Flying Qualities and
Flight Testing of
the Airplane

Frontispiece   Now Bede has produced the kit-built, jetpowered BD-10 two seater ... with features resembling a miniature McDonnell Douglas F-15 Eagle, flying qualities and flight testing of such homebuilts take on a new dimension. (Courtesy of Bede Jet Corporation)

Flying Qualities and Flight Testing of the Airplane

Darrol Stinton
Past Senior Visiting Fellow
Loughborough University of Technology
Leicestershire. UK

Copublished by
American Institute of Aeronautics and Astronautics, Inc.
1801 Alexander Bell Drive, Reston, VA 20191
and Blackwell Science Ltd, Osney Mead, Oxford, OX2 0EL, UK

American Institute of Aeronautics and Astronautics, Inc.
1801 Alexander Bell Drive, Reston, VA 20191

Library of Congress Cataloging-in-Publication Data

Stinton, Darrol, 1927–
     Flying qualities and flight testing of the airplane / Darrol
  Stinton.
       p. cm.
     Includes bibliographical references and index.
     ISBN 1-56347-274-0 (paper)
     1. Airplanes—Flight testing.   2. Airplanes—Handling characteristics.   I. Title.
TL671.7.S75   1998   629.134′53—dc21   97-43741 CIP

Set by Setrite Typesetters Ltd. Printed and bound in Great Britain.

Two Irish Setters:

BONNIE,
quiet companion of many pages, and

LIZZIE,
her mother, who always told me when to stop.

Contents

Foreword

Flying Qualities and Flight Testing of the Airplane by Darrol Stinton is published jointly by the American Institute of Aeronautics and Astronautics (AIAA) and Blackwell Scientific Publications in the United Kingdom. In this text, the author discusses the flying qualities of an aircraft, including stability, controllability, performance characteristics, functioning of flight controls, stability augmentation and autopilots, ergonomics of the flight deck, ground and water handling, cross-wind handling, buffeting, and cockpit visibility, as well as external influences such as exhaust emissions. The verification and test of these qualities are presented in this volume in a comprehensive description of the flight testing processes based on the author's extensive practical experience as a test pilot and design engineer. This text is a companion volume to the author's earlier texts, *The Anatomy of the Aeroplane* and *The Design of the Aeroplane,* also available from AIAA.

The text begins with chapters on aircraft configuration and flight characteristic examples of six different aircraft from the *Wright Flyer* (1903) to the *Rutan Voyager* (1986). The subsequent three chapters deal with aerodynamics, response, stability, maneuverability, performance, man-machine interface, and flight testing procedures. Separate chapters are devoted to airfield performance, general performance, stalls, control and maneuvers, longitudinal stability, spinning, ground and water handling, and powerplants. The text should be of interest and importance to a great number of readers: engineers, pilots, flight engineers, airplane owners, home-built aircraft enthusiasts, and aeronautical engineering students. To this latter category this book is particularly recommended, since the text provides an excellent background knowledge on the practical aspects of flying.

The AIAA Education Series embraces a broad spectrum of theory and application of different disciplines in aerospace, including aerospace design practice. The Series has now been expanded to include defense science, engineering, and technology and is intended to serve as both teaching texts for students and reference materials for practicing engineers and scientists.

J. S. Przemieniecki
Editor-In-Chief
AIAA Education Series

Preface

'To design one is nothing;
to build one is easy;
to fly one is everything.'

Otto Lilienthal (1848–96)

Let me assume that you have picked up this book either because you fly, or you are so enthusiastic about flying that you devour what anyone has to say authentically about how aeroplanes behave and feel in the air, on the ground or on water. This allows me to slip into using personal pronouns as if you and I were actually talking over a problem, which is what the many examples contained are about. Do not expect to be handling a conventional textbook on the flight testing of *aeroplanes* (i.e. *airplanes* west of 30° W) – aircraft which are heavier than air, mechanically propelled, and have their supporting surfaces fixed for flight, unlike autogyros and helicopters (which some today mistakenly refer to as aeroplanes).

The book presents selected parts of a tool kit gathered over the 50 years since I became tangled in aviation in World War II. Of these, 35 included test flying, first as an experimental test pilot at the (then) Royal Aircraft Establishment, Farnborough, and later as the airworthiness certification test pilot for light aircraft with the Air Registration Board and Civil Aviation Authority. All started with an old-fashioned apprenticeship on the factory floor, leading to aircraft design, before side-tracking into RAF squadron service, test flying, accident investigation, operational requirements, consultancy, tutoring and teaching.

Flying Qualities and Flight Testing of the Aeroplane is a book about the special **airmanship** of the test pilot. Airmanship is being 'street-wise', on the ground and in the air, in all matters affecting the conduct of an effective flight with the highest level of safety for its intended purpose. Here the pilot carrying out the test is tasked with finding out as much as possible about the properties and nature of an aeroplane in one relatively short flight. It looks at raw flying qualities, assessed manually, with hand on stick, wheel or control column, and foot on rudder bar or pedal, all connected by push rod, torque tube or cable to the flying controls. Another title might be: *The Quick-Look Flight Testing of Raw Aeroplanes*. In one direction it covers the ground between the featherweight microlight and heavier light and small aircraft, up to the 19 000 lb (8600 kg) nineteen-seat, public transport commuter.

In another direction of importance to a vast number of people – other than just display pilots, owners and their engineers – are historic, classic and vintage machines, especially 'warbirds'. This is no indulgence. My personal opinion is that aeroplanes fitted with propellers (rotors and fans) are more interesting than pure jets, because they suffer from more raw dynamic and aerodynamic asymmetries and quirks, and so provide more intricate, extensive and therefore valuable lessons – in handling qualities especially. Many warbirds were once front-line aeroplanes – World War I scouts and trainers with rotary and liquid-cooled engines – more recent heavy piston-engined fighters and training aeroplanes – jet fighters and even V-bombers. A few are wholly original, with rare and irreplaceable engines and parts, and totally inadequate 'product support'. Many are replicas to a degree, either full, semi-scale or miniature, with make-do engines. Original, rebuild, refurbished, replica or 'look-alike', each brings with it unique problems of control, stability, degraded or dubious performance, traps, flaws and shortfall. All are linked by the common flight test thread of **philosophy**, **principle** and **practice**. The thread is what binds all of the widely differing examples in the text.

If the aircraft you built after reading my second book, *The Design of the Aeroplane*, does not fly quite as you hoped, the third one tells you what remedies might be tried, before you burn it, break it up, or hang it on the wall.

This book has been written for dipping into or nibbling at, rather than for reading from cover to cover, which accounts for its structure and the reason why some topics are approached more than once from quite different angles. The arrangement is to suit pilots and their overworked engineers and mechanics especially, because they are active people who get on with things. They only stop long enough to find out about what puzzles them, instead of slogging through a text. This is why the ground-clearing Introduction which follows, together with Chapters 1 and 2, are there for scene-setting.

The lay-out is not that of a classic textbook. Some repetition is inevitable. For example, the danger of the spiral dive is mentioned when discussing wing dropping at the stall. It also appears in the context of lateral and directional motions, as a consequence of excessive weathercock stability and a deficiency in corrective roll with sideslip (dihedral effect). Similarly, it is necessary to skip backwards and forwards again when considering total drag and rolling moment. Wing-tip forms affect favourably and also adversely the lift dependent (induced) drag and, therefore, performance in terms of the lift/drag ratio, high speed and *Everling number*. They also affect lateral and directional handling qualities. There are other examples.

Chapter 3 has been a problem. This book started as a handful of chapters at the end of *The Design of the Aeroplane* – except that there was then too much to squeeze between two covers for the price. *Flying Qualities* needs a mass of definitions, and some technical 'flute-music'. Chapter 3 had to go somewhere. Had it appeared at the beginning it could have been wrongly off-putting. Whichever form I tried lacked elegance. Bear with it, because it is there to be tasted before leaping into what follows. If, on the other hand, you are happy – and there is no reason why you should not simply jump over it – you can always return to it later.

If some of the drawings appear 'busy' this could not be avoided. All is relevant. If some appear crowded it is because publication costs must be contained, limiting pages available for drawings.

For literary convenience, 'he' includes 'she'. Flight testing has involved women continuously from at least the time of the Wright Brothers, as test pilots, or observers and flight test engineers. Hanna Reitsch, Jacqueline Cochran and Jacqueline Auriol were all awarded Honorary Fellowships by the Society of Experimental Test Pilots (SETP) in the USA. Others include Melitta Gräfin von Stauffenberg, Amelia Earhart, Joan Hughes, and a plethora of Russians, some of whom became astronauts. As a profession it is neither sexist nor machismo. Indeed the men and women involved are, more often than not, quiet and unassuming, revealing nothing to link them and their coolly experimental science with an occupation embarassingly misrepresented by film makers.

For those of you who have asked the publisher already, do not be disappointed to find nothing here about handling and testing what are called *high order* (electronic) *flight control systems*. I concede that fly-by-wire (FBW) and now fly-by-light (FBL); and the integration of flight and propulsion control systems are of profoundest importance. But they point the way into the acronym-laden future and a totally different book in which the flying and power controls are operated *via* a computer.

With the aid of the computer one can now fly an aircraft that is otherwise uncontrollable and unstable, making it carry out manoeuvres which were once impossible. But the computer prevents one appreciating the aeroplane itself or assessing its basic flying qualities, warts and all. Instead the pilot is assessing the veracity of a computer, the ergonomics and reliability of the circuits and the switches which bring them to life, the utility and reliability of the way in which information is presented to him or her in a 'glass cockpit', and the quality of the head-up display (HUD). These advanced devices have freed engineers from the constraint of having to design aircraft which possess good manual controls, optimized performance, manoeuvrability and inherent stability. The ability to make even a brick fly up-down-forwards-backwards-sideways and turn on a point, by means of electronic signals from pilot to computer and thence to engine and other controls, distances this combination of science and art from the 'gritty-gutty' subject of this book: the hands-on flying of an aeroplane. It has also brought with it a new class of accident – that of *computer-induced pilot error*.

This book is the third in a trilogy. *The Anatomy of the Aeroplane* surveyed a broad landscape of civil and military aeroplanes of various shapes, sizes and weights, from light to heavy, subsonic and supersonic. *The Design of the Aeroplane* is a compendium of design notes – all of the things I now find hard to remember – focused upon one area of the landscape, namely single- and multi-engined aeroplanes weighing less than 5700 kg (12 500 lb) – the upper limit for civil single-pilot operation – and all of them subsonic. *Flying Qualities...* is related to both and is an amalgam of those parts of the others which I consider to be of interest and importance for the greatest number of readers: engineers, pilots, owners, homebuilders, students and enthusiasts. Like its predecessors, it includes seaplanes because they are practical and useful aircraft – and not only for water-bombing fires.

Many flight test methods are as old as the experimental flying machines of Cayley, Lilienthal and the Wright Brothers but, like aircraft, they have been refined by skill and science. While their elements remain unchanged, improvements in precision and accuracy of techniques and complexity of test equipment have added a specialist crew member to that of the test pilot – the ubiquitous and irreplaceable Flight Test Engineer. The subject of flight test has grown with aviation and computer technology. To deal with it in depth would occupy space far beyond the scope of these pages.

Often tests concentrated on just one aspect: a rerun of a rate of climb – a closer look at the direction of roll, stick-free, in a straight steady sideslip – following up a report of an excessive wing drop at the stall – or pursuing evidence revealed in the aftermath of an accident which needed further investigation in flight. Airborne accident investigation provides some of the most rewarding test flying. It is too easy to jump to conclusions about responsibility for and the apportioning of blame in the wake of an accident. The simplest flight test often produces results which cast doubt on preconceptions and hasty judgements.

The information contained has a variety of sources, starting in 1960 with my lectures to the Empire Test Pilots' School on the design of the airframe inside the skin of an aircraft and how that affects what goes on outside it. These spawned *The Anatomy of the Aeroplane* and then developed over the next 30 years into a discussion of the interaction between configuration design and flight testing – and the role of the certification test pilot especially.

All aircraft chatter ceaselessly about their ailments and the job of the test-pilot is akin to that of a doctor, using all of his senses to diagnose what it is suffering from and why some feature or flying quality is wrong, deficient or flawed. Pressing the analogy further, this book has more in common with a doctor working in the Australian outback, or the jack-of-all-trades Chinese 'barefoot doctor', than with one in London's Harley Street. The test pilot, particularly the one concerned with airworthiness, is properly part of the process of rectification – often in the field – with the burden of responsibility for knowing exactly what he is talking about before opening his mouth.

Other sources are papers and lectures I have given to the Royal Aeronautical Society; the Royal Institution of Naval Architects; the Historic Aircraft Association; the Experimental Aircraft Association (EAA) Convention (or Fly-In), at Oshkosh, Wisconsin; to students at Loughborough University of Technology and others. Loughborough lectures in particular are shaped to meet the requirements of a syllabus, outlined in *An Engineer's Responsibility for SAFETY*, which is a document of the Hazards Forum – this is an organization which exists to provide a national focal point in the UK in which engineering features in the mitigation and reduction of man-made (and natural) hazards and disasters. The syllabus aims at developing in students particularly, by means of lectures and case studies:

- awareness as future engineers of their duty of care and responsibilities to their fellow employees, their organization and the general public in the field of safety;
- awareness of the analytical techniques and the legal, financial and human factors relating to safety.

The Empire Test Pilots' School, the oldest of the very few test pilot schools, was founded a little more than 50 years ago. In 1959 it was located at the Royal Aircraft Establishment, the RAE, at Farnborough in Hampshire (now the Defence Research Agency, DRA). Today ETPS is back at Boscombe Down, at what is now the Defence Test and Evaluation Organization (DTEO), formerly the Aircraft and Armament Evaluation Establishment (A & AEE). Both Establishments are steeped in aeronautical history, as may be seen from the copy of the historic War Office letter of 18th June 1913 (Fig. Pre.1), which approved the proposal to form an organization within the Military Wing of the Royal Flying Corps for the purpose of carrying out experimental work connected with aviation. A&AEE lineage descends directly from that approval.

Experimental work at Farnborough (South Farnborough it was) goes back even further than the 90 years since the Wright Brothers teetered into the air at Kittihawk. The tree in Fig. Pre. 2 (Ref. 0.1) shows the development of Government Control of British Aviation from 1878, and that both the Royal Aircraft Establishment and Royal Air Force (via the Royal Flying Corps) grew out of the same trunk – His Majesty's Balloon Factory and Balloon Section, Royal Engineers – when they separated into different branches in December 1909.

Technical truth is paramount and it is easy to lose sight of it when appearances and a catalogue of recorded facts are held to be the reality. Nearly every historic aircraft today is a look-alike replica to a degree. Flying qualities might be far from representative. Yet, it is in their domain that the success or failure of an aircraft lies. Unrepresentative flying qualities can lead an inexperienced and underqualified pilot into a hole. It is too easy to accept inadequacy when one does not know what adequate flying qualities were when the original aeroplane was around.

Myths appear and are hard to dispel. Dispelling them is the proper task of the great museums, and of operational organizations like the HAA and Confederate Air Force. It should be possible for a properly qualified pilot to fly a Bleriot XI, or a Lancaster, or at least an F-4 Phantom or a Mig-31 'Foxhound' in the year 2035 and know how they handled, performed – and what they smelt like for that matter. Not for nothing has care been taken to

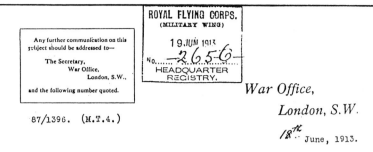

ROYAL FLYING CORPS.
(MILITARY WING)

19.JUN 1913

No. 2656

HEADQUARTER REGISTRY.

Any further communication on this subject should be addressed to—

The Secretary,
War Office,
London, S.W.,
and the following number quoted.

87/1396. (M.T.4.)

War Office,

London, S.W.

18th June, 1913.

Sir,

In reply to your letter of March 21st 1913, on the subject of Experimental work to be carried out by the Military Wing of the Royal Flying Corps, I am commanded by the Army Council to inform you that they approve of your proposals to form an organization within the Military Wing for the purpose of carrying out experimental work connected with aviation on the understanding that this organization will require no addition to the Establishment already authorized.

As regards your proposal that you should have authority to initiate experiments, I am to say that you should submit monthly for approval of the Army Council a statement of the experiments you propose should be initiated with an estimate of the cost of these experiments. On receipt of this statement the Council will decide which of the experiments shall be undertaken by you, and which shall be carried out elsewhere. In the case of the former the necessary funds will be placed at your disposal.

I am,

Sir,

Your obedient Servant,

(signature)

The Officer Commanding,
Royal Flying Corps,
(Military Wing),
South Farnborough.

Fig. Pre. 1 The historic War Office letter of 18 June 1913, which approved experimental work in connection with aviation in Great Britain, and led to the creation of the A&AEE (now DTEO); and later to the founding of the ETPS, the world's first test pilot school, in 1943. (Reproduced by kind permission of the DTEO, Boscombe Down)

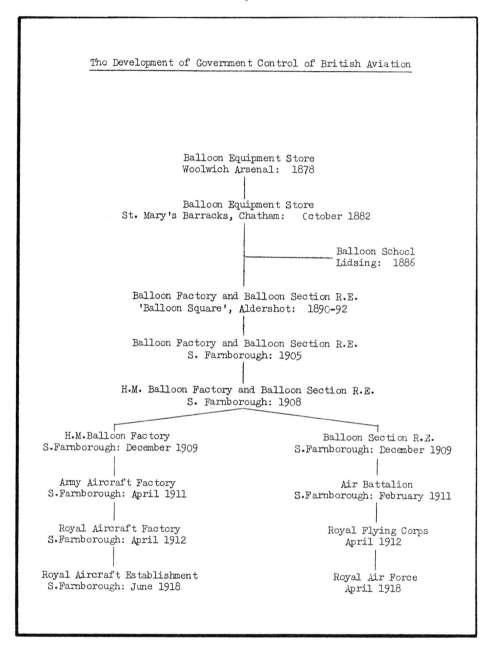

The Development of Government Control of British Aviation

Balloon Equipment Store
Woolwich Arsenal: 1878

Balloon Equipment Store
St. Mary's Barracks, Chatham: October 1882

Balloon School
Lidsing: 1886

Balloon Factory and Balloon Section R.E.
'Balloon Square', Aldershot: 1890-92

Balloon Factory and Balloon Section R.E.
S. Farnborough: 1905

H.M. Balloon Factory and Balloon Section R.E.
S. Farnborough: 1908

H.M.Balloon Factory
S.Farnborough: December 1909

Army Aircraft Factory
S.Farnborough: April 1911

Royal Aircraft Factory
S.Farnborough: April 1912

Royal Aircraft Establishment
S.Farnborough: June 1918

Balloon Section R.E.
S.Farnborough: December 1909

Air Battalion
S.Farnborough: February 1911

Royal Flying Corps
April 1912

Royal Air Force
April 1918

Fig. Pre. 2 Historic tree showing the origins of the British Royal Aircraft Establishment (RAE), Farnborough, and the Royal Air Force, from the same common trunk in 1909 (Ref. 0.1). (Reproduced by kind permission of the former RAE (now Defence Research Establishment))

re-rig *Victory* (Viscount Nelson's flagship, and already old at the Battle of Trafalgar in 1805) in her final dry berth in Portsmouth Dockyard with rope made from local West Indian hemp, with comparable oak and tar, to evoke something of the truth of life on board. One must get past what has been called the screen of recorded action and into the inner world of belief, motive and perception where history really begins. Only then do we begin to see those who went before us as people with whom we can identify, and from whom we start to learn for the good of our own survival.

A knowledge of history – the chronological record of causes and their effects – is of practical value to people who design and test-fly aircraft. To take an example from a sister-discipline, D.K. Brown, M Eng, FRINA, a consultant naval architect and historian, with a background in the Royal Corps of Naval Constructors, said in his paper on *History as a Design Tool* (RINA, 1992):

> 'Historical methods can help the modern designer in five main areas, though there is considerable overlap between them.
>
> - Design policy and decison-making reviewing past requirements and decisions; their successes and failures.
> - Teaching and training, by examples from the past.
> - Lessons and disasters, including damage in war.
> - Testing new theories against records of past events.
> - Avoiding the repetition of mistakes.'

Often something old, less developed, simpler, highlights basic truths that become obscured (for an infinite variety of reasons) with development. Study of these older forms can be, therefore, very instructive not only because they 'speak' more clearly to the discerning eye and mind, but they also plant knowledge and experience from which intuition can suddenly emerge – perhaps in a crisis and under stress.

It is often in the historic context that understanding comes in a flash, as occurred when I read in a 1939 technical journal an article by Sidney Barrington Gates (1893–1973), who was also at Farnborough, from the middle of World War I until after I left, at the end of 1963. His explanation of longitudinal stability and the part played by the elevators enabled me to see at once something I had not seen clearly in years of struggling with the classical and purely mathematical approach to stability and control. In his article, written for designers as World War II started, one drawing showed simply what it was all about. Dozens of Greek and other squiggles of mathematical flute-music previously used by way of explanation fell away. We shall come to Gates, that diagram, his work and that of others like him, particularly in Chapter 10.

Both of the previous books started from a technical viewpoint: that of the designer and engineer, faced with a set of operational and other requirements, tasked with providing an aircraft with features which would satisfy them as completely as possible. Even though the *Anatomy of the Aeroplane* was fashioned out of notes on aero-structures, written for student test pilots, it orientated to the view of the person seated at a desk, or at a drawing board, looking beyond his windows to the airfield outside. In this book we start the other way around to look in the opposite direction: how do the flying qualities anticipated or discovered affect a rethink, redesign and modification, and possibly design in the future?

This direction of approach is thought to be of more value when dealing with a subject like flying qualities. Of all readers, pilots are compulsively interested in the ways in which aircraft behave. They argue over and question the physical principles involved. Those who will cajole and weedle any owner in an attempt to get their hands on his aircraft, just to see how it handles and performs.

An airworthiness test pilot is a critic. He is trained and employed professionally to look for the worst and most dangerous features of an aeroplane, not for the best. For a person like that, tackling the subject from such a direction can lead to difficulties. Even when he finds good features he wonders if it is a trick and what is the probability of this proving to be so? In one respect he is a gamekeeper, in another a hired gun, both of which are unpopular. Because he almost always finds some shortfall or inadequacy which affects certification and regulation, there is often an objector armed with the lamest of excuses for preventing him flying and so discovering an adverse fact about this or that particular aeroplane. The best way of dealing constructively with such obstruction before it occurs is to invite the owner, or his/her pilot or engineer to come along too – NOT as a passenger but as part of the functional crew.

The project test pilot is confronted with the ideas of others, in the form of an aeroplane which he has to explore on the ground or in water as well as in flight. *Every* aeroplane is flawed with some faults at least, so that it represents compromise, on an increasingly grand scale. Not one is perfect, even among those which have been the most successful. Dealing with raw assessments it would be too easy, but wholly improper, to criticize poor, or arguable design features of any existing aeroplane that I happen to have flown, beyond pointing out here and there what appear to have been fixes, incorporated to rectify something which may have been found to be inadequate, or wrongly fitted afterwards, or to have an adverse effect upon safety.

Therefore, in line with the two previous books, several hypothetical aeroplanes are introduced in the text. Each has a rational basis for its shape and features, regardless of how wild or extreme they might appear to be. The appearance of three lifting surfaces will be recognizable as representing several modern forms and applications. Each example is a compromise to be assessed critically, just as it would be during the project stage before much material has been cut, or when presented for type or other certification. The drawings are early project sketches. As such they are neither clean nor elegant. They are accompanied by scrap notes, comments and calculations – just as they would be in a project, drawing or flight test office – and their purpose is for cutting teeth. They are here not to be sold but to be analysed, criticised and pulled apart.

Some drawings are of poor quality because they are original or just old. They are representative of actual material with which we must work, because either their authenticity and relevance are beyond doubt, or because they are the best that could be assembled at the time.

Examples, whether historic or not, are used where there is a point to be made. Some examples of mistakes and errors are my own, because there are lessons in them. The scientific approach is reductive, seeking to generalize from sets of data – *one* is never a statistic. Pilots and people like them who work on the cutting edge cannot risk living life on such terms. There is no time to wait for the statistical second example. Anecdotal evidence is vital on the grounds of '*Been there, done it* (or '*tried to do it. Couldn't because …*)'. If someone comes back with a story, listen. There may be no second chance.

Anecdotal evidence – anathema to scientists and academics working with generalities – is vital to test (and other) pilots. To paraphrase Professor Trevor E. Kletz, talking to the Hazards Forum in London on 28 April 1994:

☐ First, there is a moral obligation to impart information which might prevent an accident or prevent someone else from making a mess.
☐ Second is the pragmatic reason that if we tell others about our errors they might come and tell us theirs and we shall then be able to avoid them ourselves. Learning from others while not giving information in return makes us 'information parasites', those birds which rely upon others to warn them of approaching enemies.

☐ Third is the economic reason that safety measures cost companies money. If we tell a competitor about the action our company took after an accident, then he might spend as much as we have done to prevent the accident happening again.

☐ Fourth, if we have a serious accident the whole movement, or industry, loses esteem. New preventive legislation might then be introduced by the regulators which affects all adversely.

☐ Fifth, to be told from the 'horses mouth' has more credibility than reading a list of detached 'Do's and Don'ts'.

Throughout the book I use the word aeroplane, not 'airplane'. While touching my forelock to American friends and colleagues, I much prefer aeroplane because it comes directly from the classical Greek '$\alpha \varepsilon \rho o \pi \lambda \alpha v o \varsigma$, air-wandering'. Similarly the word enthusiast is used in the classical sense of $\varepsilon v \theta v \sigma \iota \alpha \sigma \tau \varepsilon \varsigma$, meaning 'possessing a god within'. Both contain that special Greek fire which, I believe, lies at the heart of the urge to fly like Daedalus and Icarus. It lives in the dancing Greek characters which leap from their alphabet into our mathematics. There is too a Greek connection in my subject. The precision of ancient Greek, which had a word for everything, enabled philosophers to discuss all things with accuracy, there was no room for untruth. *'Truth above all'* is honoured in the test flying profession. As John Farley, Past Chief Test Pilot of British Aerospace Defence Ltd, Dunsfold said at the ETPS Jubilee celebrations in 1993, *'It marks the difference between a test pilot and someone who merely tests aeroplanes'*.

Flying Qualities and Flight Testing of the Aeroplane began as overspill from *The Design of the Aeroplane*. **It is not and never was intended as a manual on how to go out and test your own aeroplane**. Not only can that be a dangerous game, it could be exceedingly stupid to think that having read a book on the subject then all is well. It isn't. You know nothing more than you would about, say, surgery by reading *Gray's Anatomy*. If you read this book as a primer and then try testing an aeroplane in some tricky corner of the flight envelope, you could end up looking like one of Gray's illustrations.

Where formal flight test procedures and explanations of qualities appear, the third person is used naturally. But where I state a personal opinion of something or other, or of what was devised to solve a problem, *and which was found to work at the time* (but might not do so for you) then I resort to the personal pronoun. Then I usually insert '**Note:**' in the text.

While I realize that many readers are put off by mathematics, let me say that when young I had a brain which stalled and went blank when confronted with such squiggles. Long after leaving school I had to kick-start it again from scratch. But mathematics, especially arithmetic, cannot be avoided entirely in a working book. Mathematics is only a language in which it is possible to make a statement with such precision that it can be picked up and used as a tool, for releasing sparkling and vital truths from what looks at first like dull ore. To quote from *Calculus Made Easy*, by Sylvanus P. Thompson (Macmillan, London, 1910):

> 'Being myself a remarkably stupid fellow, I have had to unteach myself the difficulties, and now beg to present to my fellow fools the parts that are not hard. Master these thoroughly, and the rest will follow. What one fool can do another can.'

Choice of units is a problem. On the European side of the Atlantic Ocean younger generations are now brought up with the metric system. Among older generations in the United States and English-speaking world, large numbers of people continue to work in foot/pound/second/Rankine units. Whereas *previously* I wrote in FPSR primary units with metric in brackets, here I turn them around and about as required, as in the ambidextrous world of aeronautics, reserving primacy for the source being quoted at the time. Equivalent units are always in brackets.

Metric SI units (Système *International d'Unités*, 1960) employ the kilogramme, metre and second as the respective units of mass, length and time. I value the metric system and a base of ten, but think that the centimetre would be one of the greatest inventions were there 2.5 in one inch instead of 2.54, and I find the foot and the pound more useful for some jobs than others. So, I tease with tongue in cheek my seriously rational, pro-metric, students that their system came about irrationally anyway, like a work of art, telling them that Napoleon wanted to mark his millenium, in science, with a unit of length greater than that of the thirty-six inch cloth-yard arrow used by the English in the turbulent events at Agincourt and Crecy. Aides measured his girth, from belt buckle to hole, and found it to be 39.3701 inches. And so the metre was born.

The *newton*, the SI unit of force, I confess is too small to have much significance for me, beyond consistency and it being the force, pressing down upon the land, of the mass of one small apple acted upon by gravity. Multiples and powers of ten are more 'user-friendly' with calculators than 12, the base of the old duodecimal system, even though 12 has more flexible factors than ten and is more practical when sharing. Extensive conversion tables are included.

Working on a word processer instead of a typewriter has lead to changes of presentation which avoid many previous stops to handwrite into the manuscript. Thus: square root of x, which is $x^{1/2}$, might more often be written $x^{0.5}$; while (1/square root of x) appears as $x^{-0.5}$. An exception is an index which is more comfortable as a simple fraction than in decimal form: e.g., cube root of $x = x^{1/3}$, which is superior to $x^{0.3333333}$, or the less accurate $x^{0.33}$.

Quotations contain valuable insights, advice and wisdom born of experience which, although long past, remains valid for all of the historic reasons given earlier.

What is here was written in good faith, because I found it worked for me at the time, with nothing better to hand. Flying qualities of aeroplanes are fascinating. Discovering the truth of what they are, and of how one machine compares with another is like finding what is on the other side of a hill. But question everything that you read and are told. Doubt is the weapon of experimental science, as Claude Bernard (1813–78), the French physiologist pointed out:

'Le doute et l'arme de la science experimentale.'

Give doubt its head before sticking your own neck out. As John Farley also said recently:

'If there is something to be learned from a dangerous exercise then it is well worth doing. If not, don't.'

There must be a ruthless balance of risk against value of what is being pursued. Flight test techniques should only be taught by pilots who have gone through the mill and know exactly what they are doing. So, if you have not, don't succumb to the conceit of believing that as you are an ace you can omit the step of being shown. You might get away with it, but then you might not. If you don't, then you will deprive me of the pleasure of sitting at your bedside and saying: '*I told you so*'.

D. Stinton
Farnham

References and Bibliography

0.1 Child, S. and Caunter, F. C., Report No. Aero. 2150, *A Historical Summary of the Royal Aircraft Factory and its Antecedents: 1878–1918*. Farnborough: Royal Aircraft Establishment (RAE), now the Defence Research Agency (DRA), 1947

Acknowledgements

In a book of this kind an author is indebted to a vast number of people, especially when trawling up experience and ideas over a span in aviation that is longer than half the time since the Wright Brothers flew. Some are no longer around to be acknowledged. All have given me guidance, advice, photographs, information, connections and opportunities, or have helped with the rough and tumble of ideas.

The following in particular are beyond reach of thanks. They were the independently-minded Gordon Corps and Nick Warner, past Chief Test Pilots of the *Civil Aviation Authority* before joining *Airbus Industrie*. To Gordon I owed one of the best test flying jobs in the world; to Nick – joint spinning and thrashing out a revision of reversed spin-recovery requirements. Hugh Scanlan, past Editor of the lamented *Shell Aviation News*, a master of English and a flying-man with a felicitous turn of phrase which captured an aircraft in essence as few others could. Mark Lambert, formerly with *Interavia* and latterly *Jane's*, a mine of information and help over many years. The Honourable Patrick Lindsay, owner of a replica of the Sopwith Triplane which prompted a line of useful research. The artist-draughtsman, Dennis Punnett who (through the good offices of William Green and Gordon Swanborough) always seemed to make available a drawing of any aircraft. Finally, Andy Coombe of the former *Norman Aeroplane Company*, whose technical approach to certification test flying, just as to design, was practical, 'streetwise' – and refreshingly artful.

Serving and retired test pilots whom I gratefully thank are David P. Davies, OBE (past Chief Test Pilot, *Air Registration Board and Civil Aviation Authority*), my 'boss' for many years, who read the text and pointed out improvements. Four Honorary Fellows of the *Society of Experimental Test Pilots*: Neville Duke, DSO, OBE, DFC, AFC (past Chief Test Pilot, *Hawker Aircraft*, and with whom I later shared civil certification flight tests) who not only read and checked the final text but provided a gracious Foreword; Newton F. Harrison, DSO, AFC, RAF (retd) (my past tutor, *Empire Test Pilots' School*, and Chief Test Pilot of *Atlas Aircraft*, South Africa); Harald Penrose, OBE (past Chief Test Pilot, *Westland Aircraft*), who as an aviation historian also provided valuable information on the original flying qualities and background to the Lysander; Jeffrey Quill, OBE, AFC (past Chief Test Pilot of *Supermarine* and *Vickers Aircraft*) who has been a source of help and shrewd advice over many years – not only when setting up the *Historic Aircraft Association* in 1979, but as the expert on the Spitfire when historic and other information was needed. Air Vice-Marshal Geoffrey C. Cairns, CBE, AFC, RAF (retd), past Commandant of the former *Aircraft and Armament Experimental Establishment* (now *Defence Test and Evaluation Organization (DTEO)*) Boscombe Down, and colleague on many subsequent civil certification test flights; Chris Turner, ex-*Royal Aircraft Establishment*, Bedford, and now in the airline industry, who dissected projects critically with an eye like a scalpel; N. (Norrie) Grove (Chief Test Pilot, *Slingsby Aviation Ltd*), with whom I was long involved with spinning and other clearances of the Slingsby T 67; John Farley (past Chief Test Pilot, *Hawker Aircraft*) for his comments and pointing me in the direction of Sukhoi, and Menahem Schmul (Chief Test Pilot, *Israeli Aircraft Industries*); countless Officers Commanding and tutors of the *ETPS* since 1959, including especially Air Marshal Sir Donald P. Hall, KCB, CBE, AFC, the response of whose courses to my lectures were an imperative in many ways to the shaping of this book – also more recently Officers Commanding, Wing Commander W. L. Martin Mayer, AFC, RAF (retd) and the present OC, Wing Commander Robert P. Radley, RAF, both for information and opportunities given; Viv. and Rod. Bellamy, and John Fairey for Hawker Fury, Sopwith Camel and Fairey Firefly replica test flying; Gus. Limbach (ex-US Marine Corps pilot), to whom I owe a great debt, over many years, for showing this Limey test pilot the American homebuilt ropes at Oshkosh. John Sproule, aviation historian, who provided information about replicas of Sir George Cayley's flying machine(s); Derek Piggott, sailplane and power test pilot, wide in generous historic experience and numerous shared test flights over many years, who flew Sproule's Cayley replica; Dick Rutan (with whom I sampled for the *CAA* his brother Burt's canards at Mojave, and who with Jeanna Yeager later flew the record-breaking *Voyager* around the World, non-stop); George Ellis (*British Aerospace*, Woodford) for his information on the DH 88 Comet; R. (Ray) and Mark Hanna, père et fils, of *The Old Flying Machine Company*, Duxford; John T. Lewis (Chief Pilot), and Peter Symes, both of *The Shuttleworth Collection*, Old Warden; and finally Wing Commander Peter Sadler, RAAF (retd) (Australia), past tutor, *International Test Pilots' School*, Cranfield, whose skills in briefing, execution of tests and debriefing students were concise models of how it should be done.

As the history of flaws, faults and failures discovered by test flying is the essence of all learning in this fascinating subject, there were those whose insights lighted dark corners – even though I cannot now put a name to a face in every case. Among those which come most easily: the books on his test flying of an exceptional number of aircraft, military and naval, allied and enemy in World War II, by Captain Eric Brown, CBE, DSC, AFC, RN (retd) (past President, the *Royal Aeronautical Society*), contain rare information. They are instructive reading for pilot and engineer, student and graduate alike. Similarly, the works of Roland Beaumont (past Chief Test Pilot, Director of Flight Operations of *British Aerospace* (*Preston Division*), and then Director of Flight Operations of *Panavia*); of Dr John W. Fozard (past Chief Designer, Harrier, and then Divisional Director, *British Aerospace*, past President, the *Royal Aeronautical Society*, Lindbergh Professor of Aerospace History at the *Smithsonian Institution's National Air and Space Museum* in Washington). Others are Richard P. Hallion (USA); Dipl. Ing. Hans-Werner Lerche (1941–45, *Luftwaffe Flight Test Centre*, Rechlin and now Honorary Fellow of the *Society of Experimental Test Pilots*); Dr Gordon Mitchell (son of the late R. J. Mitchell, Chief Designer of among many other aircraft *Supermarine's Schneider Trophy* winning seaplanes and the Spitfire which grew from them); the late L'ingenieur general Louis Bonte, Grand Officier de la Legion d'Honneur, long involved with French aviation and test flying, and author of *L'histoire des Essais en Vol*; and Bill Gunston, an unfailing source of information. Last in this group, but not least, is the *Society of Experimental Test Pilots* (USA), and its published *Proceedings*.

In applied science and history I am grateful to Dr H. H. B. M. Thomas, OBE, formerly of the *Royal Aircraft Establishment*, Farnborough, for advice on controls and a valuable paper on the subject for the *Historic Aircraft Association*, from which I have quoted. Among other staff of the former *RAE* are Brian Kervell, with whom I carried out experimental flying years ago, and who until recently was in charge of the *RAE* museum. Past Chief Librarians (and their always helpful staffs) include Richard Searle, Gwyneth Davies and Marlene Wright; also John Bagley, formerly of Aerodynamics Department, *RAE*, before joining the staff of the *Science Museum*, who has always helped with information and advice on all matters historical, and to whom I am perhaps above all indebted for drawing my attention to the works of the late G.H. Bryan and S.B. Gates. Among the staffs of museums are Dipl. Ing. Werner Heinzerling (Abteilungsleiter Luft und Raumfahrt, *Deutsches Museum*, Munich); Dr Michael Fopp, Director of the *Royal Air Force Museum*, Hendon; C. Graham Mottram, Curator and Deputy Director and David Richardson, Research Officer, of the *Fleet Air Arm Museum*, Yeovilton; the *Imperial War Museum*, Duxford; and Ronald K. de B. Nicholson, Press Relations Officer, *A&AEE*.

Others I thank for generous advice, ideas, photographs and/or wisdom, are Mrs Ann C. Welch OBE; Desmond Norman, CBE; Terence Boughton; R. (Bob) W. Chevis of *United Turbine (UK) Ltd*; Peter J. C. Phillips of *Speedtwin Developments Ltd*; Mike Badrocke of *Aviagraphica* for drawings; J. (Jim) R. Bede of *Bede Jet Corporation* (USA); Dr Paul Franceschi (France) and Richard Goode (UK), aerobatic pilots; Professor D. Howe, *College of Aeronautics, Cranfield University*; Professor Stanley J. Stevens and Dr Lloyd Jenkinson, *Loughborough University of Technology*; Ing. A. Janda and Ing. T. Poruba of *JaPo*, Czechoslovakia; Dr Oliver L.P. Masefield of *Pilatus Aircraft Ltd*; Graham Maxwell and Andre Power of *PowerChute Systems International Ltd*; a long-serving colleague, flying instructor, test and demonstration pilot, Peter F.C.A. Thorn, ex-A.V. Roe and RAF, now of the *Historic Aircraft Association*; Ian Chichester-Miles and B.J. Cunnington, photographer of *CMC Ltd*; Angus M. Fleming, *AMF Microflight Ltd*; Jeremy K. Flack of *Aviation Photographs International*; Jack Beaumont and Simon Watson, photographer and staff member of the encyclopaedic *Aviation Bookshop* (*Beaumont Aviation Literature*); *Pilatus Britten-Norman*; James Haworth, photographer on behalf of the *Royal Air Force Museum*; Nigel Price of *British Aerospace, Regional Aircraft Ltd*; John Fack of *Solar Wings Aviation Ltd*; Allan Winn (Editor) and Douglas Barrie (Defence Aviation Editor) of *Flight International*.

In the area of historic archives, collections and sources may I thank the following aviation authors, historians and archivists for their unfailing generosity and help: Philip Jarrett; Jack M. Bruce; William Green and Gordon Swanborough. Also Arnold W.L. Nayler (Technical Manager and Chief Librarian) and Brian L. Riddle (Deputy Librarian) of the *Royal Aeronautical Society*.

Airworthiness: with the *CAA* are former flight test and performance engineers, 'handlers' and requirement writers, W. Ralph B. Bryder, John L. F. Denning, W. (Bill) H. Horsley, Charles M. Prophet, Graham J. R. Skillen, John R.W. Smith, and test pilot R. (Bob) D. Cole (who replaced me), to whom special thanks are due for their salutary sharpening of ideas and amiable guidance. Without the willing help, over more years than I can remember, of Malcolm Pedel and Mrs Margaret A. Syers of Flight Department much would not have come together as easily as it did. And finally to the former *Air Registration Board*, now the *Civil Aviation Authority*, for its tolerance and many opportunities richly given over a span of two decades.

List of Plates

List of Symbols

Symbol	Meaning	Symbol	Meaning
A	Moment of inertia in roll (about O–X axis)	C_{DX}	Intercept on CD axis = approx. C_{Dp}
		C_{Dform}	Form drag coefficient
A	Area (also of cross-section)	C_{Dfric}	Friction drag coefficient
A	Aspect ratio	C_{Di}	Induced drag coefficient
A	Number of test runs	C_{Dmin}	Minimum drag coefficient
A	A constant	C_{Dp}	Parasite drag coefficient
A_s	Aspect ratio of stabilizer	C_F	Thrust coefficient
A_w	Parasite wetted area	C_H	Hinge moment coefficient
a	Acceleration, deceleration (also f)	C_L	Lift coefficient (three-dimensional)
		C_{LA}	Lift coefficient of wing of aspect ratio, A
a	Horizontal distance between CG and main wheel axle	C_{Lf}	Lift coefficient of foreplane
		$C_{D(L/D)}$	Drag coefficient at $(L/D)_{max}$
a	Lift slope, $dC_L/d\alpha$	C_{Lmax}	Maximum lift coefficient
a	Speed of sound	C_{Ls}	Lift coefficient of stabilizer
a_A	Lift slope of surface of aspect ratio A	C_M	Pitching moment coefficient
a_1	Lift slope of tail stabilizer	C_{Mac}	Pitching moment coefficient about ac
a_2	Lift slope due to elevator deflection	C_{MCG}	Pitching moment coefficient about CG
a_3	Lift slope due to tab deflection	$C_{MNP} = C_{Mac}$	Pitching moment coefficient about NP (where NP = ac)
ac	Aerodynamic centre		
$(a_1)_\beta$	Lift slope of fin with angle of attack (yaw), β	C_{M0}	Moment coefficient at zero lift
		C_p	Pressure coefficient
$(a_2)_\zeta$	Lift slope due to rudder deflection, ζ	C_R	Crossforce coefficient
		c_r	Chord length at aerofoil root, or in the plane of symmetry
B	Moment of inertia in pitch (about O–Y axis)		
		c_t	Chord length at aerofoil tip
B	Number of lift-offs	c_δ	Chord length of control (or flap) surface under consideration
b	A constant		
b	Wingspan	$\bar{c}$	Standard mean chord, SMC
b_1	Rate of change of elevator hinge moment coefficient with change in stabilizer angle of attack	$\bar{\bar{c}}$	Aerodynamic mean chord, MAC
		c'	Power specific fuel consumption
b_2	Rate of change of elevator hinge moment coefficient with elevator deflection	DG	Descent gradient
		D	Total drag
		D_F	Zero-lift drag
b_3	Rate of change of elevator hinge moment coefficient with tab deflection	D_{fric}	Frictional drag
		D_{form}	Form drag coefficient
		D_i	Induced drag
C	Centre-line	D_L	Lift dependent drag
C	Moment of inertia in yaw (about O–Z axis)	D_p	Parasite drag
		D_0	Zero lift (profile) drag of wing
CG	Centre of gravity	d	Diameter (e.g., propeller)
CG_{test}, CG_{fwd}	Other CGs	d_{lam}	Depth of laminar boundary layer
C	A constant	d_{turb}	Depth of turbulent boundary layer
C_D	Total three-dimensional drag coefficient		
C_{DF}	Zero-lift drag coefficient	E	Everling number
C_{DL}	Lift-dependent drag coefficient	e	Base of exponential logarithms
C_{D0}	Zero-lift and profile drag coefficient	e	Oswald's efficiency factor

Symbol	Meaning	Symbol	Meaning
F	Froude number	M_{CG}	Pitching moment about CG
f	Acceleration or deceleration	m	Factor applied to stall speed
f	Frequency of sound	m	Mass of an element or part
		m	Metre
G	Gap of a biplane	m	Tab gear-ratio
H	Cruising altitude	N	Number of occupants
H	Hinge moment	N	Yawing moment (about O–Z axis)
H_{mech}	Hinge moment due to mechanical arrangement of control system	N_1	Low-pressure compressor rotor speed
		NP	Neutral point, stick-fixed
H_m	Manoeuvre margin, stick-fixed	NP'	Neutral point, stick-free
H_n	CG margin stick-fixed	$N, N_0\ N_1,$	Various neutral points, NP, on wing
H'_m	Manoeuvre margin, stick-free	$N_2, \bar{N}$	
H'_n	CG margin, stick-free	NP_L, NP_D	Lateral and directional NPs
H_s	Service ceiling	$N_p, N_r, N_v,$	Yawing moment derivatives
H_α	Hinge moment due to angle of attack	N_ξ, N_ζ	
H_β	Hinge moment due to tab movement	n	Configuration factor
H_η	Hinge moment due to elevator movement	n	Manoeuvring load factor
		n_a	Applied acceleration
h	Fraction of SMC which locates the CG	$n_p, n_r, n_\upsilon,$	Yawing moment derivative coefficients
h_m	Manoeuvre point, stick-fixed, on SMC	n_ζ, n_ξ	
h'_m	Manoeuvre point, stick-free, on SMC		
h_n	Neutral point, stick-fixed, on SMC	P1, P2	First (command) and second pilot
h'_n	Neutral point, stick-free, on SMC	P	Rated power
h_0	Location of ac of wing-plus-fuselage on SMC	P	Stick force
		P_A	Probability of an event (also P_0)
$-\Delta h_0$	Forward shift of ac of wing-plus-fuselage along SMC	P_F	Probability of fatality
		P_{G-N}	Probability or equalling or bettering nett performance
I_{xx}	Moment of inertia about O–X axis = A	P_0	Probability of an occurence (also P_A)
I_{yy}	Moment of inertia about O–Y axis = B	P_t	Thrust horsepower
I_{zz}	Moment of inertia about O–Z axis = C	p, p_1, p_2	Different, related, static pressures
i_s	Tail stabilizer incidence	p	Rate of roll about O–X axis
i_w	Wing incidence	p_η	Powerplant effectiveness
K	Control response factor	Q	Torque (sometimes T)
K	Induced wing (vortex) drag factor, $1/e$	q	Dynamic pressure
K'	Induced drag factor of complete aeroplane	q_{ws}	Dynamic pressure (in wake) at stabilizer (tail)
K	Overturning coefficient		
K	Seaplane load coefficient	R	Resistance in water
K_n	Static margin, stick-fixed	R	Reynolds number
K'_n	Static margin, stick-free	R_l, R_x	Reynolds numbers at stations l and x
k	A constant	r	Yaw rate about O–Z axis
k	Critical roughness of a surface		
k_x, k_y	Original drag and lift coefficients ($\frac{1}{2}C_D$ and $\frac{1}{2}C_L$ respectively)	SD	Standard deviation
		SMC	Standard mean chord
k_{xx}, k_{yy}, k_{zz}	Radii of gyration about O–X, O–Y and O–Z axes	S	Wing area
		S_f	Area of fin-plus-rudder, also of foreplane
L	Lift	S_r	Area of rearplane
L	Rolling moment about O–X axis	S_s	Stabilizer area
L_s	Lift of stabilizer	S_w	Wing area when it is necessary to make a distinction
L_w	Lift of wing (when distinction is needed)	S_δ	Area of surface under consideration
$L_p, L_r, L_v,$	Rolling moment derivatives	s	Semi-span of wing, $b/2$
L_ξ, L_ζ		s	Take-off distance, TOD
(L/D)	Lift/drag ratio	s_F	Landing distance from 50 ft
$(L/D)_{max}$	L/D for best glide	s_G	Deceleration distance on ground
$(L/D)_R$	L/D for maximum range		
l	Litre	T	Temperature
$l_p, l_r, l_v, l_\xi, l_\zeta$	Rolling moment derivative coefficients	T	Torque (alternative to Q)
		T_n	Total number of cycles
M	Mach number	t	Temperature, when distinction is needed
M_{crit}	Critical Mach number		
MAC	Mean aerodynamic chord	t/c	Thickness/chord ratio
MCP	Maximum continuous power		
MP	Manoeuvre point, stick-fixed		
MP'	Manoeuvre point, stick-free	u	Linear velocity (along O–X axis)
M	Pitching moment (usually about O–X axis)		
		V	True airspeed, TAS
M	Total mass of parts, (W/g)	V_A	Design manoeuvring speed
$\bar{M}$	Figure of merit	V_{AT}	Original target threshold speed

Symbol	Meaning	Symbol	Meaning
V_{AT0}	Target threshold speed, all engines operating	W_E	Empty weight (sometimes zero-fuel weight)
V_{AT1}	Target threshold speed, one engine inoperative	W_F	Weight of fuel consumed
		W_{Fuel}	Weight of fuel loaded
V_{AT2}	Target threshold speed, two engines inoperative, and so on	$W_{MAX\,dry}$	Weight limit for a given dry take-off distance
V_B	Design speed for maximum gust intensity	$W_{MAX\,wet}$	Weight limit for a given wet take-off distance
V_C	Design cruising speed	W_{Pay} or W_p	Weight of payload carried
V_D/M_D	Design diving speed and Mach number	W_{ZF}	Zero-fuel weight
V_{DF}/M_{DF}	Demonstrated flight diving speed and Mach number	W_0	Gross or all-up weight
		$W_1, W_2, \ldots,$	Incremental weights
V_{EF}	Critical engine failure speed	W_n	
V_F	Design flap-DOWN speed	w	Wing loading, W/S
V_{F1}	Procedural design flap-DOWN speed		
V_{FC}/M_{FC}	Maximum speed and Mach number for required stability characteristics	X	Force in direction O–X
		x_{max}, x_{min}	Largest and smallest observations in a set of data
V_{FE}	Maximum speed with flap extended	x_n	Moment arm of an item of weight
V_{FT0}	Final take-off speed, one engine inoperative		
V_H	Maximum level speed, MCP	Y	Force in direction O–Y
V_i	Equivalent airspeed, IAS	$Y_p, Y_r, Y_v,$	Sideforce derivatives
V_{LE}	Maximum speed, landing gear extended	Y_z, Y_x	
V_{LO}	Maximum landing gear operating speed	$y_p, y_r, y_v,$	Sideforce derivative coefficients
V_{LOF}	Lift-off speed	y_z, y_x	
V_{MBE}	Maximum brake energy speed	y_1, y_2	Complementary fractions of wing semi-span
V_{MC}	Minimum control speed		
V_{MCA}	V_{MC} in the take-off climb		
V_{MCG}	Minimum control speed on or near the ground	Z	Force in direction O–Z
		$Z_p, Z_r, Z_v,$	Normal force derivatives
		Z_z, Z_h, Z_x	
V_{MCL}	Minimum control speed, approach and landing	$z_p, z_r, z_v, z_z,$	Normal force derivative coefficients
		z_η, z_ξ	
V_{MO}/M_{MO}	Maximum operating limit speed and Mach number		
V_{MU}	Minimum demonstrated unstick speed, all engines	α	Angle of attack
		α_s	Stabilizer angle of attack
V_{NE}/M_{NE}	Never-exceed speed and Mach number	α_{si}	Stabilizer setting (rigging) angle
V_{NO}/M_{NO}	Normal operating speed and Mach number	$\alpha_{0.9}$	Angle of attack for $0.9\,C_{Lmax}$
V_R	Rotation speed on take-off	β	Angle of attack of fin (also yaw angle)
V_{RA}	Rough air speed	β	Tab deflection
V_{REF}	V_{AT} (USA)		
V_S, V_{S1}	Stall speed or minimum steady flight speed (V_{S1} refers to a configuration other than for landing)	γ	Ratio of specific heats
		γ_{min}	Minimum angle of glide
V_{S0}	Stall or minimum steady flight speed, landing	Δ	An incremental change in a quantity: e.g., $\Delta C_L, \Delta a$
V_{S1g}	The 1.0 g stall speed when $L = W$	Δ	Incremental induced drag factor, $[(1/e) + \Delta]$
V_{SSE}	Speed for one-engine-inoperative training	δ	An alternative incremental change in a quantity
V_T	Maximum aero-tow speed		
V_{Tmax}	Generalized maximum threshold speed	ε	Angle of downwash
V_X	Best gradient or angle-of-climb speed		
V_Y	Best rate-of-climb speed	η	Efficiency (e.g., of a propeller)
V_{YSE}	V_Y with an engine failed	η_K	Pazmany efficiency factor
V_{app}	Approach speed	η_{trim}	Elevator angle to trim
V_i	Indicated airspeed, IAS	η_s	Stabilizer efficiency (in wake)
V_1	Decision speed on take-off		
V_2	Take-off safety speed	Λ	Angle of wing-sweep
V_3	Steady initial all-engines climb speed	λ	Taper-ratio, tip chord/root chord
$\bar{V}_f$	Fin-volume coefficient		
$\bar{V}_s$	Stabilizer or tail-volume coefficient	μ	Braking, friction or rolling coefficient
$\bar{V}_\zeta$	Rudder-volume coefficient	μ_{dry}, μ_{wet}	Braking coefficients, dry and wet surfaces
$\bar{V}_\eta$	Elevator-volume coefficient		
$\bar{V}_\xi$	Aileron-volume coefficient	μ'	Thrust specific fuel consumption
v_c	Rate of climb	μ_1	Longitudinal relative aircraft density
		μ_2	Lateral relative aircraft density
W	Weight of aircraft		
W_{Disp}	Disposable load ($= W_{Fuel} + W_{Pay}$)	v	Coefficient of kinematic viscosity, μ/ρ

Symbol	Meaning	Symbol	Meaning
v_0	Coefficient of kinematic viscosity at sea level	ψ	Angle of yaw (general notation)
		ψ_1	A variable function affecting the stick-fixed static margin, H_n
ξ	Aileron deflection		
ξ	A factor on take-off distance due to N_1	Ω	Rate of rotation about spin axis
		ω	Angular velocity (or rate) of rotation
π	Circumference/diameter of a circle $= 3.141593\ldots$		
		AIRFIELD TERMS	
ρ	Ambient density of air	EDA	Emergency distance available
ρ_0	Ambient density at sea level	EDR	Emergency distance required
		LDA	Landing distance available
σ	Relative density, ρ/ρ_0	LDR	Landing distance required
σ	Size of a population	TODA	Take-off distance available
σ_x	Size of a population sample	TODR	Take-off distance required
ϕ	Angle of bank	TORA	Take-off run available
ϕ	Included angle of bevelled trailing-edge	TORR	Take-off run required
χ	Wing-tilt angle		
χ_1	A variable function affecting the static margin, H_n', stick-free	**COMMONPLACE TERMS**	
		$\$$	Dollar (US)
		£	Pound sterling (UK)

Equivalent Anglo-American terms

English	American
accelerate-stop	— RTO (rejected takeoff)
aerobatics	— acrobatics
aeroplane	— airplane
airworthiness requirements BCAR	— FAR
allowable deficiency	— no-go item, despatch deviation or minimum equipment item
amateur-built	— homebuilt
boost pressure	— manifold pressure
coaming	— glareshield
clamshells	— buckets
compressor	— spool
EAS	— CAS (at sea level only)
engine acceleration	— spool-up
flick roll	— snap roll
inter-com	— inter-phone
Mach trimmer	— pitch-trim compensator
maximum continuous	— METO (maximum except take-off; piston engines only)
microlight (or minimum aircraft)	— ultralight
relight	— flight start
spectacles	— yoke
tailplane	— stabilizer
throttle	— thrust lever
undercarriage	— gear
unstick	— lift-off
V_{A1}	— V_{ref}

A comprehensive source of terms is to be found in: Gunston, W., *Jane's Aerospace Dictionary*, London: Jane's Publishing Company Limited, 1980

FLYING QUALITIES AND AIRWORTHINESS

Introduction

'The ideal type of pilot for production testing is a man of many parts. Basically he must be an average, or better, pilot. He must have great patience in learning the causes of faults in aircraft and all their equipment; this entails training his mind to be analytical and to have a great capacity for small detail. The successful test pilot should possess a sound practical knowledge of structures, aerodynamics, heat engines, hydraulics, electricity, and physics.'

J. A. Crosby Warren, *The Flight Testing of Production Aircraft* (Pitman, 1943)

– half a century after Lilienthal (whose work had inspired the Wright Brothers, who had flown at Kittihawk, forty years earlier, on 17 December 1903)

This quotation shows the magnitude of the strides taken to understand thoroughly the how and why of an aircraft in flight in just the handful of decades which followed the successes of Lilienthal and the Wrights. There is a need here to prepare the ground before attempting to separate and assess the mass of causes and effects we now call flying qualities. Civil aircraft are the bulk of those dealt with in the examples, so that airworthiness and risk assessment need an early mention because it is impossible to fly a civil aircraft which does not comply to a degree with airworthiness criteria set by the regulatory airworthiness authorities.

What are Flying Qualities?

The scope of what is meant by flying qualities of a large aeroplane is shown in Table 1–1, while here we simplify and concentrate upon the following, omitting what is asterisked in the table:

☐ Performance.
☐ Handling qualities.
☐ Functioning (of systems).

Of these three there is little to say about systems beyond those linking, simply and mechanically, the hand and foot of the man to the machine. Performance is dealt with in two chapters: one, field performance and the way in which despatch rules are devised; the other with climb, glide and high speed.

An aircraft in motion has six degrees of freedom, three in translation (forward, sideways, upwards or downwards) and three in rotation (roll, pitch and yaw). A pilot has four main

TABLE 1–1

Scope of flying qualities

Handling	Stability. Controllability. Stalling. Ability to trim.
Performance	Field distances. Climb rates and gradients. Descent rates. Range and endurance*.
Functioning	Flight controls. Stability augmentation*. All other systems*.
Flight deck / Cockpit	Ergonomics.
Automatics*	Flight control systems (FCS/FBW/FBL) and autopilots.
Ground and water handling	Steering. Brakes. Operation on and off the step.
CG-position	Effect on handling qualities.
Cross-wind handling	Drift. Rolling tendencies.
Turbulence*	Motion. Instrument indications. Wake effects of helicopters and larger aeroplanes.
Buffeting	Stall. High Mach number. Manoeuvring. Airbrakes, gear and flaps.
Noise*	Aerodynamic. Engines and propellers. Gear, flaps. Air-conditioning. Audio warnings.
Motion	True. Apparent.
Visual	Day. Night. Dusk. Clear/direct-vision panels.
Exhaust emissions	Carbon monoxide (CO).

* Omitted here.

controls: aileron, elevator, rudder and throttle(s) (or power lever(s)) – and gravity. To cope with six degrees of freedom using four main controls is complicated.

Those responsible for recommending aircraft for certification are sometimes forced to argue with applicants for Cs of A. When a deficient flying quality reveals a non-compliance, a frequent response is: *'So what, does it matter – anyway, I can cope with it?'* – implying that the requirement is wrong, or silly, or out of date. First, I cannot think of any airworthiness flight requirement that does not have its origin in blood or misery or heartbreak. Second, and these things have happened: when later the same applicant has had a heart attack/wrapped his car around a tree/lost his medical category/gone into receivership, his dependents or the receivers will want to sell the aeroplane. What if the next owner, who is not as skilled as the applicant, finds too late that he is unable to handle it?

What is 'Airworthiness'?

Airworthiness is defined as 'the continuing capability of an aircraft to perform in a satisfactory manner the flight operations for which it was designed'. By flight operations we mean:

- ☐ Flight crew workload.
- ☐ Flight handling characteristics.
- ☐ Performance within the design envelope.
- ☐ Safety margins.
- ☐ Welfare of occupants.
- ☐ Punctuality (which is a consequence of reliability).
- ☐ Economics.

The passage of time affects none of these things. All that has changed is the size, the complexity and level of aircraft performance, material technology and design techniques.

The Chicago Convention and the ICAO

The international scale and scope of civil aviation necessitates international agreement among nations about air safety standards. The Chicago Convention of 1944 set up the International Civil Aviation Organisation (ICAO), which is an organ of the United Nations Organization. One task of ICAO is to adopt standards and recommendations on airworthiness, in the form of published national and international codes.

Considerable work is involved in demonstrating that an aircraft designed to such a code satisfies it in every respect. The indicator of credit-worthiness in the eyes of the uninvolved public (the innocent third party) lies primarily in the award of Type Certification, marked by a C of A.

Certificates of Airworthiness

The award of a C of A is akin to issuing a passport to go abroad. It permits an aircraft to fly in the sovereign airspace of any country which is a signatory of the Chicago Convention.

Test flying is not a matter of 'throwing an aircraft about'. It is a cool, meticulously painstaking process of examining, bit by bit and step by step, the nature of a piece of flying machinery which can be lethal and, more often than not, costs the earth. Thus, in civil matters one finds that the Convention spells out the scope of what is required. It is published by the ICAO and contains 18 Annexes, of which Annex 8 is concerned with civil flight testing (*see* Appendix A). The Military also has its equivalent sets of specified standards to Annex 8, plus specialized military specifications and defence standards in the form of US Mil Specs and UK Def Stan. Flowing from Annex 8 on the civil side are the US Federal Aviation Requirements (FARs), European Joint Aviation Requirements (JARs) and the older British Civil Airworthiness Requirements (BCARs). Airworthiness covers every material aspect of an aircraft: airframe, engine, systems, avionics and the operational carriage of a payload of passengers and freight.

The eighteen ICAO topics are:

(1) Personnel licensing.
(2) Rules of the air.
(3) Meteorological service for international air navigation.
(4) Aeronautical charts.
(5) Units of measurement to be used in air and ground operations.
(6) Operation of aircraft.
(7) Aircraft nationality and registration marks.
(8) Airworthiness of the aircraft.
(9) Facilitation.
(10) Aeronautical telecommunications.
(11) Air traffic services.
(12) Search and rescue.
(13) Aircraft accident investigation.
(14) Aerodromes.
(15) Aeronautical information services.
(16) Environmental protection.
(17) Security – safeguarding international civil aviation against acts of unlawful interference.
(18) Safe transport of dangerous goods by air.

A fundamental principle of the Chicago Convention is that all states should be able to participate in air transportation as equals. Such a requirement presupposes good faith on the part of states in their dealings with one another, and to a regard for fair play in the matter of equal opportunity. Where the principle falls down is in the invocation of sovereignty of any state over its own airspace, which the Convention recognized that states retained, according to the ancient Roman adage:

'Cujus est solum, ejus est usque ad coelum et ad inferos.'

i.e. '*The owner of the land owns everything above it (to the sky) and also everything beneath it (to the depths)*'. Governments seek advantages for their flag carriers and other national airlines, and impose major limitations on foreign carriers. These may affect the numbers of passengers to be carried, frequency of flights, and other vital matters, and can take the form of sly and covert extensions of foreign policy.

There is a burden of responsibility in determining whether or not an aircraft is fit for the purpose intended and worthy of a C of A. In the event of an accident one of the first questions is '*Did the aircraft have a valid C of A, or Permit covering the operations, and if so on what basis was it issued?*' This brings us back to the matter of flight testing, because a certification test pilot approaches an aircraft to find out three things:

☐ What is the *nature* of the machine coming on to the national Register?
☐ What *numbers* are needed to establish a measure against which to issue a Type Certificate, a C of A, or a Renewal next time (and to assess other examples of the type for certification)?
☐ Does the aircraft have any *dangerous features*?

It is surprising how many potentially dangerous features can be found. It is a consequence of being human, and it is the best reason for having an impartial certificating organization tasked with a duty of care for the innocent and uninvolved general public.

Risk and Safety Assessment

In recent years the mathematical theory of statistical probability has appeared in what looks at first sight like a new scientifically-mounted application of airworthiness requirements to the safety of aircraft. The mathematical approach, in which quantitative numerical techniques are used is new, but the philosophy is not only old, it is reflected in every field of human endeavour. Every day, in many situations, at home and at work, we try to choose courses which reduce risk to a minimum. In aeronautics, mathematics is enlisted to help ensure that the more hazardous an event the lower will be the probability of it occurring. One has only to look back at the history of aircraft design and the different trends set since the Wright Flyer of 1903, to realize that hazards force changes of direction and technical advance – and these have occurred fastest and most dramatically in the crucible of war.

Experience in war brought new developments, innovations or simply in-built redundancy. Examples are the modern aileron, because wings that could be warped by the pilot were not strong enough at higher speeds and generated too much adverse drag on the down-going side. As aeroplane performance and potential increased, the need for swift response to control inputs by the pilot led to the introduction of aerodynamically balanced control surfaces to reduce the increasing forces to be overcome. This is discussed initially in Chapter 2 when we examine the different flying qualities of two aeroplanes from World War I which are favourites with the public at air shows.

Another example was the fitting of engines with two magnetos, because one frequently failed and that meant a forced landing or a crash. Not only that, but quite early it became conventional for electrical switches to be UP for ON, so that they could be knocked OFF with a downward sweep of the hand to remove one fire hazard. Sir Thomas Sopwith (1888–1989), early aviator and manufacturer, interviewed not long before he died said: '*we had lots of lovely crashes*', adding that speeds were so low that one rarely got badly hurt.

The pursuit of speed led to the monoplane – and rapidly to its reappraisal – because of structural failures which resulted in the wings folding, leading-edge downwards at high speed and shallow angle of attack. The cabane bracing, which was believed to be needed only to hold the wings in place on landing, was lowered in height to reduce its drag without it being realized that the bracing also opposed the leading edge-down pitching moment caused by the aft movement of the Centre of Pressure on the wing. This improperly understood 'weakness' lay behind a succession of monoplane failures. This led in turn to a policy of continuing development of the stronger, stiffer and lighter military biplane, particularly in Britain, well into the 1930s. Even so, high loads on wings caused designers to double the lift-bracing wires of biplanes and monoplanes at an early date, a wholly natural belt-and-braces solution.

To avoid what is now known as *common mode failure* designers and engineers introduced dissimilar redundancies. A hand pump to lower flap and landing gear in an emergency was an early example. Now a ram air turbine (a RAT) is used on high performance aeroplanes. Alternative air sources, manually controlled, for the carburetter system of a piston engine, emergency batteries, manual over-rides for electric flying control systems, are further examples.

People are fallible and two pilots are needed for civil transport category aeroplanes weighing more than 5700 kg (12 500 lb). Duplicate inspections are carried out by separate, independent, inspectors after flying control systems have been broken down and reassembled.

The requirements governing the strength and stiffness of structures are to ensure that the airframe is durable enough to resist forces and excitations by accelerations, gusts and turbulence, even though these are thought to be remotely probable.

The Concorde supersonic transport aeroplane is an example of statistical failure assessment being used throughout. This aeroplane first flew more than 20 years ago and has been one of the most successful proof-of-concept civil aircraft ever built. The same techniques are used increasingly by manufacturers, for example, Boeing in the USA, Airbus Industrie in France, British Aerospace and SAAB in Sweden.

Failures can be hazardous in more ways than one. Safety margins may be reduced, as when a loss of power of an engine erodes the rate of climb and obstacle clearance. There may be a reduction in strength of the airframe caused by, say, an uncontained turbine burst. A control system failure almost invariably degrades the handling qualities and increases workload on the crew. Yet another may result in injury or death. One catastrophic failure may cause all of them.

A catastrophic event results in the loss of the aeroplane and/or more than a small proportion of the occupants being killed. The failure of a single window might result in nothing more serious than shock and painful ears through rapid decompression. But a window failure which results in the captain of the aeroplane being sucked half-way out, or a passenger sucked out completely (both of which have happened) have the potential for catastrophe, but are not catastrophies in themselves. However, if the roof of the cabin had opened causing a number of passengers to be sucked out; or if the failure of bolts in an engine pylon causes the engine to become detached, knocking away the one alongside, degrading low-speed lateral and directional control enough to result in the heavily laden aeroplane hitting a block of flats, then multiple fatalities would convert what started as an apparently simple initial failure into a catastrophe.

In Chapter 6 we look at the use of risk assessment only in the matter of field performance. Risks are of two kinds:

☐ *Individual risk*, in which a person takes his life in his own hands. More often than not it is a matter for the person concerned to decide what level of risk he accepts as tolerable in a given set of circumstances, and that level is higher than:

☐ *Societal risk*, in which one's life is in the hands of others in situations which result in multiple fatalities when things go wrong. The level quantified as being publicly acceptable is a matter of mass perception and depends upon many social and other factors, including the size (capacity) of aircraft which have featured in recent accidents, the nature of an accident and the attitude of the press. The loss of a large ship and its crew, sinking quietly beyond the horizon, often attracts no more than a line. But aircraft accidents, which tend to happen in public, are the subject of infinitely more attention by the press, even though there may have been fewer fatalities.

There must be a limit to a book of this kind, which is that in general it is confined to discussion of aircraft with manual control systems. Modern military aeroplanes, with fly-by-wire, and fully powered control systems can make an aerodynamically unstable military aircraft feel stable by artificial means. Having said that, military aeroplanes are not dismissed entirely, because they often possess features which are useful when making a point. The same applies to large transport aeroplanes, which are not dealt with for the same reason, beyond noting a feature here or there which serves to illustrate something fundamental.

Among civil aeroplanes light twins, defined as having not more than nine passengers, with a maximum take-off weight of 5700 kg (12 500 lb), are a possibility. But more interesting as an upper limit is the civil commuter category aeroplane, with seats for nine to 19 passengers and a maximum weight of 8600 kg (19 000 lb), as defined in the US FAR 23. It has the advantage of linking the light and heavy ends of the spectrum of manually-controlled aeroplanes. For purposes of certification commuters are treated like much larger public transport aircraft, because they are capable of carrying and affecting the lives of a significant number of people.

Commuter risk assessment is at a societal level. Whereas at the other end of the scale light aeroplanes, especially one or two-seat aerobatic and microlights, do not warrant such treatment because they rarely involve anyone who, leaving child passengers aside, is not aware of the individual risks involved.

That is not to say that the individual matters less than the crowd. Even the largest crowd is no more than a mass of individuals. As far as regulatory authorities are concerned they are constrained by public perception of what is acceptably safe, and the public expects tighter and higher standards of safety than does the individual taking his own life in his hands.

Permits to Fly and Experimental Cs of A

In the following sections we look in two other significant directions, towards aircraft which, in the UK, are given Permits to Fly, and in the USA an Experimental Certificate of Airworthiness.

In the UK a Permit is a passport to fly in sovereign airspace, with certain permissions and, therefore, restrictions. Unlike a C of A, to fly abroad it is necessary to negotiate with a foreign government to accept the terms of the Permit. Because Permit aircraft are not given

Cs of A it would be inconsistent to cause them to comply with a published code of airworthiness requirements. However, they are treated with a shrewd eye for good airworthiness practices. Although the letter of airworthiness requirements is lacking their flavour and underlying philosophy is at the heart of surveryor's airworthiness inspections and flight tests (which are airborne inspections) for the award of Permits to Fly.

Around the world there is growing interest in the sizeable and wealthy airshow industry and the historic, classic and vintage aeroplanes (both original, full-scale and scale replica) upon which it relies. In the UK the Historic Aircraft Association (HAA) maintains a register of pilots who have undertaken, by the adoption of a code of conduct, to display such valuable historic aircraft (and not themselves) with the degree of sympathy and care which are their due. Elsewhere numerous organizations exist to own and operate such machines. In the USA the independent Confederate Air Force (which has charitable status), also the Antique/Classic Division, and Warbirds of America (both of which belong to the Experimental Aircraft Association, EAA), provide considerable resources and expertise for the reconstruction and maintenance of vintage, classic and historic aircraft, in both static and working order.

Much of the research needed for *The Design of the Aeroplane* was useful in work for the CAA when it came to supporting recommendation of Permits to Fly for a number of historic, vintage and classic machines. It led into historic corners of aircraft design, of control and stability, of engines and their handling, of aircraft operation, and advice in pilot's and other notes.

For this and the reason connected with the RAE Farnborough, given earlier, this book draws heavily upon history, because it contains many valuable lessons which often must be relearned. They are as relevant today as they were yesterday, or 100 years ago when Lilienthal was flying; or in 1809 when Sir George Cayley, experimenting, then built a full-size glider, which Gibbs-Smith called the first tentative glider to carry a man in history. Quoting Cayley – who did not himself fly – it would *'frequently lift him up, and convey him several yards together'*. Accidents and their outcome keep on repeating themselves for the same physical and human reasons, while only aeronautical technology differs, generation by generation.

Similarly, the highly motivated homebuilder of (amateur-built) light aircraft is enabled to fly by the award of a Permit to Fly in the UK. In the USA such flying is enabled by means of the Experimental C of A of the FAA. The permissions and restrictions are similar, only the titles differ. In the UK the award of a Permit implies no such burden. In the face of much past argument and criticism levelled at the CAA, and the ARB before it, for not adopting the FAA title, the UK (rightly) preferred to reserve the title C of A for the document which bore a burden of airworthiness, by signifying compliance with published airworthiness requirements.

It is with experimental aircraft produced by enthusiastic amateurs that small but significant advances are often made. In the USA the FAA and the military keep an eye open from the sidelines, to see what is produced in garages and spare corners of hangars. One has only to mention Burt Rutan in the USA, for example, to conjure images of sleek tail-first aeroplanes for homebuilders, and the ultimate long-range record-breaking Voyager (a tail-first tandem-engined twin) flown around the world non-stop in December 1986 by his test pilot brother Dick, and Jeanna Yeager.

In the late 1960s and early 1970s Rutan was carving a path in the direction of efficient tail-first aeroplanes. A plethora of canards culminated in the two-seat around the world non-stop Voyager. About the same time another American, Jim Bede, caught the imagination of home-builders with his diminutive and slightly more conventional BD-5, with a tail at the back and a pusher propeller – followed later by a jet-powered variant. Now Bede has

Plate 1–1   McDonnell Douglas F-15. (Courtesy of McDonnell Douglas and Philip Jarrett Collection)

produced the transonic, kit-built, jet-powered BD-10 two-seater, with swept wings, twin fins and features resembling a miniature McDonnell Douglas F-15 Eagle. At an estimated $600 000 with avionics in 1993, the BD-10 is likely be out of reach of many who once afforded the BD-5. This is an indication of the changing 'ball-game' at the upper end of the range.

Plate 1–2   Homebuilt BD-5B with turbocharged 1200 cc Honda
'Civic' engine. A little aeroplane with quick reactions due to the
authority of aerodynamics over inertia. (Courtesy of Howard Levy)

The BD-5, which was for me a skittish joy to fly, suffered from the failure of engine
manufacturers to meet the enormous public demand for kits at that time. Impatience to get
airborne with the wrong two-stroke engines caused a number of fatal accidents. Bede fired
the imagination of the public in a similar way to Henri Mignet in France, around half a
century earlier, with his Pou du Ciel, or 'Flying Flea'. Although subsequently banned in its
original form, because of initial dangerous features which caused fatal accidents (Ref. 1.5),
derivatives of the Flea are delightfully simple and relatively safe, when designed and used
properly.

Flying qualities are not constrained by weight and size. Push the stick forward and down-
elevator is applied to a three-axis microlight, an advanced fighter, a heavy transport, or a
historic replica. When rigged, the only difference in flying qualities between a microlight
(responding more like a feather than a dynamic brick), a light or a small aeroplane, or one
which is potent and heavy, are the control forces, rates of response, and levels of performance, not their nature. What is said here applies to a far wider range of aeroplanes than those
which need only one pilot to fly them.

SECTION 2

CONFIGURATION AND CHARACTERISTICS

CHAPTER 1

The Shape of the Aeroplane:
What the Drawings Reveal

'Form follows function.'
Attributed to Leroy Randle of Grumman

The *shape* of an aeroplane is determined by its intended purpose. Some aircraft are elegant and sleek, others are chunky and ungainly. Some have squared corners and gaping apertures, which can be opened and closed in different phases of flight. Others have fixed landing gear; while those with critically high performance have relatively high aspect ratio wings, slots, flaps (maybe leading as well as trailing-edge) and retractable landing gear, to give them wide speed-range while operating from airfields which are as short as possible for the weight lifted. The shorter the airfield, the greater the utility and marketability of the aircraft. Each physical feature is blended with as much artistry as the designer and his team of aerodynamicists, propulsion, systems and structural engineers can muster. The way in which the aircraft flies at any particular point in the flight envelope is the result of its *configuration* at the time. If it does not fly as well as it should, then the configuration must be fettled and 'tuned' by modifications, called 'fixes'.

Configuration is a precise word meaning far more than what the aircraft looks like when body, engine nacelles, wings and tail or foreplane surfaces have been assembled. It includes changes in flap, wing-sweep, landing gear position, propeller pitch, throttle and power-setting, deployment of air brakes, tail parachutes or drag-chutes, lift dumpers, thrust reversers, use of reheat. It also includes physical changes to suit alteration of role (e.g., bomb and other doors open for despatching stores or parachutists).

In the world of civil airworthiness certification configuration means especially *loading* and *location* of the Centre of Gravity, CG, and power settings, all of which profoundly affect flying qualities. However, in the world of military test flying configuration tends not to include location of the CG, which is left aside as a separate item. The reason for this is that configuration is seen as driven by the phase of a mission, while CG is determined by less directly controlled activities, like fuel-burn, in-flight refuelling and the release of weapons or stores.

Learn from National Differences in Problem Solving

Every aircraft represents a set of solutions to a problem with many interrelated facets. These facets are not always apparent when looking at a three-view drawing alone. There is no substitute for getting close to an aircraft, gathering as many photographic and other details as possible, and watching it intently in *manoeuvring* flight. A case in point has been the revelation of Russian fighter technology since about 1989. Mikoyan's MiG-29 (NATO codename 'Fulcrum') and Sukhoi's Su-27/Su-35 ('Flanker') are fighters with an edge over those of the West in demonstrated post-stall agility, needing trainers, like Sukhoi's projected Su-54, with closely similar flying qualities.

The Russians (and the Israelis) listen to their test pilots, who have the freedom and authority to say what *nature* of aircraft they think should be developed for a particular purpose. The limitations and boundaries within which their aircraft operate originate from the sharp end, with no excess of technology beyond that needed to achieve what is required. The Russians still use wind-tunnels where many in the West now resort to computer graphics. Run your hand over flying and other surfaces, right down to the tiny lateral vortex-generating strakes at the root of the nose pitot tube of the MiG-29, and you realize that Russian aeroplanes are uncompromisingly built to do a job. In the West we seem more concerned with producing aircraft as elaborate vendors of ever more advanced technology packages, so that the pilot at the sharp end is there to do as he is told with a product and limitations defined by someone who, no matter how brilliant, may never have been there and actually done it. It is said that Israeli test pilots wrote their own software for their fighter, Israeli Aircraft Industries' Lavi – and that the top role-model for youth in Israel is the fighter pilot.

There is much to be said for the Russian and Israeli approach: producing test pilots who are effective in the cockpit and in judgement, with voices which are authoritative in the design office and boardroom. This is the no-nonsense way to improve one's company and its products and, of course, the effectiveness and relevance of one's aircraft industry. Those who seek savings to the economy by cutting, say, an establishment like a test pilot school should realize this.

Take, for example, the relevant observations of Valery E. Menitsky, Chief Test Pilot of the Mikoyan Design Bureau, when talking about development of the (albeit fly-by-wire) MiG-29M (as reported in the *1994 Proceedings of the Society of Experimental Test Pilots*):

> '... we have introduced limitations considering many factors such as Mach and AOA (*angle of attack*). There are two possible approaches: (1) to develop an aircraft with good flying characteristics in a limited flight envelope; or (2) to develop an aircraft with a wider performance envelope and to intentionally accept some degradation of flight qualities at higher angles of attack.
>
> The first approach is easier. Our opinion, not always shared by the Soviet Air Force, is that the second approach offers more potential, since we consider an aircraft and its pilot to be an integral system ... Although the latter approach requires greater efforts in terms of increased flying hours, more dedicated training in free air combat manoeuvring, and flight into critical handling regimes, it guarantees a significant increase in the overall quality and capability of the aircraft–pilot combination. This method also allows a pilot to use his intellect and initiative to their fullest extent.'

With this purpose in mind, now let us run an eye over the external shapes of three aeroplanes, to see what bearing these might have upon their possible flying qualities and the relationship between aircraft and pilot. Technical terms are used without much clutter of formal definition, because most readers are already used to aeronautical terminology through the wide range of popular publications now available. But, as in Ref. 1.5, there is a tying-together in Chapter 3, which defines and links them more precisely. That chapter is inserted at a point in the text where it may be used either as a refresher or a reference.

Unfortunately aircraft are vastly expensive to build, and the tendency (especially where military projects are concerned) is to call for as much role-flexibility as can be compressed within one airframe by a committee. Too often the penalty is reduced performance. Historically, many successful aeroplanes have been conceived for one primary role. Later they have been modified for other originally unforeseen roles, sometimes with surprising, at other times mixed, success.

The value of a test pilot in a company must be to provide an input of streetwise experience from the world outside. His value as the voice of reality is especially high today when much technical design is electronic. Without practical leaven a consequence can be '*garbage in, garbage out*', resulting in mistakes and later accidents.

When confronted with a technical committee it is worth attempting to concentrate decisions in favour of optimum configuration and performance from an aircraft which is good technically, even though it might not be the best for the main job for which it is intended. There is or was a plaque at Bristol Engines, made in 1939 for Sir Roy Fedden who, according to his biographer Bill Gunston, regarded these words by James H. 'Dutch' Kindelberger, of North American Aviation, as his favourite text:

> 'It may not always be the best policy to adopt the course that is best technically, but those responsible for policy can never form a judgement without knowledge of what is right technically.'

Avoid being diverted. It is relevant in a technical project to apply a cardinal one of ten *Principles of War* defined by Karl von Clausewitz (1780–1831), a Prussian staff officer when Napoleon was roaming Europe:

> 'Selection and maintenance of aim.'

Disregard of that principle on two occasions in particular had disastrous consequences for German control of the air in World War II. First was the Luftwaffe's change to night bombing from attacks upon the RAF fighter force, which was fast running out of pilots. Second, the Fuhrer-Befehl of Adolf Hitler, was fatal for Germany. Hitler caused the advanced Messerschmitt Me 262, a twin-jet fighter to be produced as a 'Blitzbomber' instead. Engines were unreliable and there was insufficient development of an aeroplane which might have tipped the balance against the Allied bombing campaign and changed the outcome of the later air war over Western Europe. Many times since that war aeronautical projects have been cancelled because costs escalated beyond reason as a result of politically-driven sponsors subsequently altering an aim by asking for too much to be incorporated in one airframe (Ref. 1.1).

Analysing the Drawings

The first sight of an aeroplane is usually through a picture or a drawing. Good drawings are of enormous value and one is worth a ton of words. For years I made it a practice to incorporate a three-view general arrangement drawing in each flight test report. This was once commented upon favourably in court by a judge who found it '*useful to know what the lawyers are talking about*'. Collect drawings, but be careful. Even the best and most

Plate 1–3a Mikoyan MiG-29 ('Fulcrum'). (Courtesy of *Flight International*)

Plate 1–3b Sukhoi Su-35/27M (Flanker). (Courtesy of *Flight International*)

Plate 1–3c Projected Sukhoi Su-54 trainer. (Courtesy of Douglas Barrie, *Flight International*)

Plate 1–3d Israeli Aircraft Industries' Lavi. (Courtesy of Menachem Schmul, IAI)

Plate 1–4   The Messerschmitt Me 262. '... that's a wicked aeroplane ... wicked ... WICKED! I'm sure glad that they screwed up the tactical use of this aeroplane.' Quoted comment by General Carl A. Spaatz, USAAF, in 1944 (Hugh Morgan and Isolde Baur, Watson's Whizzers, *Aeroplane Monthly*, January 1995). (Courtesy of Philip Jarrett Collection)

beautiful of published drawings can be wrong in detail. Not all manufacturers produce ideal drawings for publication. Not all published drawings are accurate.

Similar advice applied to the homebuilders and kit-builders, who are increasingly one and the same. It is now possible to buy exotically tailored kits for nearly as much as a second- or third-hand commercial product (which has the advantage of having most of the bugs removed) already. Whereas the slinky and elegant, as well as the apparently basic and safe, kitted aeroplane may still possess design features and flying qualities which range from unpleasant to downright dangerous. Examples which occur time and again are inefficient tail configurations, excessive control circuit friction and aileron drag.

Personal satisfaction for the amateur-builder lies in performance coupled with appearance for least cost. Many choose to build miniature semi-scale 'look-alikes' of their favourites, to judge by what appears at the EAA Fly-In and Forum at Oshkosh each August. One of the sleekest semi-scale piston-propeller aeroplanes there, painted midnight blue and white, was designed by a former US Navy flight surgeon who '*wanted something that looked good in Navy colours*'.

In recent years the RPV (remotely piloted vehicle) powered by turbojet engines has made possible the development of small turbines for homebuilders.

In Ref. 1.5 (1983) two small homebuilt jet aeroplanes, based upon French twin Microturbo TRS 18 engines each producing 225 lbf (1.0 kN) thrust, were used as design examples. On paper they had the potential to reach M 0.67 to 0.7 and suffer compressibility effects. An actual example of an experimental small-jet with engine-ancestry of that kind is the Chichester-Miles Consultants Ltd, CMC Leopard (1981), under development as a high performance four-seat light business aeroplane, with a production aircraft weight of 3750 lb (1700 kg). It was powered initially by two Noel Penny turbines of 300 lbf (1.33 kN), with the production version to be fitted with NPT 754 turbofans of 750 lbf (3.34 kN), based upon a development of an RPV engine designed by NPT, giving it a M_{NE} of M 0.81 (according to *Jane's All the World's Aircraft*). Since the demise of NPT there are plans to use other engines.

In the USA the Bede BD-10 appeared in 1993 at the EAA Convention in Oshkosh. As was noted earlier, it resembles a miniature modern fighter. The aeroplane has a gross weight of 4400 lb (2000 kg) and a single General Electric J-85 turbojet of 2950 lbf (13.12 kN), which is claimed to give it the potential to reach M 1.4.

Plate 1–5   Marking a significant trend, like the BD-10 in its own field, is the Chichester-Miles Leopard business jet. Potential lags engine technology too often, and each mating of a new engine with an existing airframe complicates flight test programmes. (Courtesy of Photolink/CMC)

Example 1–1: Semi-scale – Sukhoi Su-35 (Su-27M), Twin-Jet

The BD-10 marks a trend to be explored. The following example in Fig. 1.1 is a scheme for a slightly smaller and less powerful semi-scale 'Twin-Jet' for the homebuilder, based upon the Russian Sukhoi Su-35 fighter, a development of the Su-27 (NATO code-name 'Flanker'). The drawing was built-up from a tracing of the plan-view of the Su-35, while using the side-view to provide a guide to the location of the mainwheels, the rear fuselage and tailpipes. The resulting semi-scale drawing gives some idea of what is involved and how careful one must be in the selection of features which, though attractive, even stimulating and known to work full-scale, might prove tricky. The planform shares common aerodynamic features and broadly comparable proportions with the Su-35: swept wings and tails; broad lifting-body with twin-fins and rudders rather like endplates aft; canard foreplanes with slab, all-moving, tail plane; and a bubble canopy. In elevation, the side-view, the Twin-Jet has room for two occupants in tandem. Powerplants for the semi-scale aeroplane were to be, like the first prototype Leopard, two turbofan NPT 754 engines, each of 750 lbf (total 6.7 kN). These are no longer available. It is the shortage of certificated and available powerplants of the right power and thrust which places some of the most severe restraints on design and manufacturing organizations.

The planform is unique, in that it is highly integrated. The fuselage, broadening aft, appears to sprout canards, wings and tailplanes in a way that makes it wise to treat wings, strakes and fuselage as one aerodynamic unit, instead of a family of discrete surfaces. In this respect it has features in common with a lifting body, augmented by aerofoil surfaces, and the aerodynamic centre of the whole must be determined, and not just that of the wings attached to a body.

The drawing is accompanied by notes and rough estimates, because one cannot be sure that the proportions of the kit-built fighter will be satisfactory, or even safe. The notes are the consequence of the designer and a test pilot getting their heads together. This and what follows pose questions dealt with in the rest of the book.

Points raised with the designer by the pilot

First is that the centre of gravity as drawn appears to be a long way aft aerodynamically, and a long way forward relative to the landing-gear. It is a small aeroplane and the location of a passenger, or second pilot so far ahead of the CG is likely to cause trim changes between dual and solo. Ballast will be needed when flown solo – and that leaves scope for error. It is important to find out where the aerodynamic centre of the aeroplane lies (a rough method is shown in Ref. 1.5). To the eye there are enormous question marks against control and stability alone.

The broad centre-section and intake region of the Su-35 configuration contributes lift, but probably not efficiently as the aspect ratio of (wing-plus-body) is low, around 2. A pessimistic estimate of lift slope without flaps suggests a minimum level speed (which is an indication of stall speed) at a projected target gross weight of 4000 lb (1820 kg), well in excess of 100 knots. However, the Su-35 canard foreplanes (often described as 'moustaches'), being ahead of the centre of gravity, marginally improve lift on take-off and agility in manoeuvre. Together with the lifting body vortex-lift should be present. If so, the landing speed might be reduced somewhat, even at a limiting angle of attack of 13°. Landing and take-off speeds increase as the square root of the increase in weight, relative to the projected target weight. Even so, the foreplanes are very small and fast touchdown will increase considerably brake and tyre wear. With the mainwheels well aft of the CG, a steerable

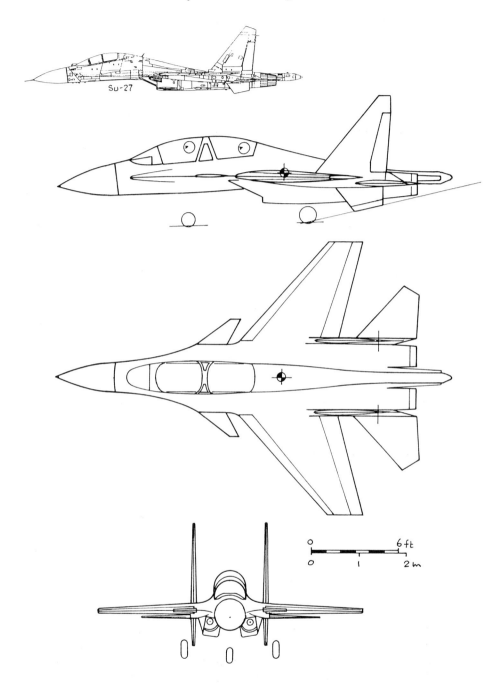

Fig. 1.1 Twin turbofan, semi-scale Sukhoi Su-35 (derivative of Russian Su-27M) kit-built for home construction.

- Weight breakdown?
- Gross weight, $W_0 = 4000$ lb (1815 kg)
- Thrust, $F = 2 \times 750$ lbf (6.67 kN) NPT turbofans
- $F/W_0 = 0.38$
- What forward and aft CG limits – limiting amounts of ballast needed in rear cockpit?

  – CG movement with fuel-burn?
- Wing area = 51.5 ft² (4.8 m²); canards = 4.2 ft² (0.4 m²); body = 109.8 ft² (10.2 m²)
- Wingspan = 17.5 ft, i.e. aspect ratio = $17.5^2/(51.5 + 4.2 + 109.8) = 1.85$

  (all areas included as aircraft is short-coupled)
- C_{Lmax} approximately 1.1 with flaps: lift slope a_A = about 0.05 (wing + body)

  = about 0.06 (wing alone)
- Take-off: maximum angle of attack without tail-strike = about 10°:

  Over-rotation (possible attitude), $\theta° = (C_{Lmax}/0.05) + 3°$ (say) = 25°
- Eyeline over nose parallel with horizon with tail bumper on ground?
- CG too far forward geometrically and too far aft aerodynamically for manual control system?
- Area of vertical tail surfaces = 2×40 ft² (7.43 m²)
- Tail moment arms too short: not enough fin volume?

  : not enough rudder volume?
- Where are the airbrakes?
- Is there a drag-chute?
- What is the location of the combined aerodynamic centre of the (wing + body)?
- What flying controls: ailerons, elevators, stabilators, trim-tabs?
- Are there ailerons on the wings, or is the horizontal tail a combined *taileron*?
- Is a leading and trailing-edge flap interconnect needed?
- Wheelbase – is this too short? How will aircraft behave on a rough surface?

The test pilot adds, looking at the side elevation, that he thinks there is a distinct risk of the aeroplane trying to change ends directionally at high angles of attack – or to suffer from severe 'Dutch-rolling' at least, especially in the landing configuration.

Fig. 1.1 Notes/commentary

nosewheel will suffer much scrubbing when countering a crosswind. The low Reynolds number of the small canard surfaces might cause them to be inefficient and stall at shallow angles of attack when manoeuvring, or in the flare. The aeroplane needs both leading and trailing-edge flaps, and for greatest advantage these must be full-span, as near as possible, leaving no room for conventional ailerons. All this must be looked at during the test programme.

With an estimated maximum lift coefficient around 1.1, and knowing that low aspect ratio shifts the angle for maximum lift to larger angles of attack, it would appear that this aeroplane might attain 25° to 30° nose-up at C_{Lmax}. Therefore, an over-enthusiastic pilot could over-rotate on take-off and scrub the jet pipes and tail on the ground, unless the aeroplane has some form of angle of attack information and limitation. Of course – and although it is less likely – the aeroplane could also be over-rotated in the flare on landing, so that the same need arises.

Therefore, choice of wing section is tricky. A high speed section with a small radius leading-edge might fail to generate enough lift on take-off at the maximum possible angle of attack – especially with a strong download (anti-lift), needed to rotate the aircraft on take-off. This must be considered. The early De Havilland Comet 1 suffered ground stalls and failure to take-off with the tail too low: one at Ciampino, Rome, in October 1952, a second, serious, at Karachi in March 1953. It was a previously unknown phenomenon and is now the subject of tests (see Example 13–4). The cure was to droop slightly the leading-edges of the wings and the upper lips of the intakes of the four engines buried in the wing centre-sections. To avoid such a situation with the Twin-Jet a solution would be either to incorporate fixed droop along the leading-edge (which increases drag at high speeds), or mechanical droop/Kreuger flaps, both of which are heavier, but do not increase drag very much when retracted.

An aerodynamic feature which might cause problems with longitudinal stability is the combination of canards and LERX (leading-edge root extensions), which fair like chines into the forebody. They bring the aerodynamic centre of the whole aeroplane for-ward, which is destabilizing because it reduces the static margin, which is aft of the centre of gravity. To move the CG forward would involve rearrangement of the landing gear. There-fore, the canards must be linked into the pitch-control system, as with the original aero-plane. Photographs show them to have wide angles of deflection, especially nose-down when the leading edge flaps are deployed. This suggests that they are also able to function like slats, or on rotating them downwards through ninety degrees after touchdown, like airbrakes/lift-dumpers on landing.

The Su-35 has leading and trailing-edge flaps. Clearly, lift-enhancing devices are needed, but what of the effects of downwash over the tailplanes immediately behind the wing? There would appear to be a need for linking canard and tailplane deflection. Would this be by an actual mechanical linkage, or by means of separate control by the pilot? It seems reasonable, therefore, to make the tailplanes into tailerons, so providing control in roll as well as pitch, freeing the wings to have only high-lift devices fitted.

Because the original Su-27 had fly-by-wire (FBW) control in pitch, might it be necessary to consider increasing the scale area of the stabilators – or are they adequate for joint control in roll as well as in pitch?

Most stabilators have fairly narrow chords when manually controlled, to reduce the stick-forces (which vary directly with chord for a given hinge/pivot position (see Equation (3–29)). The chord of the stabilators in this case is relatively broad and one might expect fairly large changes in stick-force with changes in strength of downwash from the wings and flaps. This suggests that the stabilators need their own balance/anti-balance tab sys-tems, mounted along their trailing-edges, or very careful optimization of pivot position if aerodynamic balance is to be maintained within close enough limits? It is not an easy de-sign problem, especially if friction is present in the control circuits.

Very small RPV engines are designed for short lives. Will there be enough rudder author-ity to cope with loss of an engine on take-off? A further consideration is that such aircraft are necessarily small, and so must be the retractable wheels and undercarriage legs. It is impracticable to think of operating with small wheels from grass, because of roughness of ride and high rolling friction, μ, which lengthens the take-off run (see Chapter 6). Flying with small wheels must be from airfields with smooth paved runways. The Su-27 and Su-35 have mud/stone-guards fitted to their nosewheel units.

From the practical point of view, flying surfaces are thin and in small scale do not leave much room for actuator rods, bell-cranks and mechanisms to operate control and flap sys-tems. The Su-35 has a complete digital fly-by-wire control system. Thus, there could be difficulties in semi-scale, because, unlike the original, there could be no auto-stabilizing and yaw-damping devices fitted to keep the aeroplane pointing in the right direction at all times.

Fin and rudder surface areas, for example, could need enlarging further, because the semi-scale aeroplane has more side-area ahead of the CG than does the Su-35. It looks too short-coupled, fin and rudder-volumes are small and directional stability could be weak. The ventral fins are reported to have been fitted to improve spin recovery. Semi-scale there is scope for enlarging them in chord and making them into tail-bumpers, as the angle of attack on landing and take-off is limited. The undercarriage units of the Su-35 are awkwardly placed for the Twin-Jet. Crosswind handling with such a large side-area ahead of the mainwheels could present a problem, through lack of rudder power and excessive brake-wear.

Roll with sideslip and the need for dihedral is something of an unknown quantity. Swept-back wings augment positive dihedral effect. The tall fins and rudders should also augment dihedral effect, although they could also cause the aeroplane to roll away from the direction of the applied rudder.

The fitting of an upper fuselage-mounted airbrake as on the Su-35 and a brake parachute in the tail-cone must also be considered. Twin fins and rudders are advantageous at transonic and supersonic speeds but the Twin-Jet could need different wing and tail sections of greater thickness ratio than the Sukhoi, to provide volume for flap and manual control system linkages.

An unhealthy feature for the homebuilder in the event of a wheels-up landing is the vulnerability of the engine-boxes and the costly and so far rare fan engines they contain.

Level speeds should be fast, and the aeroplane may be expected to encounter compressibility. It has swept surfaces to reduce compressibility effects. But the problem is not simple.

Also, what of flight for best range and endurance at high altitude? That will need oxygen. A comprehensive and expensive set of instruments, radio and nav-aids are needed in the event of finding oneself above the weather, short of fuel. These could cost at least $14 000 in 1994.

With uncertificated RPV engines, and a configuration fraught with potential problems, how might one get a Permit to Fly in the UK? Could it be done through the amateur Popular Flying Association, or expensively through the Civil Aviation Authority? As this is written, on past experience the probable answer is that it would have to be the CAA. There would be tight restrictions to ensure that it flew only where it could not injure the innocent third party, and flight test results would only be accepted from a pilot the Authority knew it could trust.

Aerofoil Sweep to Counter Compressibility Effects

The semi-scale Su-35 raises the problem of compressibility. Swept wing and tail surfaces are commonplace for high speed flight. The beneficial effects of wing sweep at high speed were discovered separately and independently in different countries, according to Theodore von Kármán (Ref. 1.2). Late in the 1920s Italy had won the Schneider Trophy against foreign competition. Flushed with victory, the Italian Government, under Benito Mussolini, arranged an international meeting to discuss the problems of high speed flight, which was attended by von Kármán as a member of an American delegation. The meeting – the Volta Congress for High Speed Flight – was held in Rome in 1935, and it was there that the German aerodynamicist, Adolph Busemann, presented his first theory of the effects of sweep in reducing the drag of a wing at supersonic speed. This advantage was also realized in the USA, late in 1944, by Robert T. Jones (Ref. 1.3) of the then National Advisory Committee for Aeronautics (NACA, now NASA).

Busemann made use of the *Independence Principle* (also called the *Principle of Superposition*), which assumes that the aerodynamic forces on a wing panel, long in the spanwise direction, can be considered as independent of any spanwise component of the velocity of the aircraft in the direction of flight. In short, supersonic drag depended instead upon the component of airflow *at right angles* to the long axis of the wing – i.e. near enough normal to the leading edge. He limited his consideration of the benefit of wing sweep to a type of flow explored earlier by J. Ackeret, in which the component at right angles to the long axis of the wing is everywhere supersonic, so causing expansion and compression wave drag (a phenomenon which occurs when the component normal to the length of the wing is faster than the speed of sound).

At airspeeds which are high enough for compressibility to occur, the distinction is made between flows which are *subcritical*, when the flow component normal to a leading or trailing edge is subsonic; and those which are *supercritical*, when the normal flow components reach supersonic velocities and expansion and compression (shock) waves are formed. Only later was it realized that wave drag could be eliminated entirely at supersonic speeds by sweeping the wing and other surfaces to lie completely within the Mach cone (Ref. 1.3.), so ensuring subcritical leading and trailing edges.

Figure 1.2 shows the broad principles involved (Ref. 1.3), and Fig. 1.3 the effect of aerofoil sweep at the $\frac{1}{4}$ chord upon fineness and, hence, critical Mach number (Ref. 1.4). Forward and backward sweep similarly reduce wave-drag, but have different low speed aerodynamic side-effects at large angles of attack, and aero-elastically at high speed.

The first practical aeroplane designed to investigate the advantages of swept wings for high speed flight was the Messerchmitt P 1101 of World War II. It was intended to have a ground adjustable three-position variable-sweep wing, but never got airborne, being overtaken by other events. Nevertheless, when modified in the USA it was later transformed into the experimental Bell X-5, which flew in 1951 and revealed clearly many practical problems, then largely unexpected but now thoroughly understood.

The Su-35, like many modern supersonic aeroplanes, appears to be 'area-ruled'. Although area-ruling is linked with the name of Dr Richard T. Whitcomb of the now NASA Langley Research Center similar solutions were being investigated elsewhere, for example at the Royal Aircraft Establishment, Farnborough – albeit from a different direction. The story of the way in which Whitcomb saved the Convair YF-102, by means of a lengthened and more tapered forebody, waisting of the fuselage in the way of the wings, and a lengthened body aft of the wings, widened by additional bulges, is now aeronautical lore. The performance of the delta-winged Convair YF-102 which, until then, had failed to exceed the speed of sound, was dramatically improved. As the renamed YF-102A it exceeded Mach 1 the day after it first flew following modification, in December 1954.

Angles of sweep, Λ, as measured from drawings, may be used to estimate the critical Mach number of wing and tail surfaces by means of the following equation (in which V is TAS and a is the local speed of sound):

$$\text{Mach number, M} = (V/a) = \text{cosec } \Lambda \tag{1–1}$$

When analysing drawings one needs good state-of-the-art data, derived from reliable sources. One source, a widely recognized and internationally respected authority, is *Jane's All the World's Aircraft*, published annually. Reliable information is also found in manufacturer's data sheets, and we shall look at a parametric means of assessing such data to test its reliability. Parameters are combinations and ratios of apparently unrelated items of raw data, e.g., weight divided by wing area to give *wing loading*, or weight divided by total power or thrust, to produce *power* and *thrust loadings*.

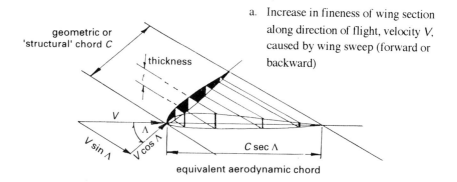

a. Increase in fineness of wing section along direction of flight, velocity V, caused by wing sweep (forward or backward)

b. Mach cone shed by source at M = 2.0

$M = \operatorname{cosec} \Lambda$ eq. (1-1)

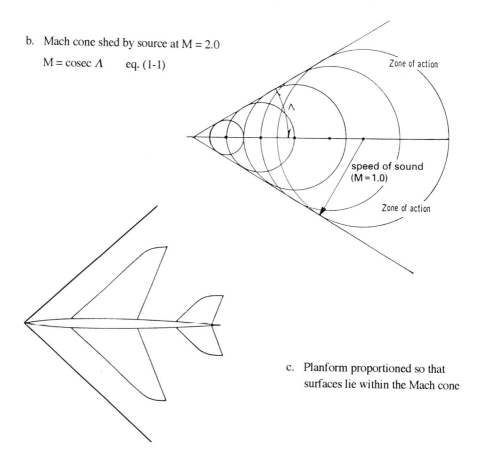

c. Planform proportioned so that surfaces lie within the Mach cone

Fig. 1.2   The reason for sweeping aerofoil surfaces at high speed (M > 0.6).

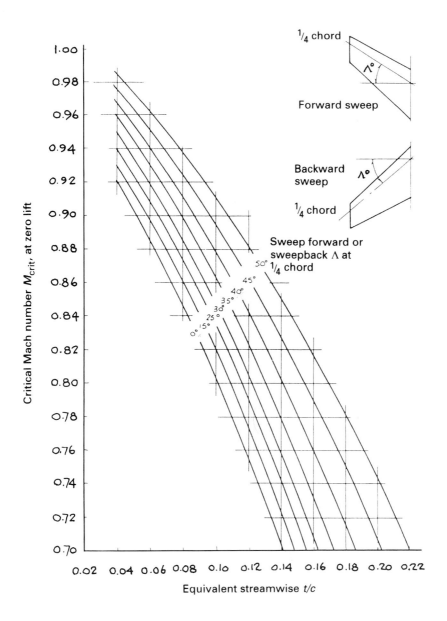

Fig. 1.3 The effect of aerofoil sweep at the $\frac{1}{4}$ chord upon fineness and critical Mach number. Leading edge and $\frac{1}{4}$ chord sweep are roughly the same. Streamwise $t/c = $ (structural t/c) cos Λ; and M_{crit} is without aspect ratio effects.

Plate 1–6   The Messerschmitt P1101 with a three-position variable-sweep wing was overtaken by events in 1945. It was developed in the USA as the experimental Bell X-5. Ref.: Example 1–1. (Courtesy of Deutsches Museum Munich)

Example 1–2: Pilatus PC-7

Some years ago we were asked to compare eight turbopropeller military training aeroplanes. Data was obtained from *Janes' All the World's Aircraft*, and a parametric method was used which, inspired by the analytical methods of an American consultant, Rolf Wild, had been devised for an earlier investigation. *Jane's* was used as the source of all of the information, because the manufacturer's technical data was not to hand for every aeroplane at that time, and it seemed the fairest way of treating all alike.

Of the eight aircraft the Pilatus PC-7 came out at the top of the survey, as an effective aeroplane. It is of value at this point because deductions from the published drawings are augmented and modified in a *Postscript,* based upon actual experience of the aeroplane in flight.

The PC-7, which is shown in Fig. 1.4, has a Pratt and Whitney Canada PT6A-25A turbo-prop engine of 550 shp (shaft horsepower). Built in Switzerland, the aeroplane was designed to meet Swiss civil airworthiness requirements, those of the US FAR 23, and the requirements of a selected group of US Military Specifications (Trainer Category), termed Mil Specs.

From the drawing the following deductions might be made:

Wings are straight tapered. The taper ratio (tip chord/root chord) is a little less than 3/5.

Plate 1–7 The Bell X-5 (1949 proposed) experimental variable-sweep Aeroplane evolved out of the German Messerschmitt P1101 which did not fly. Ref.: Example 1–1. (Courtesy of Bell Aerosystems and Philip Jarrett Collection)

The purpose of taper is to reduce wing structure weight, by concentrating area (and lift) inboard, while maintaining a long span (high aspect ratio) across the air mass being worked on, so reducing the downwash velocity and acceleration imparted, and with it the associated lift dependent (induced) drag. Induced drag is directly proportional to span loading, i.e. the weight of the aeroplane divided by the wing span. Also, too much taper (less than tip chord/root chord $= \frac{1}{2}$, say) and a wing becomes prone to dangerous stalling of one tip before the other, and wing dropping which may be followed by either a spiral dive or a spin.

The drawings suggest that the wings are assembled in three parts: outer planes plus parallel chord centre section, with what look like two transport joints. If correct, then separate flap linkages will be needed in between the planes and centre section.

The flaps appear to be plain and to extend beneath the fuselage. Their locations are indicated by lines representing skin structure in the plan and side views.

Wings have marked dihedral outboard of what we have assumed to be transport joints. Tips are raked, with chamfer underneath, while wing trailing edges are longer than the leading edges. Chamfer makes a small additional contribution to favourable roll

Plate 1–8   Area-ruled according to the principles of Dr Richard T. Whitcomb of (now) NASA Langley, the Convair F-102A with longer waisted fuselage and bulged lateral fairings abaft the wing trailing-edges. It went supersonic the day after it first flew in December 1954. Ref.: Example 1–1. (Courtesy of Convair (General Dynamics) and Philip Jarrett Collection)

with sideslip (the tendency of the upwind wing to rise). Properly designed raked tips are the most efficient for reducing the induced drag. Long trailing edges trap the maximum working mass of air before it is able to free itself from the wing surfaces.

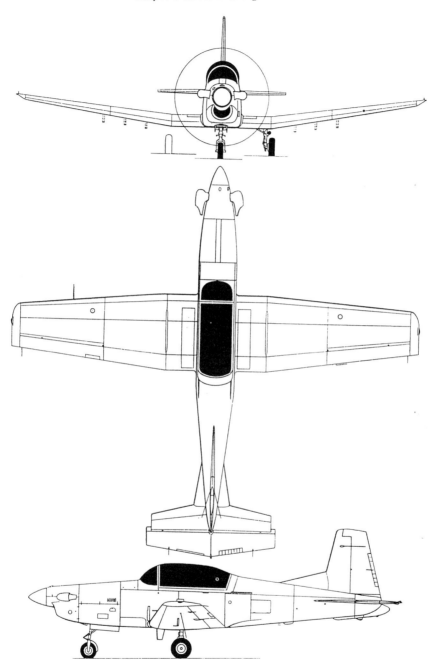

Fig. 1.4 Pilatus PC-7 Turbo-Trainer. (Courtesy of Greenborough Associates Ltd)

The walkways on each side of the wing centre-section in plan view confirm that the cockpit is approached from either side, and that the canopy must therefore slide aft.

☐ *Ailerons* have an inset trim-tab (pilot operated) to port, and what looks like a small fixed tab to starboard. The line forward of one parallel with the leading edge of each aileron is possibly a shroud, and not a spoiler, of which there is no mention.

☐ *Propeller thrust-line* is almost in line with the tailplane and elevator. This arrangement generally produces the least longitudinal trim change with power.

☐ *Fuselage and canopy* have carefully blended and harmonious lines. The canopy has a separate windscreen, is blown in one piece and slides back on rails, confirming the earlier deduction from the twin walkways.

A sliding humped canopy suffers from high aerodynamic suction, forwards, and an increasing pressure recovery down the slope at the back, both of which can hold it shut against the windscreen arch. This might present a problem in an emergency. Some aircraft have needed mechanical assistance, in the form of a hatch to ventilate and break the adverse pressure-differential aerodynamically, or a spring, or an explosive canopy release. Today, turboprop training aeroplanes can reach speeds high enough to justify fitting ejection seats. These have 'hood-crackers' to enable the pilot to eject through the hood; or the canopy itself may be fitted with an explosive device to destroy the transparency before seat and pilot penetrate it.

The risk of a birdstrike at high speed must be considered.

The fuselage itself is nearly symmetric in elevation, especially where it tapers aft of the wing. The fin and rudder are located forward relative to the tailplane and elevator. This is beneficial in assisting spin recovery as it minimizes rudder and fin blanketing at high tailplane angles of attack, although fin and rudder area, working on a shorter moment arm about the CG, must be made slightly larger in area to achieve adequate tail volume – the product of surface area and moment arm (to which directional stability and control are proportional).

The pleasant lines of the fuselage are marred somewhat by the addition of dorsal and ventral strakes, which have played a part in the reshaping of the tail surfaces during development test flying.

Note: Dorsal and ventral fin surfaces, strakes added to tailplane leading edges, and a plethora of other devices, indicate that the designer and his team encountered trouble.

☐ *Fin and rudder surfaces* appear to have been shaped originally to correspond with the proportions of the other flying surfaces, and to blend into the lines of the tailcone. But during development two modifications were incorporated:

(1)   The addition of a straight dorsal strake at the base of the fin.
(2)   The addition of a small ventral fin surface beneath the tailcone.

Thus it is reasonable to suppose that the aeroplane suffered either a problem with spin recovery, or prestall buffeting of the fin at large angles of sideslip.

The rudder is horn-balanced, to lighten the foot forces and to provide volume for the mass balance ahead of the hinge (which appears to lie aft of the line marking the leading edge of the rudder – this line probably indicates the trailing edge of the rudder shroud). There is also an articulated rudder tab for trimming the aeroplane directionally, which is needed with changes in power.

Note: The presence of dorsal and ventral strakes is often a consequence of such turbo-prop aeroplanes having been designed by 'rules of thumb' commonly applied to piston-propeller aircraft, or when converted from piston to turbine. Piston engines are heavier per horsepower and kilowatt than turboprops, and noses are shorter when engines are mounted in front. Tail sizes, calculated by means of standard tail volumes, then tend to be smaller than is needed, because lightweight turboprops must be hung further ahead of the wing to balance the CG, which is arranged to lie near the $\frac{1}{4}$ mean aerodynamic chord (i.e. $\frac{1}{4}$ MAC or simply the 'quarter chord'). This lengthens the nose, which introduces more body surface area forward of the CG, which is destabilizing in pitch as well as in yaw; while the tractor propeller, which is also destabilizing, now has an even longer moment-arm ahead of the CG than it had before. All is exacerbated by the extra power of turboprops because of *P-effect* which, as we shall see, reduces aerodynamic damping, making the aeroplane livelier in response to control. This is discussed in Chapter 10.

☐ *Tailplane and elevator* reveal the greatest changes. Whatever horn and mass balances once existed appear to have been altered subsequently. Horns have been cut away, are unshielded and small, giving the tailplane and elevators 'stepped tips'. Strakes have been added at the roots of the tailplane, hinting at difficulty in meeting spin recovery requirements.

On the elevators two tabs are shown. The one on the starboard side resembles the rudder tab. If correctly illustrated then this could have been the original tab for trimming in pitch. One might only guess the function of the high aspect ratio (long and narrow) tab on the port side of the elevator. It might be an anti-balance tab intended to increase the stick-force felt by the pilot (a reason for cutting away elevator horn length ahead of the hinge). A heavier stick-force discourages the pilot from overstressing the aeroplane in manoeuvres, or breaking it, or hastening a structural failure later in its life for someone else. Although we can only speculate, modifications (fixes) of this kind always indicate that there has existed a problem which demanded attention.

☐ *Landing gear* is a retractable nosewheel tricycle arrangement, with the nose-leg mounted aft of the engine firewall. In flight nosewheels, legs and doors ahead of the CG are destabilizing directionally, while all undercarriage arrangements appear to be destabilizing laterally, in the landing configuration. They are a factor in the determination of fin and rudder sizes, and in the amount of favourable roll with yaw (dihedral effect) needed to meet sideslip requirements.

Postscript to the PC-7

Figure 1.5 is a manufacturer's specification drawing and this postscript is added to show how three-view drawings can differ, while additional information, culled in connection with flight testing of the aeroplane, modifies some of our earlier deductions.

In fact the wings are built in one piece for lightness and there are no transport-joints. They are assembled in a jig before being mated with the fuselage. Flaps are split, are built in three parts, two outer and one across the centre-section, beneath the fuselage. They are electrically operated, as is the landing gear. This makes the assembly and subsequent break-down of components much easier than when a purely mechanical, pneumatic or hydraulic system is fitted.

Drag of the walkways on the wings is said to cost 1 to 2 knots.

The right aileron has no fixed tab. The aeroplane is adjusted to fly wings-level in the cruise by altering the gap of the ailerons with washers, and filing away some of the depth of a small fixed strip working in the airflow on the underside of the aileron trailing edge. The strip causes upfloat of the aileron. Varying the depth of the strip by filing away material enables one to trim out the forces, wings level, between 120 and 210 KIAS (knots, indicated airspeed) in the clean configuration. Residual out of trim forces in the landing configuration are then negligible. Ailerons are adjusted after the first test flight, and checked during pro-duction testing.

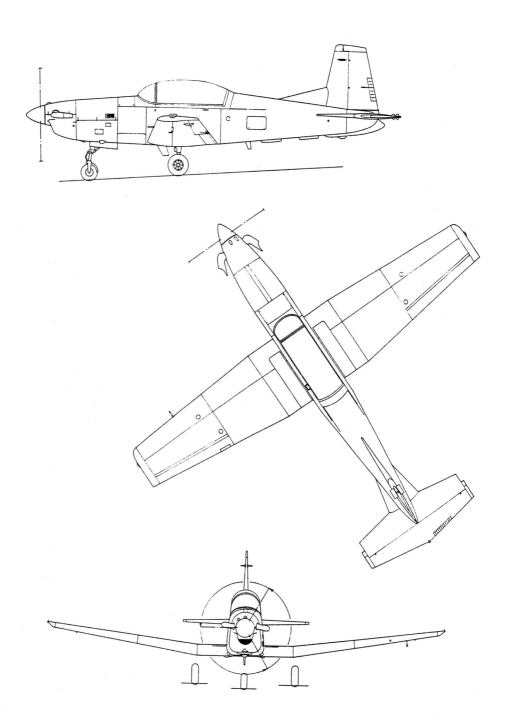

Fig. 1.5 Manufacturer's drawing of Pilatus PC-7. Note the differences between this and Fig. 1.4?

There is no offset to the propeller thrust line.

The canopy can be motored aft in flight by about 30 cm (11.8 in) at speeds up to 135 KIAS. It is jettisoned by releasing the rails, with the canopy leading edge slightly raised so as to make it fly clear of the tail.

The fin and rudder are direct copies of the Pilatus P-3 and were found to be too small for the directional stability and spin recovery requirements without strakes. Tailplane strakes were made out of wood and their shape determined by trial and error, primarily to achieve a stable spin with good recovery.

Originally the rudder showed signs of locking over in sidelips, and this led to the addition of the dorsal strake.

The rudder trim tab serves as an anti-balance tab. Proportions differ between Figs 1.4 and 1.5. Differences between the drawings need checking with photographs, or examination of the aeroplane at first hand. The strakes ahead of the tailplane are not the same length in Fig. 1.4 as in Fig. 1.5. The *Jane's* drawing correctly shows two elevator tabs, one of which could have been added during development and simply overlooked by the company draughtsman. The trim tab on the port side of the elevator is fixed. The elevator trim tab has end-plates which increase its aerodynamic efficiency (in terms of the lift slope, i.e. the lift generated per degree of deflection) and consequent authority needed to meet the minimum trim speed requirement at forward CG.

The elevator circuit has both a down-spring to increase stick-force (so providing an apparent improvement in longitudinal stability) and a *negative* bob-weight (which *reduces* stick forces with applied g) which is an unusual combination. The effect of the latter can be felt through the control column in turbulence.

Both of these examples show that a considerable amount of 'ballpark' (approximate, but broadly acceptable) information can be gleaned from drawings alone. Every line on a drawing tells something of what is inside, under the skin. Rigging angles (of incidence) between aerofoil surfaces and fuselages, thrust-lines, differences in rigging angles between wings, tailplanes and elevators, rigging angles of fins relative to the fore and aft axis of the aircraft, joints between panels and lines of dots, representing fasteners, biplane wings of unequal size (when the lower plane is much smaller the arrangement is called a *sesquiplane*), all point to this or that particular route for further inquiry.

Example 1–3: Aerobatic project – Système 'D'

As a third example an aeroplane which is awkwardly monstrous and graceless to the eye, which the company test pilot sees on a sketch-pad when passing through the Project Office. Nothing is known of the actual flying qualities of such a shape, beyond those rumoured of X-winged radio-controlled model aeroplanes, which appear to be indifferent to which way up they are. The designer is thinking of Olympic class aerobatics in his tea-break, and has sketched an aeroplane with one purpose in mind: *agility*, with the added ability to stand still and hover.

Test pilots, especially those in companies, see the aircraft they will eventually fly on drawing boards, as forms and sketches, long before the design is finalized. In principle and in practice they must have the opportunity to say what they think, and not only in terms of view, cockpit design and ergonomics.

The shape of the aeroplane, and the title came about some years ago when a French friend, an aerobatic pilot, drove us both at high speed on a devious route around a traffic jam *en route* to the airfield. He dismissed my nerves as an Anglo-Saxon 'malade'. His solution to the traffic problem was an example of French trouble-shooting he called *Système 'D'*

Plate 1–9   Typical of the competition in advanced aerobatics, the Sukhoi Su-31. A spinning clearance is needed for aeroplanes certificated for aerobatics in the UK. This involves comprehensive flight tests. Ref.: Example 1–3. (Courtesy of Richard Goode)

(which derives from the transitive verb '(*se*) *debrouiller*', to unravel, to extricate oneself, to manage).

Whereas an Englishman or an American, when faced with official obstruction, might shrug and accept the situation, the strong revolutionary spirit in a Frenchman looks immediately for a way under, over or around it. With aerobatics in mind, conversation turned to the possibility of designing an aeroplane which incorporated this French characteristic: which could ravel and unravel itself in (hopefully) unlimited aerobatics.

Designer's reasons for shape

By 'unlimited aerobatics' is meant the ability to explore sequences within the international competition box of air, shown in Fig. 1.6 (Refs 1.5 and 1.6), of which even the best of current Olympic standard aerobatic aircraft are not yet capable. To hover, if necessary. To spin either erect, inverted, or on one side or the other, in 'knife-flight'. Similarly, to carry out loops, rolls and flick (snap) manoeuvres erect, inverted, and banked relative to the plane of symmetry. To execute in either direction rotary stall turns or straight tailslides; and *lomçovaks* (in which the flying controls are used to set up conditions during which the gyroscopic properties of the propeller take over).

☐ *Power requirement* to carry these out in either direction, and to hover when required, counter-rotating propellers or rotors are needed, otherwise torque effects will be asymmetric. For manoeuvring in the hover thrust must exceed weight by about 20%. A turbopropeller engine is envisaged, driving rigid rotors. The turbine may then be run at

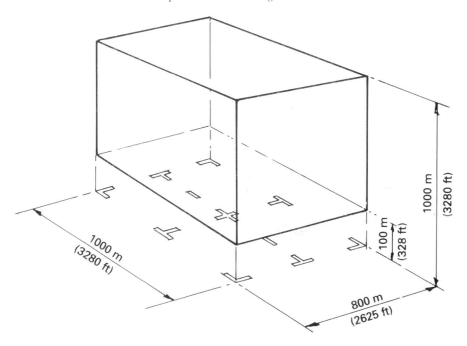

Fig. 1.6 The international competition box of air within which a sequence of aerobatic manoeuvres must be accomplished (Refs 1.5 and 1.6).

constant speed, with thrust changes by means of the rotor pitch controls. The immediate problem is that a single engine driving contra-rotors will need a very heavy and complicated gearbox if the pilot is to control torque at will, by controlling power by means of each rotor separately. The alternative is to fit coupled engines, each driving its own rotor.

Using Equation (3–71) with a single engine of 1000 eshp one must design down to $(1000 \times 2.6)/1.2$, i.e. a gross competition weight of only 2150 lb (982 kg). Yet the engine, gearbox and rotors will weigh together around 46% of this, leaving little or nothing in hand for structure, fuel, systems and pilot. A major difficulty is making the device balance properly, because of the length of its nose.

Fitting coupled engines, the length of the nose is much the same as for the single, while the engine installation weight, gearbox and rotors are much heavier, in which case one is looking at a much bigger and heavier aeroplane, with 2000 eshp and grossing around 4700 lb (2136 kg).

As the aeroplane is not intended for public transport operations, design and operation avoids the difficulties encountered when attempting to meet airworthiness codes written to safeguard fare-paying passengers.

The *configuration* shown in Fig. 1.7 is a form of triplane. Historically, triplanes have been amongst the most agile aerobatic aircraft ever built. But they have not been fast, nor is this one. It is no more intended to fly from A to B than is a Formula 1 racing car intended for roads with ordinary traffic. Wing and tail surfaces have a multiplicity of junctions at their roots, all of which are sources of turbulence and drag. The unusual wing and tail arrangement is an attempt to achieve aerodynamic and dynamic symmetry about the CG, with adequate ground clearance for the rotors.

The wings are thick, for structural strength and lightness. Sections are formed from a truncated ellipse, with straight tangents to the trailing-edge. Maximum lift coefficients

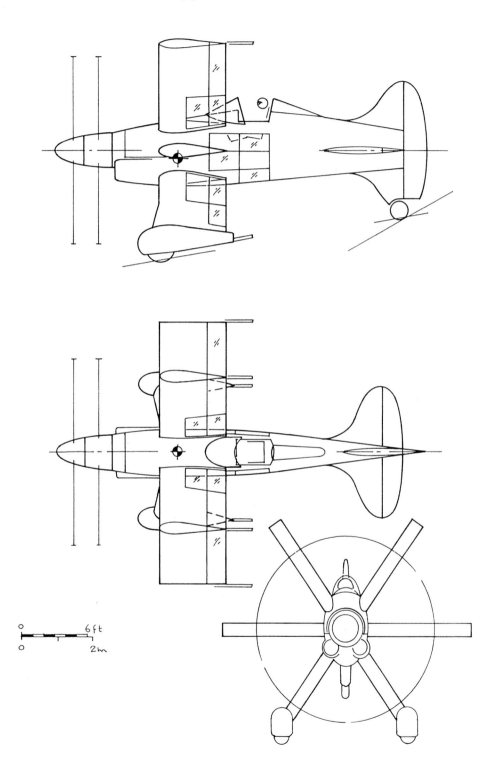

Fig. 1.7 Système 'D' aerobatic triplane.

- 2 × 1000 eshp turboprops.
- Design weight estimate 4700 lb (2140 kg).
- Weight breakdown, position and range of movement of the CG?
- What are the projected areas of the mainplanes, throughout 360° roll?
- How will stall speeds change: power-OFF and power-ON: erect, inverted knife-flight, 55° bank, 125° bank?
- What is the design airspeed for aerobatic manoeuvres?
- What is the effect of 55° dihedral on the lift slope?
- What power and rotor-pitch controls? What flying controls? Any flap/aileron droop?
- What nose-UP attitude might be reached with power on take-off? (with such low aspect ratio about 30° seems possible).
- How will one rotate the aeroplane nose-UP on take-off with the tailwheel on the ground?
- What are the tail areas and volumes?
- In what direction is the aeroplane likely to roll with yaw?
- Is the eye-line over the spinner parallel with the ground, tail-down?
- Do the control surfaces need horns?
- How much fuel is needed for an aerobatic sequence, and how much reserve?
- What is the size of the static and manoeuvre margins?
- What stick-force per g might be expected?
- Has the arrangement of the exhaust pipes been sorted out?
- How will the heat of the exhausts affect the transparencies?
- How does the pilot enter and leave the cockpit?
- What will be the effect upon the contra-rotor system of the failure of one engine?
- How long will it take to feather one rotor in the event of an engine failure – will it be manual or automatic? Will an ejection seat be fitted?

The test pilot doubts the ability of any pilot to cope with a significant emergency in a competition box, at an average altitude of 550 m (1800 ft). He insists that there must be a sure way of abandoning such a complicated aeroplane, ideally by an ejection-seat.

Fig. 1.7 Notes/commentary

will not be large – say 1.0 to 1.2, even with power. With a lift slope averaging, say, 0.065 to 0.07/degree one might expect extreme take-off and landing attitudes around 1.2/0.065, say, 15 to 20 degrees – or more – touching down tailwheel first, with power.

Wings have lightweight rods fitted at their tips, with the ends in the plane of the pilot's eye, to provide visual cues.

The landing gear is relatively heavy and durable, with large tyres for soft ground, and a heavy duty tailwheel unit for landing tailwheel first. Mainwheel spats are shaped to provide some lift, and they are large to make them easy to keep clear of mud, grass or snow.

- *Compactness* is needed and a triplane with a comparatively short span has the advantage of small rotary moments of inertia (i.e. 'flywheel' effects about the CG). But inertia in pitch (B) is expected to be larger than in roll (A), affecting significantly the (B/A) ratio and in particular the required spin characteristics (see Chapter 12). Short wings enable the aeroplane to be lighter than otherwise. Short span is no disadvantage because the aeroplane need not fly for range. Provided that it can be dismantled and transported the fuel capacity is determined by training needs and competition sequences, plus a reserve.

☐ *Aerodynamic damping,* in moderation, is an advantage for an aeroplane intended to be more of a circus pony than a flying machine. It lengthens time of response to application of a control, otherwise everything could happen too fast, causing pilots to be imprecise in manoeuvre. Slowing down the rates of manoeuvres also helps the judges. But:

☐ *Stability* need be little better than neutral about any axis. Such an aeroplane must be all control, but with adequate strength and stiffness for precision, and to prevent it from being broken too easily. It may sound dubious to those who are not aerobatic pilots, to suggest that there is no need for conventional stability. But there is a practical precedent in the evolution of the agile and effective fighting scout of World War I, discussed in Chapter 10. One example in particular, the Sopwith Camel of 1916, epitomized agility (Ref. 1.7). Small rotational inertia in pitch, roll and yaw implicit in the configuration of the Camel, revealed Sopwith's philosophy, which is as relevant to the subject of agility today as it was then, as recalled in the words of Cecil Lewis, a former Royal Flying Corps Pilot:

> 'Like all Sopwith productions, it was a bit on the light side; but for actual flying, next to the Triplane it took first place with me.'   (Ref. 1.8)

Plate 1–10   The Sopwith Camel. Described as a 'popping firecracker', it epitomized agility in World War I. Pilots either hated or loved it. There was no in-between. Ref.: Example 10–4. (Courtesy of the Imperial War Museum)

Plate 1–11 The Schneider Trophy-winning Supermarine S-6B (1931) opened a new chapter in the development of high-speed technology and test flying. Ref.: Example 9–10. (Courtesy of Philip Jarrett Collection)

☐ *Strength and stiffness* is essential for maximum aerobatic agility, without risk of breaking the aeroplane, or of diminishing control authority through aero-elastic distortion. While not being able to envisage everything that competing pilots might attempt, it would be wise to design the Système 'D' structure for an ultimate load factor of, say, 8 in the first instance. This is then multiplied by a factor of 1.5 for safety.

☐ *All-around view* is essential, which is why the pilot is seated high, and being slow there is no need for a canopy. The height of the pilot's eye is dictated by his view over the nose, tail-down, and by slant lines to either side of the cockpit section which is narrow at the top, to improve the view. There must be extensive transparent areas of wing and fuselage skin, in an attempt to provide clear views of the ground from inside the cockpit, in almost any attitude. It is obstruction of arcs of view by wing structure that could be one of the most damning criticisms of the aeroplane.

Note: Transparent panels are shown, but are a disadvantage for the judges, tasked with deciding which way up the aircraft is in a given aerobatic sequence. All surfaces need colour-coding.

☐ *Aerodynamic controls* are six ailerons, one on each wing, functioning also as flaps; and a cruciform tail arrangement, with tailplanes and elevators set forward of the rudder post. Elevators and rudder all serve as controls in pitch and yaw, as required. Up-elevator (be it the elevator or the rudder in use as an elevator at the time) lowers just sufficiently those ailerons which work as flaps in the relevant plane.

Note: For even higher rates of roll there are cogent arguments for dispensing with conventional ailerons and replacing them with wings which rotate in pitch on mainspars which are, in effect, spindles.

At this stage the designer does not know whether control horns are needed to relieve the pilot's work-load during an aerobatic sequence.

Rotor torque is to be controlled in *lomçovaks* by operating power levers asymmetrically. Although the aeroplane should hover in a vertical position, it is incapable of landing other than on its wheels. The tailwheel is large, strong and heavy to take the load of landing first.

The first reaction of the test pilot is that he is appalled by the complexity, and has reasons for believing intuitively that the aeroplane will not work. We shall deal later with some of his morc factual misgivings.

The test pilot as one of a team

To be able to cope with the ideas of engineers and others can be demanding. There is in the nature of a test pilot more than a little of the loner. Taken to an extreme, of course, it makes him a misfit. But on the other hand a pilot who is highly gregarious could find it emotionally difficult to probe the extreme behaviour of an aeroplane in the lonely isolation of a cockpit, if he or she does not understand in technical depth what the design team is attempting to achieve.

Aviation has developed to the point where design, construction and flight testing are team efforts. Today solo flight testing – other than for the purest research – is rare because many operational military and civil aeroplanes have seating for a test crew of at least two. Then one flies with a Flight Test Engineer, a person who is the highly trained product of a college or university, the industry, and probably a school, like the ETPS (which runs FTE courses); or with an observer who is an engineer, who knows the aircraft and represents the owner.

All the best designers have an ear for pilots. For example, it is said of R. J. Mitchell, designer of the Schneider Trophy-winning Supermarine S.5, S.6 and S.6B floatplanes and, later, the Spitfire that:

> 'Mitchell got on very well with the RAF pilots, for whom he had the greatest respect, and when they were flying one of the S.5s he was always down at Calshot, anxiously watching every performance. He soon became known as 'Mitch', a familiarity never permitted at Supermarine.'   (Ref. 1.9)

Jeffrey Quill was for a time assistant to Captain J. 'Mutt' Summers (then Chief Test Pilot of Vickers Supermarine) and knew Mitchell from 1936 to his death in 1937. Writing later in his autobiography Quill said:

> 'He (Mitchell) made me see that the most valuable contribution that I could make was to concentrate upon becoming a better and better pilot and try to absorb a natural understanding of the problems of engineers and designers by maintaining close personal contacts with them and doing my best to interpret pilots' problems to them, and vice versa.'   (Ref. 3.3)

Engineers must listen to pilots – the people who have been there and done it, or at least tried to do it (some have not listened and have made a mess). By the same token, a pilot must never be a prima donna, too proud to listen to an engineer or a mechanic (it has happened). Those who refuse to do so make their own kind of mess.

References and Bibliography

1.1 Wood, D., *Project Cancelled*. London: Jane's Publishing Company, 1975

1.2 von Kármán, T., *Aerodynamics*. Ithaca, New York: Cornell University Press, 1957

1.3 Jones, R. T., *Wing Theory*. Princeton, New Jersey: Princeton University Press, 1990

1.4 Stinton, D., *The Anatomy of the Aeroplane*. Osney Mead, Oxford: BSP Professional Books, 1966

1.5 Stinton, D., *The Design of the Aeroplane*. Osney Mead, Oxford: BSP Professional Books, 1983

1.6 Williams, N., *Aerobatics*. Shrewsbury: Airlife Publications, 1975

1.7 Yeates V. M., *Winged Victory*. London: Jonathan Cape, 1934

1.8 Lewis, C., *Sagittarius Rising*, London: Peter Davies, 1936

1.9 Mitchell, G., *R. J. Mitchell, World-Famous Aircraft Designer, Schooldays to Spitfire*. Olney, Bucks: Nelson and Saunders, 1986

CHAPTER 2

Flight Characteristics:
What the Test Pilot Tells Us

'Historic aircraft are those generally regarded as possessing such technical features, flying-qualities, operational and historic associations as to warrant uncommon knowledge or skill when preserving them for display and use in representative working order.'
British *Historic Aircraft Association*, founded in 1978

Many people now have an interest in (re)constructing and operating originals and near-accurate flying replicas of aeroplanes belonging to earlier periods of aeronautical history. Their numbers extend from a mixed bag of enthusiasts at one end of the scale, to hard headed business men at the other. The Air Show industry is big, and growing, with hot competition for the rotting remains and wreckage of aircraft long since abandoned in jungles, on the sea bed, or buried in snow. Rebuilt, they pose problems: *'What handling differences were there between the Rolls Royce Merlin-engined Spitfire and one with a Rolls Royce Griffon?'. 'Did all Bristol Mercury engines cut like that when the throttle was opened rapidly?'. 'What modifications are needed to locate the CG of a full-scale replica of a Sopwith Triplane in the **right place**, when fitted with a fixed radial engine instead of the heavier rotary of the original? And what is the **right place** – the drawing does not show it?'. 'Why does the inverted Ranger engine in the Albatros DV replica, keep on choking and then failing?'. 'They want to stick a gunner out in front to make it look like Gunbus in a film. Isn't there too much side area ahead of the wings?'. 'Might not the wheels be too far back?'. 'Does the fin need some more offset with that engine fitted?'.*

To find and make available the right information and safest answers the British Historic Aircraft Association was founded in 1978. Its definition of an historic aircraft heads this chapter, and the essential words are *'in representative working order'*, i.e. with unadulterated (original) flying qualities. There was also founded a Register of pilots who could be relied upon to display such aircraft as they deserved, and not to use them to display themselves.

There are numerous sources of historical information. The Royal Aeronautical Society in London is one; as is the Library of the former RAE (Royal Aircraft Establishment) at Farnborough (now the DRA (Defence Research Agency)). The Royal Air Force Museum at Hendon, in London, is another. While the National Air and Space Museum in Washington, and the Deutches Museum in Munich are two among many others outside of the UK.

Engines play an important part in the maintenance of flying qualities. A World War I aeroplane which originally had a rotary engine (propeller and engine rotating on a common shaft set in the fuselage structure) cannot be reproduced with accuracy when powered by a fixed radial engine, 20 years younger. There will be no gyroscopic effects of engine rotation. The CG will almost certainly come out in the wrong place i.e. too far aft, because the younger engine could be lighter per horsepower, providing insufficient weight ahead of the wings. With later aeroplanes heavy operational equipment, like radio, guns, turret mechanisms, will be missing. One must dig through records or find actual equipment (even if it no longer works), to manoeuvre the centre of gravity to where one thinks it should be.

The answer to where the CG should lie, and between what limits, is not always easy to find, especially when the aircraft is venerably old, and the records have long since disappeared. Another aspect is the need to discover characteristics of the original wing section – and the sections of the tail surfaces (when these were designed to support a considerable part of the load in flight). This information is needed to determine the relationship between CG and CP of the aerofoil surfaces so that a proper and safe flight test may be carried out.

It is not a safe rule of thumb to say that the CG of an old aeroplane should lie '... *about* $\frac{1}{3}$ *chord back from the leading edge*'. Some were much further aft. And anyway, which leading edge of which plane, in the case of a biplane or a triplane?

Example 2–1: Biplane Fighter Replica – Hawker Fury 1 (1929)

Of all historic British aeroplanes one of the most elegant was the Hawker Fury biplane fighter, the side-view of which is shown in Fig. 2.1. One of the most interesting projects has been the building of a full-scale replica using an original Rolls Royce Kestrel II S, engine, carefully inhibited and in working order, complete with an original two-blade Watts wooden propeller.

The only drawing to hand for our present purpose is a three-view GA drawing in a journal for aeromodellers (which is a typical starting point for many such projects).

Note: A similar drawing together with bits and pieces of scrap parts provided by the manufacturer (and many more handcrafted) enabled a replica of the Westland Lysander, designed in the 1930s, to be constructed in the 1970s. It was subjected to a similar cross-examination before being test flown for the issue of a Permit to Fly. It had to be proved that what looked like a Lysander also flew like one. Fortunately, reprinted copies of Air Ministry Pilots' Notes had been published in recent years. These enable one to check sensibly the anticipated flying qualities. Such notes state basic drills, procedures, airspeeds, and provide valuable advice on handling.

The lines on the drawing of the Hawker Fury show it to be mixed metal and fabric covering. It is a single-seater, and the gun-troughs confirm that it was an interceptor fighter. It happens also to be the father of the Hawker Hurricane which, because of the form of construction pioneered by Hawkers, was repaired and returned to service faster than the all-metal Spitfire in World War II, enabling Hurricanes to shoot down more enemy aircraft than did the more elegant and better remembered Spitfire on the British side.

There are metal panels between cockpit and propeller spinner. The rear fuselage, from the end of the checkerboard marking aft, has lines revealing the existence of stringers covered with doped fabric, which is also on the fin and rudder. Double lines marking the fin and rudder ribs are an indication of the fabric strips covering stringing, through from side to side, to hold fabric onto the ribs. A control cable is drawn emerging from a hole in the fabric just above the second digit in the registration: K2050. There are two struts beneath the

Plate 2–1 A modern replica of the Hawker Fury (origin 1927) with an original Rolls Royce Kestrel engine. Ref.: Example 2–1. The tough tube and fabric airframe led directly to that of the Hurricane. (Author)

tailplane, which indicate the presence of a substantial piece of structure – a longeron – at the bottom of the fuselage, to which they must be attached. A straight member, the top longeron, runs forward from immediately beneath the tailplane to a vertical line running downwards from the connecting point of the forward vee-struts with the fuselage. This group is termed the cabane structure, which links the top plane to the fuselage. From the drawing we deduce that:

☐ the straight member is the top longeron, therefore:
☐ above the top longeron the fuselage probably has a curved 'turtle-deck' section, and it must be almost flat between the bottom longerons (this can usually be confirmed by looking elsewhere on the drawing);
☐ the vertical line, running downwards from the top longeron and forward cabane vee-struts has fasteners shown on each side. This suggests that it is substantial, not only because it forms the anchorage of the forward legs of the fixed undercarriage; but it also marks the frame which ties the engine to the rest of the aeroplane, while supporting the firewall between the engine bay and the fuselage;
☐ the 'box' beneath the fuselage is the radiator for the in-line liquid-cooled Kestrel;
☐ a teardrop-shaped bump beneath the top line of the cowling, just behind the propeller, is a fairing where each forward corner of the in-line vee-engine would otherwise protrude. The six oval holes are for the ejector exhausts, which discharged straight to atmosphere

Plate 2–2 A Hawker Hurricane with the Rolls Royce Merlin engine. This refurbished example, seen at a 1980s air display at Biggin Hill, in night fighter black, shows clearly the ventral strake, with recess for the tail wheel, introduced to improve quality of spin recovery. (Author)

(unlike the modified flame-trap exhausts, developed much later for other aircraft);

- the markedly humped fuselage raised the eye-line of the pilot, so reducing the blind arc through the top plane, confirming that we are indeed looking at an interceptor fighter (guns, which were fitted on the top decking of the fuselage ahead of the pilot, are indicated by lines in the panel immediately ahead of the cockpit, which gave access to the breeches and ammunition feeds, and also by the hint of long channels ahead of the muzzles, which show clearly in other views);
- the aerial wires running from a mast on top of the rudder to the top plane and to the fuselage turtle deck, behind the pilot, indicate that the aircraft had radio. The point of attachment of the aerial to the radio (heavy by modern standards) is aft of the wings and well behind the CG – which is somewhere between the wings, but one cannot yet be quite sure where. So, with no radio of the period available, one might expect the Fury to be nose-heavy and to need ballast aft (which was found on a first test flight in the replica Fury);
- if the machine is nose-heavy, then with a collapsed oleo (shock-absorber) leg and flat tyre the aeroplane will not need to tip forward very far to snag on the ground an expensive, hand-carved, propeller; and
- there may be insufficient up-elevator authority to get the tail down into the three-point attitude for a slow and short landing. There could be bouncing, stalling or a wing dug

into the ground (and only a new set of teeth if the pilot is lucky). Thus, it is of critical importance to sort out early in the project where the CG was originally located, before charging bull-like into the design of a replica which, looking right, in the end flies wrongly.

There is no insult in cross-checking what an engineer has calculated. Two heads are better than one – and such advice cuts both ways!

Figure 2.1a is marked with a number of questions to which answers had to be found before carrying out test flying for the CAA. Figure 2.1b is a copy of my calculations for finding the equivalent monoplane wing, the test CG and its location as a fraction of the mean chord of the equivalent plane.

Postcript: Summarized flight test findings – full-scale replica

The Fury replica, tested on a warm September day, combined potent performance (comparable with a number of competition aerobatic machines) with amiable and docile handling qualities. The airspeed indicator was calibrated in statute miles per hour (MIAS).

Taxying the aeroplane was lazy in response to the effective control by rudder and toe brakes. The aeroplane leapt into the air on takeoff, in what could only be likened to three strides, around 50 MIAS (80 kph). In flight the controls were effective, but friction was unexpectedly high.

The liquid-cooled engine tended to overheat in the climb. Even so, at 105 MIAS (169 kph), where cooling was better, a rate of 1250 ft/min (6.35 m/s) was achieved at 2500 ft (762 m). Elevator trim was fully nose up at 90 MIAS (145 kph) with throttle set for a 5% descent gradient (about 300 ft/min (0.15 m/s)).

Stick force/g was about 8 lbf (36 N) at 150 MIAS and 2500 RPM.

Straight stalls, power off (i.e. throttle closed) were docile occurring at 50 MIAS (80 kph). There was a slight tendency to drop either wing, which could be prevented by full rudder against the direction of wing drop (left rudder as the right wing went down). With throttle set at 2500 RPM, the wing dropped vigorously, but this could be checked by rudder. Finally a docile nose drop occurred at 52 MIAS (84 kph). In turning stalls within 30° bank and 2500 RPM there was a sedate nose drop at 65 MIAS (105 kph) to the left; and a gentle roll out of the turn at 50 MIAS (80 kph) to the right, with some aerodynamic buffet.

Never Exceed Speed, V_{NE}, of 225 MIAS (362 kph), was not checked, as wing strength calculations had not yet been substantiated. High speed handing to 170 MIAS was satisfactory.

Directional stability and corrective roll with sideslip were not worse than neutral. Full rudder sideslips produced no fin buffet, nor tendencies to lock, at 72 MIAS (116 kph), although with throttle closed, the foot force flattened beyond half rudder deflection. With throttle set for 5% descent gradient at the same IAS, with feet off the rudder pedals sideslip continued.

Longitudinally the aeroplane was stable, but nose-heavy, with minimum trim speeds which were far too high. This pointed to a CG too far forward, which was confirmed when attempting to land. It was hard to reduce speed with throttle closed. The aeroplane floated, touched down three-point around 55 MIAS (88 kph), but ballooned back into the air. Fast feet were needed with the rudder and care with the toe-brakes, because of nose-heaviness.

The defects were nose-heaviness (a CG coefficient of 0.27 (27%) of the mean equivalent monoplane chord was clearly too far forward); and the engine running over-rich (black smoke from the exhausts). To cure the first, lead was later added in the rear fuselage.

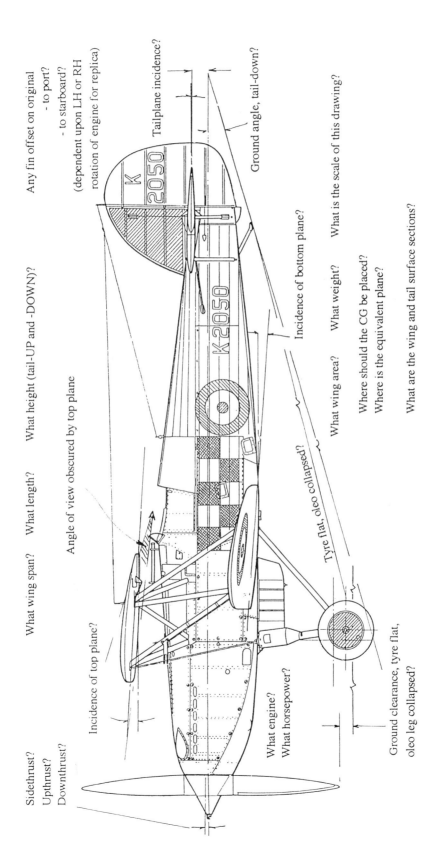

Sidethrust?
Upthrust?
Downthrust?

What wing span?

What length?

What height (tail-UP and -DOWN)?

Any fin offset on original
    - to port?
    - to starboard?
(dependent upon LH or RH
rotation of engine for replica)

Tailplane incidence?

Ground angle, tail-down?

Angle of view obscured by top plane

Incidence of top plane?

Incidence of bottom plane?

What is the scale of this drawing?

What wing area?

What weight?

Where should the CG be placed?
Where is the equivalent plane?

What are the wing and tail surface sections?

Tyre flat, oleo collapsed?

What engine?
What horsepower?

Ground clearance, tyre flat,
oleo leg collapsed?

Fig. 2.1a Pre-World War II Hawker Fury interceptor fighter showing some of the most fundamental questions to be answered when thinking of building and flying a full-scale replica. (Courtesy of *Aeromodeller*, drawing used as a guide)

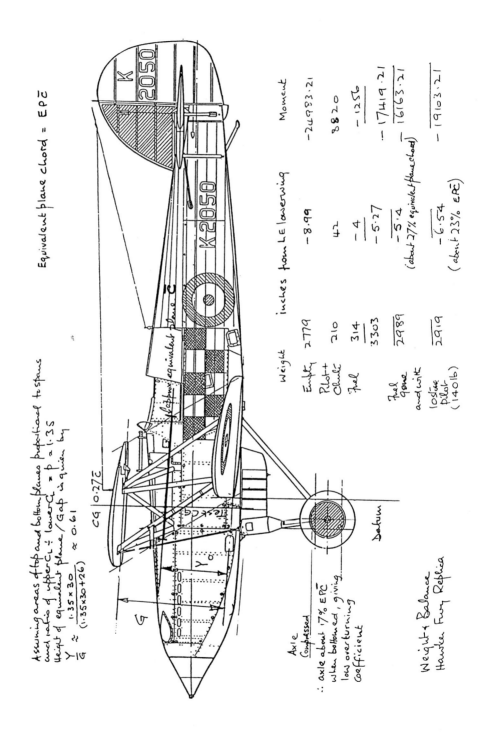

Fig. 2.1b Elevation of the Hawker Fury used in a Flight Test Report with calculation of the equivalent plane and CG, with full fuel, fuel gone, and 210 lb and 140 lb pilots. (Courtesy of *Aeromodeller*)

Emergence of Flight Dynamics

Response of the Fury to control in pitch and yaw altered dynamically the original 'static' (initial) stick and rudder forces felt by the pilot. This was to be expected with any manually controlled aeroplane. In the early days it had attracted attention enough to bring about the study of flight dynamics as a subject in its own right. The need for such disciplined approach was anticipated before the Wright Brothers flew. Afterwards, the subject gathered momentum, with performance and handling qualities being investigated in depth, as flight dynamics in Germany, at the Göttingen model test institute, before the outbreak of World War I in 1914. The subject was also studied at the Royal Aircraft Factory in England, the USA National Advisory Committee for Aeronautics, and at the Russian Central Aerohydrodynamics Institute (TsAGI), formed on 1 December 1918, by Professor Zhukovsky.

Note 1: The British Royal Aircraft Factory (the originator, *inter alia*, of a series of aerofoil sections, including the RAF 15 and RAF 32) was founded in April 1912 and became the Royal Aircraft Establishment (RAE) in June 1918. The Royal Air Force (RAF), formed in April 1918, was the direct descendant (via the Royal Flying Corps of 1912 and later integration with the Royal Naval Air Service) of the Balloon Section, Royal Engineers, which had separated from His Majesty's Balloon Factory in December 1909. The latter organization became the Army Aircraft Factory in April 1911, and then the Royal Aircraft Factory one year later.

Note 2: In the USA the National Advisory Committee for Aeronautics (NACA, now NASA), also a government organization, was created almost by default by a rider to the Navy Appropriations Bill, passed by Congress in the Winter of 1915.

The study of technical advances from World War I onwards is relevant to the design and flying-qualities of modern light and sporting aeroplanes. Significant changes were taking place, and a substantial lead in wing and control surface design was held by the Germans. Two historically significant fighting aeroplanes, one from each side, exemplify in the words of a British pilot who flew both types the marked differences in handling which stemmed directly from German research into control design. Not only that, but what follows is typical of the investigation required when considering (re-)building and flying such aircraft today.

Example 2–2: The Fokker DVII (1918) and Bristol F2B Fighter (1918)

In his book *Flying Minnows* the South African author Roger Vee, who had joined the RFC just before it became the RAF (on 1 April 1918), wrote of flying two-seat Bristol Fighters in combat (Fig. 2.2 and Ref. 2.2). That aeroplane was designed by Captain Frank S. Barnwell (1890–1938). The Fokker DVII is included because, following the Armistice, Vee also managed to fly a captured example and recorded his impressions (Fig. 2.3 and Ref. 2.2). Fortunately, contemporary performance figures for both types are available (Refs 2.3, 2.4, 2.5 and 2.7) and these are used here to make a rudimentary technical comparison. The DVII had a reputation as the most redoubtable of all German fighting scouts, and was mentioned specifically in the Armistice conditions, which demanded immediate surrender of all aircraft of the type.

The DVII was designed by Reinholdt Platz (1886–1966) who (though somewhat eclipsed by the personality of his employer, Anthony Fokker) deserves nothing short of admiration for the fertility of his ideas, and his skill in bringing them to practical fruition.

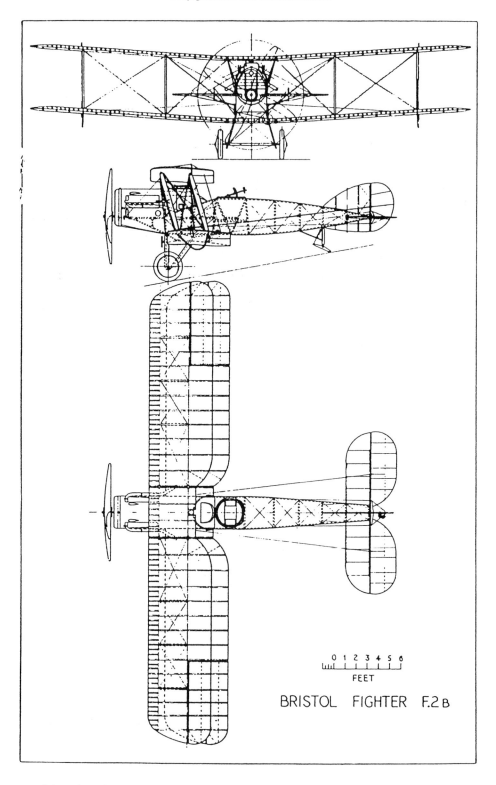

Fig. 2.2 The Bristol F2B, designed by F. S. Barnwell (Ref. 2.6). Although larger and heavier than the Fokker DVII, note that it has no control horns ahead of the hinges. (Courtesy of J.M. Bruce)

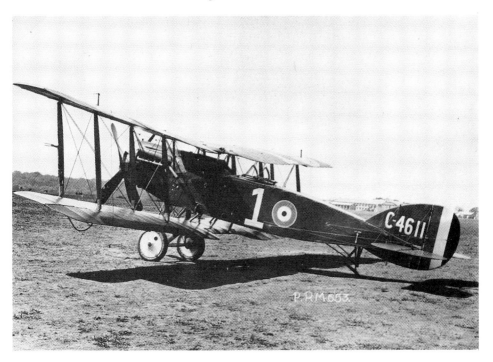

Plate 2–3 Frank Barnwell's Bristol F2B (1916) with Rolls Royce Falcon engine. Performance and handling of the F2B and Fokker DVII are compared in Example 2–2. (Courtesy of the Imperial War Museum)

Vee's comparison between the two aeroplanes is caught in these extracts from Ref. 2.2 (italics are mine):

'... each *Bristol* used for war-shows carried a 264 hp engine. The short exhaust pipes from this, one on either side, ended just next to the pilot's cockpit. The noise from the engine was thus deafening ....'

'A *Bristol Fighter* in the air is a glorious sight. It is like a flying tiger and it is an inspiring thing to look upon ....'

'A formation of E.A.* gave one a strong impression of deadly insects crawling through the sky. Because of their "balanced" controls they moved jerkily, with none of the graceful swing of the British machines ....'

* Enemy aircraft.

And on flying a DVII:

'The single seater was very small and light after our heavier machines. I got in, ran up the engine and waved the chocks away. Just before I left, the Flight Sergeant advised me to open the throttle slowly as the engine choked easily. The pitot-tube and other instruments had been removed, so as I began pushing forward the throttle I intended to take good care to get up sufficient speed before taking the *Fokker* off the ground. When the throttle was about half-open I glanced over the side and, to my surprise, found that we were already in the air. Immediately I opened the engine fully and eased forward the stick. The little *Fokker* answered beautifully to

Plate 2–4   Replica of the World War I Fokker DVII with a Ranger engine mounted inverted. Careful research is needed to ensure that look-alike replicas may be flown for display in representative working order. Replica Albatros DVa behind. Ref.: Example 2–2. (Origin of plate unconfirmed, author's collection)

her controls, but was very, very sensitive to them. I realized then why the *Fokkers* moved so jerkily compared to our own graceful machines.'

Vee has distinguished for us two handling features of what were quite different types of aeroplane (even though the Bristol Fighter, although larger and heavier than the Fokker DVII, was still a fighter – of a specialized kind compared with a single-seat scout):

☐ the 'balanced controls' of the Fokker linked with its moving 'jerkily' (we would now describe this as a response);
☐ the 'graceful swing' of the Bristol Fighter (another type of response).

Figures 2.2 and 2.3a and b are not to the same scale. Note the 'horns' on the control surfaces of the DVII (not present on the F2B) which, being ahead of the control hinges, lighten the forces felt by the pilot, so quickening response in pitch, roll and yaw. Note too the compact wing arrangement, with deep strong spars which reduced bracing and attendant parasite drag, providing strength with lightness, thereby improving performance.

Figures 2.3a with 2.3b were chosen deliberately. Their reproduction is of poor quality from a battered copy of the technical magazine, *Flight*, for 3 October 1918 (Ref. 2.5), which reproduces an official report by the Technical Department, Aircraft Production, Ministry of

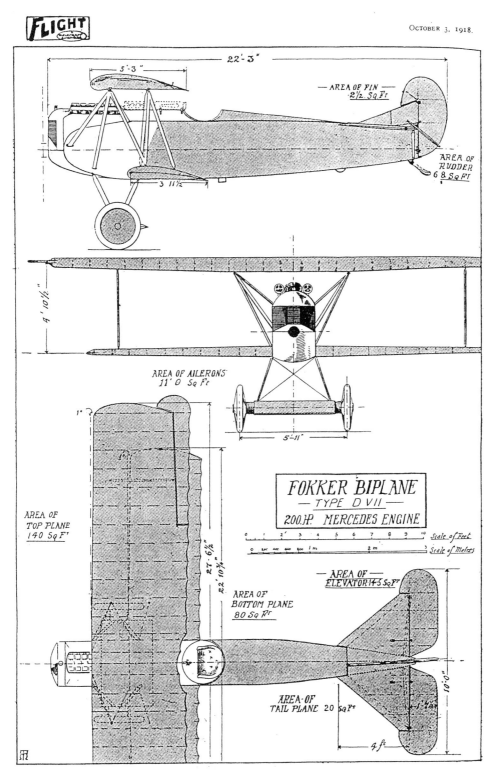

Fig. 2.3a Typical research material, pre-flight test, copied from *Flight* for 3 October 1918.

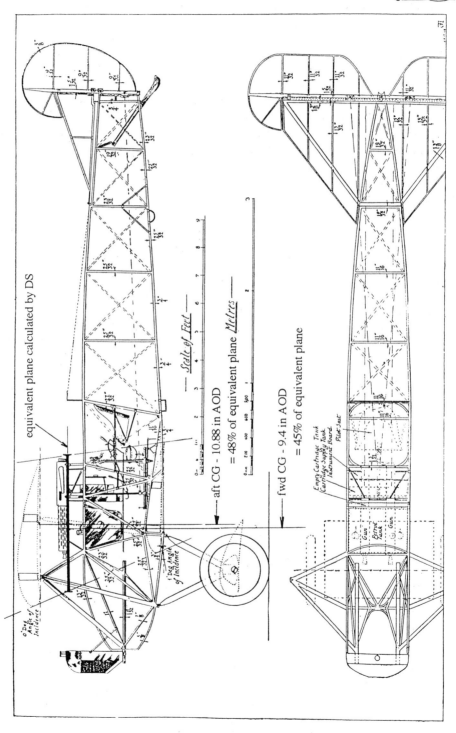

Fig. 2.3b   Companion drawing of Fokker DVII from Ref. 2.5, showing construction of the equivalent plane and CG locations, for the test flight of a full-scale replica in 1989.

Munitions. *Flight* shows details which differed from the official report. It highlights one of the many problems encountered in historical research. Which source is the most credible? Details can be missed, because when drawings are worn, or printed on war-time quality paper, they can be hard to reproduce cleanly.

Flight discusses details, not in the official report, which are significant where flying-qualities are concerned. First is that there is one degree of sweepback shown on the drawing. Second, the triangular fin is offset to port (a detail which might be overlooked, but in photographs the propeller is a Right Hand tractor, propwash of which, spiralling aft, would push the tail to starboard, needing a lift force from the fin to port, to reduce the pilot's footload, hence the offset). Third, the upper surface of the top plane is flat, unlike some other drawings, which shows that it has slight dihedral (confirmed by photographs). This aeroplane has a 200 hp Mercedes engine, but lacks performance details.

Quoted performance differs between examples of the same type, every bit as much as the recorded weights. For example, Ref. 2.3 gives the DVII as having an all-up weight of 955 kg (2100 lb) and a BMW IIIa engine which produced 185 hp (138 kW). Other variants had a 170 hp (basic rating 160 hp) Mercedes DIII, and weights varying between 1870 lb (850 kg) and 1936 lb (880 kg) (Refs 2.4, 2.5 and 2.7).

The Bristol F2B is quoted in Ref. 2.5 as weighing 2590 lb (1177 kg) with a 275 hp (205 kW) Rolls Royce Falcon III engine (whereas Vee says 265 hp). Using 275 hp for calculations, Table 2–1 shows the weight, wing and power loadings for both aeroplanes in imperial and metric units, of the period. Here FPSR are the primary units, with metric in brackets (i.e. 'weight' is in old fashioned kilogrammes).

An old measure of how 'good' the performance of a light aeroplane might be was given in terms of (wing loading) times (power loading). We do not know for sure that the weight given is the gross or maximum take-off weight, W_0, we will generalize and use:

$$(\text{weight}/\text{wing area}) \times (\text{weight}/\text{hp}) = (W_0/S)\,(W_0/P) = W_0^2/SP \qquad (2\text{–}1)$$

TABLE 2–1

Technical comparison

Item	Bristol F2B Rolls Royce	Fokker DVII Mercedes	BMW
Weight lb	2590	1936	1936
(kg)	(1177)	(880)	(880)
Power hp	275	160	185
(kW)	(205)	(119)	(138)
Wing area ft^2	405.6	231.4	
(m^2)	(37.7)	(21.5)	
Power loading lb/hp	9.42	12.1	10.46
(kg/kW)	(5.74)	(7.36)	(6.38)
Wing loading lb/ft^2	6.39	8.37	
(kg/m^2)	(31.22)	(40.93)	
W_0^2/SP lb/ft^2 hp	60.19	101.23	87.55
(kg^2/m^2 kW)	(179.20)	(301.24)	(261.13)

in which the smaller the product the better the performance. Equation (2–1) is an example of the parametric approach to assessing probable flying qualities.

The term (W^2/SP) applied to very basic, simple, aeroplanes (fixed gear, split or plain flaps (if any), coupled with low wing loading). Agility and performance would be 'good' when:

(wing loading) × (power loading), i.e. W^2/SP

satisfies: $90 < W^2/SP < 135$ in FPSR units   (2–2a)

or: $270 < W^2/SP < 400$ in SI units   (2–2b)

There appears to be some justification for the formula because both aeroplanes were rated as good and had values less than 90 (FPSR). Indeed, considering the Bristol F2B was the larger, heavier and, potentially, the more sedate of the two, nevertheless it had both lower wing and power loadings.

Note 1: The formula does not have to use the gross (or take-off) weight, W_0, exclusively.

Note 2: The Système 'D' aeroplane in the previous chapter (Fig. 1.7) has a value of:

$W_0^2/SP = 25.8$ in FPSR units

which is much less than 90 and points to perhaps unmanageable agility.

Table 2–2 tells us little until we start to tease the results into surrendering evidence of characteristics they might otherwise conceal. Starting with Fig. 2.4, we have a rough plot of rate of climb, in feet per minute, from Refs 2.3, 2.4, 2.5, 2.6 and 2.7. It is crude because now we can find little or nothing about the ambient conditions at the time. The altitudes are probably pressure altitudes measured by altimeter, but we know nothing of errors or corrections made, or of the change of ambient air temperature, density and pressure with altitude on the day the tests were carried out, or by whom. In short, the results are *'not proven'*, in the old Scottish legal sense, nor are they proveable at this remove.

Even so, there are several things which might be said before looking at climb performance in a later chapter. The first is that, as a general rule, it is not good practice to draw a mean line through only three points. Second, in Fig. 2.4 we ignore this stricture on the principle of the flying instructor's advice: *'Do as I say not as I do'*, by resorting to the coarse rule of thumb: that a normally aspirated piston engined aeroplane takes, roughly, 45 minutes or a little more to reach its absolute ceiling. So, for the Bristol Fighter and the Mercedes-engined Fokker DVII fair curves are used to join four points: start, two interim, and absolute ceiling, estimated at 45 minutes after takeoff. The captured DVII, with only two points, is an exception.

Measuring the slopes of tangents to both curves over 10 minute intervals, at different altitudes, it is now possible to 'eyeball' a local rate of climb at each point, where:

local rate of climb = change in altitude between ends of each

tangent (ft)/time taken (min)   (see Equation (3–69))

The result is the approximate average rate of climb at each point at which a tangent touches the climb curve. This should produce a straight line in practice, showing rate of climb to decrease with altitude, because of the fall in air density, and pressure. Figure 2.5 is not far out, considering that the curves are sketched in. So, with a combination of three-view drawings and fairly sparse technical data, we find it possible to make a broad comparison between quality of performance of typical examples of two aeroplanes which were opponents three-quarters of a century ago.

If the curves are a fair indication, then the climb performance of the heavier two-seat Bristol F2B was better than the single seater DVII scout with a Mercedes engine, but not as

TABLE 2–2

Comparison of climb times

Climb time in minutes*	Bristol F2B	Fokker DVII		
		BMW	Mercedes	'Captured'
1000 ft (305 m)	0.83			
3200 ft (1000 m)		2.50	3.8	4.25
10 000 ft (3048 m)	11.25			
16 400 ft (5000 m)		16.00	31.5	
18 000 ft (5486 m) Service ceiling**			44.0	
Absolute ceiling ft**	22 000			22 900
(m)	6706			6980

* Climb times are plotted in Fig. 2.4, with reduction to rates in Fig. 2.5.
** Exceptionally hard to measure, because it takes something like 45 minutes to reach an absolute ceiling which cannot be determined accurately, even when there. It is much better to measure the time to a service ceiling, at which the rate of climb is 100 ft/min (0.51 m/s), and then to extrapolate with rates of climb established on the way to altitude (see Fig. 2.5 and Chapter 7).

good as the BMW-powered DVIIF. The Fokker had a reputation for good climb performance, and the steeper slopes of both DVII climb lines in Fig. 2.5 tends to confirm this. Assuming that the airspeeds in the climb of the F2B and the DVII were similar, then the angle of the climb path, i.e. the *gradient* would have given the impression of the DVII being able to climb like a lift, when seen from the cockpit of an F2B. This is the possible origin of impressions brought back by two-seater crews, that the new German scout had enviable speed and rate of climb, for:

> 'it appeared to be able to stand still in the air with its nose up, in which attitude it would pour a stream of bullets into the two-seater from the blind-spot.'   (Ref. 2.8)

You may ask why go to the trouble of calculating climb performance when speed, the ability to catch or run away from one's opponent must be of greater importance? An answer is that speed is one point at one given altitude, measured by an airspeed indicator, the instrument and position errors of which, as this is written, one knows nothing. We do not know if the speeds quoted are *indicated* (instrument) or *true* (over the ground in still air) speeds. We can only make a reasonable guess. To compare values and ascribe full significance to them we must know that each instrument was calibrated and comparable with the other. With this in mind, Table 2–3 gives unqualified, albeit published, maximum speeds for the F2B and Mercedes-engined DVII (Refs 2.6 and 2.7 respectively).

The ability to climb higher faster than one's opponent bestows a tremendous tactical advantage; and height can always be converted into airspeed. No maximum speed has been found for the DVIIF, with the BMW engine, but knowing that for a given drag coefficient, C_D (see Equation (3–28)) which depends intimately upon the configuration of an aeroplane, then:

$$\text{power, } P, \text{ varies as } C_D \, (V_{max})^3 \tag{2–3}$$

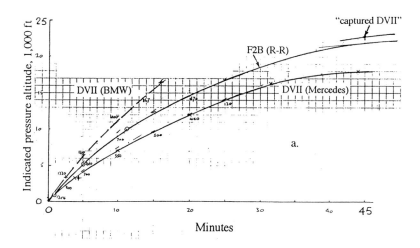

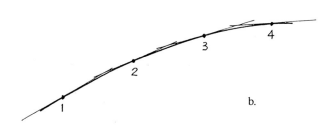

Fig. 2.4 Data in Table 2–2 plotted and 'eyeballed'. This has been reduced in scale so that in the lower figure the method of constructing tangents at convenient points can be shown. This enables rough local rates of climb to be calculated.

so that: $(V_{max}, \text{BMW})/(V_{max}, \text{Mercedes}) = (P, \text{BMW}/P, \text{Mercedes})^{1/3}$ (2–4a)

i.e. $V_{max}, \text{BMW} = (185/160)^{1/3} \times 110 = 1.05 \times 110 = 115$ mph (185 kph)   (2–4b)

This corresponds with values of 116.6 mph (186.5 kph) and, 114.1 mph (182.5 kph), quoted in Ref. 2.4, but without saying which engine.

Postscript: Test of a DVII replica with 200 hp Ranger engine (1989)

The replica weighed 2257 lb (1026 kg), making it about 17% heavier that the 185 hp BMW-engined aeroplane in Table 2–1. The CG was at 48% of the equivalent monoplane chord for take-off, moving forward as fuel was consumed to 45% (see Chapter 10, and the discussion surrounding Example 10– 4 for significance of this).

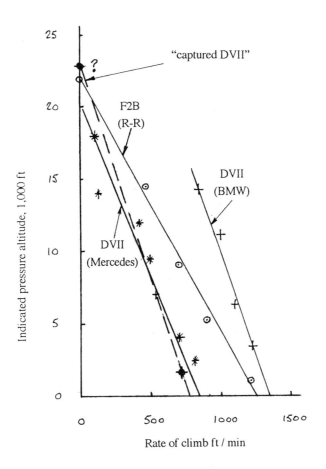

Fig. 2.5 Local rates of climb calculated using the method in Fig. 2.4. The results are necessarily crude at this remove because nothing is known of the accuracy of the instruments, or of the ambient conditions at the time (see Chapter 5). Even so, it is better than having nothing at all against which to compare a flight test result of a replica.

TABLE 2–3

Comparative speed		
Maximum speed	Bristol F2B	Fokker DVII (Mercedes)
10 000 ft (3048 m)	113 mph (182 kph) (98 knot)	110 mph (177 kph) (96 knot)

The tail was lifted at 45 MIAS and the main wheels at 50 MIAS. During the climb at 70 MIAS the ailerons were heaviest, with: A : E : R = 3 : 2 : 1 (see Equation 2–5), so that in spite of horns, the replica exhibited force ratios the reverse of those rated as harmonious by World War II standards. A rate of climb of 490 ft/min was measured at 2400 ft (731 m), well below the rates calculated subsequently in Fig. 2.5, in air colder than ISA. Stalls were sedate at 50 MIAS power OFF, and 42 MIAS with 75% power. Longitudinally the aeroplane was untrimmable below 70 MIAS and beyond 95 MIAS, needing then to be flown all of the time. Roll with sideslip (dihedral effect) was neutral, as was directional stability.

The ailerons suffered backlash and were relatively weak, although adequate, compared with rudder and elevator. Response in roll was such that I felt disinclined to try rolling aerobatic manoeuvres. Thus, in spite of a more powerful engine and modern materials, the replica was pleasant to fly but sedate. It could not be rated in the same class as its 'jerky' scouting predecessors, banned under the terms of the 1918 Armistice.

Agile Versus Sedate – Opposed Flying Qualities

The last example, taken from an historic period of rapid technical development, separates machines that are *agile*, from those which are *sedate*. These two opposing qualities can be developed in several different directions. At one remove we may gather and discuss the flying-qualities of lively aerobatic, agricultural, trainer and aerial-work aircraft. At the other are the more dignified air-taxi, air-ambulance, commuter and passenger-friendly transport and freight-carrying aeroplanes.

History proved useful in another way, because of a book written by the same Captain Barnwell, together with W. H. Sayers in 1916 (Ref. 2.9), which gave the needed clue to the way in which aircraft designers, not only of World War I but long after, set about arranging engine position and thrust line, conjunctions of flying surfaces, CP and CG, and disposition of the undercarriage. We now understand their effect upon flying qualities. The difference in terminology is that we now use aerodynamic centre, ac, of the wings (and body combinations) of the complete aircraft, instead of the Centre of Pressure, CP.

Example 2–3: Tail-first Rutan Voyager (1986)

In 1986 an event in aviation took place which was as great in its way as the flight by the Wright Brothers at Kittihawk, on 17 December 1903. The Voyager, a 'tail-first' or canard monoplane, with twin-booms and twin-engines, front and rear in a central nacelle containing the crew of two, flew around the world non-stop. The configuration provided every structural, aerodynamic and propulsive advantage needed to succeed. The wings were exceptionally long and of high aspect ratio for low lift-dependent drag. The nacelle and twin booms were slender (the second crew-member occupied a prone position in a space not many inches deep next to the pilot). The wings and booms contained fuel. The front engine, used on takeoff, was essentially a spare, shut down and feathered in the cruise, which was carried out on the more efficiently located rear engine. The aeroplane was designed by Burt Rutan, of Mojave in California. It was flown by his brother Dick Rutan and Jeana Yeager, both of whom were part of the team which built it.

The aerofoil sections were designed by aerodynamicist John Roncz. The aeroplane needed a high lift/drag ratio for long range, and this meant smooth laminar flow profiles. If the relative airflow over the surfaces of an aircraft becomes turbulent, then wing profile drag (caused by skin friction and the form (bluffness) of the shape being pushed through the air), and the parasitic zero-lift (extra to wing) drag is increased. Laminar flow is easily destroyed

Plate 2–5 Around the world non-stop, the Rutan Voyager (1986) at Oshkosh, showing long span, taxying on the rear engine with the front engine stopped. Ref.: Example 2–3. Note: the front (No. 1) engine was shut down for cruising and the propeller feathered. (Author)

by bugs splattering against leading edges, and by rain, which forms discrete blobs, like drops of water on a clean and highly polished surface. In the words of Dick Rutan and Jeana Yeager, after experiencing severe loss of lift from the canard, which put them into an uncontrollable dive when they entered a shower of rain (Ref. 2.10, and my italics):

'For the canard, in particular, he (*Roncz*) had worked out a shape that had a huge amount of lift – 132-to-1 lift-to-drag ratio as compared with the 20-to-1 ratio of a standard wing. It may have been the "most lifting" airfoil in the history of aviation, and in Breguet's formula it translated straight into range ....
'What we found was that the rain disturbed the smooth passage or "laminar flow" of air over the top of the canard. Each drop, at the moment it beaded on the canard, had the same effect as if it had been a solid bump on the smooth surface. "Nature", John Ronz said, "sees each drop as a rock." Sure enough, we lost 65 percent of the lift on the canard when it was wet. No wonder we were falling out of the sky in that shower.'

The cure suggested by Dr Gregorek was vortex generators:

'... 210 tiny angled tabs, which looked like little teeth or shark's fins, glued to the canard with *Hot Stuff* glue. These kept the air swirling over the aerofoil and stopped the drops from appearing. As an added precaution, John later had us sand the leading edge down so the water didn't bead up. The vortex generators added about 2% drag, but as John put it, that was a very inexpensive premium for the life insurance they provided.'

Note: Having tested numerous tail-first aeroplanes with manual flying controls, I must confess that they are not among my favourites. To me, each felt strange and back-to-front. All were less agile and coordinated in response to control, with limited manoeuvrability compared with conventional aeroplanes having similar wing and power loadings. I cannot pass an opinion on the qualities of high performance tail-first jet aeroplanes, with expensive computers between pilot and flying-controls.

During World War II a qualitative rule for judging handling to be 'sweet' was that for a given rate of response in roll, pitch and yaw, the control forces of a conventional aeroplane with a tail at the back should be:

$$\text{aileron : elevator : rudder} = 1 : 2 : 4 \tag{2–5}$$

To my hands and feet comparable tail-first aeroplanes have felt, very roughly:

$$\text{aileron : elevator : rudder} = 4 : 2 : 1 \tag{2–6}$$

Even so, where there is a requirement to fly long distances, efficient cruise performance is needed and tail-first aeroplanes can be shaped to generate better overall lift/drag ratios than conventional aircraft. The reason, simply, is that the foreplane and rearplane both lift in the same direction – upwards. Whereas a download is produced by a conventional tail – or, if not a download, then not as much proportional upload as a foreplane. The canard configuration operates more efficiently at one design point in the flight envelope. But, unlike a conventional aircraft, it is less efficient and handy off-design.

Note on the Breguet Range Equation –Tail-First Aeroplanes

The shape of tail-first Voyager provided an efficient solution to the *Breguet range equation* (Louis Breguet, pioneer French aircraft designer (1880–1955)). This equation shows that the joint contribution to range-flying ability made by the three primary elements of the configuration of an aircraft: *propulsion*, *aerodynamics* and *structure* is, qualitatively:

$$\text{range} = \text{a constant} \times (\text{propulsive efficiency}) \times (\text{aerodynamic efficiency})$$
$$\times (\text{structural efficiency}) \tag{2–7}$$

The equation is adaptable to propeller-driven and jet aeroplanes as follows:

☐ *Piston and turboprop aeroplanes* (in FPSR units)

$$R \text{ nm} = 750 \, (\eta/c')(L/D)_{\text{max}} \, (\log (W_0/W_E) \tag{2–8}$$

in which propulsive efficiency is (thrust efficiency/specific fuel consumption), (η/c'); aerodynamic efficiency is equated to the hard-to-achieve, theoretical, maximum (lift/drag), $(L/D)_{\text{max}}$; while structural efficiency is measured by the weight of fuel carried, expressible in terms of the (gross weight lifted per unit of empty weight) as $\log (W_0/W_E)$. Thrust efficiency, η, is the propeller efficiency, which appears in the equation linking thrust horsepower with rated, or installed, horsepower and, hence, thrust, F, and drag, D (which are equal and opposite in steady, level, cruising flight); true airspeed is V KTAS:

$$\text{thrust horsepower, } P_t = \eta P$$
$$= FV/326 = DV/326 \tag{2–9}$$

As a general rule, η is around 85% for a well finished metal propeller, and is nearer 75% for one made of wood. These values can drop markedly with wear and tear (nicks, bites, bug-splat, erosion by rain, hail, sand and gravel, hangar-rash) to nearer 70% and 62% respectively, spoiling rates of climb and range efficiency.

Equation (2–8) tells us that for long range we need to combine a high mileage per unit of fuel consumed, with high lift and as little drag as possible, while carrying the maximum amount of fuel in an airframe structure of least weight. If the last term in the equation looks odd, don't worry about it. The gross weight, W_0, comes into the picture only because the airframe structure must also be able to carry its own weight in addition to everything else.

☐ *Turbojet and turbofan aeroplanes* (in FPSR units)

$$R \text{ nm} = 2.3 \, (V/\mu') \, (L/D)_R \log (W_0/W_E) \tag{2–10}$$

in which the meanings are substantially as before, except that we employ V KTAS, μ' is thrust specific fuel consumption in lb/h/lbf, and $(L/D)_R$ is the best lift/drag ratio for range (at a speed somewhat faster than that at which the maximum (L/D) occurs).

We know from Equation (1–1) that Mach number $M = V/a$. Therefore, knowing too the local speed of sound at a given air temperature and pressure, only a slight adjustment is needed to Equation (2–9) to change $V(L/D)$ into $M(L/D)_R$. This combination of terms has a more powerful effect upon achievable range (which is an operational requirement and as such the subject of penalty in most contracts) than the product of specific fuel consumption and structural efficiency as a fuel-carrier in the remainder of the equation (the value of which is around 1.0, rising to near 2.0 only at high Mach numbers). So, for simple practical purposes, we may say that for jet and turbofan aircraft as indicated in Fig. 2.6:

$$\text{range varies as } M \, (L/D)_R \tag{2–11}$$

and replacing lift and drag with weight and thrust given in the brochure:
we have a further indicator of aerodynamic and propulsive efficiency of a jet aeroplane

$$W_0 \, M/F \tag{2–11a}$$

which is useful in Chapters 5 and 13 (see Fig. 13.7).

Equations (2–11) and (2–11a) enable designers to use configurations (like highly swept deltas) which are capable of achieving high Mach numbers in spite of apparently low lift/drag ratios in the cruise. As jet aircraft fly high, true airspeed increases with altitude at constant Mach number, with longest ranges being achieved at or near the tropopause.

This is the fundamental difference between Voyager, which was piston-propeller driven and a long range jet aeroplane. Whereas the jet flies fast and farthest at high altitude, Voyager, like other piston-engined aeroplanes was constrained to low altitudes where the static pressure of the air squeezed into the cylinders is highest, and power output greatest. It flew at low airspeeds where aerodynamic forces, which depend upon speed squared, are low and structure weight light in consequence – the weight of material which would have been needed at a higher cruising speed, being traded for a heavier fuel load.

The light and floppy structure of Voyager caused handling problems:

'As we flew at heavier weights, we found out even worse things about the airplane: we discovered that, at a point just above eighty-two knots, we had a major pitch-problem – pitch porpoising … in an updraft the wings bow, the tips go up, and the wing roots go down, carrying the fuselage with them. This in turn pitches up the nose and canard. The motion becomes a wave, repeating itself and growing in magnitude – a heaving or "porpoising" … we were facing the danger of an airplane that had a basic aerodynamic flaw … that could come apart around our ears ….

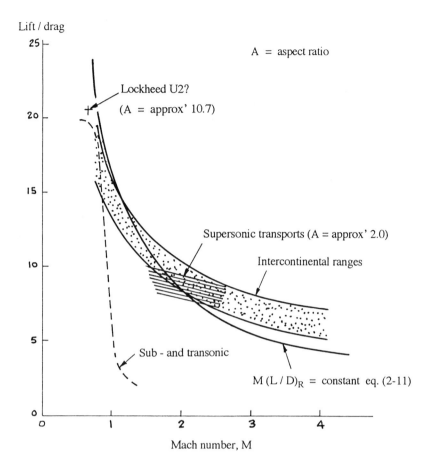

Fig. 2.6 Turbojet configurations which provide 'good' lift/drag ratios for long ranges. Each aspect ratio fits only one part of the operational envelope, defined by range and Mach number.

... People would ask how far the wings could bend, and we would answer jokingly, "Until they touch". The composite construction did give the wings a remarkable amount of resilience, and the fifteen or so feet wingtip movement we often saw was not in itself a concern ....

When we flew at the heavier weights, the flapping became serious. The flights were showing that at exactly eighty-two and a half knots of speed, the flapping of the wings went out of control – the oscillations, in technical terms, "became divergent". The flapping began to increase in height, doubling with each cycle. Unless the pilot or autopilot stopped it, the wings would literally flap themselves off in about ten or twelve seconds. If you let go of the stick, this airplane would come apart ....

If you attack the porpoising wrong, it gets worse. It can easily turn into "pilot-induced oscillation." The only way to control it is to lead the wave by ninety degrees. When the nose is at the bottom of the curve, you give it forward stick, even the stick as it passes the horizon, then, as it reaches the top of the cycle, pull back ...'   (Ref. 2.10)

Foreplanes, being mounted ahead of the CG, are destabilizing. As described in the case of Voyager, wing tips rose and the wing roots and fuselage went down, and there would be an increase in angle of attack of the canard foreplane, to the changed relative airflow. That would increase the lift which, acting ahead of the CG, then pitched the nose upwards, driving a further cycle, like an oscillating spring. Inertia of the masses ahead of the mainplane roots would exacerbate pitch-up and increasing angle of attack, by trying to resist being forced downwards as the wing tips began to rise.

Note: As far as I know there was no 'fix'. The pilot rode it out.

Example 2–4: Wright Flyer (17 December 1903)

Pitch-up and longitudinal instability caused by canard surfaces is not a new phenomenon. It is a direct consequence of configuration. The Wright Brothers experimented with foreplane pitch controls. Their gliders, and their aeroplane, the Wright Flyer of 1903, all began with a tethered kite, to which they later added a forward 'rudder' (elevator) for control in pitch. The control was in front where it provided much needed lift in the upward sense. Their gliders of 1900 and 1901 had no vertical rudders. Lateral and directional control was by wing warping. Their earliest aircraft, built in 1899, had wings which moved fore and aft, to adjust the CP relative to the CG. A moving foreplane interconnected with the wings and acted as an automatic trimming elevator. Longitudinal control by adjustment of the CP and CG, which is now commonplace with hang-gliders and weight-shift microlight aeroplanes, was employed by Otto Lilienthal (1848–96), one of the giants in the history of flight. F. W. Lanchester (1878–1946) alleged that the Wright's machine could:

'... metaphorically speaking trace its ancestry back to the gliding apparatus of Otto Lilienthal.' (Ref. 2.11)

The Wrights' aircraft were initially flying wings – biplane cellules, to provide maximum strength for low weight and large lifting area. They were all control and no inherent stability, like the birds which they observed closely and copied. Their Flyer had anhedralled wings (tips lower than the roots), because birds arched their wings downwards in strong and gusty conditions. Earlier experiments with dihedral (tips higher than roots) had caused them to bank out of control when they fitted a directional rudder at the back. The machine had then crashed after rearing up and dropping a wing. Today we would describe the rearing up as a form of pitch-up, nodding sagely, because we know with benefit of hindsight that the lateral rolling moment with sideslip and directional stability and control are mutually interactive. Neither is isolated from the other. Hence the interconnected behaviour of the aircraft in lateral response to dihedral and directional response to rudder.

The Wrights flew successfully on a knife-edge. Eventually they gave up their forward 'rudder' (elevator), the change being marked by a letter from Orville to Wilbur on 24 September 1909 (Ref. 2.12):

'The difficulty in handling our machine is due to the rudder* being in front, which makes it hard to keep on a level course .... The machine is always in unstable equilibrium .... I do not think it is necessary to lengthen the machine but to simply put the rudder behind instead of before ....'
* Elevator.

Perhaps the aeroplane would have been easier to handle had there had been a fixed foreplane to support the elevators, this could have slowed the motion by aerodynamic damping in pitch.

Example 2–5: North American X-15 (NACA Committee Decision, June 1952)

Porpoising, reported by Yeager and Rutan in Example 2–3, is not confined to tail-first aeroplanes. Other configurations suffer, and not necessarily for the same reasons. A. Scott Crossfield, a NACA test pilot (who was the first to exceed Mach 2.0), is reported to have experienced uncontrolled porpoising during an early test flight in the North American X-15, after being dropped from a Boeing B-52 at 38 000 ft (11 582 m). The landing approach was made after jettisoning the bottom of the ventral fin, but on doing so the aeroplane began to pitch up and down as the power control system took every pilot input and magnified it, leaving him out of phase with the motions. In Crossfield's referenced words and italics:

'The X-15 was porpoising wildly, sinking towards the desert at 200 miles per hour. I would have to land at the bottom of an oscillation, timed perfectly; otherwise, I knew, I would break the bird. I lowered the flaps and gear.
  My mind was almost completely absorbed in the tremendous task of saving the X-15, of getting it on the ground in one piece. But I could not push back a terrible thought that was

Plate 2–6   A North American X-15 A2 with overload fuel tanks, designed to probe to M8. This aircraft is of the type described in Example 2–5. (Courtesy of Philip Jarrett Collection)

forming. *Something was dreadfully wrong. We had pulled a tremendous goof. The X-15 in spite of all our sweat and study, our attempt at perfection, had become completely unstable.'* (Ref. 2.13)

Fortunately, the nose came up at the same time as the two tailskids hit the ground. The nose pitched down, bringing the nose gear smartly on to the ground without damage, and the landing was completed safely.

Remembering that flying qualities embrace performance, handling and functioning, this account reveals that often they are neither comfortably (nor comfortingly) separate. Short-fall in one can trigger even worse shortfall in another. In this case a valve in the powered flying control system needed adjustment. Even so, this was only found out after it had induced a sudden, uncontrollable, and almost catastrophic change in a handling quality, which manifested itself to the pilot as longitudinal instability. Further, the 'instability' did not appear until the ventral fin was jettisoned, which would immediately suggest an aerodynamic causal connection to any pilot.

In fact, Crossfield is reported as saying that the X-15 did not have any serious design flaws. Bridging the gap between conventionally supersonic air-breathing flight, and hypersonic research, including simulated lifting atmospheric re-entry from space, it has been one of the most productive experimental aircraft ever built.

Note: As a measure of the changes that take place as a design progresses from a paper submission to reality, the X-15 in its final forms operated at gross weights in excess of 56 000 lb (over 25 000 kg). With external and internal rocket fuel exhausted it weighed less than 19 000 lb (some 8600 kg). The original project design submitted to NACA by North American had a launch weight of a little under 18 000 lb (8200 kg) and a landing weight around 10 000 lb (about 4600 kg).

The first flight of a new design is more often than not a simple affair – certainly, it should be. It is unwise at first encounter to push an aircraft further than calculation and prudence advise. Longevity in test flying lies in taking half a step (or less) at one time, on the principle of: '*Softly, softly catchee monkey*'. For a pilot to lose or damage a new aeroplane through showing off is proof that he is untrustworthy and unsuitable. Sack him, if he survives.

Example 2–6: Focke-Wulf Fw 190 (1938–1945)

Kurt Tank (1897–1983), designer of the German Focke-Wulf Fw 190, served in the cavalry and infantry in World War I. That experience no doubt influenced his approach to the design of the Fw 190, which he conceived as a *Dienstpferd*, a cavalry horse, unlike the racehorse Messerschmitt Bf 109 and Supermarine Spitfire. The rare Fw 190 is probably the most sought-after aeroplane among collectors and those in the air show industry. The later Ta 152 is unobtainable as this is written.

Every modification or 'fix' tells a story, which is the reason for using Tank's Fw 190 and the Ta 152, shown in Fig. 2.7.

In the period between the wars Tank made his name as a design engineer and practising test pilot. His combined skills enabled him to fly himself to Luftwaffe fighter units during World War II, to talk with the young pilots. Over beer and schnaps, out of the language of fighter pilots (enhanced in every air force by movements of hands) he translated their requirements into the exemplary flying qualities of an aggressive family of piston-engined fighters, two of which are shown in Fig. 2.7 (Refs 2.14, 2.15 and 2.16). His ability to mix with and learn from squadron pilots gave him an edge which paid dividends for the Luftwaffe.

Plate 2–7   Messerschmitt Bf 109-E1: a small light aeroplane with a powerful engine. (Courtesy of Deutsches Museum, Munich)

The first prototype of the Fw 190, the V1, flew in May 1939. The test pilot, Hans Sander reported:

> 'Aerodynamically she handled beautifully. The controls were light, positive and well balanced and throughout the initial flight I never once had to make use of the tailplane trim. I suppose most test pilots would have made at least a roll in the new aircraft, but I did no aerobatics during the maiden flight of the Fw 190; I was quite happy to leave such fancy flying until later in the test programme, when I knew a little more about her. At this stage my task was merely to 'taste' the handling characteristics of the new fighter.'   (Ref. 2.16)

Air Marshal Sir Sholto Douglas, C-in-C Royal Air Force Fighter Command, writing to the Under Secretary of State for Air in July 1942, is quoted as saying:

> '... the Fw 190 is the best all-around fighter in the world today.'   (Ref. 2.16)

So highly was Tank regarded, that the Ta 152, which reached operational status shortly before the war ended, bore the prefix of the initials, Ta, even though it was manufactured by a company which, unlike Messerschmitt, Heinkel and Junkers, did not bear his name.

Interpreting what pilots say

In every accident or incident we find numerous *predisposing factors* and, acting as the trigger, one cardinal *precipitating factor*. To diagnose and report an event accurately, so as to separate the factors, pilots and engineers must understand the language and especially the concepts each employs. The foregoing examples were chosen to raise question marks in your mind. While those questions cannot be answered yet, that they should exist is

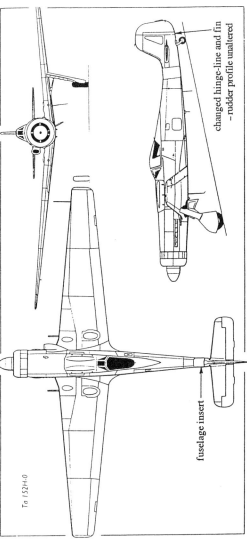

Ta 152H-0

changed hinge-line and fin
– rudder profile unaltered

fuselage insert →

Fig. 2.7a Kurt Tank's Dienstpferden (cavalry horses): the BMW radial engined Focke Wulf Fw 190 A-3 and the Jumo in-line engined, high altitude Ta 152H-0, bearing initials of his name instead of Fw. (Courtesy of William Green)

Fig. 2.7b The modification of a basic airframe to incorporate an engine of higher power, and a wing of longer span requires additional tail-volume if handling qualities are not to be degraded. Modifications to tail-volume (Equation (3–53) applies to all tail surfaces) may properly be regarded as 'fixes'. In the case of the Ta 152 a parallel-sided section was added to the rear fuselage to increase the tail-moment arm (see Ref. 2.20). This 'plug' provided spare volume for compressed air, oxygen bottles, and possibly ballast. The plug was in effect a squashed cylinder suggesting that it was a quick-fix for low cost and ease of manufacture at a time when Germany was hard-pressed by daylight raids of high-flying USAAF B-17 Fortresses. Fin and rudder volume was also increased by moving the rudder hinge-line aft and increasing the fin area ahead of it. No change seems to have been made to the rudder profile. (Courtesy of Ing. Aleš Janda and Ing. Tómás Poruba, Czechoslovakia)

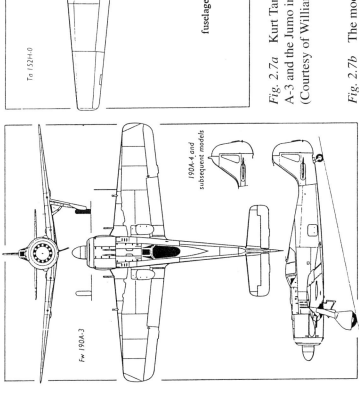

190A-4 and
subsequent models

Fw 190A-3

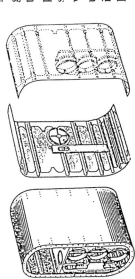

important. A question implies the emergence of a distinction. A distinction marks the first step on the road to discovering and understanding technical truth.

As Tank knew, listening to pilots is instructive. Listening to qualified test pilots bestows the advantage of hearing unvarnished truth, because of the ethics of discipline and technical integrity instilled by their training. David P. Davies, Past Chief Test Pilot and Head of Flight Department of the UK Air Registration Board and the Civil Aviation Authority put it this way:

> 'You must also stand by the technical truth, however great the commercial or political pressures.' (Ref. 2.17)

Davies did not believe that committees would do this, unless there were men among them who would dissent on points of principle. He had been surprised how many would not do so, *'not when the chips are down'* (Ref. 2.17). The test pilot must be his own man, *'vera prae ceteris'* (truth above all), because technical truth is paramount. I would add that he must also be practical and a realist, because reality is the killer. Ignore it at your peril.

HRH The Duke of Edinburgh, an experienced flying-man, observed in an essay on test pilots that:

> 'In all this conflict of new and old, strange and familiar, and in the confusion of human emotions, the test pilot has somehow to keep a grip on reality. He has got to find the best and most practical without allowing himself to be held back by doubt or urged on by optimism. He has got to be prepared to be unpopular and, if his reason dictates it, he must be ready to be unfashionable as well. The designer can have faith and enthusiasm, the manufacturer can have conviction and determination, but the test pilot must have common sense.' (Ref. 2.18)

The other side of the coin, as Wing Commander Norman Macmillan said of test pilots lost while probing the unknown:

> 'From their sacrifices it became possible for other men to fly in safety where they had flown in peril; theirs was the greatest love, for they gave their lives for their friends.' (Ref. 2.19)

Given common sense, how is what the pilot says translated into numbers and technical language useable by the engineer? Kurt Tank had the advantage of direct contact with flyingmen, as an engineer and a test pilot himself. Many engineers do not have that advantage today. But there are techniques which pilots can employ for recording and conveying intelligibly their impressions and results, and these are discussed in Chapter 4.

In the Introduction and these two chapters we have encountered many technical terms. Most rate as common useage in the popular press and in aviation journals. Others may not do so to the same extent, yet all will tend to recur in one way or another throughout the remainder of the book. So, to ease your memory, and to avoid jumping around too much, the next chapter gathers language and concepts together, to make reference easier, as required.

References and Bibliography

2.1 Cox, G. A. G., *Famous Biplanes No 23, Hawker Fury*. Plan No 2727. Aeromodeller Plans Service, 1959

2.2 Vee, R., *Flying Minnows*. London: John Hamilton Limited, 1920

2.3 Kosin, R., *The German Fighter since 1915*. London: Putnam, 1983 (German) and 1988 (English)

2.4 Gray, P. and Thetford, O., *German Aircraft of the First World War*. London: Putnam, 1962

2.5 The Fokker Biplane, Type DVII, *Flight*, 3 October 1918

2.6 Bruce, J. M., The Bristol Fighter (No 2 of the Series), *Flight*, 7 November 1952

2.7 Nowarra, H. J. and Kimborough, S. B., *von Richthofen and the Flying Circus*. Letchworth: Harleyford Publications, 1958

2.8 Weyl, A. R., *Fokker: The Creative Years*. London: Putnam, 1965

2.9 Barnwell, F. S., *Aeroplane Design* and Sayers, W. H., *A Simple Explanation of Inherent Stability*. London: McBride, Nast and Co., 1916

2.10 Yeager, J. and Rutan, D. (with Patton, P.) *Voyager*. New York: Alfred A. Knopf, 1987

2.11 Lanchester, F. W., The Wright and Voisin types of flying machine, a comparison, *The Aeronautical Journal*. London: RAeS January 1909

2.12 Gibbs Smith, C. H., *The Wright Brothers ... A Brief Account of their Work 1899–1911*. London: HMSO (also obtainable from the Science Museum), 1963

2.13 Hallion, R. P., *Test Pilots, the Frontiersmen of Flight*. Washington: Smithsonian Institution Press, 1981

2.14 Nowarra, H. J. *Focke-Wulf Fw 190 & Ta 152*. Sparkford, Somerset: FOULIS, Haynes Publishing Group, 1988

2.15 Wagner, W. *Kurt Tank – Konstrukteur und Testpilot bei Focke-Wulf*. Munich: Bernard & Graefe Verlag, 1988

2.16 Price, Dr A., *Focke Wulf's Dienstpferd*. Stamford: Air International, May and June 1992

2.17 Ramsden, J. M., Interviews D. P. Davies, Test pilot's test pilot, *Flight International*, 24 April 1982

2.18 HRH The Duke of Edinburgh, *Men, Machines and Sacred Cows*. London: Hamish Hamilton, 1984

2.19 Macmillan, Wing Commander N., *Great Aircraft*. London: G. Bell and Sons, 1984

2.20 Janda, Ing. A. and Poruba, Ing. T., *Focke-Wulf Fw 190D & Ta 152*, (Hradec Kralove: JaPo, M Horakove 273, 500 06, and Na drahach 456, 500 09, Czechoslovakia), c1990

2.21 Green, W., *Warplanes of the Third Reich*. London: Macdonald and Jane's, 1970

SECTION 3

PREPARATION

CHAPTER 3

Language and Concepts

'The scientists explore what is, engineers create what has never been.'
Theodore von Kármán (1881–1963)

In a little over 50 years aviation has turned the English language into the modern equivalent of Latin, with many aeronautical words now so commonplace that one might ask if definition is really necessary. The answer is that definitions are needed, because of the abuses which creep in to useage, transforming it to mis-usage. One common example is calling a helicopter an aeroplane. While acronyms like HUD (head-up-display) and V/STOL (vertical/short takeoff and landing) economize effort, they mask with ease the concepts they are meant to define, becoming instead a wretched nuisance when the reader is unsure of their translation. As an aid to clear expression, (Sir) Winston Churchill (1874–1965), a British prime minister and master of English, gave crisp advice to all who speak or write for the benefit of understanding (and those who construct flight test reports please note): *'Pray tell me on one side of one sheet of paper.'* ... *'Short words are better than long words.'* ... *'Old words are best.'* Similar if somewhat grittier advice was offered by a Cambridge don (a fellow of a college), giving a BBC Reith Lecture on living language, many years ago: *'Good English is effective English, and there are times when a bargee speaks better English than an Oxford don.'*

There is no neat and easy way of structuring this chapter to make it 'user-friendly' for all. It falls under the following broad headings:

> Aerodynamics and aeroplane geometry
> Control and handling
> Response of aircraft
> Stability and manoeuvrability
> Performance
> Power terms
> Weight and balance
> Tables and other conversion factors

under each of which the relevant terms are arranged alphabetically. If this gives an impression of bittiness, bear with it. Other arrangements were much worse.

The following definitions are not academically pure. Aeronautics is a mixture of precise science, discipline, accuracy in telling it as it is (or was), gut-feeling, artistry, and specialist forms of expression. Mathematics is a handmaiden, a shorthand language. Aircraft in motion (on the ground or water, as well as in the air) are obdurate, wilful and can be skittish. In a spin an aeroplane behaves like a horse which has bolted and is 'doing its own thing'. It can take cool and apparently irrational actions by the pilot to regain control. This chapter is to help the reader to visualize and feel what is involved in those interactions and responses to control that we call flying qualities.

Aerodynamics and Aeroplane Geometry

Aerodynamics is the scientific study of the way in which air moves, and of reactions felt by surfaces to its motions.

Aerodynamic coefficients are non-dimensional coefficients for aerodynamic forces and moments.

Aerodynamic centre (ac) is a point about one quarter of the distance from the leading edge to the trailing edge of an aerofoil section, about which the pitching moment coefficient of all

Plate 3–1   Aerodynamics is the scientific study of the way in which air moves and of the reactions felt to its motions. The later cancelled British TSR-2, trailing vortices from wing and flap tips, can be seen on an early test flight, around 1964. (Courtesy of British Aerospace (Warton))

aerodynamic forces acting upon it is constant within the working range of angles of attack. Consequently the rate of change of pitching moment about the ac, M_{ac} (and therefore the pitching moment coefficient, C_{Mac}) with change in angle of attack is zero. Using the Greek symbol, Δ, meaning a small change, we write:

$$\Delta C_{M_{ac}} / \Delta a = 0 \tag{3-1}$$

The whole aircraft (wing-plus-body-plus-nacelles) has an ac, which is the point through which the resultant lift *increment* acts when there is a small change in angle of attack from its value in steady straight and level flight. The ac is, therefore, the *neutral point*, which is dealt with in due course:

$$ac = NP \tag{3-1a}$$

Tail and canard stabilizer surfaces also have their own aerodynamic centres, as do all bodies.

Aerodynamic force is the reaction felt by the surface of a body moving through air to changes in pressure and friction as the air slides past:

pressure × area over which it acts = force

i.e.

$$pA \text{ or } pS = F \tag{3-2}$$

In aeronautics we use:

wing area = wingspan × mean chord

written:

$$S = b\bar{c} \text{ (or } b\bar{\bar{c}}) \tag{3-2a}$$

in which $\bar{c}$ and $\bar{\bar{c}}$ are the *standard mean* and *aerodynamic mean chords* respectively.

General predictions of lift, drag and hydrodynamic resistance, are made in terms of products of wing area, fluid density, speed relative to air or water, and pure numbers like *lift coefficient*, C_L, and *drag coefficient*, C_D (C_{Mac}, the pitching moment coefficient about the aerodynamic centre in Equation (3-1) is another example). The terms are tied together in the form of Equations (3-9) and (3-10) using dynamic pressure, q, and wing area, S.

The two mean chords in Equation (3-2a) are defined as:

☐ *Standard mean chord*, SMC, $\bar{c}$, which is the ratio of the gross wing area, S, divided by the wingspan, b:

$$\text{or:} \quad \bar{c} = S/b = \frac{\sum_{-b/2}^{+b/2} c\Delta y}{\sum_{-b/2}^{+b/2} \Delta y} \tag{3-3}$$

in which $b/2$ is the semi-span of the wings, often denoted s. For a straight tapered wing:

$$\bar{c} = (\text{root chord} + \text{tip chord})/2 \tag{3-3a}$$

☐ *Aerodynamic mean chord*, MAC, $\bar{\bar{c}}$, is the chord length usually drawn in the plane of symmetry and lying along the zero lift line of the wing, where:

$$\overline{\overline{c}} = \frac{\sum_{-b/2}^{+b/2} c^2 \, \Delta y}{\sum_{-b/2}^{+b/2} c \, \Delta y} = \frac{\sum_{-s}^{+s} c^2 \, \Delta y}{S} \tag{3-4}$$

It is the equivalent chord of an aerofoil surface having the same span, area and aerodynamic properties as a wing or other relevant aerofoil surface of a real aircraft. It is used as the datum length for calculating pitching moment, M, and pitching moment coefficient, C_M. It is the datum chord to which the ac of the wing is referred.

Figure 3.1 shows the relationship between both chord lengths geometrically. In terms of the semi-span, s, which is split into fractions, y_1 and y_2, and λ, the taper ratio of the tip chord, c_t, to the root chord, c_r. These enable it to be shown that if:

taper ratio, $\lambda = c_t / c_r$ $\hspace{2cm}$ (3–5)

then:

$y_1 / y_2 = (1 + 2\lambda)/(2 + \lambda)$ $\hspace{2cm}$ (3–6)

and in terms of semi-span:

$y_1 / s = y_1 /(b/2) = (1 + 2\lambda)/3 \, (1 + \lambda)$ $\hspace{2cm}$ (3–7)

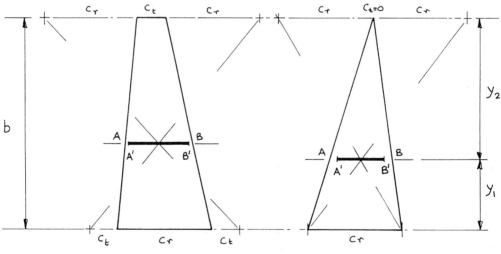

A - B $= \overline{\overline{c}} =$ mean aerodynamic chord, MAC

A' - B' $= \overline{c} =$ standard mean chord, SMC

$= S / b = (c_t + c_r) / 2$ $\hspace{0.5cm}$ eq. (3-3a)

Fig. 3.1 Geometric construction of mean chords, MAC and SMC. The MAC is the one to which forces and moments are applied. For most practical purposes: $\overline{\overline{c}} = \overline{c} =$ wing area/span.

Manipulating further:

$$\text{MAC}/\text{SMC} = \bar{c}/\bar{\bar{c}} = (4/3)\,(1 - \lambda/(1 + \lambda)^2) \tag{3–8}$$

A wing with a pointed tip, although impracticable (unlike a flexible *Bermudan sail*), has a taper-ratio, $\lambda = 0$ and the MAC $= (4/3)$ SMC. But when the wing is of parallel chord resembling a 'plank', then: $\lambda = 1.0$ and MAC $=$ SMC.

In practice one should never have a taper ratio less than about $\frac{1}{2}$ if tip stalling is to be avoided. When $\lambda = 0.5$ the MAC is only about 4% larger than the SMC, in short, they are near enough equal and are often treated as one and the same without further distinction.

Lift, L, and drag, D, are commonly expressed in terms of wing area, S, (as in Equation (3–2a)) and dynamic pressure, q (see Equation (3–26)):

$$\text{lift, } L = C_L q S \tag{3–9}$$

$$\text{drag, } D = C_D q S \tag{3–10}$$

Dynamic pressure is explained when we come to Equation (3–26).

Aerodynamic mean chord, $\bar{\bar{c}}$, was defined in Equation (3–4).

Aerofoil is a body (often referred to as an 'aerofoil surface') shaped to produce more lift than drag. Wings, tails and canard foreplanes, fins and rudders are all examples. The forces generated are the same whether the body moves through stationary air, or air flows steadily past the stationary body – Sir Isaac Newton (1642–1727) showed that this law applies to any fluid.

Note: All bodies generate lift when in motion and inclined to the air at an angle of attack. A fuselage designed to lift is called a *lifting body*.

Aerofoil section is the profile of an aerofoil when sliced from leading to trailing edge – usually, but not necessarily, parallel with the normal flight path (Fig. 3.2).

Aeroplane as we saw earlier is a scholarly word from the Greek meaning literally 'air-wandering'. The Oxford English Dictionary and the Glossary of terms of the British Standards Institution conspire to define it as '*a mechanically-driven heavier than air flying machine, with supporting surfaces fixed for flight*'. Thus, rotary winged *aircraft* are never *aeroplanes*.

Airbrake is a high-drag surface opened into the airflow to slow down.

Aircraft is a generic term used to describe any kind of flying machine (Ref. 3.1).

Airflow is the relative flow of air past any surface when the surface is imagined to remain fixed. The effect is the same if the eye moves with the surface.

Airspeed (see *Performance terms*).

Ambient conditions are the physical conditions of the atmosphere at a particular place, altitude, date and time.

Angle of attack is the angle of inclination of a surface to the relative airflow, generally measured between the flight path and the chord line of an aerofoil section, Fig. 3.3.

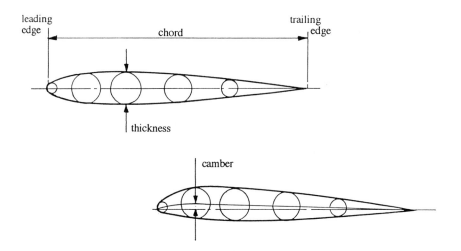

Fig. 3.2   Aerofoil section geometry.

Angle of incidence is, strictly, the rigging angle between an aerofoil datum, like the chord line and the datum line of the aircraft (see Figs 3.3 and 3.6). However, it is sometimes loosely used as a synonym for angle of attack.

Aspect ratio is the geometric description of how long is the span compared with the (mean) chord length of an aerofoil. The average distance between the leading and trailing edge is called the *standard mean chord* (SMC), $\bar{c}$. When an aerofoil surface resembles a rectangular plank of constant chord, then the aspect ratio is the ratio of the (span/SMC). But when it is not like a plank one must resort to Fig. 3.4 and the following:

(geometric) aspect ratio, A = wingspan/SMC = $b/\bar{c}$ (3–11a)

and as area = span × chord in Equation (3–2a), the SMC = area/span, such that:

aspect ratio, A = wingspan^2/area = b^2/S (3–11b)

The *aerodynamic aspect ratio* is in the form of Equations (3–11a) and (3–11b), but uses the *aerodynamic span* of the wings between the cores of the trailing (horseshoe) wingtip vortices, and the MAC. It is smaller than the geometric aspect ratio and accounts for losses in theoretical lift/drag ratios.

Bernoulli's law, so named because it is usually ascribed to Daniel Bernoulli (1700–82), was discovered by Leonard Euler (1707–83). To demonstrate it we consider a small element of air or fluid flow and ignore the frictional effects, so that the sum of the *static pressure* and the *dynamic pressure* is then constant along a streamline. The static pressure is equivalent to the weight of a column (or head) of air pressing down on unit area of a surface with a value equal to that measured at the streamline. Dynamic pressure is the force per unit area that would be exerted by the flow being dammed and brought to rest against a flat surface

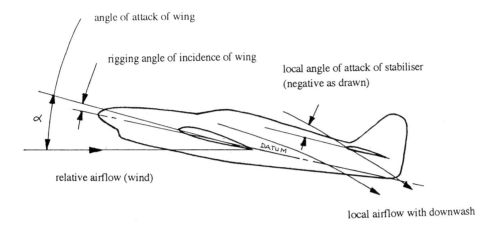

angle of attack of wing

rigging angle of incidence of wing

local angle of attack of stabiliser
(negative as drawn)

α

DATUM

relative airflow (wind)

local airflow with downwash

Fig. 3.3   Angles of attack and incidence.

perpendicular to it (and without leakage around the edges). The pressures at any point in the flow are summed to give:

total pressure = static pressure + dynamic pressure
$$p_0 = p + q = \text{a constant, } C \qquad\qquad (3\text{--}12)$$

A physical explanation is suggested in Fig. 3.5a and developed further in Ref. 1.5. Figure 3.5b shows patterns in flows at different airspeeds and Reynolds numbers (Equations (3–30a) and (3–30b)).

Biplane is an aeroplane with one wing arranged more or less above the other (see also *gap* and *stagger*), Fig. 3.6.

Boundary layer is the region of fluid (air, next to the skin of an aircraft) through which there is a change of relative velocity, from zero at the skin to the full speed of the flow some way from it. A boundary layer is either *laminar* (in which fluid particles slide in smooth stream-lined layers), or *turbulent* (in which they are disturbed and tumble over one another). A turbulent layer is thicker than one which is laminar (Fig. 3.7). A laminar layer is more vulnerable and can be caused to *separate* more easily from a surface (i.e. it is replaced by a wedge of stagnating fluid clinging to and being carried along with the aircraft, which under-cuts the main flow, forcing it away from the surface). The boundary layer has the effect of an added layer of skin of finite thickness, called the *displacement thickness*, which is deep-est where the boundary layer is turbulent, and which has the effect of changing the basic contours of a body.

It has been useful at times to be able to estimate boundary layer and displacement thick-nesses, δ and δ^*, when deciding how deep strakes or vortex generators should be in differ-ent locations. Depths can be calculated approximately for laminar and turbulent flows in

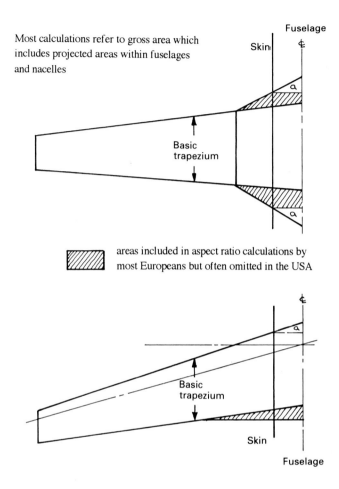

Fig. 3.4   Wing areas used in aspect ratio calculations.

terms of distance downstream, x, and the Reynolds number, R_x, at distance x (see Equations (3–30a) and (3–30b)).

laminar thickness, $d_{\text{lam}} = 5.5\,R_x^{-1/2}$, approx.   (3–13)

laminar displacement thickness, $\delta_{\text{lam}*} = 0.35\,\delta_{\text{lam}}$, approx.   (3–14)

turbulent thickness, $\delta_{\text{turb}} = 0.38\,R_x^{-1/5}$, approx.   (3–15)

turbulent displacement thickness, $\delta_{\text{turb}*} = 0.14\,\delta_{\text{turb}}$, approx.   (3–16)

When $x = l$, the length of a body:

$R_l = R_x$ (see Equation (3–30b)).

The critical roughness of a surface, k, needed to trip a laminar boundary layer into turbulence must be large enough to protrude through the laminar sub-layer (see Fig. 3.7):

$$\text{critical roughness, } k = 100 \; \upsilon/V \qquad (3-17)$$

where:

$$\text{kinematic viscosity, } \upsilon = \mu/\rho \qquad (3-18)$$

in which μ = dynamic viscosity, such that if v is the velocity of the fluid at some distance, z, from the skin, then frictional shear stress between the layers of fluid, sliding over one another:

$$\text{shearing force/unit area} = \mu \, (\mathrm{d}v/\mathrm{d}z) \qquad (3-19)$$

and ρ = density of the fluid, be it air or water.

Transition from laminar to turbulent flow is sensitive to vibration of the surface on which it occurs. Engine and air noise, transmitted through an airframe, can trip a laminar flow into turbulence. The frequency at which this is most likely to occur is given by:

$$\text{vortex frequency, } f = 2 \; V^2/10^5 \; v \; \mathrm{c/s} \qquad (3-20)$$

Note: Physically, air and water are elegant sisters and they obey a number of similar laws. Racing yachtsmen insist on silence when 'ghosting' in light airs. A sudden noise, causing vibration, can trip a slow laminar flow along the hull into one that is turbulent and much thicker, with greater resistance.

Behaviour – really, *mis*behaviour – of the boundary layer marks it as the prime mischief-maker. It mocks the careful calculations of aerodynamicists and the fairness of forms draughted by designers. It needs 'fixes': warts and distortions: vortex generators, fairings, fences, slats and slots, wash-in and wash-out of wing tips to control and curb its excesses.

Buffet is felt when the airframe and control surfaces are shaken and thumped by local changes in pressure caused by vortices and turbulently agitated air being shed and left behind in the wake. Most commonly it is a low speed phenomenon associated with large angles of attack, but it is also caused at high speeds when the airflow is slowed violently through shock waves.

Buffet boundary, for a given set of conditions, is the line of speed and altitude combinations at which buffet will be experienced.

Buffet margin, for a given set of conditions, is the amount of 'g' which can be imposed for a given level of buffet.

Buffet threshold, for a given set of conditions, is the point at which buffet (either low or high speed) is first felt.

Buoyancy is the property of being able to float, either in air or water, the weight of fluid displaced being equal to the weight it supports (the Principle of Archimedes (287–212 BC)).

Camber is the curvature of the geometric mean line of an aerofoil section, measured in terms of percentage of and distance along the chord line (see Fig. 3.2).

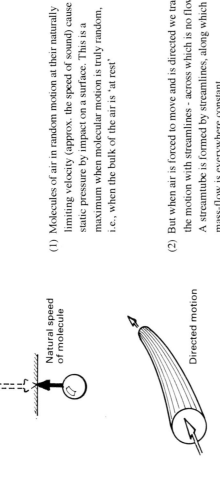

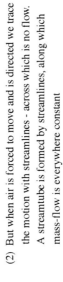

(1) Molecules of air in random motion at their naturally limiting velocity (approx. the speed of sound) cause static pressure by impact on a surface. This is a maximum when molecular motion is truly random, i.e., when the bulk of the air is 'at rest'

(2) But when air is forced to move and is directed we trace the motion with streamlines - across which is no flow. A streamtube is formed by streamlines, along which mass-flow is everywhere constant

(3) The addition of a component of directed motion cannot alter the natural limiting speed of a molecule (as long as the temperature of the air remains unchanged), so that the component of velocity with which it impacts against a surface, or another molecule, is now reduced, and the static pressure falls

Fig. 3.5a Air when caused to flow. Fluids and gases behave in similar ways.

For constant size of body, air density and viscosity:

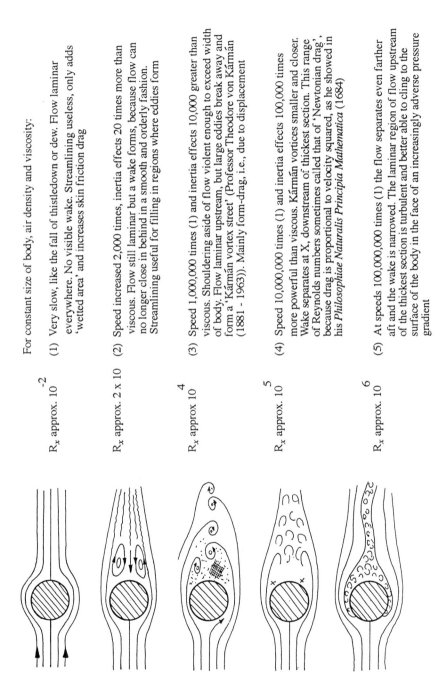

R_x approx. 10^{-2}

(1) Very slow, like the fall of thistledown or dew. Flow laminar everywhere. No visible wake. Streamlining useless, only adds 'wetted area' and increases skin friction drag

R_x approx. 2×10

(2) Speed increased 2,000 times, inertia effects 20 times more than viscous. Flow still laminar but a wake forms, because flow can no longer close in behind in a smooth and orderly fashion. Streamlining useful for filling in regions where eddies form

R_x approx. 10^4

(3) Speed 1,000,000 times (1) and inertia effects 10,000 greater than viscous. Shouldering aside of flow violent enough to exceed width of body. Flow laminar upstream, but large eddies break away and form a 'Kármán vortex street' (Professor Theodore von Kármán (1881 - 1963)). Mainly form-drag, i.e., due to displacement

R_x approx. 10^5

(4) Speed 10,000,000 times (1) and inertia effects 100,000 times more powerful than viscous. Kármán vortices smaller and closer. Wake separates at X, downstream of thickest section. This range of Reynolds numbers sometimes called that of 'Newtonian drag', because drag is proportional to velocity squared, as he showed in his *Philosophiae Naturalis Principia Mathematica* (1684)

R_x approx. 10^6

(5) At speeds 100,000,000 times (1) the flow separates even farther aft and the wake is narrowed. The laminar region of flow upstream of the thickest section is turbulent and better able to cling to the surface of the body in the face of an increasingly adverse pressure gradient

Fig. 3.5b Air in laminar and then turbulent motion. Reynolds number R_x (see Equations (3–30a) and (3–30b). (4) references (Sir) Isaac Newton (1642–1727), born in the year Galileo died.

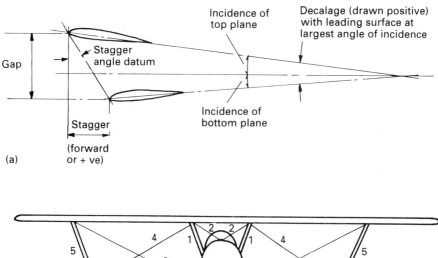

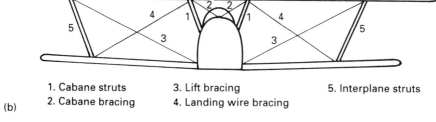

1. Cabane struts   3. Lift bracing   5. Interplane struts
2. Cabane bracing   4. Landing wire bracing

(b)

Fig. 3.6 Basic rigging and bracing terms. Decalage describes the difference in wing, foreplane or tailplane incidence. Stagger is stated as an angle, a distance, or as a percentage of the gap.

Centre of pressure, CP, is the point at which the resultant pressure on a body is taken to act. The sum of the moments of all the pressure forces about the CP is zero.

Chord is the length of the straight *chord line* joining the leading and trailing edges of an aerofoil section (*see* Fig. 3.2).

Circulation is the motion of a fluid along a curved path about an instantaneous centre of rotation. A flow marked by curved streamlines shows that a centripetal acceleration is being imparted to the fluid, either by curvature of a surface around which it is flowing (e.g., *Coanda effect*), or by a region of rising pressure which is deflecting its path. Curvature of a flow reveals that circulation is present. The centripetal acceleration is reacted by centrifugal resistance of the fluid to deflection so that pressure, which is lowest on the inside of the curve, increases as one moves radially outwards from the centre. The radial pressure differential produces a cross-force on a curved surface. Splitting the cross-force into mutually perpendicular components, one normal to the flight path, the other along it, produces lift and drag (and sometimes thrust i.e. an anti-drag force tending to draw the surface forward).

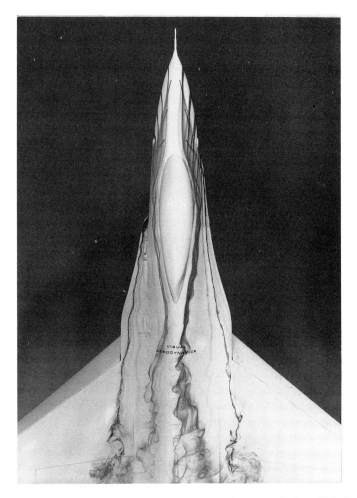

Plate 3–2   The breakdown in flow from laminar to turbulent. Eidetics
F-16 strake modification in a water-tunnel. (Courtesy of *Flight
International*)

Coanda effect is the tendency of air or water to follow a curved surface. It is called after
Henri Coanda (1886–1972), the Rumanian engineer who discovered its importance. Coanda
effect can de demonstrated by running a tap and bringing the curved back of a spoon into
contact with the jet of water. The tip of the handle should be held lightly so that the spoon
can swing freely. The spoon will be sucked sharply into the jet.

Compressibility is the ability of a gas to change volume and density when squeezed.

Compressibility effects embrace the formation of shock and expansion waves in a flow
when air is forced to move faster than the natural limiting speed of its molecules (called the
speed of sound). The speed at which shock waves are first apparent is the critical Mach

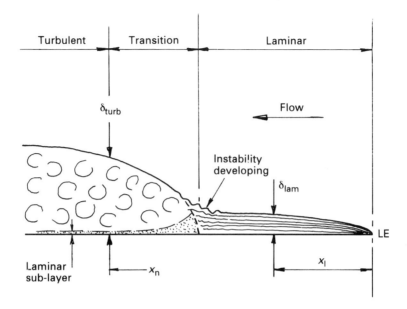

Fig. 3.7 Change of boundary layer from laminar to turbulent at a distance *x* downstream of the leading edge (in this case of a sharp-edged flat plate).

number, M_{crit}.

Configuration describes the geometry of an aeroplane when various elements (flaps, air-brakes, landing gear, nozzle setting, power and propeller settings, weight and CG) which make up its shape are set for a particular phase of flight.

Datum (line) is generally accepted as a straight line drawn through a *datum point* from which measurements are made.

Decalage is the angle subtended between the chord lines of biplane wings, and between wings and stabilizer (*see* Fig. 3.6). It is positive when the upper (or leading) aerofoil is set at a larger angle of incidence than the one which follows, and/or is set lower. In the case of wings and stabilizers, decalage is often called *longitudinal dihedral*, which must be positive for longitudinal trimability and, hence, stability.

Dihedral and *anhedral* (originally *cahedral*) is the change in height of an aerofoil from root to tip. When the tip is higher we have dihedral, subtending a *dihedral angle* inboard. When the tips are lower anhedral (negative dihedral) is present.

Downwash is the downwards motion (no matter how slight initially) of the mass of air, accelerated mainly by the wings, which produces an equal and opposite (upwards) lifting force as soon as an aircraft begins to move forwards. Bodies also lift and, when they do so, contribute to the overall downwash (and upwash, i.e. negative downwash).

Drag, D (see also Equation (3–10)), is the sum of the components of all of the aerodynamic forces acting on an aircraft, when resolved along the flight path. Such forces are caused by pressures normal (perpendicular) to the skin at any point, and by friction in the boundary layer as it slides past. Drag represents lost energy and it has numerous sources.

A useful break-down is shown in Fig. 3.8. There are two views of drag of which pilots should be aware, depending upon whether it is an aerodynamicist or a performance engineer who is involved:

☐ The **aerodynamicist** splits drag into *wing drag* and *extra-to-wing drag (parasite drag)*. *Wing drag* has two parts: *lift-dependent*, or *induced* drag, D_L (or D_i), caused by the act of lifting; and *profile (zero-lift)* drag of the wing alone, D_0:

$$D_{wing} = D_0 + D_L = D_0 + D_i \qquad (3\text{--}21)$$

in which:

$$D_0 = D_{form} + D_{fric} \text{ of the wing alone} \qquad (3\text{--}22)$$

Form drag, D_{form}, is due to the shape or bluffness of the wing, shouldering aside the particles of air.

Skin friction drag, D_{fric}, is the drag caused by friction, retarding the air as it slides over the surface of the wing.

Note: *Extra-to-wing* (commonly called *parasite*) *drag*, D_p, is the (form + friction + interference + all other bits and pieces of drag) of the remainder of the aircraft.

☐ Drag as seen by the **performance engineer**, is split into 'lifting' and 'non-lifting' elements:

Lift-dependent or *induced drag*, D_L or D_i, as we have seen. However, at sonic and supersonic speeds, when local flows reach the critical speed of sound on parts of the airframe, then wave drag, caused by expansion (low pressure) and shock (high pressure) waves, appears and this must be included in the lift-dependent term, hence D_L.

Strictly speaking the term induced drag, D_i, should be reserved for flight at subcritical speeds when there is no compressibility present. Induced drag is the ever-present resistance attending the lifting circulation (mainly around the wings) which forms the trailing (horseshoe) vortex system, producing *downwash*.

Zero-lift, D_F, or *parasite drag*, D_p, can be summed as follows:
no compressibility:

$$\text{total drag, } D = D_i + D_p \qquad (3\text{--}23)$$

compressibility present:

$$D = D_L + D_F \qquad (3\text{--}24)$$

Thus:

$$D_L = D_i + (\text{wave drag caused by lift in a compressible flow}) \qquad (3\text{--}25a)$$

and

$$D_F = D_p + (\text{wave drag caused parasitically}) \qquad (3\text{--}25b)$$

Dynamic pressure, q, in Equations (3–9), (3–10) and (3–12), is the limit force felt per unit frontal area of an aircraft – or any other shape – impacting (like a bulldozer) fluid of density,

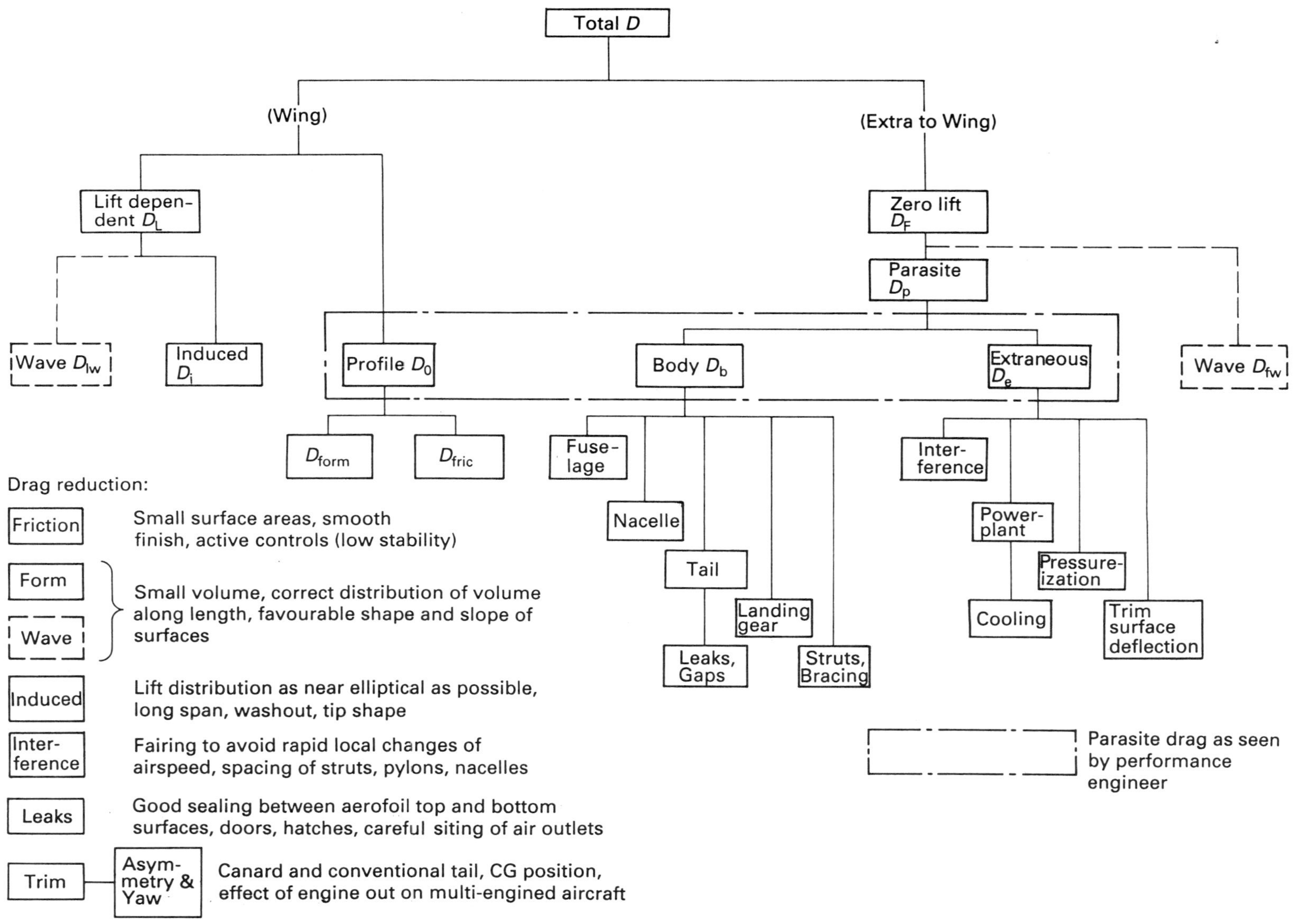

Fig. 3.8 Drag breakdown (wave-drag is not significant when M < 0.6).

ρ, at velocity V. By fluid we mean both liquids and gases. Force per unit area is, of course, pressure. It may be shown that dynamic pressure is equivalent to the *kinetic energy* imparted per second to a unit volume of air of density, ρ, when it is accelerated to velocity, V, such that:

dynamic pressure $= \frac{1}{2} \times$ density of air $\times$ relative true airspeed^2

$$q = \frac{1}{2}\rho V^2 \tag{3-26}$$

'Relative' is used because the pressure obtained is the same whether it is caused by the aircraft moving through still air, or by the air flowing past a stationary aircraft, as in a wind tunnel.

The equation is not used in this form, because it is extremely difficult to determine the true airspeed in flight, TAS. Instead we resort to equivalent airspeed, EAS (see Equations (3–63) to (3–66)).

Empenage is the collective French noun for the tail surfaces.

Flap is that part of the surface of an aerofoil that can be deflected to produce a change in effective camber and, hence, lifting pressure distribution.

Free stream is the relative air flow close to, but undistarbed by, the aircraft.

Gap is the vertical distance between planes, for example of a biplane, or a triplane.

Geometric mean chord, GMC (see *standard mean chord*, SMC, $\bar{c}$).

Hinge moment is the magnitude of the twisting exerted at a control or flap surface hinge by the action of the aerodynamic force on the surface times its distance from the hinge (see also Fig. 3.19).

Incidence (see *angle of incidence*).

ISA (see *Performance terms* (describing ambient conditions)).

Lift, L, is, primarily, the component sum of all aerodynamic forces acting on an aircraft when resolved at right angles to the flight path. However, powered lift components should also be included, where they exist. In steady straight and level flight the lift is equal and opposite to the weight of the machine.

Lift, *drag* and *pitching moment coefficients*, C_L, C_D and C_M (see Fig. 3.9). Equations (3–9), (3–10) and (3–29) are 'dimensionless coefficients' (pure numbers) used to calculate forces and moments in general form. They are useful because they remain largely independent of the conditions, except in so far as they are affected by Reynolds number and Mach number. For example, the wing loading, w, which is the ratio of the (weight/wing area), has the dimensions of a pressure: the 'footprint pressure' of the wing upon the air. Knowing that the aircraft is in flight at true airspeed V in air of density ρ, and that both are uniquely linked by the value of dynamic pressure q (Equation (3–26)) then as (pressure/pressure) is dimensionless:

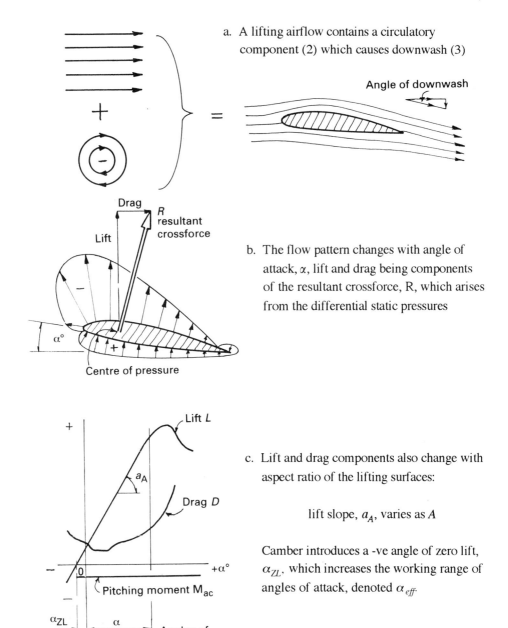

a. A lifting airflow contains a circulatory component (2) which causes downwash (3)

Angle of downwash

b. The flow pattern changes with angle of attack, α, lift and drag being components of the resultant crossforce, R, which arises from the differential static pressures

c. Lift and drag components also change with aspect ratio of the lifting surfaces:

   lift slope, a_A, varies as A

   Camber introduces a -ve angle of zero lift, α_{ZL}, which increases the working range of angles of attack, denoted α_{eff}.

Fig. 3.9 The way in which lift, drag and pitching moment are caused at subsonic (subcritical) speeds. α_{eff} is the 'working-range' of angles of attack over the straightest part of the curve.

wing loading/dynamic pressure $= w/q = C_L$ (3–27)

i.e.

$C_L = W/qS$ (3–27a)

when lift, L, and weight, W, are equal. The higher (larger) the maximum lift coefficient, the slower an aircraft of a given weight and wing loading can fly before it stalls.

Drag and its components are also generalized as dimensionless coefficients by means of wing area and dynamic pressure:

(drag/wing area)/dynamic pressure $= C_D$ (3–28)

Similarly, $D_F, D_L, D_i, D_{form}, D_{fric}$, for example, when divided by qS, become $C_{DF}, C_{DL}, C_{Di},$ C_{Dform}, C_{Dfric}, respectively. The same treatment is given to any other aerodynamic terms which are dynamic pressure-dependent, like pitching, yawing and rolling moments, see Equation (3–29), making them independent of ambient conditions.

Lift (*coefficient*) *slope*, a, (see Figs 3.9 and 3.10) is the rate of change of lift with angle of attack. Formulae for lift slope with aspect ratio, a_A, are given in Equations (3–41) to (3–43).

Pitching moment is a torque, a twist applied to a body by a system of forces acting in pitch about a chosen centre. The pitching moment on an aircraft is dealt with in exactly the same way as lift and drag, by turning it into a dimensionless coefficient. Knowing that the *aerodynamic moment* is numerically equal to a *force × lever arm*, for convenience we substitute the mean aerodynamic chord, or the standard mean chord. As these are virtually one and the same for our kind of aircraft::

total pitching moment = moment coefficient × dynamic pressure × wing area × SMC

i.e.

$M = C_M qS\overline{c}$ (3–29)

This is the general form. In practice we distinguish between moments about the CG and ac and their respective coefficients by suffixes: $M_{CG}, M_{ac}, C_{MCG}, C_{Mac}$.

Pressure altitude (see *Performance terms*).

Pressure coefficient is the ratio of the pressure (or pressure difference) at some point on a surface to the dynamic pressure of the free-stream. A pressure coefficient is negative when the local velocity of the flow is higher than the free-stream velocity (thereby increasing the dynamic pressure, q, and reducing the static pressure, p, see Equations (3–12) and (3–26)).

Rake (of tips of wings and tails) is to cut them at an angle, so that they are not parallel with the plane of symmetry of the aircraft, so making leading and trailing edges of differing lengths.

Reynolds number, R (often R_l or R_x), is a pure number named after Osborne Reynolds (1842–1912), Professor of Engineering in the University of Manchester, who in 1883 carried out systematic experiments on the transition of flows from laminar to turbulent. Turbu-

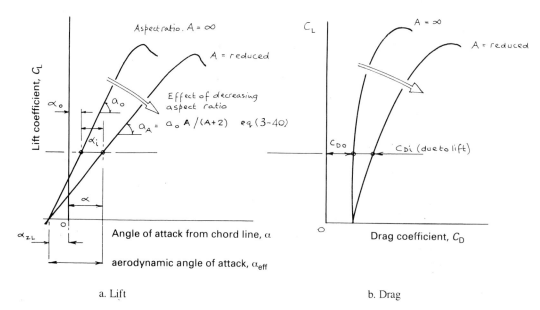

a. Lift   b. Drag

Fig. 3.10a and b   Effect of aspect ratio upon lift and drag, showing that decreasing aspect ratio reduces lift and increases drag-due-to-lift (i.e. lift-dependent or induced drag). The lift slope a_A is as given in Equation (3–42).

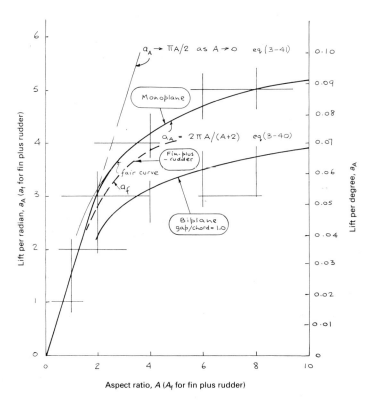

Fig. 3.10c   Effect of aspect ratio upon lift slope a_A. The effect of taper is ignored. The lift slopes are as given in Equations (3–42) and (3–43), depending on the aspect ratio.

lent flows had already been analysed by a German engineer, Gotthilf Heinrich Ludwig Hagen (1797–1894) (Ref. 1.2).

Reynolds number is a measure of *scale effect* and is of vast importance, because as long as two bodies, or surfaces of similar shape even though differing in size, have the same Reynolds number, then the flows around them remain geometrically similar.

$$\text{Reynolds number, } R_x = \frac{\text{inertial factors which excite a disturbance}}{\text{viscosity which quietens and sooths}} \qquad (3\text{--}30\text{a})$$

☐ Factors which agitate and excite fluid particles:

Scale: size of body (*l*) or length of surface (*x*) in contact with the fluid (see Fig. 3.7). In the case of an aerofoil section we commonly use local chord length, *c*, for localized effects, and SMC, $\bar{c}$, for the whole aircraft.
Density of fluid, ρ, and, hence, the crowding of particles which can jostle one another in a given volume in contact with a length of surface. The denser the fluid the more easily is turbulence spread.
Velocity of the body relative to the fluid, *V*.

☐ Factors which sooth and quieten motion:

Dynamic viscosity, μ, which measures the 'treaclyness' of the fluid (see Equation (3–19)). Hence:

$$R_x = \frac{V\rho x}{\mu} = \frac{Vx}{v} \qquad (3\text{--}30\text{b})$$

in which $v = \mu/\rho$, the coefficient of kinematic viscosity.

At low Reynolds numbers, less than about $R_x = 5 \times 10^5$ on an aerodynamically smooth surface (critical roughness less than the thickness of the laminar sub-layer), the flow is laminar and relatively thin. Thereafter those factors which spread excitement cause transition. Smooth laminar conditions break down and by about $R_x = 5 \times 10^6$ the boundary layer becomes thick and turbulent. Transition from laminar to turbulent flow may oscillate, varying at a point on a surface by as much as ±30% of the distance *x*.

Scale effect (see *Reynolds number*).

Slat and *slot* (see Table 8–2) are aerodynamic devices which help to maintain attachment of the airflow to an aerofoil surface at low speed and large angles of attack. (Later Sir) Frederick Handley Page (1885–1962) in the UK and Gustav V. Lachman (1896–1966) when a German pilot in World War I, had both conceived the slat and slot independently, in 1919 and 1918 respectively (Ref. 3.2).

The slat is a small auxiliary aerofoil close to the leading edge of the main surface, which opens and forms a slot between them. Alternatively, a fixed slat is shaped out of the wing or tailplane leading edge, so forming a permanent slot. Air, squeezed through the slot with increased airspeed, re-energizes the boundary layer, preventing separation and a subsequent stall.

Slug is the unit of mass based upon the British Imperial System. Mass is the 'quantity of matter' contained within a body. One slug is the mass of a body weighing *g* pounds (i.e. 32.174 lb, or 14.59 kg). All masses press down on the ground under the effect of gravitational acceleration, *g* ft/s² (9.8 m/s²), with forces called *weight*. The relationship between force and acceleration was correctly formulated by Newton, whose second law states:

force = mass × acceleration

$$F = ma = Wa/g \qquad (3\text{–}31)$$

(see also *applied normal acceleration* under *Handling Terms* and Equation (3–31a) *et seq.*).

The FPSR weight of air is 0.076 474 lb/ft³ (1.225 kg/m³) at sea level in the International Standard Atmosphere, ISA. So, transposing Equation (3–31) for mass, which is denoted ρ_0 at sea level:

$$\begin{aligned}
\text{mass, } \rho_0 &= \text{weight of one cubic foot of air}/g \\
&= 0.764\ 74/32.174 \\
&= 0.002\ 378 \text{ slugs/ft}^3 \ (1.225 \text{ kg/m}^3), \text{ ISA} \qquad (3\text{–}32)
\end{aligned}$$

Looked at another way, a force of 1 pound will accelerate a mass of 1 slug at 1 ft/s².

Span is, in general, the geometric distance between the tips of wings and similar aerofoils, measured normal to the plane of symmetry of an aircraft. However, the *aerodynamic span* (when such a term is used), is the distance measured between the tip vortices, and this depends upon shaping of the wing tips. The object of good aerodynamic design is to make the aerodynamic and geometric spans equal.

The *structural span* of an aerofoil is the sum of the (swept) lengths of both halves, measured from root to tip of each *structure*, regardless of sweep, dihedral or anhedral.

Stagger is the amount by which one wing of a biplane is set ahead of the other (see Fig. 3.6). Stagger is positive when the top wing is ahead of the bottom, and vice versa.

Stagnation point is the point on the section of skin surface of an aircraft at which the local relative airflow is brought to rest. A *stagnation line* joins a number of stagnation points.

Stall is the gross loss of lift caused by breakdown and separation of the relative airflow over a wing or similar aerofoil surface, in the face of an adverse pressure gradient downstream. Stalling is usually associated with large angles of attack, but not always. Formation of a shock wave on an aerofoil, through which a sharp deceleration from supersonic to subsonic flow occurs in a fraction of a millimetre, accompanied by a violent rise in static pressure, provides the adverse gradient needed to precipitate a *shock stall*.

Standard mean chord, SMC, $\bar{c}$, (also *geometric mean chord* (see under both *aerodynamic force* and *aspect ratio*, also Equation (3–3))).

Static pressure, p, is the average force exerted upon unit area of surface by the bombardment due to random thermal motion of molecules of air (or any other fluid). See also *Bernoulli's law* (Equation (3–12)).

Streamline and streamtube (see Fig. 3.5a) are imaginary (although streamlines can be simulated by smoke filaments in air and dye in water to mark the direction of a flow). A streamtube is a filament of flow bounded by streamlines. It follows that, by definition, there can be no flow across a streamline; and there is neither gain nor loss of fluid along a streamtube, so that the mass flow past any cross-section is constant:
i.e.

cross-sectional area × velocity × density = a constant

$$AV\rho = C \qquad (3\text{–}33)$$

Thus, provided that the flow is subsonic and incompressible, so that ρ does not change, a contraction in cross-sectional area (a squeeze) increases the velocity, and vice versa.

When the streamlines in a fluid are steady (i.e. successive imaginary snapshots look identical) the flow is called laminar. See also *turbulence*.

Subsonic (*transonic*) and *supersonic* are descriptions of flows which are slower and faster than the local speed of sound. Transonic is the mixed, in-between, state when the relative flow over some parts of the skin has been forced to supersonic velocities, while over other parts it remains subsonic.

A distinction made between Mach number of the aircraft and that of the relative airflows over the skin is:

low speed $M < 0.2$
subsonic $M = 0.2$ to 0.9
transonic $M = 0.7$ to 1.5
supersonic $M > 1.0$

Sweep, Λ, is the backwards or forwards sweeping of the wings. They are truly swept when a spanwise line between, say, 15% and 70% of the chord back from the leading edge is *not* at right angles to the plane of symmetry of the aircraft.

System of axes and notation, see Fig. 3.11 and Table 3–1.

Turbulence describes the behaviour of a fluid when it moves in a chopped up and lumpy way, with particles tumbling over one another in a disorderly fashion. In contrast with laminar flow, no streamlines are present.

Volumes of *ailerons, tails, elevator, fin, rudder, stabilizer* (see Equations (3–53) and (3–58) to (3–61)).

Vortex is the circulatory motion of a fluid about a centre, like water rotating as it flows down a drain-hole in a bath.

Wash-in is the building into the wing of a twist along the span, such that the incidence of the tip is larger than that of the root. The reverse is:

Wash-out, in that the built-in twist gives the tip less incidence than the root, inhibiting a premature tip-stall.

Control and Handling

Aircraft motion involves six degrees of freedom: three in translation (*forward, sideways, and up or downwards*), and three in rotation (*pitch, roll* and *yaw*).

The pilot has three flying controls: *elevator, aileron* and *rudder*; and subsidiary flying controls in the form of trim and other *tabs*. He also has a fourth primary control: of *power* and *thrust* by means of the power lever, or a throttle.

It is not immediately obvious how a pilot with four controls manages to fly an aircraft with six degrees of freedom.

Aircraft equilibrium and *trim* are essential in practice if the aircraft is to fly steadily, i.e.

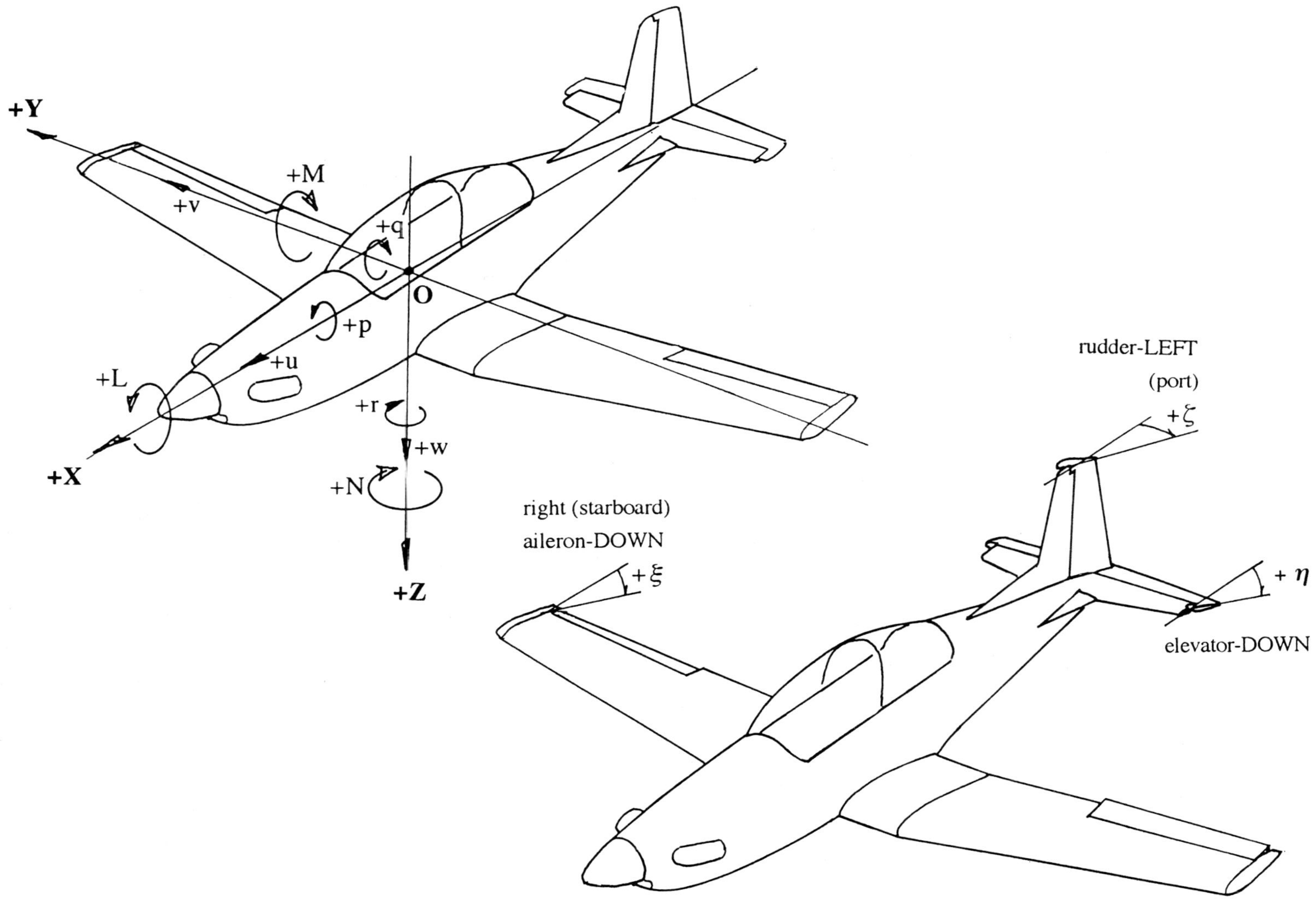

Fig. 3.11 System of axes and notation (UK).

TABLE 3–1

General notation*

Axes	OX	OY	OZ
Aerodynamic forces	X	Y	Z
Angular motions	ϕ (bank)	θ (pitch)	ψ (yaw)
Linear velocities	u	v	w
Angular velocities	p	q	r
Moments of forces	L (roll)	M (pitch)	N (yaw)
Moments of inertia	A	B	C

(Note: $A + B \approx C$)

* This notation is not used exclusively in this book.

without transient changes in applied accelerations, so that we may describe it as being in steady level flight: steady climb and descent, steady pull-outs and steady turns. In such conditions the overall forces of lift, drag and sideforce balance thrust and weight. Rolling moments in pitch, roll and yaw are zero. Thus, in steady flight all forces and moments about the centre of gravity balance, and the aircraft is in equilibrium (see *Weight and balance* and *CG*, at the end of this chapter).

In such a state of equilibrium the aircraft is said to be *trimmed*. But here we run into a semantic and conceptual problem, between theory on the one hand and practice on the other. The definition of trim as mathematical equilibrium suits the scientist and the engineer. The pilot puts it simply and more grossly. When he says he has trimmed his aircraft he means that if it is disturbed, by say a gust, then it should return to the trimmed condition without his having to touch the controls. If it does not, then either his trimmer is ineffective, or everything is on so much of a knife-edge that to talk of the aircraft being trimmable, or in trim, is academic.

Note: (see below) *Trim tabs* of various kinds are used (some fixed, most adjustable by the pilot) to enable the controls to be released in flight, while leaving the aircraft in trim.

Control and *stability* are complementary. Control is needed to make an aircraft manoeuvrable, either in some transient trajectory or from one steady state to another. Stability is provided to make it behave during those moments when controlling authority of the pilot is relaxed or diverted. However, control power is needed to overcome stability: the greater the inherent stability the more effort that is required. In the end, the degree of control applied becomes a measure of aircraft stability.

Handling qualities are the subjective opinions of pilots on the ease of manoeuvring, and assessing the workload involved in maintaining a steady flight condition, especially in turbulence. An aircraft may be stable and controllable, but unpleasant to fly. Another may be a sheer delight on a short sortie, even though stability is almost non-existent. Assessment of handling and the associated terms is a disciplined way of describing subjective qualities, like sweetness and liveliness, objectively.

Control terms

Aileron is a simple flap-like surface set at the trailing edge of a wing to provide control in roll. Usually mounted outboard, they may also be inboard at the trailing edge of high speed aeroplanes for structural reasons. Aileron control may also be combined in the function of a combined stabilizer and elevator (stabilator), making it a *taileron*.

Backlash is the free-play in cockpit control, i.e. any 'lost motion' which does not immediately produce the required mechanical effect at the other end of the circuit.

Breakout force is the effort a pilot must apply to move a control from a given position. Such forces usually involve circuit friction, but may also be caused by a badly designed linkage.

Elevator controls an aircraft in the pitching plane. It is usually a simple flap-like surface attached to a tailplane, or a canard foreplane. It can form an integral part of a dual-function all-moving *stabilator* (*monoblock elevator* in France) which has no separate tailplane and elevator surfaces.

Float or *trail* of a control surface is shown in Fig. 3.12. It is the tendency of a control surface to move with changes in the direction of the relative airflow (wind) – either with the wind or against it.

Friction is resistance in a circuit to the pilot's effort to move a cockpit control.

Gearing is the relationship between control surface movement and the movement of the pilot's control in the cockpit.

Horn balances are fitted to aileron, elevator and rudder surfaces, as required to reduce control forces felt by the pilot (see Chapter 9 and Fig. 9.10).

Locking can occur when a control surface moves to full deflection (due to overbalance caused by an over-powerful horn). It is often hard to overcome locking without considerable redesign.

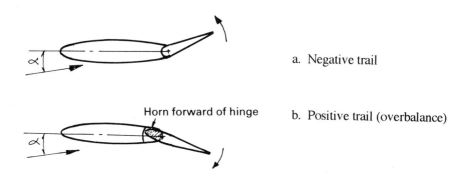

a. Negative trail

Horn forward of hinge   b. Positive trail (overbalance)

Fig. 3.12   Control trail or float. See also Figs 2.3a and 2.3b, and Fig. 9.6 for hinge-moment convention.

Monobloc stabilizer or *elevator* (see *elevator*).

Overbalance is the opposite of self-centring. A control is overbalanced when a force must be applied to prevent further deflection. If the force needed to deflect a control decreases with increasing deflection (but does not reverse) a *tendency to overbalance* exists.

Rudder is a simple flap-like surface usually, but not always, attached to a fin at the tail, giving control in yaw. Some aeroplanes employ monobloc fin and rudder surfaces. Originally the elevator control was called a rudder (see Example 2–4: letter of Orville to Wilbur Wright).

Stabilator (see *elevator*).

Tab is a minor flap-like control at the trailing edge of an aileron, elevator or rudder. *Trim tabs* enable the controls to be released in flight. They fix the angle at which a control surface trails or floats when the pilot releases his control. Others are used for *balance*, to reduce the force needed to deflect a control. *Anti-balance* tabs have the opposite effect, they increase control forces when these are too light for the amount of deflection achieved, and are intended to prevent inadvertent overstressing of the aircraft. *Geared* and *servo* tabs are special variants of the basic balance tab.

Trail (see *float*).

Handling Terms

Pilot assessment

Applied normal acceleration (noticeable through the seat of the pants) is the measure of an apparent increase in weight when manoeuvring. Equation (3–31) and Fig. 8.10 show that in a level turn, angle of bank, ϕ, the lift and apparent weight are increased by an amount $(n_a)\,g$ over the 1.0 g case in straight and level flight. Thus:

$$\text{applied normal acceleration} = (n_a)\,g = (n-1)\,g \qquad (3\text{–}31a)$$

and the angle of bank, ϕ, is obtained from:

$$W/\cos\phi = n\,W$$

i.e.

$$\cos\phi = 1/n \qquad (3\text{–}31b)$$

The centripetal acceleration, V^2/R, acting on all of the masses causes a centrifugal reaction:

$$(W/g)\,V^2/R = n\,W\sin\phi \qquad (3\text{–}31c)$$

in which R is the radius of the turn and V the true airspeed, in compatible units.

Control effectiveness means is used to indicate that the control works, or that it does not. There are no degrees of effectiveness. Thus, never say that the control was **'*very* effective'**.

Compensation (see Table 5–4 for this and related definitions).

Cooper–Harper Rating is a disciplined method by means of which pilots may assess aircraft handling qualities (see Chapter 5).

Harben shorthand (*also: Stone spin shorthand* (see Table 4–5 (Refs 4.6 and 4.7)) is a shorthand which is useful during briefing before a flight, providing a glance-reference to cockpit control positions during a particular exercise or test, recording on a knee-pad, and as an *aide-mémoire* for later debriefing.

Harmonization describes the relationship between control forces and deflections. They are said to be well harmonized when forces, deflections and responses of the aircraft are such that the pilot is not aware of having to exert too much or too little effort. A rule passed down from World War II is that for 'sweet handling' the aileron : elevator : rudder force relationship needed to produce a common rate of response about each axis should be:

$$A : E : R = 1 : 2 : 4 \qquad\qquad\qquad (2\text{–}5)$$

Performance of a control (see Chapter 5, *Cooper–Harper* and other rating scales).

Response is a term open to misuse, because it must be related to time. It is wrong to talk of 'control response' and correct to say '*aircraft response* to control'. Responsiveness depends upon rotational moment of inertia about an axis, aerodynamic damping and stability. Large aeroplanes tend to be sluggish compared with small, which are as quick as cats – or 'squirrelly' (as quick-reacting as a squirrel).

Sensitivity describes a control which produces a rapid response of the aircraft, having due regard for smallness of control deflection, or control force, or both.

Spongy describes a control which feels as though it is connected to a spring, which must be compressed or extended before the control surface begins to move.

Task (see Chapter 5, Table 5–4).

Workload (see Chapter 5, Table 5–4).

Response of Aircraft (see *Response*)

Dutch roll, a regular short period oscillation in roll and yaw.

Flutter, a high frequency oscillation resulting from interactions between aerodynamic forces and the dynamic frequencies of oscillation and vibration of control surfaces and other masses.

Heave is (a sometimes fluctuating) motion of the CG along the Z-axis (which lies in the plane of symmetry). It is associated with changes of lift with angle of attack.

Pilot induced oscillation, PIO, is a driven oscillation, sometimes divergent, arising out of an adverse interaction between the pilot and the aircraft. Aircraft response to control can be so fast that slower corrective actions by the pilot get out of phase, sometimes disastrously.

Pitch-up is a nose-up pitching motion which may be uncontrollable, and may be especially violent during steep turns at high Mach numbers.

Pitching is an angular motion about the lateral Y-axis (spanwise, through the CG).

Porpoising is a fairly regular pitching motion of low frequency, accompanied by some 'heave'.

Post stall gyration, PSG, uncontrolled motion(s) about one or more aircraft axes following departure from controlled flight. Large fluctuations in angle of attack are usually encountered.

Rocking is a lateral oscillation, predominantly in roll, but often accompanied by a directional component and 'sway'.

Roll is an angular motion about the longitudinal (X-axis) through the CG.

Sideslip is flight with a component of airflow from one side. *Skid* is centrifugal, away from the centre of the turn. *Slip* is centripetal, towards the centre of the turn.

Sinking is a marked increase in rate of descent with the aeroplane maintaining a more or less constant attitude, increasing the angle of attack. The condition is frequently met in the approach to the stall.

Skid (see *sideslip*).

Snaking is a lateral and directional oscillation, predominantly in yaw.

Surge is (a sometimes fluctuating) motion of the CG along the longitudinal X-axis. The same word is used to describe the effects of transient changes in thrust (especially of a jet engine, when it is associated with irregularity of the airflow due to compressor stalling), which causes the aircraft to respond as described.

Sway is (a sometimes fluctuating) motion of the CG along the spanwise Y-axis.

Tightening is the condition in a turn or a pull-out such that a push force is needed on the stick to hold a constant '*g*' loading on the aircraft.

Vibration is a high frequency structural oscillation, usually more than 1 cycle per second (1 Hz).

Wallowing is an uncommanded regular or irregular motion about all three axes at once.

Wandering is the description of slow and steady uncommanded changes of heading, not necessarily in the same direction.

Yawing is angular motion about the normal (Z-axis).

Stability and Manoeuvrability

When we talk of stability we tend to mean longitudinal stability in the pitching plane. Directional stability or weathercock stability is in the yawing plane. It is usual to couple lateral and directional stability together, because they are interdependent – except that there is no such thing as *lateral stability*. What is called lateral stability is the *rolling moment due to sideslip*, otherwise called *dihedral effect*, which is positive when the lower wing (upwind or to windward) tends to rise. The term lateral stability is inexact but maintains its dignity as a semantic convenience.

The *definition of stability* by Jeffrey Quill (Supermarine Spitfire test pilot) was put simply as:

> 'The tendency of an aircraft when disturbed from a condition of steady flight to return to that condition when left to itself; conversely instability is the tendency of the aircraft to diverge further away from the condition of steady flight if once disturbed.'   (Ref. 3.3)

$$\text{measure of stability} = \text{magnitude of restoring moment}/\text{magnitude of disturbance} \qquad (3\text{--}35)$$

The magnitude of the restoring moment, $-M_{CG}$, is measured about the CG of the aircraft, which acts as a fulcrum in pitch, roll and yaw (*see* Fig. 3.13, which shows M_{CG} in the positive (+) sense). When the aircraft is disturbed in pitch, e.g., by an upgust, a stabilizing force, L_s, is produced by the stabilizer (or, L_r in the case of a rearplane) which in an ideal world acts like a weather-vane in pitch and tips the nose back into the new relative wind. The magnitude of L_s depends upon the new angle of attack at the stabilizer, α_s (Fig. 3.14b).

If the restoring moment is zero, stability is *neutral* and the aircraft remains in the disturbed state. A nose-up restoring moment is positive and nose-down negative. So, taking moments about the CG in Fig 3.13 we can write:

$$M_{CG} = (\text{approx}) \, L_w x - L_s l_s \qquad (3\text{--}34)$$

which can be turned into a moment coefficient, C_M, by means of Equation (3–29):

$$C_{MCG} q S_w \bar{c}_w = C_{Lw} q S_w x - C_{Ls} q_s S_s l_s \qquad (3\text{--}35)$$

giving approximately:

$$C_{MCG} = C_{Lw}(x/\bar{c}_w) - C_{Ls}(q_s/q)(S_s l_s/S_w \bar{c}_w)$$
$$= 0 \text{ for trim} \qquad (3\text{--}36)$$

The suffix w for wing terms is normally omitted, it is included in Equations (3–35) and (3–36) merely to distinguish between the different contributions. Note that two new expressions in brackets have crept into the equation:

(q_s/q) and $(S_s l_s/S_w \bar{c}_w)$, which is normally written $(S_s l_s/S\bar{c})$.

The former is the 'tail efficiency factor' (Equation (3–50)) and the latter the 'stabilizer (or tail) volume coefficient' (Equation (3–53)), both of which are dealt with in due course.

Fundamentally there are two kinds of stability:

☐ *Static stability*, which determines whether or not the aircraft will initially tend to return to its trimmed condition.
☐ *Dynamic stability*, which governs the subsequent behaviour after the initial response of the aircraft to the static restoring moment.

Both are affected in turn by the position of the control surfaces. If these are held rigidly by the pilot throughout the disturbance they are said to be *stick-fixed*, and the restoring moment has a related value. But if during a disturbance the control surfaces are allowed to trail or float, as when the pilot releases the stick (hence, *stick-free)* then the stick-fixed restoring

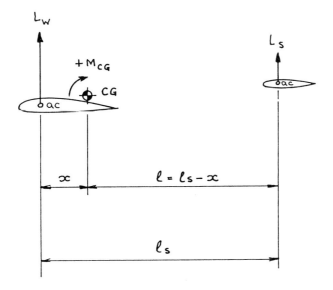

Fig. 3.13 The measurement of longitudinal stability is taken about the centre of mass, the CG, the fulcrum about which the aircraft pitches, rolls and yaws. Aerodynamic pitching moments about the aerodynamic centres, ac, of the flying surfaces have been neglected. The ac is, roughly, the quarter chord of each surface. L_w = wing or foreplane lift, L_s = tail stabilizer or rearplane lift. For simplicity the drag of each surface has also been neglected.

moment may be reduced or even increased, depending upon the direction of float and the position in which the surface settles (see Fig. 3.12).

Centre of gravity (CG) *margins*, stick-fixed and -free (H_n and H'_n), are the distances between the position of the CG and the *neutral points* (NP and NP′), expressed as percentages or fractions of the standard mean chord of the wing. They are guides to the permissible CG-ranges. The larger a CG margin in relation to the SMC, the faster the return response to a trimmed condition after disturbance. Thus, the larger the CG margin the more stable the aircraft. Figure 3.15 shows physically what are meant by the terms Neutral Point, NP, and CG (static) margin.

   Changes in control force and deflection with changing airspeed of real aircraft are used as measures of 'stick-movement' and 'stick-force' stability, in terms of:

   stick travel to change speed = stick-fixed static margin   (3–37a)

   stick force to change speed = stick-free static margin   (3–37b)

(for static margins see Equations (3–47) and accompanying definitions).

Directional stability is the same as 'weathercock-stability' in *yaw*.

Lateral stability is in *roll*. It does not exist, being roll with sideslip or 'dihedral effect'.

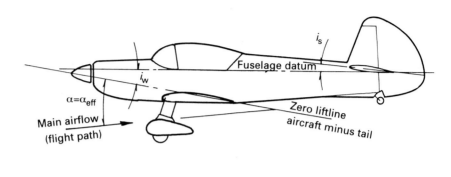

(a)

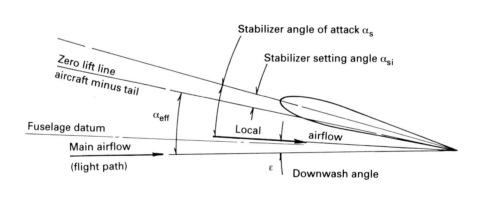

(b)

Fig. 3.14   Factors affecting stabilizer angle of attack and, hence, lift L_s, in Fig. 3.13.

Longitudinal stability is in *pitch*. Historically this was the earliest mode of stability to be analysed by, for example, mathematicians like G. H. Bryan and W. E. Williams (Ref. 3.4). The principles had been studied and were understood by Sir George Cayley, as far back as 1809, when he had discovered the importance of dihedral (Fig. 3.16):

'This angular form, with apex downward, is the chief basis of stability in aerial navigation ....'
(Ref. 3.5)

With tandem surfaces: wing in front and separate tail behind, or vice versa, the leading surface must fly at a larger angle of attack than the rear. Even tailless configurations, which lack a separate stabilizer, employ a longitudinal vee between those portions ahead of the

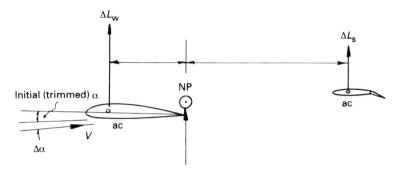

Neutral Point
centre of pressure and resultant of ΔL_w and ΔL_s:
the 'fulcrum' of the aerodynamic forces

Disturbing moment   Righting moment
from wing   from stabiliser

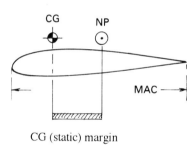

CG (static) margin

Fig. 3.15   The amount of longitudinal stability is proportional to the CG- (or static-) margin which is the distance measured between the CG and the neutral point, NP. The distance is expressed as a percentage of the mean aerodynamic chord of the *complete* aircraft. The drag forces have again been neglected.

CG and those behind (see Fig. 3.17 (Ref. 1.5)).

Magnitude of a disturbance is measurable in terms of the change in angle of attack, $\Delta\alpha$, which in itself is useless. More usefully we may resort to the change in lift coefficient, ΔC_L within the 'working range of angle of attack', α, so that the slope of the lift (coefficient) curve (Figs 3.9b and 3.10a), denoted a is given by:

lift slope, $a = \Delta C_L / \Delta\alpha$ \hfill (3–38)

Fig. 3.16 Cayley's simple experiment conducted in 1809 with a cone made of writing paper and weighted at its apex. It led to his deduction that the vee-form, apex downwards 'is the chief basis of stability in aerial navigation.'   (Ref. 3.5)

Figures 3.10a to 3.10c show how lift (and drag coefficient) slopes change with aspect ratio. Thus we may rewrite our statement of stability in Equation (3–35) in terms of the restoring moment about the CG (with a minus sign in front to show opposition) when the aircraft is disturbed by, say, a gust which changes the angle of attack of the wing by an amount $\Delta\alpha$ (which we can then replace with ΔC_L):

$$
\begin{aligned}
\text{measure of stability} \quad &= -\Delta M_{CG}/\Delta C_L \\
&= \text{CG } \textit{margin}, H_n, \text{ stick-fixed} \tag{3–39a} \\
&= \text{CG } \textit{margin}, H'_n, \text{ stick-free} \tag{3–39b}
\end{aligned}
$$

in which the disturbance is restated as:

$$
= \Delta C_L = a\Delta\alpha \tag{3–40}
$$

Lift slope, a, is for an infinitely long wing with no losses at the tips. In the real world of finite aspect ratio, A, it becomes a_A. The higher the aspect ratio, i.e. the longer the span of a wing across the relative wind, the more nearly does lift slope approach the theoretical maximum for an infinitely long wing (Fig. 3.10c):

$$
\text{lift slope for an infinite span wing, } a_o = 2\pi/\text{radian } (0.11/\text{degree}) \tag{3–41}
$$

But as aspect ratio is reduced, lift slope decreases by a factor $A/(A+2)$, so that for real aspect ratios, far removed from infinity:

$$
\begin{aligned}
a_A &= 2\pi A/(A+2) \\
&= a_o A/(A+2)/\text{radian} \tag{3–42}
\end{aligned}
$$

While for very low aspect ratios of 2 or less:

$$
a_A = 2\pi (A/4) = a_o (A/4) \tag{3–43}
$$

It follows that a wing of low aspect ratio, which has a shallower lift slope, has less gust-response and is better suited to high speed flight in turbulence than one of higher aspect ratio. The British TSR.2, designed for transonic speeds at low altitudes, but cancelled in 1965, had a low aspect ratio wing for that purpose.

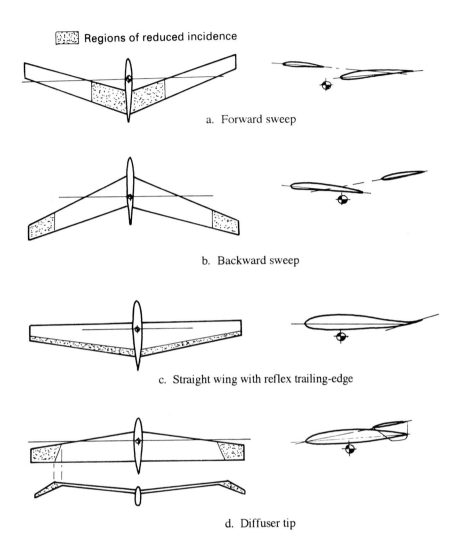

Regions of reduced incidence

a. Forward sweep

b. Backward sweep

c. Straight wing with reflex trailing-edge

d. Diffuser tip

Fig. 3.17 Stabilization of tailless aircraft by means of the vee-form, apex downwards, which confirms Cayley's observation in Fig. 3.16 and Ref. 3.5.

Manoeuvrability is the ability to change direction: the greater the change in a given time, the more manoeuvrable is the aircraft. Because a stick movement by the pilot changes the flight path in pitch and so introduces an acceleration, '*g*', normal to the flight path, we have two criteria:

$$\textit{stick-movement/applied 'g'} = \text{stick-fixed manoeuvre margin, } H_m \qquad (3\text{--}44a)$$

$$\textit{stick-force/applied 'g'} = \text{stick-free manoeuvre margin, } H'_m \qquad (3\text{--}44b)$$

The first is rarely measured. The second is more directly related to 'feel', because the stick-force felt by the pilot is that needed to restrain elevator movement.

Manoeuvre margins, stick-fixed and -free (H_m and H'_m) are the distances between the CG and the manoeuvre points, expressed as percentages of the SMC, and give a measure of the manoeuvrability of the aircraft. The smaller the manoeuvre margin the 'twitchier' or more 'squirreley' is response to control.

Manoeuvre points, stick-fixed and -free (h_m and h'_m), are in effect aft centre of gravity positions which must never be exceeded. The CG must always lie forward of an MP:

h_m stick-fixed = CG position at which stick movement per 'g' is zero   (3–45a)

h'_m stick-free = CG position at which stick force per 'g' is zero   (3–45b)

They are physically similar to the neutral points, h_n and h'_n (see Fig. 3.19 and Table 3–2), except that the aircraft is rotating in pitch (as when pulling out of a dive) which increases the local angle of attack and lift of the tail, this in turn causes the manoeuvre points to move aft of the neutral points, making the manoeuvre margins larger than the static margins.

Neutral points, stick-fixed and -free (h_n and h'_n) are the centre of gravity positions for neutral static stability (Fig. 3.18 and Table 3–2). The neutral point is in effect the same as the aerodynamic centre, in that:

$$\Delta C_{MNP}/\Delta\alpha = \Delta C_{Mac} / \Delta\alpha = 0$$ (3–46)

and:

h_n stick-fixed = CG position at which stick movement to change speed is
zero   (3–46a)

h'_n stick-free = CG position at which stick force to change speed is zero   (3–46b)

Figure 3.18 is exaggerated for clarity. The neutral point actually lies around 0.35 to 0.45$\bar{c}$. It is worth recording that S. B. Gates defined the neutral point as:

'... that point on the wing chord the pitching moment coefficient about which does not change when the incidence* changes.'
* Here, angle of attack.

which is what Equation (3–46) says mathematically.

Static margins, stick-fixed and -free (K_n and K'_n) are the measures of longitudinal stability with real aircraft, and are proportional to the CG margins H_n and H'_n which are the static margins of theoretical and infinitely stiff aeroplanes. In the real case we have, e.g.:

$$K_n = \psi_1 H_n = -dC_{MCG}/dC_L$$ (3–47a)

in which ψ_1 is the factor of proportionality by which the stick-fixed CG margin is altered by airspeed (hence aerodynamic loading, propwash effects, movements of passengers and crew, and aero-elastic bending and twisting of the airframe). It is easier to make a structure strong than to make it stiff. Structures of larger aircraft are heavier than the ideal for strength because of the need for stiffness.

Similarly,

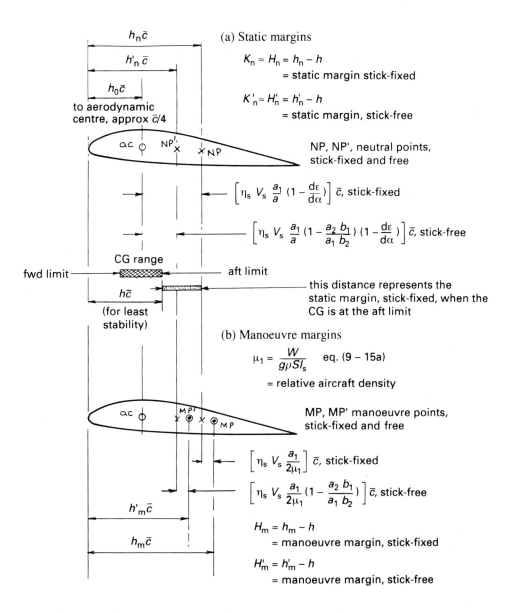

Fig. 3.18 The relationships between the aerodynamic centre of aircraft-minus-stabilizer, the neutral points stick-fixed and -free, and the manoeuvre points stick-fixed and -free.

$$K'_n = \chi_1 H'_n = -dC_{MCG}/dC_L \qquad (3\text{--}47b)$$

involves the factor of proportionality, χ_1, in the stick-free case. Pilots need only be aware that such factors exist, and that they account for the differences between paper and real aeroplanes.

Table 3–2

Summary of definitions and notation

Name	Stick condition	Symbol	Meaning	Used to determine
Neutral point	Fixed	h_n	CG for neutral static stability	CG for zero stick travel to change speed
	Free	h'_n	As above	CG for zero stick force to change speed
CG margin (theoretically rigid aeroplane)	Fixed	H_n	$h_n - h$	Guide to permissible CG range
	Free	H'_n	$h'_n - h$	Ditto
Static margin (real, flexible aeroplane)	Fixed	$K_n = H_n$ ψ_1	Measures stick-fixed static stability	Stick travel to change speed
	Free	$K'_n = H'_n$ χ_1	Measures stick-free static stability	Stick force to change speed
Manoeuvre point	Fixed	h_m	CG position for zero manoeuvre margin as above	CG for zero stick travel per g
	Free	h'_m		CG for zero stick force per g
Manoeuvre margin	Fixed	H_m	$h_m - h$	Stick travel per g
	Free	H'_m	$h'_m - h$	Stick force per g

Static and manoeuvre margins

Many of the apparently tortuous calculations in this chapter are deployed to help the pilot to understand what engineers and aerodynamicists are talking about, so that he or she is not misled into imprudent action by lack of understanding. Not knowing the right question to ask when the time comes is a quick way of being trapped by the mistakes of others. And if there is an accident the first and easiest person to blame will be the pilot. So, 'CHECK – SIX' (in fighter jargon, watch your tail, your six-o'clock position, treating this as armour). The pilot's neck and reputation are at stake long before those of anyone else.

In the following equations which compare static and manoeuvre margins, there are several strange terms, some of which may be made clearer by a diagram than in words. Figure 3.19 shows tail (stabilizer) effects upon the component hinge moments and lift coefficients, first by the whole surface, then the elevator alone, and finally the tab acting by itself. *Hinge moments* are caused by the centres of pressure of the moving surfaces acting at a distance from their hinges (see Fig. 9.4 and Equation (9–14)).

There are several important items to be borne in mind before summarizing. The first is

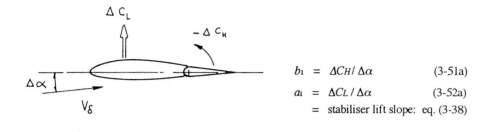

$$b_1 = \Delta C_H / \Delta \alpha \qquad (3\text{-}51a)$$

$$a_1 = \Delta C_L / \Delta \alpha \qquad (3\text{-}52a)$$

$$= \text{stabiliser lift slope: eq. (3-38)}$$

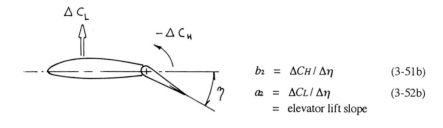

$$b_2 = \Delta C_H / \Delta \eta \qquad (3\text{-}51b)$$

$$a_2 = \Delta C_L / \Delta \eta \qquad (3\text{-}52b)$$

$$= \text{elevator lift slope}$$

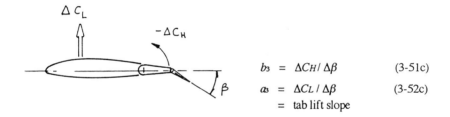

$$b_3 = \Delta C_H / \Delta \beta \qquad (3\text{-}51c)$$

$$a_3 = \Delta C_L / \Delta \beta \qquad (3\text{-}52c)$$

$$= \text{tab lift slope}$$

V_δ is the local velocity of the airflow affecting a control surface of chord length, c_δ and area S_δ. A change in angle of attack, $\Delta \alpha$, or deflection of the control surface through an angle, δ, causes an incremental change, ΔH in the hinge moment, H. If q_δ is the local dynamic pressure, then the hinge-moment coefficient is given by:

$$C_H = H / q_\delta S_\delta c_\delta \qquad (3\text{-}49)$$

In the case of a stabiliser the deflection δ becomes η for an elevator and β for a trim tab, so that using the terms defined in the diagrams:

total hinge-moment coefficient, $C_H = b_1 \alpha + b_2 \eta + b_3 \beta \quad (3\text{-}51a)$

total lift coefficient, $\qquad C_L = a_1 \alpha + a_2 \eta + a_3 \beta \quad (3\text{-}51b)$

Fig. 3.19   Definitions of stabilizer and control surface hinge-moment and lift coefficients.

that the angle of attack of the relative airflow at the tail, α_s, depends upon the nature of the wake and downwash shed by the wing ahead of it (see Fig. 3.14b, downwash angle ε). It may be shown from this diagram that when a disturbance is small, as measured in terms of $\Delta\alpha$ at the wing, the rate of change of angle of attack at the tail, $\Delta\alpha_s$ with change in angle of attack at the wing is:

$$\Delta\alpha_s/\Delta\alpha = (\Delta\alpha - \Delta\varepsilon)/\Delta\alpha \text{ (neglecting stabilizer setting angle, } \alpha_s)$$

As Δ grows smaller and smaller and approaches zero, then this equation becomes:

$$d\alpha_s/d\alpha = [1 - (d\varepsilon/d\alpha)] \qquad (3\text{–}48)$$

☐ *Stabilizer* (tailplane + elevator + tab in fixed relation one to another) produces a lift slope a_1 due to an increase in angle of attack (or incidence) at the tail. The elevator and tab each attempt to float or trail and together contribute a moment about the elevator hinge; but we usually ignore the small tab contribution in what follows. The hinge moment, H, transmits a force to the pilot's hand (the stick-force). In general terms the hinge moment coefficient is:

$$C_H = H/q_\delta S_\delta c_\delta \qquad (3\text{–}49)$$

The term q_δ is the dynamic pressure recovery at the surface to which the equation applies in the wake effects of dirty airflows over surfaces and junctions upstream. In the case of a tail, $q_\delta = q_s$ in a flow velocity, V_s, which is less than the free stream value, V (see Equation (3–26)). Indeed, we found q_s appearing first in Equation (3–36):

$$\text{tail (or stabilizer) efficiency, } (\eta)_s = q_s/q \qquad (3\text{–}50)$$

this, as $\rho/2$ is constant:

$$= (V_s/V)^2 \qquad (3\text{–}50a)$$

enabling us to substitute $(\eta_s q)$ for q_s.

Note: Tail efficiency is exceptionally hard to calculate, but it must not be overlooked. Designers often produce, quite unintentionally, tails which are too small in some corner or another of the flight envelope, because they have no figures for tail efficiency. Pilots should be aware of this. It is a cause of inadequate longitudinal and directional stability with kit-built designs especially, and I spent one visit to Oshkosh photographing tail surfaces which looked too small, and/or inefficiently arranged because of it.

The same tail efficiency term in Equation (3–50) can be used without much error for vertical tail surfaces (fins and rudders). On finding during flight tests that the tail area is too small, the obvious solution, and the hardest, is to suggest that it needs enlarging in the ratio:

$$\text{practical tail area} = (1/\eta_s) \text{ theoretical tail area} \qquad (3\text{–}50b)$$

With real values around $0.7 < \eta_s < 0.85$, many deficient tail areas should have been 15 to 40% larger than drawn. When a fuselage is bluff and 'draggy' and the wing junction is a bit of a mess, then the value of η_s may drop as low as 0.5, requiring a tail twice as big (or twice as effective) as that provided.

A high-winged aeroplane usually needs larger fin and rudder surfaces than one which is low winged, because the surfaces downstream of high wing and body junctions are closer to the plane of the wake and suffer more interference than when a wing is low-set on the fuselage.

The hinge moment coefficient can be written in terms of the component surfaces:

$$C_H = b_1\,\alpha + b_2\,\eta + b_3\,\beta \qquad\qquad (3\text{--}51)$$

while the overall lift coefficient, in terms of the component lift slopes and angles is:

$$C_L = a_1\alpha + a_2\,\eta + a_3\,\beta \qquad\qquad (3\text{--}52)$$

These terms are set out in Fig. 3.19, Equations (3–51a) to (3–51c) and (3–52a) to (3–52c).

Note: A similar approach may be used for the fin, rudder and tab, and the ailerons (Ref. 1.5).

Second:

☐ *Stabilizer (tail) volume coefficient, $\overline{V}_s$, appeared in Equation (3–36) and is defined as:*

$$\overline{V}_s = (S_s/S)\,(l_s/\overline{c}) \qquad\qquad (3\text{--}53)$$

in which S_s and S are the stabilizer and wing areas; l_s is the moment arm of the aerodynamic centre, ac_s, of the stabilizer about the ac of the wing; and $\overline{c}$ the SMC of the wing.

Note: One may use the same method with fins and rudders, using the semi-span of the wing, $(b/2)$, wing area, S, fin and rudder moment arm, l_f, such that fin (and rudder) volume coefficients are as described in Chapter 9, Equations (9–11a) and (9–11b). From these we see that an increase in wingspan, b, diminishes the volume coefficient of the fin and rudder, making the wing dominant and the original tail combination relatively less effective. In Example 2–6, the long-winged Ta 152, developed out of the shorter winged Fw 190, needed major modifications to the fuselage length and the fin and rudder-hinge arrangement to restore tail volume and authority for high altitude control and stability.

Third, we have the:

☐ *Effect of float or trail* upon *lift coefficient, stick-free.* When the pilot releases the stick then under the influence of the pressure field at the tail (the cause of the hinge moment) the elevator will float to a new angle, $\Delta\eta$, at which the moment is reduced to zero. But, having moved, the elevator has also changed the lift of the tail, altering the tail loading and, by changing too the effective camber of tailplane-plus-elevator, it has altered the lift slope to a new equivalent value, a_1. The change is calculated as follows, the minus sign showing that trail occurs in the normal sense:

$$b_1\,\Delta\alpha_s = -b_2\,\Delta\eta$$

i.e.

$$\Delta\eta = -(b_1/b_2)\,\Delta\alpha_s \qquad\qquad (3\text{--}54)$$

Thus, the change in stabilizer lift coefficient, stick-free, becomes:

$$\begin{aligned}
\Delta C_{Ls} &= a_1\Delta\alpha + a_2\,\Delta\eta \\
&= a_1\Delta\alpha_s\,[1 - (a_2 b_1/a_1 b_2)] \\
&= \overline{a}_1\,\Delta\alpha_s \qquad\qquad (3\text{--}55)
\end{aligned}$$

in which the *equivalent lift slope*, $\overline{a}_1 = a_1\,[1 - (a_2\,b_1/a_1 b_2)]$ $\qquad\qquad$ (3–55a)

Within this equation, (b_1/b_2) determines the resultant direction in which the elevator will float, while:

$$(a_2/a_1) = (dC_{Ls}/d\eta)/(dC_{Ls}/d\alpha) = d\alpha_s/d\eta \qquad (3–56)$$

which is the effective rate of change of angle of attack of the stabilizer with elevator float.

Why is this so? Simply because with elevator upfloat $\Delta\eta$ is negative and stabilizer lift is reduced. Upfloat makes b_1 negative. The lift slopes a_1 and a_2 are always positive. The hinge moment due to elevator movement, b_2, must also be negative if the control is to feel right and in the correct sense to the pilot. Of course, in all of this $\Delta\alpha_2$ is assumed to be positive. Inserting these conditions in Equation (3–55a) the term inside the bracket must be less than 1, making the equivalent lift slope, $\bar{a}_1$, less than the stick-fixed value, a_1. But when b_1 is positive, causing the elevator to float downwards, then stabilizer lift is increased and the equivalent lift slope, $\bar{a}_1$, becomes greater than a_1 stick-fixed. Thus, the degree of static stability stick-free is increased when b_1 is positive, and decreased when it is negative. Elevator-fixed stability is considered to be the measure of basic static stability, in exactly the same way as with any free-flight model aeroplane. Releasing the stick so as to free the elevator, adds to or detracts from the basic stick-fixed stability.

Summarizing

Changes in control force and deflection with given changes in airspeed are important measures of stability, enabling a pilot to make a swift qualitative assessment:

☐ Stick-fixed stability = *stick movement needed to change speed*   (3–57a)
☐ Stick-free stability = *stick force needed to change speed*   (3–57b)

It was S. B. Gates who, apparently with a flash of insight, realised that *stick-force per applied 'g'*, felt by the pilot, is a direct measure of static longitudinal stability, and an indicator of the size of the static margin (Fig 3.20). Although the remaining equations for the static margin stick-free and for both manoeuvre margins look formidable and complicated at first, they are not. See them as growing by elements added to the original expression for static margin stick-fixed (free-flight model aeroplane) case:

$$K_n = (h_0 - h) + \eta_s \bar{V}_s (a_1/a) [1 - (d\varepsilon/d\alpha)]$$
$$= \text{static margin stick-fixed (basic case)} \qquad (3–58)$$

$$K'_n = (h_0 - h) + \eta_s \bar{V}_s (\bar{a}_1/a) [1 - (d\varepsilon/d\alpha)]$$
$$= \text{static margin stick-free} \qquad (3–59)$$

$$H_m = (h_0 - h) + \eta_s \bar{V}_s ((a_1/a) [1 - (d\varepsilon/d\alpha)] + a_1/2\mu_1)$$
$$= K_n + \eta_s \bar{V}_s a_1/2 \mu_1$$
$$= \text{manoeuvre margin, stick-fixed} \qquad (3–60)$$

$$H'_m = (h_0 - h) + \eta_s \bar{V}_s ((a_1/a) [1 - (d\varepsilon/d\alpha)] + \bar{a}_1/2 \mu_1)$$
$$= K'_n + \eta_s \bar{V}_s \bar{a}_1/2\mu_1$$
$$= \text{manoeuvre margin, stick-free} \qquad (3–61)$$

in which:

$$\mu_1 = W/g\rho (l_s S)$$
$$= \text{longitudinal relative aircraft density, i.e. the 'density' of the aircraft}$$
$$\text{relative to the surrounding air} \qquad (3–62)$$

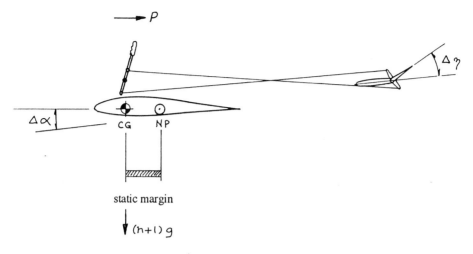

static margin

$(n+1)\,g$

$\Delta\alpha$ = change in angle of attack to apply 1.0 g by entering a turn, or tightening a manoeuvre

$\Delta\eta$ = change of elevator angle to achieve $\Delta\alpha$

P = pilot's stick-force to deflect elevator through $\Delta\eta$

P/g = stick-force per g

Note: the larger the (static margin / chord), the heavier the stick-force per g

Fig. 3.20 Stick-force per g as an indicator of longitudinal stability – a flash of insight of S. B. Gates.

Remember that $\bar{a}_1$, the equivalent lift slope, is given in Equation (3–55), and is the effect of the pilot releasing the stick to fly 'hands-off' so letting the elevator float freely.

Performance

The following terms are of most use to pilots. Several quantities, such as ambient temperature and airspeed, cannot be measured directly by means of instruments, but must be derived from raw material of the instrument readings, using theoretical relationships. In this section we assume that in all cases an 'instrument correction' is applied to instruments to remove any inaccuracy.

Airspeed basics

ASIR, airspeed indicator reading.

CAS, calibrated airspeed – the indicated airspeed of the aircraft corrected for position and instrument error. CAS is the true airspeed (TAS) in the standard atmosphere at sea level.

EAS, equivalent airspeed, V_i, is CAS corrected for *scale altitude effect* (the effect of compressibility as altitude is gained).

GS, ground speed is that of an aircraft over the ground.

IAS, indicated airspeed shown on the airspeed indicator (ASI). Published IAS values assume zero instrument error. When scale altitude effect is more or less zero and there are no instrument errors, then EAS and IAS are nearly equal.

M, Mach number, after Ernst Mach (1838–1916) the Austrian philosopher and physicist who first noted its significance. The ratio of the TAS, V, to the speed of sound, a, for the ambient conditions:

$$M = V/a \tag{1–1}$$

TAS, V, true airspeed relative to the undisturbed air. It is related to calibrated airspeed corrected for altitude, temperature and compressibility, and to equivalent airspeed by the *relative density*, where:

$$\text{relative air density, } \sigma = \rho/\rho_0 \tag{3–63}$$

and:

$$TAS = EAS/(\sigma^{0.5}) \tag{3–64}$$

i.e.

$$V = V_i/(\sigma^{0.5}) \tag{3–65}$$

Calibrated and equivalent airspeeds (CAS and EAS)

KCAS/MCAS, calibrated airspeed and Mach number in knots and miles per hour respectively. When CAS and EAS are almost the same (at low speed and altitude) then approximately:

$$V = CAS/(\sigma^{0.5}) \tag{3–66}$$

Figure 3.21 shows the relationship between KTAS and KEAS in the International Standard Atmosphere, ISA; and Fig. 3.22 the theoretical relationship between airspeed and Mach number.

Pitot head, the sensor which measures the total static-plus-dynamic pressure (Equation (3–12)).

Position error (PE) is the error in readings caused by the position of the pitot head and the static sources (where the static pressure, p, in Equation (3–12) is measured) due to variations in the pressure field surrounding the aircraft. The field changes with Reynolds number, R, and attitude to the flight path (angle of attack). In the absence of instrument error:

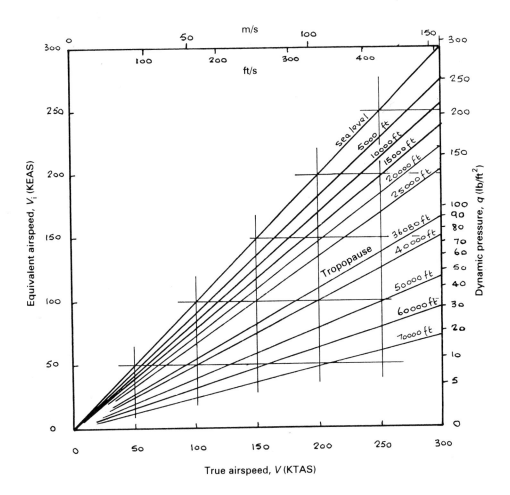

Fig. 3.21 The relationship between true and equivalent airspeed and dynamic pressure in the International Standard Atmosphere.

calibrated airspeed = indicated airspeed + position error
$$CAS = IAS + PE \qquad\qquad (3\text{--}67)$$

PEs are presented as tables, or graphically. For our purposes EAS = CAS (at sea level).

V_A, the design manoeuvring speed, a structural limitation, at which full application of control will not break the aircraft.

V_{app}, approach speed.

V_{AT}, originally the target threshold speed, now replaced by V_{REF}, of which it is a function (e.g., $V_{AT} = V_{REF} + 5$, etc.).

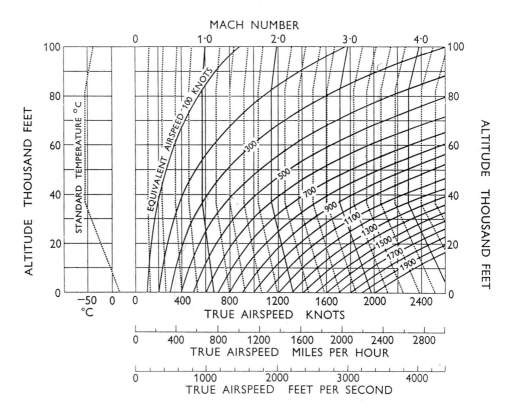

Fig. 3.22   Theoretical relationship between airspeed and Mach number (Ref. 1.4).

V_{ATO}, target threshold speed, all engines operating.

V_{AT1}, target threshold speed, one engine inoperative.

V_{AT2}, target threshold speed, two engines inoperative, and so on.

V_B, the design speed for maximum gust intensity.

V_C, the design cruising speed.

V_D/M_D, design diving speed and Mach number, respectively.

V_{DF}/M_{DF}, *demonstrated* flight diving speed and Mach number, respectively.

V_{EF}, the speed, selected by the applicant for a C of A, at which the critical engine is assumed to fail. It must not be less than 1.05 V_{MC}.

V_F, the design flap (down) limit speed.

V_{F1}, the design flap speed for certain procedural conditions.

V_{FC}/M_{FC}, the maximum speed and Mach number for required stability characteristics.

V_{FE}, the maximum speed with flap extended.

V_{FTO}, the final take-off speed of the aeroplane that exists at the end of the take-off path in the *en-route* configuration with one engine inoperative.

V_{H}, the maximum speed in level flight with maximum continuous power (MCP).

V_{i}, equivalent airspeed, EAS.

V_{LE}, the maximum speed with landing gear extended.

V_{LO}, the maximum landing gear operating speed.

V_{LOF}, lift-off speed.

The next five apply to multi-engined aeroplanes:

V_{MBE}, the maximum brake energy speed.

V_{MC}, the minimum control speed in accordance with (US) Federal Aviation Regulations and (European) Joint Aviation Requirements. In the case of a reciprocating engine-powered aircraft the conditions are, broadly: the critical engine inoperative (propeller windmilling, or auto-feather when applicable) and not more than 5° of bank towards the live engine; at maximum take-off weight, or any lesser weight necessary to show V_{MC}; the most unfavourable CG; with maximum available power on the live engine(s); flaps in the take-off position and landing gear retracted. Hence we also use V_{MC} (F_{TO}) and V_{MC} (F_{L}), i.e. with flaps for take-off and landing.

V_{MCA}, is substantially the same as V_{MC}, but in the take-off climb.

V_{MCG}, the minimum control speed on or near the ground.

V_{MCL}, the minimum control speed, approach and landing as an alternative to V_{MC} (F_{L}).

V_{MO}/M_{MO}, the maximum operating limit speed and Mach number.

V_{MU}, the minimum demonstrated unstick speed with all engines operating.

V_{NE}/M_{NE}, the never exceed speed and Mach number, where:

$$V_{NE} = 0.9\ V_{DF} \tag{3–68}$$

V_{NO}/M_{NO}, the normal operating speed and Mach number are the structural maxima and should not be exceeded, except in smooth air, and then only with caution.

V_{R}, the rotation speed on take-off.

V_{RA}, the rough air speed.

V_{REF}, a term which originated in the USA. Although it is not used officially in the UK, it is increasingly entering common useage to replace V_{AT}.

V_S or V_{S1}, the *stall* speed, or *minimum steady flight speed* at which the aircraft is controllable (V_{S1} specifies a configuration, *other than for landing*).

V_{S0}, the *stall* or *minimum steady flight speed* at which the aeroplane is controllable *in the landing configuration*.

V_{S1g}, the 1.0 g stall speed at which the aeroplane generates a lift force (normal to the flight path) equal to its weight.

V_{SSE}, the intentional one engine inoperative speed is a minimum speed selected by the manufacturer for intentionally rendering one engine inoperative in flight for pilot training. In the UK one engine should not be shut down intentionally for any reason other than an emergency, below 3000 ft (910 m) above ground level.

V_T, the maximum aero-tow speed.

V_{Tmax}, a generalized maximum threshold speed term.

V_X, the best angle-of-climb airspeed which delivers the greatest gain in altitude over the shortest horizontal distance (also called the best gradient-of-climb speed).

V_Y, the best rate-of-climb airspeed, which delivers the greatest gain in altitude in the shortest time.

V_{YSE}, V_Y with an engine failed on a multi-engined aircraft.

V_1, the *decision speed* on take-off. Up to V_1, if an engine fails, it should be possible to stop in the remaining length of runway. Beyond V_1 the take-off is continued.

V_2, the *take-off safety speed* established by the manufacturer to add a safety margin to the minimum control speed, taking into account such items as: failure of the critical engine (propeller windmilling, or auto-feather when applicable); maximum take-off weight, or any lesser weight necessary to show V_{MC}; the most unfavourable CG; full power on the live engine(s); landing gear and take-off flaps extended; average pilot strength and ability; and an element of surprise (having been caught unawares by failure of the engine).

V_3, the steady initial climb speed, all engines operating.

Terms involved in performance measurement

Accelerate–stop distance, the distance needed to accelerate an aeroplane to a specified speed and, assuming failure of an engine at that moment the speed is attained, to bring the aircraft to a stop.

Altitude and height are not quite the same. Altitude is 'height' measured by an altimeter above a pressure datum. Being a calibrated and sensitive pressure-measuring device, with a scale which presents changes in pressure as changes in 'height' above a datum, and so changes with the weather, it is only accurate under one set of precise conditions. *Height*, on the other hand, means 'tape-measure height', which is almost impossible to measure accurately other than by radar altimeter within an aircraft, or a theodolite on the ground.

Ceiling, of which there are two:

☐ *Absolute ceiling* is the altitude at which the rate of climb is zero. Most normally aspirated aeroplanes seem to take about 45 minutes to reach somewhere near their absolute ceilings, as we noted in Chapter 2, when constructing Fig. 2.4. In practice, absolute ceiling is unobtainable.
☐ *Performance and service ceiling*, are civil and military ceilings, respectively:

(1) *Performance ceiling* is the altitude at which rate of climb = 150 ft/min (0.762 m/s).
(2) *Service ceiling* is the altitude at which the rate of climb is 100 ft/min (0.508 m/s).

Climb gradient, the demonstrated ratio of the change in height during a portion of a climb in still air, to the horizontal distance traversed in the same time interval, usually expressed as a percentage, or in degrees.

Climb rate (rate of climb), the demonstrated ratio of the change in height during a portion of a climb in still air, to the time taken for that change:

$$\text{rate, } v_c = \text{change in height/time, m/s (ft/min)}$$
$$= (\text{change in height/horizontal distance}) \times$$
$$(\text{horizontal distance/time taken})$$
$$= \text{climb gradient} \times \text{true airspeed} \qquad (3\text{-}69)$$

(see Fig 3.23).

For constructing *en-route* flight paths the following simple formulae are acceptable:

$$\text{Height gained in feet per nautical mile flown} = \frac{60 \times \text{rate of climb}}{\text{ground speed}} \qquad (3\text{-}68a)$$

$$\text{Distance travelled horizontally in nautical miles per 1000 ft height change} = \frac{17 \times \text{ground speed}}{\text{rate of climb/descent}} \qquad (3\text{-}68b)$$

Descent gradient and descent rate (rate of descent), v_d, are worked out in exactly the same way as climb gradient and rate of climb. In the case of an approach to land, this is expressed as, e.g.,

$$5\% \text{ descent gradient (DG)} = 5/100 = 3 : 60 = 3° \text{ glide slope} \qquad (3\text{-}68c)$$

Demonstrated crosswind velocity, is the velocity of the crosswind component for which adequate control of the aeroplane during take-off and landing was actually demonstrated during tests. Crosswind handling must be demonstrated. Calculation is no substitute.

ISA, International Standard Atmosphere, defined by the ICAO (Ref. 3.6):

the air is a perfectly *dry gas*;

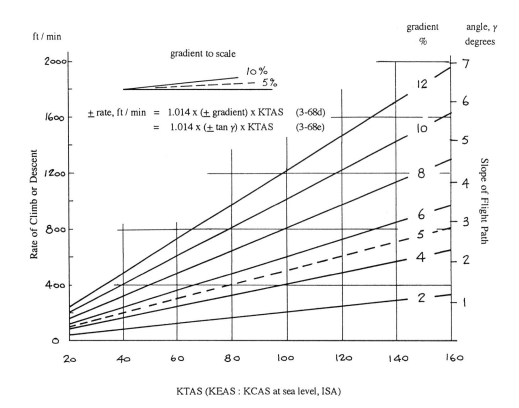

Fig. 3.23   Conversion of rates of climb and descent to gradients and angles.

temperature at sea level 15° Celsius (59° Fahrenheit);
pressure at sea level 1.013 250 × 10⁵ Pa (29.92 inHg) (1013.2 mb);
temperature gradient (lapse rate) from sea level to the altitude (called the
tropopause, at which the temperature stabilizes at −56.5°C) is 3.25°C per 500 m
(1.98°C/1000 ft);
density at sea level, ρ_0, 1.225 kg/m³ (0.002 38 slugs/ft³).

Additionally:

speed of sound at sea level, 340.3 m/s (1116.4 ft/s);
kinematic viscosity, $\upsilon = \mu/\rho = 1.4606 \times 10^{-5}$ m²/s (1.5723×10^{-4} ft²/s);
ratio of specific heats, $\gamma = 1.4$.

Density altitude, the altitude in the standard atmosphere at which the prevailing air density
occurs. Density altitude exceeds the *pressure altitude* (measured by the altimeter) by roughly
33.5 m/°C (110 ft/°C) that the ambient temperature exceeds the standard temperature for
the altitude, and vice versa.

Indicated pressure altitude is the altimeter reading when the subscale has been set at
1013 mbar (29.92 inHg).

OAT, the outside air temperature is the free static (ambient) temperature obtained either from in-flight temperature indications (adjusted for instrument error and compressibility effects) or ground meteorological sources.

Performance Group is the present classification of a public transport aeroplane for a British Certificate of Airworthiness, by weight and minimum level of performance (Ref. 3.1). However, large parts of Ref. 3.1 will almost certainly become redundant around 1995, when the operational requirements of the Joint Aviation Authorities in Europe, JAR-OPS, are fully implemented. The present proposals for JAR-OPS, as this is written in mid-1992, include a system of *Performance Classes*, but the full detail for smaller aircraft has yet to be decided and significant differences from existing UK Performance Groups may yet emerge. For the present, Fig. 3.24 and Table 3–3 show the proposed new JAR-OPS Performance Classes A, B, and C and their correspondence with existing CAA Performance Groups, which are listed as follows:

☐ *Group A:* Propeller and turbine-engined aeroplanes certificated to BCAR Sect D, or JAR 25, with a maximum certificated weight exceeding 12 500 lb (5700 kg) and with a performance level such that if a power-unit fails at *any* moment, a forced landing should not be necessary. Basically, this group contains large aeroplanes.

Note: Although *Group A* also includes jet aeroplanes with more than nine seats or with an MTOW exceeding 12 500 lb (5700 kg), it is the intention of the JAA to treat all passenger-carrying jet aeroplanes as large, regardless of size. As this is written it remains an open item awaiting a policy decision.

☐ *Group B*: Commuter or small twin-engined aeroplanes certificated to BCAR-23, with a certificated performance level such that a forced landing is unlikely to be necessary if an engine fails at any point in a flight.

Note 1: *Commuter Category* means a piston or turbopropeller-driven multi-engined aeroplane of 19 000 lb (8600 kg) or less, having not more than 19 passenger seats. Propeller-driven aeroplanes with no more than nine seats must, if their MTOW exceeds 12 500 lb (2700 kg), also be certificated as Commuters.

Note 2: *Performance Group B* falls a little short of the standard of existing *Group A*, embodied in JAR-25 for large aeroplanes. In due course *Group B* aeroplanes will be required to meet, when operating for Public Transport, the operating rules applicable to *Group A* aeroplanes.

☐ *Group C*: Aeroplanes certificated to BCAR Sect K, with a maximum certificated take-off weight not exceeding 12 500 lb (5700 kg) and a level of performance which enables them to continue a flight with a failed engine, but only after the take-off and initial climb are completed. This group will no longer be an option for any new application for type certification.

☐ *Group D*: Basically single-engined aeroplanes certificated to BCAR Sect K, with a maximum certificated take-off weight not exceeding 12 500 lb (5700 kg) and no provision for performance if an engine fails. The value of V_{s0} appropriate to the maximum weight shall not exceed 60 knots (111 kph), and there are limits of cloud ceiling and visibility for operations, and climb performance after take-off for obstacle clearance.

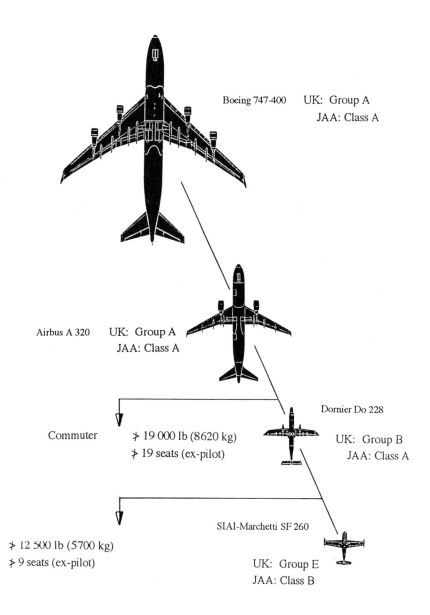

Boeing 747-400   UK: Group A
                   JAA: Class A

Airbus A 320   UK: Group A
                JAA: Class A

Commuter   ≯ 19 000 lb (8620 kg)
                ≯ 19 seats (ex-pilot)

Dornier Do 228

UK: Group B
JAA: Class A

SIAI-Marchetti SF 260

≯ 12 500 lb (5700 kg)
≯ 9 seats (ex-pilot)

UK: Group E
JAA: Class B

UK PERFORMANCE GROUP

JAR-OPS CLASS

Fig. 3.24   UK Performance Group JAR-OPS Class.

TABLE 3–3

Proposed new JAR-OPS Performance Classes

Level of safety acceptable to public = sum of safety regulation

= airworthiness rules + operational rules

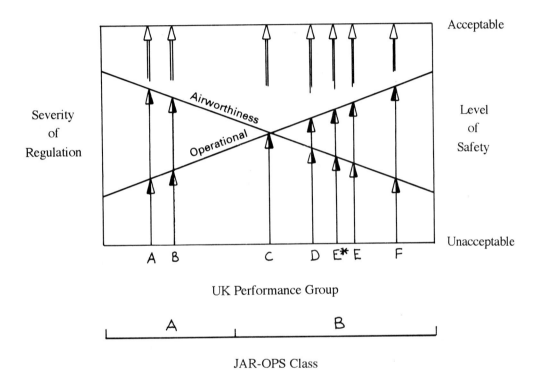

UK Performance Group

JAR-OPS Class

UK Performance Group	JAR-OPS Class
A, B	A
C, D, E and F	B
Unclassified and X	C

Note: E* is an aeroplane normally expected to lie in Group E, but with a good engine-out performance

Note 1: The required limit on V_{S0} of 60 knots has caused many arguments between the UK and USA (where the limit is 70 mph and corresponds with the kinetic energy in a surviveable accident on a freeway or motorway, when seat belts are properly fastened). The conversion factor was read as producing 60 knots in the UK, but 61 knots in the USA. Such a small difference might seem insignificant. It was not for UK manufacturers faced with foreign competition from an aircraft certificated to FAR 23. Nor was it fair for a foreign manufacturer seeking Type Certification from the UK Authority attempting to maintain fair play between commercial competitors – and with a duty of care for the innocent general public. When I left the CAA 61 knots was beginning to be accepted for aircraft certicated to FAR 23.

Note 2: As this is written *Group D*, like *Group C*, will almost certainly no longer be an option for any new application for type certification.

☐ *Group E*: Aeroplanes for which there is limited scheduling of performance. In reality the paperwork has been simplified, but the required level of performance is raised in consequence to at least the levels of *Groups C* or *D*. For example, *Group E* aeroplanes are limited to a maximum take-off weight of 6000 lb (2730 kg); and with a corresponding V_{S0} that shall not exceed 60 knots (111 kph), but see **Note 1** above).

Note: Propeller-driven and jet-engined small aeroplanes, previously limited to 6000 lb MTOW (see Equation (3–78)) for *Group E* certification, may now be certificated to the slightly more severe *Performance Group F* requirements, up to 12 500 lb MTOW, provided that they have no more than nine passenger seats.

☐ *Group F*

    (1)   Small twin-engined aeroplanes having not more than nine passenger seats, with a performance level such that a forced landing may be necessary following engine failure shortly after lift-off, or shortly before touchdown.

    (2)   Small single-engined aeroplanes, meaning those with a maximum certificated take-off weight of 12 500 lb (5700 kg) or less, having not more than nine passenger seats.

Note: To be certificated, propeller-driven aeroplanes with more than nine passenger seats, even if their MTOW does not exceed 12 500 lb, must have more than one engine and be treated as Commuters. There are those who argue that this is unfair: e.g., to the 10 000 lb (4545 kg), 14-seater. Others reply that more than nine seats constitutes serious Public Transport operations which must be certificated as such.

☐ *Group X*: Appears to have no formal definition. It was introduced to cover large piston-engined aeroplanes certificated to the old US CAR code, and with appropriate AFM (Aircraft Flight Manual) data acceptable to the UK Authority.

Large aeroplane means an aeroplane of more than 12 500 lb (5700 kg) maximum certificated take-off weight. It does not include aeroplanes in the Commuter category.

Note 1: *Big transport aeroplanes* weigh more than 30 000 lb (13 620 kg).

Note 2: *Ex-military aircraft* operating on the civil register are classified as follows:

Single piston engine	5000 kg (11 000 lb)
Multi-piston engine	>5000 kg
Single turbine engine	8000 kg (17 640 lb)
Multi-turbine engine	>8000 kg

Pressure altitude is the altitude measured from standard sea level pressure (1013 mbar) by a pressure or barometric altimeter. It is the indicated pressure altitude corrected for position and instrument error. Here we assume that instrument errors are zero (but see Figs 5.8 and 5.9).

Scale altitude effect is so called because of the increase in air density when air is displaced bodily and compressed by an aircraft at (or near) the speed of sound. This, the acoustic speed, is, near enough, the average limiting speed of air molecules trying to get out of the way of an approaching body in their own time. It is determined by their ambient temperature, pressure and density. The ASI, on the other hand, is calibrated on the assumption that the air is incompressible, and so it over-reads in the same way as it would if the aeroplane were to fly at the same TAS, but at a lower altitude.

Screen height, the height above ground level at which a flight path is assumed to pierce an imaginary erect screen on takeoff and landing. Screen heights are 35 ft (10.7 m) and 50 ft (15.2 m), depending upon the performance group and case.

Station pressure, the atmospheric pressure at airfield level.

Power terms

General

Auxiliary power unit, APU, any gas turbine-powered unit delivering rotating shaft power, compressor air, or both which is not intended for direct propulsion of an aircraft.

Beta control, a system whereby the propeller can be operated at blade angles selected manually, normally used during the approach and ground handling.

Brake horsepower, the power delivered at the propeller shaft (main drive or main output) of an aircraft engine.

Critical engine is the engine whose failure would most adversely affect the performance or handling qualities of an aircraft.

Detent, a mechanical arrangement which indicates, by feel, a given position of an operating control. Once the operating control is placed in this position the detent will hold the lever there and an additional-to-normal force will be required to move the operating control away from the position.

Engine, is a prime-mover. It consists of at least those components and equipment necessary for the functioning and control, but excludes the propeller.

Feathered pitch, is the propeller pitch setting, specified in the appropriate propeller manual, which in flight with the engine stopped, gives approximately the minimum drag, and corresponds with a windmilling torque of approximately zero.

Propeller (fixed pitch), is one for which the pitch cannot be changed, except by processes constituting a workshop operation.

Propeller (reverse pitch), means a propeller with a range of blade angles used for producing reverse thrust.

Propeller (variable pitch), is one for which the pitch setting changes or can be changed when the propeller is rotating or stationary. It includes:

☐ A propeller, the pitch setting of which is directly under the control of the flight crew (controllable pitch propeller).
☐ A propeller, the pitch setting of which is controlled by a governor or other automatic means which may be either integral with the propeller or a separately mounted equipment, and which may or may not be controlled by the flight crew.
☐ A propeller, the pitch setting of which may be controlled by a combination of the above methods.

Rated (*power* or *thrust*) *output*, is the approved shaft power or thrust that is developed statically at standard sea level atmospheric conditions for unrestricted use.

Piston-propeller engine

Manifold pressure, the absolute static pressure measured at the appropriate point in the induction system, usually in inches or millimetres of mercury.

Maximum climb power, maximum power permissible during climb.

Maximum continuous power, MCP, the maximum power permissible continuously in flight.

Maximum cruise power, maximum power permissible during cruise.

Turbine engine

Maximum contingency power and/or *thrust rating* means the minimum test bed acceptance power and/or thrust, as stated in the engine type certificate data sheet, of series and newly overhauled engines when running at the specified conditions and within the appropriate acceptance limitations.

Maximum continuous power, MCP, and/or *thrust rating* means the power and/or thrust identified in the performance data for use during periods of unrestricted duration (clearly determined between the constructors and certification authority).

Total equivalent static power, means:

☐ Total equivalent static power, kW = propeller shaft power + $\dfrac{\text{static jet thrust, N}}{15}$ (3–70)

☐ Total equivalent static power, hp = propeller shaft hp + $\dfrac{\text{static jet thrust, lbf}}{2.6}$ (3–71)

Weight and Balance

It is essential in flight testing to know the weight of the aircraft and the location of the centre of gravity, before, during and after flight. Knowing the range of movement of the CG as fuel is consumed during the flight enables us to avoid it moving dangerously out of limits. The weight and location of the CG affect stall speed and trim drag. Forward CG increases stability and control forces needed to manoeuvre – i.e to overcome the enhanced stability – while producing a lower rate of climb than when the CG is further aft, at the same weight. This is because more up-elevator is needed to trim the aircraft which, by increasing the download at a conventional tail, decreases the overall lift of wings-plus-tail. An aft-CG, on the other hand, makes an aeroplane twitchier and less stable, by shortening the moment arms of the tail surfaces about the CG. But the reduced trim drag enables it to fly further at a given airspeed, because it then requires less power and fuel to do so.

Centre of gravity

Arm, the moment arm of an item of weight: i.e. the distance between its CG and the reference CG-datum for calculating the CG of the aircraft.

Centre of Gravity, CG and *CG-datum*. The CG of an aircraft (or any item of equipment) is the point at which it would balance if suspended. Its distance from the reference datum (the CG-datum) is found by dividing the total moment by the total weight.

Figure 3.25 (including Equations (3–72) and (3–73) in Fig. 3.24) refreshes in simple terms why the CG is determined by multiplying weights by moment arms, so as to take moments about a datum – either a point or an axis – called the CG-datum. This datum is chosen to make calculations as easy as possible. It may lie right outside the system if that is the most convenient arrangement. In Fig. 3.25 the fulcrum is used as the datum, with moments taken to right and left. Or, knowing that the reacting force at the fulcrum is (W_1 + W_2), the datum could as easily have been at the weight, W_1, distance x_1 from the fulcrum – or at the distance of weight W_2 if preferred. It does not matter what datum is used, as long as we are consistent and define beforehand in which directions arm measurements are to be taken as positive and negative. The moments about a datum are then summed as follows:

$$W \bar{x} = W_1 x_1 \pm W_2 x_2 \pm W_3 x_3 \pm ... \pm W_n x_n \qquad (3\text{--}74)$$

in which $\bar{x}$ is the distance of the centre of gravity of the total weight, W, i.e. the sum of W_1 to W_n from the datum. When the CG is also the datum (see Fig. 3.23) $W \bar{x} = 0$.

In exactly the same way we may calculate a change in the moment arm, $\Delta \bar{x}$, caused by a change, ΔW_1, in a component weight located at station x_1:

$$W \Delta \bar{x} = \Delta W_1 x_n \qquad (3\text{--}75)$$

But when the weight W is on the limit and cannot be changed, the CG is in the wrong place and we have to adjust it by exchanging the weight of one item for another (e.g., an increase in weight of ballast countered by a decrease in the weight of fuel on board, while keeping both fixed at their stations x_1 and x_2), then:

$$\pm W_1 = \mp W_2$$

so that:

$$W\Delta \bar{x} = \Delta W_1 (x_1 - x_2)$$

and

$$\pm \Delta W_1 = W \Delta \bar{x} / (x_1 \mp x_2) \qquad (3\text{--}76)$$

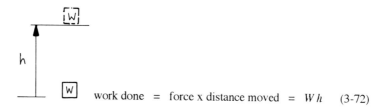

work done = force x distance moved = $W h$ (3-72)

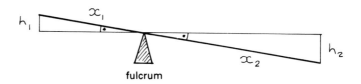

fulcrum

The beam is rigid, so that heights h_1 and h_2 at each
end are directly proportional to arms x_1 and x_2

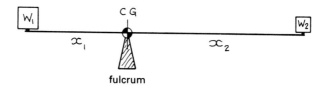

fulcrum

For equilibrium, work done by W_1 must equal work done
by W_2, which is directly proportional to moments:

$$W_1 \, x_1 \; = \; W_2 \, x_2 \qquad\qquad (3\text{-}73)$$

The fulcrum, the point of balance, is their centre of gravity

Fig. 3.25   Work done, moments and balance.

When it is necessary to move the CG by a shift of ballast, or by putting a crew-member in
another seat, distance Δx_n away from the original station, this is accomplished by manipulating the change of moment and transposing as required:

$$\Delta \bar{x} = (W_n / W) \, \Delta x_n \qquad\qquad (3\text{–}77)$$

These last two equations are most useful when it is necessary to move the CG between tests.

CG limits are the extreme locations of the CG within which the aircraft must be operated at
a given weight.

Note: Flight at weights and CG limits which are beyond those specified in the C of A or in
the Permit to Fly can invalidate the insurance in the event of an accident.

Weight definitions (see also Chapter 6)

AUW, W_0, the all-up (gross) weight is the maximum weight at which flight requirements must be met. It is substantially:

$$maximum\ take\text{-}off\ weight = \text{gross (all-up) weight} = \text{MTOW}$$
$$= \text{operating empty weight} + \text{disposable load} \qquad (3\text{--}78)$$

in which operating empty weight and disposable load are built up as follows:

$$operating\ empty\ weight = \text{basic empty weight} + \text{crew} \qquad (3\text{--}79)$$

$$basic\ empty\ weight = \text{standard empty weight} + \text{optional equipment} \qquad (3\text{--}80)$$

which also:

$$= APS\ weight \text{ of the aircraft when prepared for service:}$$
$$\text{i.e. a fully equipped operational aeroplane, without}$$
$$\text{crew, usable fuel, or payload.}$$

$$standard\ empty\ weight = \text{weight of the standard aircraft (as}$$
$$\text{manufactured)} + \text{unusable fuel} + \text{full operating}$$
$$\text{fluids} + \text{full engine oil} \qquad (3\text{--}81)$$

$$disposable\ load = \text{payload} + \text{useable fuel} (+ \text{any necessary ballast}) \qquad (3\text{--}82)$$

$$maximum\ ramp\ weight = \text{maximum weight approved for ground manoeuvre}$$
$$= \text{maximum take-off weight} + \text{start, taxi and run-up}$$
$$\text{fuel}$$
$$= \text{basic empty weight} + \text{useful load} \qquad (3\text{--}83)$$

$$maximum\ landing\ weight = \text{maximum weight approved for touchdown} \qquad (3\text{--}84)$$

$$maximum\ zero\ fuel\ weight = \text{the maximum weight approved} - \text{usable fuel} \qquad (3\text{--}85)$$

Tare weight is the weight of chocks, blocks, stands and other items of equipment included in the reading on the scale when the aircraft is weighed. Tare weight is deducted from the scale reading to obtain the actual (net) weight of the aeroplane.

Fuel capacity and weight conversions

Care must be taken when working in fuel capacities and weights to avoid confusion between US and Imperial gallons, litres, kilogrammes and pounds. Figure 3.26 shows two charts which have been useful for many years, but I cannot remember where they came from.

Tables and Other Conversion Factors

Tables 3–4 to 3–9 are reproduced from Ref. 1.5 to assist in handling a wide range of awkward quantities, and in finding one's way around.

The next step is to look at the relationship between the pilot and the aeroplane, when operating as a man–machine system into which a pilot who is mentally and physically *average* should fit.

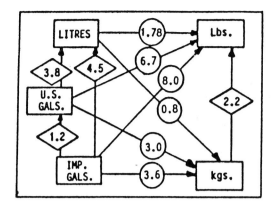

FOR JET 'A'
CALCULATIONS
TURBINE S.G. 0.8

• Follow the arrow and multiply.
• Backtract the arrow and divide.

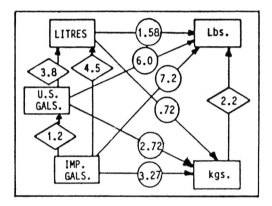

FOR AVGAS
CALCULATIONS
PISTON S.G. 0.72

• Follow the arrow and multiply.
• Backtract the arrow and divide.

Fig. 3.26 Fuel quantity and weight conversions (used for many years, but origin unknown).

TABLE 3–4

Greek letters used as symbols

Capital	Small	English name	Capital	Small	English name
A	α	Alpha	N	υ	Nu
B	β	Beta	Ξ	ξ	Xi
Γ	γ	Gamma	O	o	Omikron
Δ	δ	Delta	Π	π	Pi
E	ε	Epsilon	P	ρ	Rho
Z	ζ	Zeta	Σ	$\sigma \varsigma$	Sigma
H	η	Eta	T	τ	Tau
Θ	θ	Theta	Υ	υ	Upsilon
I	ι	Iota	Φ	ϕ	Phi
K	κ	Kappa	X	χ	Chi
Λ	λ	Lambda	Ψ	ψ	Psi
M	μ	Mu	Ω	ω	Omega

TABLE 3–5

Prefixes

The following prefixes may be used to indicate decimal fractions or multiples of basic or dervied metric SI units (*Système International d'Unités,* 1960) which employs the kilogramme, metre and second as the respective units of mass, length and time

Fraction	Prefix	Symbol	Multiple	Prefix	Symbol
10^{-1}	deci	d	10	deca	da
10^{-2}	centi	c	10^2	hecto	h
10^{-3}	milli	m	10^3	kilo	k
10^{-6}	micro	μ	10^6	mega	M

TABLE 3–6

Units in BCAR flight requirements

Quantity	Unit	Symbol	Remarks
Length	metre	m	Heights, altitudes and relatively
	foot	ft	short distances quoted in this unit.
Mass	kilogramme	kg	The term 'weight' has been retained
	pound	lb	when 'mass' is strictly correct. Equivalent existing unit not quoted where the weight represents a legal discriminant, e.g. 5700 kg is maximum AUW for aeroplanes certificated to Section K.
Time	second	s	
Temperature	degree Celsius	°C	The Kelvin scale is not used. Equivalent temperature in °F is
	degree Fahrenheit	°F	not quoted if it has always been the custom to use °C in the particular context.
Angle	degree	°	Radians are not used, even as equivalent.
Speed	kilometre per hour	km/h	Metres per second not used.
	knot	knot	The knot is a speed equal to one nautical mile per hour. The British nm is 6080 ft (1853.18 m). The International nm is 1852 exactly. Aviators tend to use 6000 ft as a crude rule of thumb. Where it is necessary to be more accurate, we shall use the International nm, for which 1 kn = 0.514 m/s (whereas 1 kn (UK) = 0.514773 m/s).
Force	newton	N	
	pounds-force	lbf	

continued

TABLE 3–6 – *continued*

Quantity	Unit	Symbol	Remarks
Rate of climb	metre per minute feet per minute	m/min ft/min	
Gradient of climb	per cent	%	Tangent of the climb angle expressed as a percentage.
Pressure (altimeter setting)	millibar	mbar	Newtons per square metre not used.

TABLE 3–7

Flight manual units

Quantity	Unit	Symbol	Remarks
Length (heights and altitudes)	foot	ft	Note conflict with BCAR primary unit.
Length (short horizontal, e.g., runway lengths)	metre	m	
Length (long horizontal)	nautical mile	–	
Length (depth, e.g., slush on runways)	millimetre inch	mm in	
Length (aircraft dimensions)	feet or inches	ft or in	
Mass	kilogramme pound	kg lb	The term 'weight' is commonly used. Equivalent not always quoted. Metric tons (tonne) not yet used but may be in the future.
Temperature	degree Celsius	°C	
Angle	degree	°	
Speed	knot	kn	
Rate of climb	feet per minute	ft/min	
Gradient of climb	per cent	%	'ft/100 m' is also used in certain contexts.
Pressure	millibars	mbar	

TABLE 3–8

Conversion factors between the British FPSR (foot, pound, second, rankine) system and SI (Système International) units

Quantity	FPSR units	Multiply by	To obtain SI units	Multiply by	To obtain FPSR units
Mass (M)	slug	1.459×10	kg	6.852×10^{-2}	slug
Length (L)	ft	3.048×10^{-1}	m	3.281	ft
Density (ρ)	slug/ft^3	5.155×10^{2}	kg/m^3	1.940×10^{-3}	slug/ft^3
Temperature (T)	°F + 460	5.56×10^{-1}	°C + 273	1.8	°F + 460
	°R		K		°R
Velocity (V)	ft/s	3.048×10^{-1}	m/s	3.281	ft/s
	mph	1.609	kph	6.214×10^{-1}	mph
	knot	1.853	kph	5.396×10^{-1}	knot
		0.515	m/s	1.942	
Force (F)	lbf	4.448	N (newton)	2.248×10^{-1}	lbf
	slug ft/s^2		kg m/s^2		slug ft/s^2
Work	slug ft^2/s^2	1.356	Nm	7.376×10^{-1}	slug ft^2/s^2
Energy (J)	BTU		(joule)		BTU
Power (W)	slug ft^2/s^3	1.356	Nm/s	7.376×10^{-1}	slug ft^2/s^3
	hp (550 ft lbf/s)	7.456×10^{2}	(watt)	1.341×10^{-3}	hp (550 ft lbf/s)
Pressure (p)	slug/ft s^2	4.788×10	N/m^2	2.088×10^{-2}	slug/ft s^2
	lbf/ft^2		(pascal)		lbf/ft^2
		4.788×10^{-4}	bar	2.088×10^{3}	
Specific energy, etc.	ft lbf/slug	9.290×10^{-2}	Nm/kg	1.076×10	ft lbf/slug
Gas constant	ft lbf/slug°R	1.672×10^{-1}	Nm/kg K	5.981	ft lbf/slug°R
Coef. of viscosity (μ)	slug/ft s	4.788×10	kg/m s	2.088×10^{-2}	slug/ft s
Kinematic viscosity (ν)	ft^2/s	9.290×10^{-2}	m^2/s	1.076×10	ft^2/s
Thermal conductivity (k)	lbf/s°R	8.007	N/s K	1.249×10^{-1}	lbf/s°R
Heat transfer coef.	lbf/ft s°R	2.627×10	N/m s K	3.807×10^{-2}	lbf/ft°R
Frequency	c/s	1.0	Hz (hertz)	1.0	c/s

Note: Various derived units are named after eminent scientists and engineers. In addition to the newton (Sir Isaac Newton), there are:

degree Celsius: Anders Celsius (1701–44), Swedish astronomer;

hertz: H. R. Hertz (1857–94), German physicist;

joule: James Prescott Joule (1818–89), English physicist;

kelvin: Lord William Thomson Kelvin (1824–1907), Scottish physicist;

pascal: Blaise Pascal (1623–62), French philosopher and scientist;

degree Rankine: William George Macquorn Rankine (1820–72), Scottish scientist and engineer:

watt: James Watt (1736–1819), Scottish engineer who invented the first really efficient steam engine.

TABLE 3–9

Conversion factors (mixed FPSR, metric and SI)

Multiply	By	To obtain
acres	0.4047	ha (= 10^4 m^2)
	43 560	ft^2
	0.0015625	miles^2
standard atmospheres (atm)	76	cmHg
	29.92	inHg
	1.01325	bar (= 10^5 N/m^2)
	1.033	kgf/cm^2
	14.70	lbf/in^2
	2116	lbf/ft^2
	101 325	N/m^2
bars (bar)	0.98692	atm
	14.5038	lbf/in^2
British thermal unit (Btu)	0.5556	CHU
	1055	J
	0.2520	kcal (kilocalorie)
centimetres (cm)	0.3937	in
	0.032808	ft
centimetres of mercury at 0°C (cmHg)	0.01316	atm
	0.3937	inHg
	0.1934	lbf/in^2
	27.85	lbf/ft^2
	135.95	kgf/m^2
centimetres per second (cm/s)	0.032808	ft/s
	1.9685	ft/min
	0.02237	mph
cubic centimetres (cm^3)	0.06102	in^3
	3.531×10^{-5}	ft^3
	0.001	litre
	2.642×10^{-4}	US gal
cubic feet (ft^3)	28 317	cm^3
	0.028317	m^3
	1728	in^3
	0.037037	yd^3
	7.481	US gal
	28.32	litre
cubic feet per minute (ft^3/min)	0.472	litre/s
	0.028317	m^3/min
cubic inches (in^3)	16.39	cm^3
	1.639×10^{-5}	m^3
	5.787×10^{-4}	ft^3
	0.5541	fl oz
	0.01639	litre
	4.329×10^{-3}	US gal
	0.01732	US qt
cubic metres (m^3)	61024	in^3
	1.308	yd^3
	35.3147	ft^3
	264.2	US gal
cubic metres per minute (m^3/min)	35.3147	ft^3/min

continued

TABLE 3–9 – *continued*

Multiply	By	To obtain
cubic yards (yd³)	27	ft³
	0.7646	m³
	202	US gal
degrees (arc)	0.01745	radians
degrees per second (deg/s)	0.01745	radians/s
erg	1.0×10^{-7}	J
feet (ft)	30.48	cm
	0.3048	m
	12	in
	0.33333	yd
	0.0606061	rod
	1.894×10^{-4}	stm
	1.646×10^{-4}	nm (international)
feet per minute (ft/min)	0.01136	mph
	0.01829	km/h
	0.508	cm/s
	0.00508	m/s
feet per second (ft/s)	0.6818	mph
	1.097	km/h
	30.48	cm/s
	0.5925	knot (international)
foot-pounds (ft lbf)	0.138255	kgf m
	3.24×10^{-4}	kcal
	1.356	N m (J)
foot-pounds per minute (ft lbf/min)	3.030×10^{-5}	hp
foot-pounds per second (ft lbf/s)	1.818×10^{-3}	hp
gallons, Imperial (Imp gal)	277.4	in³
	1.201	US gal
	4.546	litre
	153.707	fl oz
gallons, US dry (US gal dry)	268.8	in³
	1.556×10^{-1}	ft³
	1.164	US gal
	4.405	litre
gallons, US liquid (US gal)	231	in³
	0.1337	ft³
	4.951×10^{-3}	yd³
	3785.4	cm³
	3.785×10^{-3}	m³
	3.785	litre
	0.83267	Imp gal
	133.227	fl oz
US gallons per acre (gal/acre)	9.353	litre/ha
grammes (g)	0.001	kg
	2.205×10^{-3}	lb
grammes per centimetre (g/cm)	0.1	kg/m
	6.720×10^{-2}	lb/ft
	5.600×10^{-3}	lb/in
grammes per cubic centimetre (g/cm³)	1000	kg/m³
	0.03613	lb/in³
	62.43	lb/ft³
hectares (ha)	2.471	acres
	107 639	ft²
	10 000	m²

continued

T<small>ABLE</small> 3–9 – *continued*

Multiply	By	To obtain
horsepower (hp)	33000	ft lbf/min
	550	ft lbf/s
	0.7457	kW
	76.04	kgf m/s
	1.014	metric hp
	745.70	Nm/s (W)
horsepower, metric	75	kgf m/s
	0.9863	hp
inches (in)	25.40	mm
	2.540	cm
	0.0254	m
	0.08333	ft
	0.027777	yd
inches of mercury at 0°C	0.033421	atm
(inHg)	0.4912	lb/in^2
	70.73	lb/ft^2
	345.3	kg/m^2
	2.540	cmHg
	25.40	mmHg
	3.386×10^3	N/m^2
inch-pounds (in lbf)	0.011521	kgf m
J (joule)	0.27778×10^{-6}	kWh
	1	Nm
	1	Ws
kilogrammes (kg)	2.204623	lb
	35.27	oz avdp
	1000	g
kilogramme-calories (kcal)	3.9683	Btu
(kilocalories)	3088	ft lbf
	426.9	kgf m
kilogramme-metre^2 (kg m^2)	3417	lb in^2
	23.729	lb ft^2
	0.7376	slug ft^2
kilogrammes per cubic metre	0.06243	lb/ft^3
(kg/m^3)	0.001	g/cm^3
kilogrammes per hectare (kg/ha)	0.892	lb/acre
kilogrammes per square centimetre	0.9678	atm
(kg/cm^2)	28.96	inHg
	14.22	lbf/in^2
	2048	lbf/ft^2
kilogrammes per square metre	2.896×10^{-3}	inHg
(kg/m^2)	1.422×10^{-3}	lb/in^2
	0.2048	lb/ft^2
kilometres (km)	1×10^5	cm
	3280.8	ft
	0.6214	stm
	0.53996	nm (international)
kilometres per hour (kph)	0.9113	ft/s
	58.68	ft/min
	0.53996	knot (international)
	0.6214	mph
	0.27778	m/s
	16.67	m/min
kilowatts	1.34	hp

continued

TABLE 3–9 – *continued*

Multiply	By	To obtain
knots (knot) (international)	1	nm/h
	1.688	ft/s
	1.1508	mph
	1.852	km/h
	0.5144	m/s
litres (litre)	1000	cm^3
	61.02	in^3
	0.03531	ft^3
	33.814	fl oz
	0.2642	US gal
	0.2200	Imp gal
	1.0568	US qt
litres per hectare (litre/ha)	13.69	fl oz/acre
	0.107	US gal/acre
litres per second (litres/s)	2.12	ft^3/min
metres (m)	39.37	in
	3.280840	ft
	1.0936	yd
	0.198839	rod
	6.214×10^{-4}	stm
	5.3996×10^{-4}	nm (international)
metre-kilogrammes (kgf/m)	7.23301	ft lbf
	86.798	in lbf
metres per minute (m/min)	0.06	km/h
metres per second (m/s)	3.280840	ft/s
	196.8504	ft/min
	1.9427	knot (international)
	2.237	mph
	3.6	km/h
microns	3.937×10^{-5}	in
miles (stm)	5280	ft
	1.6093	km
	1609.3	m
	0.8690	nm (international)
miles per hour (mph)	44.704	cm/s
	0.4470	m/s
	1.467	ft/s
	88	ft/min
	1.6093	km/h
	0.8690	knot (international)
millibars	2.953×10^{-2}	inHg
	0.1	kN/m^2
millimetres (mm)	0.03937	in
millimetres of mercury at 0°C (mmHg)	0.03937	inHg
international nautical miles (nm)	6076	ft
	1.1508	stm
	1852	m
	1.852	km
newtons (N)	0.2248	lbf
ounces, fluid (fl oz)	8	dr fl
	29.57	cm^3
	1.805	in^3
	0.0296	litre
	0.0078	US gal

continued

Preparation

Table 3–9 – *continued*

Multiply	By	To obtain
ounces, fluid per acre (fl oz/acre)	0.073	litre/ha
pounds (lb): mass	453.6	g
	0.453592	kg
	3.108×10^{-2}	slug
pounds force (lbf)	4.4482	N
	0.45359	kgf
pounds-feet (lbf ft)	1.356	Nm
pounds-feet^2 (lb ft^2)	0.421	kg m^2
	144	lb in^2
	0.0311	slug ft^2
pounds per acre (lb/acre)	1.121	kg/ha
pounds per cubic foot (lb/ft^3)	16.02	kg/m^3
pounds per cubic inch (lb/in^3)	1728	lb/ft^3
	27.68	g/cm^3
pounds per hour per pound force (lb/h/lbf)	28.3	mg/N s
pounds per hour per horsepower (lb/h/hp)	169	μg/J
pounds-force per square foot (lbf/ft^2)	0.1414	inHg
	4.88243	kgf/m^2
	4.725×10^{-4}	atm
	0.048	kN/m^2
pounds per square inch (psi or lbf/in^2)	5.1715	cmHg
	2.036	inHg
	0.06805	atm
	0.0689476	bar
	703.1	kg/m^2
	6.89476	kN/m^2
quarts, US (qt)	0.94635	litre
	57.750	in^3
	3.342×10^{-2}	ft^3
radians	57.30	deg (arc)
	0.1592	rev
radians per second (radians/s)	57.296	deg/s
	0.1592	rev/s
	9.549	rpm
revolutions (rev)	6.283	radians
revolutions per minute (rpm or rev/min)	0.1047	radians/s
revolutions per second (rev/s)	6.283	radians/s
rods	16.5	ft
	5.5	yd
	5.0592	m
slugs	14.594	kg
	32.174	lb
slug feet^2 (slug ft^2)	1.3559	kg m^2
	4633.1	lb in^2
	32.174	lb ft^2
square centimetres (cm^2)	0.1550	in^2
	0.001076	ft^2
square feet (ft^2)	929.03	cm^2
	0.092903	m^2
	144	in^2
	0.1111	yd^2
	2.296×10^{-5}	acres

continued

TABLE 3–9 – *continued*

Multiply	By	To obtain
square inches (in²)	6.4516	cm²
	6.944×10^{-3}	ft²
square kilometres (km²)	0.3861	stm²
square metres (m²)	10.76391	ft²
	1.196	yd²
	0.0001	ha
square miles (miles²)	2.590	km²
	640	acres
square rods (rod²)	30.25	yd²
square yards (yd²)	0.8361	m²
	9	ft²
	0.0330579	rod²
tons	2240	lb
	1016	kg
	1.016	t (tonne)
ton-force (tonf)	9.964×10^{3}	N
tons per square foot (tonf/ft²)	107.252×10^{3}	kN/m²
tonnes	2204.62	lb
	1.10231	short ton (2000 lb)
	0.984207	ton
watts (W)	1.34×10^{-3}	hp
	10^{-3}	kW
yards (yd)	0.9144	m
	3	ft
	36	in
	0.181818	rod

References and Bibliography

3.1 CAP 393 *Air Navigation: The Order and Regulations*. London: Civil Aviation Authority, 1991

3.2 Lachman, G. V., ed., *Boundary Layer and Flow Control*. Oxford: Pergamon Press, 1961

3.3 Quill, J. K., *Spitfire, a Test Pilot's Story*. London: John Murray, 1983

3.4 Bryan, G. H. and Williams, W. E., The longitudinal stability of aerial gliders. *Proceedings of the Royal Society*, Vol. 73, 1903

3.5 Gibbs-Smith, C. H., *Sir George Cayley's Aeronautics 1796–1855*. London: a Science Museum Handbook, 1962

3.6 *ICAO Document 7488/2*. International Civil Aviation Organisation, Document Sales Unit, 1000 Sherbrooke Street West, Suite 400, Montreal, Quebec Canada H3A 2R2

3.7 Swatton, P. J., *The Operator's Handbook of Scheduled Aircraft Performance*. Author's draft, c1987

CHAPTER 4

The Man–Machine Interface

'Do not let yourself be forced into anything before you are ready.'
Letter from Wilbur to Orville Wright, August 1908

The classification, *man*, describing the human species applies equally to both sexes. Both suffer the same professional stresses in the same working environment, and there is not a great deal of difference between men and women in the ways that they respond. It has been observed that men are physically stronger than women (a point forgotten by designers of some aeroplanes flown by women), but that they 'break' more easily when stressed emotionally. Women survive longer. Both are, fundamentally, creatures designed for a two-dimensional existence, who will insist on exploring in three dimensions. For this, special equipment, training and other facilities are needed.

Pilots are as prone to making mistakes as anyone else. The circumstances under which they operate conspire to overload them at times, and we must look first in a mirror at this average, malleable, often under-confident and over-anxious mistake-maker which the test pilot and engineer has to protect from himself and others like him. It is not easy to deal rationally at all times with the problem, being like carving a lump of butter with a knife made of butter.

Stress and human limitations

In the early 1960s I was employed to use a method I had devised much earlier (out of interest as an Instrument Rating Examiner on a night all-weather squadron) in a test programme run by the Royal Air Force Institute of Aviation Medicine (IAM) at Farnborough. It involved testing a number of pilots who had reported disorientation (a loss of spatial reference) in flight.

Disorientation can follow swiftly upon the heels of a quite simple mistake. I developed the method out of curiosity, to investigate mistake-making in a practical and rudimentary way among experienced night fighter pilots (including myself) under artificial (procedural) instrument rating tests (IRTs) during the late 1950s (Ref. 4.1). Some pilots were more prone to making mistakes than others under the same test conditions – and there was no clear correlation that one could see between training, experience or type of man.

The IAM disorientation tests were carried out with pilots described as being in a high risk group. The methods outlined were broadly the same as those developed for the IRT and Ref. 4.2, which includes later work and summarizes a number of findings.

In Chapter 2 we noted that in accidents and incidents one detects a multiplicity of *predis-posing* and usually one *precipitating* factor (which is often nothing more than the trigger which sets in motion a train of events). In flying accidents roughly two-thirds are found to have been caused directly by human error; about two-thirds of which occur on take-off and landing, with most on landing. Thus four-ninths, or 44%, of pilot error accidents occur in the launch and recovery phases of flight. Although this is an old rule-of-thumb, it is still clearly borne out by a finding in an article by Learmount, on *Human Factors* in world airline safety in 1992 (Ref. 4.3), in which 21% of the accidents were on take-off and 23% on landing: a total of 44%.

The most significant cause of accidents is the stall. Stall quality is the keystone in the arch of air-safety. The airspeeds specified for demonstrating compliance with airworthiness requirements are based upon the stall speeds in the relevant configurations, multiplied by factors like: 1.2, 1.3 and 1.4, to produce 1.2 V_{S1}, 1.3 V_{S0}, 1.4 V_{S0}.

Most everyday mistakes are small, like putting on a switch or carrying out a check in the wrong order, letting fuel get somewhat out of balance, fumbling a radio transmission. Usually you '*catch yourself on*' (as an Irish flying instructor would urge a slow student), sorting it out before it becomes significant. But if the mistake is not small in scale, then the conse-quences might rapidly get out of hand.

For sound practical reasons the test pilot accepts that mistakes will be made, approaching an aircraft, and the cockpit/flight-deck layout in particular, with this in mind. It is for this reason that audio stall warning, stick-shakers and pushers, warning-lights and blinking doll's-eyes exist. The Certification Test Pilot looks for dangerous features, on the principle of 'Murphy's Law', that: '*If something can be assembled or done wrongly, then it will be*'; and 'Sod's Law', that: '*If something goes wrong, it will rapidly get worse.*' Both of these are real. They are aspects of the 'General Law of Selective Gravity', one version of which is that '*Buttered toast hits the floor butter-side downwards*'!

So, at one extreme we take into account human ergonomics and the extreme range of dimensions of those who fly, male and female: size, reach of arms and legs, and of course weight. At the other, we assume that pilot-error is inevitable and do everything possible to reduce, or neutralize its effects. Aircraft should be designed for average pilots of average skill and ability. Not only must crews be given enough to do to keep them awake and active; they must be kept sharp and skilled enough to cope with the emergency should a fuse blow (as it will).

In this last respect, technology which produces public transport aircraft designed with high-order flight control systems and full automation from take-off to touchdown, with computers and no manual-reversion between a minimum flight-deck crew and the flying control surfaces, might well produce fewer accidents and incidents in future; but opera-tional experience with existing automated aircraft already points towards a potential for catastrophe which, though less frequent, will be larger in scale when it does occur. The reason is that computer software cannot discriminate, make judgements and weigh odds as effectively as can the deficient human being. Without manual-reversion which enables the pilot(s) to override and command the flying and other controls, crew and passengers could find themselves in the hands of a piece of mentally-handicapped machinery which lacks the gift of foresight.

For many reasons, military strike and specialized reconnaissance aircraft are a different matter.

When stressed we might become mentally overloaded, whereas on the previous day, with the same workload, we were not. Overloading is either self-induced, or is caused by cir-cumstances around us: the actions both direct and indirect of others. Usually it derives from a mixture of causes. A healthy amount of stress keeps us on our toes. Too much kills, and from physical side effects as well as directly.

Table 4–1 lists stress factors found to affect people adversely. Table 4–2 shows effects of life-style changes. Both are from US Navy sources (Ref. 4.2, originating in Refs 4.8 and 4.9). What they show is that many factors are capable of causing significant risk of accident, if not illness. For example, it was found that scores between 150 and 199 points could lead to a 37% probability of health being affected within two years. Over 300 points could result in accident, injury or illness among 4/5 ths of those investigated. High and low risk groups were revealed.

Figure 4.1 is a typical 'arousal curve', modified by varying workload and task, which shows changes in pilot efficiency with stress. No two curves are ever identical. They change with time, and they are different for every individual. When we have an 'off-day' we can be vastly less efficient than we should be, all else being equal. There are studies of biorhythms which seek to explain variations for which obvious causes are less apparent.

Curves like those shown are common in all walks of life. The left-hand side of the curve is the shape associated with low arousal, when there is too little stimulation. Stimulation should not be confused with either physical exertion or with discomfort. One significant danger of automation in the cockpit is understimulation. Alright, the workload and the number of flight deck crew can often be reduced, but it is also human to doze off betimes, especially when cruising on long-haul legs with nothing much going on. Then there is the risk of the flight-deck crew operating on the backside of the arousal curve.

Figure 4.2 shows what happens in general to someone with the same job, but with adversely changed circumstances. Superimposed is the efficiency demanded by, say, an aircraft which is suffering various failures. Here pilot efficiency, $\eta = \dot{I}/\dot{O}$, with a dot over each letter, is the ratio of the *rate* at which information, I, fed to the pilot is used and translated by him into a *rate* of output, O, needed to operate the aircraft and stay ahead of its increasing demands. Handling mistakes are rarely single. They become increasingly cumulative. Furthermore, under increasing stress one notices a tendency to confuse left with right: hands, feet, direction in which to look out when told that another aircraft is close, direction of turn, which engine to shut down or propeller to feather.

Symptoms of Pilot Stress Under Test Conditions

Symptoms of increasing stress, revealed during instrument flying tests in Ref. 4.1 were progressive. They are not confined to instrument flying alone:

☐ *Poor scanning* of the array of instruments. This is the most common failing, which shows as inability of the pilot to maintain completely a predetermined flight path all of the time.

☐ *Roughness* is of two kinds, depending upon the degree of overcontrolling by the pilot and inability to remain calm or to settle down. The distinction arose out of the circumstances of Ref. 4.1, for which tests had been carried out in the tandem two-seat Meteor T7 trainer. The Instrument Rating Examiner sat in the front, with the pilot being tested out of sight in the back. By placing an index finger lightly on top of the stick, and using the seat of the pants and his instruments, the IRE (who was not flying blind) could correlate between the flight path and what the pilot behind him was doing.

(1) *Roughness on the controls*, felt by the examining pilot, was the term used to distinguish small and rapid stick movements as the pilot being tested hunted for feedback of changes in control force. It showed that, although under some stress, the pilot was managing to cope well enough, because the ride was smooth.

TABLE 4–1

Stress factors
(US Navy study: Refs 4.2, 4.8 and 4.9)

Rank	Life event	Mean value	Rank	Life event	Mean value
1	Death of a spouse	100	23	Son or daughter leaving home	29
2	Divorce	73	24	Trouble with in-laws	29
3	Marital separation	65	25	Outstanding personal achievement	28
4	Jail term	63	26	Wife begins or stops work	26
5	Death of a close family member	63	27	Begin or end school	26
6	Personal injury or illness	53	28	Change in living conditions	25
7	Marriage	50	29	Revision of personal habits	24
8	Fired at work	47	30	Trouble with boss	23
9	Marital reconciliation	45	31	Change in work hours, conditions	20
10	Retirement	45	32	Change in residence	20
11	Changes in family member's health	44	33	Change in schools	20
12	Pregnancy	40	34	Change in recreation	19
13	Sex difficulties	39	35	Change in church activities	19
14	Gain of new family member	39	36	Change in social activities	18
15	Business readjustment	39	37	Mortgage or loan under $10 000	17
16	Change in financial state	38	38	Change in sleeping habits	16
17	Death of a close friend	37	39	Change in number of family get togethers	15
18	Change to different line of work	36	40	Change in eating habits	15
19	Change in number of arguments with spouse	35	41	Vacation	13
20	Mortgage over $10 000	31	42	Christmas	12
21	Foreclosure of mortgage or loan	30	43	Minor violations of the law	11
22	Change in work responsibilities	29			

TABLE 4–2

Lifestyle changes during deployment
(US Navy Study: Refs 4.2, 4.8 and 4.9)

Marital Separation	65
Change in responsibility at work	29
Change in living conditions	25
Revision of personal habits	24
Change in working hours or conditions	20
Change in residence	20
Change in recreation	19
Change in social activities	18
Change in sleeping habits	16
Change in eating habits	13
Total	249

(2) *Roughness of ride* revealed that the task was getting harder and the pilot was out of phase with the responses of the aircraft. Motion is akin to that described as pilot-induced oscillations, which result in the pilot becoming further stressed as he fights to maintain his (increasingly imagined) flight path. Control movements coarsened

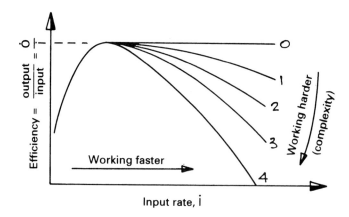

Fig. 4.1 Efficiency of the mind when carrying out a task.

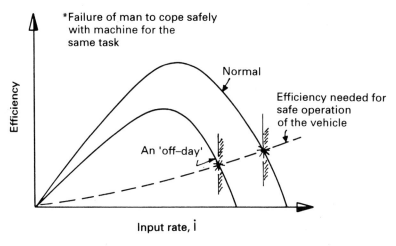

Fig. 4.2 Same task, same person, but different circumstances arising from change in life events (Ref. 4.2).

as he lagged further and further behind the response of the aircraft. The pilot showed signs of 'losing the place' as he ceased to think as well as he should, because:

- ☐ *Sluggish forethought and anticipation* leads to poor coordination of aerodynamic and power control movements needed to adjust and smooth the flight path.
- ☐ *Fatigue*, or symptoms very like it, began to set in. Only by observation of attitudes over a period of time could one distinguish between the pilot who was becoming flustered and lacked stamina, another who was genuinely fatigued (through having insufficient rest), and yet another one who was simply lazy, or bored, and having an 'off-day'. Snap judgements can be faulty.

Scott Crossfield, a former NACA test pilot (Chapter 2, Example 2–5) did not believe in pilot fatigue. At the fourth General Assembly of the Advisory group for Aeronautical Research and Development (AGARD) of the North Atlantic Treaty Organization (NATO), early in 1954, he was reported as saying:

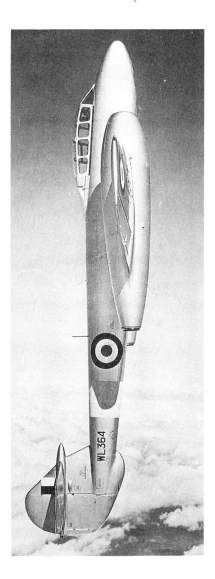

Plate 4–1   The Gloster Meteor T7, a superb trainer. There are (Flettner) heavying strips on the rudder trim tab trailing edges. (Courtesy of the former Gloster Aircraft Company)

'I am still a young man and I have little truck with the idea of pilot fatigue for a sedentary worker ....

'Most of the problems that arise that I observed in my service in the United States Navy and since then, that were attributed to fatigue were, to my mind, largely a result of boredom. They had taken so many operations away from the pilot in the modern airplane that he didn't have anything to do unless he was fighting for his life or doing some specific manoeuvres gathering data or some specific tactic. But any time he wanted to go from here to there, or was flying home, he went to sleep on the job, as many of us do when we are sitting.'   (Ref. 4.4)

What Crossfield had to say nearly 40 years ago has an echo in the modern advanced cockpit, designed to reduce pilot workload to a minimum. It does not apply to combat fatigue, or the fatigue experienced by a pilot or crew struggling with one or more failures of engines and equipment.

☐ *Disorientation*, as revealed by increasing loss of reference in space.

☐ *Trial and error activity*. Under intense stress, a well learned recent skill degenerated into an earlier less skilled form. One result is that a pilot, unable to think clearly and effectively, resorted to a procedure which no longer applied. Control movements became rapid and bore little or no relation to those required. At this stage it was easy to get the hands and/or feet mixed up. This is often apparent when an out of practice pilot is spinning, or has suffered an engine failure unexpectedly, and cannot decide which way to push the rudder pedals, and may also feather or shut down the wrong engine.

Note 1: I have also had a civilian flying instructor, demonstrating to me spin-entry-and-recovery procedure under test conditions (during exploratory tests following a fatal accident in another aircraft of the same type) run accurately through the 'patter' quite unaware that his hands and feet were not doing as he said.

Note 2: The modern flight simulator (which hardly existed at the time the tests recounted here were being run) is useful for keeping pilots procedurally sharp.

☐ *Capitulation*. Fortunately this was rare. It marked the moment of truth when the pilot, beaten by the aeroplane, gave up.

Note: Sometimes a pilot, unable to cope and in a panic has locked on to the controls and commited himself to the charity of his Maker. This is a dangerous situation for the other pilot. Always be prepared for it.

A flying instructor friend of mine, having a student freeze on the controls in the front seat of a Gloster Meteor 7 in a spin, '*unfroze him*' by firing the contents of one of the two fire extinguishers at the back of his neck.

In another case, a fatal spinning accident, there was evidence of the large and strong student having locked on to the control column with it pulled fully back (which has led to a high rotational spin in that type of aircraft). His right hand showed none of the expected crash damage, but was extensively bruised – apparently caused by the small instructor, sitting on his right, attempting to make the student release his grip.

Human vulnerability is something that must be accepted and defended against. Never be taken in by the pilot who says that he is not afraid – or is contemptuous of an aircraft (as I have had said to me by one instructor who wished to assert his competence and superiority). The best is to remain silent.

There are a number of techniques and devices by means of which we may equip ourselves, and a number of basic rules, based upon experience, which apply when carrying out a first flight in an unproven or previously unexplored configuration.

Approach to the game

☐ *Take small steps*. Approach no part of a test so vigorously that you cannot move backwards out of danger. Test flying consists of taking cautious and tiny steps (like putting a toe into possibly hot water, so as to take a full step backwards if required).

☐ *Boldness* is not a virtue.

☐ *Acquire a sympathy for machinery*.

☐ *Listen to your aircraft*. Rather like a rather boring acquaintance that wants to talk about itself it will chatter away to all of your senses about its ailments. Have the wit to pay heed, using:

> *sight, sound, touch* (including the seat of the pants), *taste* and *smell*.

☐ *Never plan or try to do too much.*

☐ *The only criterion of truth is the word of the pilot who has done it, or tried to do it.* Here is advice of David P. Davies in a lecture to other test pilots:

> 'Don't believe other people, prove it for yourself.
> Stick to what you have proved believable.
> Don't be overawed by other more senior people.
> Don't ignore the feelings in your bones.'

☐ *Never go beyond gliding distance* from an airfield on a first flight, particularly when carrying out stalling, in case the engine stops (as it can do if idling RPM are too low).

☐ *Never use the aircraft to show anyone how good a pilot you think you are.* If you feel the need to do so, then you are not that good.

Cockpits and Vital Statistics

Figure 4.3 shows some measurements needed in the ergonomic design of cockpits and crew-stations. The letters of the alphabet refer to Table 4–3A–L. It is essential to know your weight and, if possible, some of your own basic measurements in test flying (e.g., sitting heights of eye and top of the head, length of leg and reach when strapped in) so that meaningful assessments can be made. We may think that we are of average height, average build, average leg length, but what do we really mean? Of those listed above, which are average? Each is of crucial importance if you complain about some feature in the cockpit which causes you discomfort. It could cost the manufacturer a lot of money to rectify matters, if it is you (like me) and not the aircraft that is non-standard in certain respects.

Note: One measurement which appears to be lacking from Table 4–3 is the downwards stretch of the arm from the shoulder to the tips of the fingers when seated.

Having said that, and finding that maybe your reach or length of leg is on the short side for your height, do not be dismayed. The careful designer builds in tolerances. For example, Figs 4.4a,b, and 4.5a,b (Ref. 1.5) show cockpit dimensions, tolerances and proportions which have been found to give reasonable comfort. Some pilots will be uncomfortable because they are cramped. Others might need a cushion at the back, or on the seat to provide additional adjustment.

Anthropometrics and percentile tables

A *percentile* is a mathematical device used in statistics, to analyse groups of things, such as people. The method lends itself to the anthropometric measurement of aircrew for cockpit workspace and functional flying clothing. Results are summarized in the form of *percentile tables*.

Tables 4–3(1) to (12) are derived from Ref. 4.5 and show the range of significant measurements A to R at the head of the table. The range of each measurement is from 1 to 99. The same number of aircrew was used for each, and the resulting distribution is shown in each column. The *mean value, standard deviation, coefficient of variation, range of results*

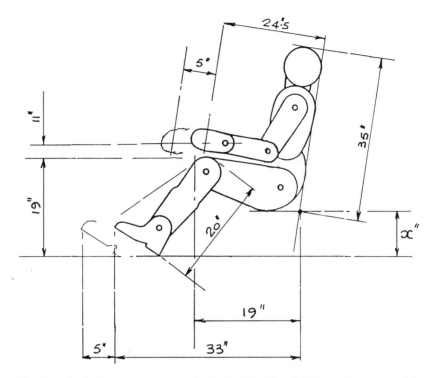

Fig. 4.3   Sketch of pilot measurements in the De Havilland Chipmunk, extracted from an ETPS Preview report. The units (pre-SI) are inches.

and *number of subjects* measured are given at the foot of each column. These terms are defined in Chapter 6, under the heading of *Handling scatter*, Equations (6–13) to (6–14a).

Of course, one must be careful when using percentiles to specify the particular element to which it refers. A pilot with an 18 percentile sitting height might easily have a 66 percentile reach and a 73 percentile buttock to heel length.

Example 4–1: Modified Spitfire Mk IX, Two-Seat Configuration

The usefulness of knowing one's own percentile measurements was made clear when testing a highly modified civilian Spitfire Mk IX, for use in airshows and training pilots. It had been converted by eye, to resemble a two-seat training version built for the Russians in World War II and was proportioned to fit the engineer carrying out the modification.

The normal cockpit had been moved forwards and a second one, with dual controls, built behind. This necessitated rearrangement of several items of controls and equipment. Landing-gear warning lights were now on the coaming of the rear cockpit, instead of being below, on the instrument panel. A standard Spitfire gear-selector was fitted to the cockpit structure, to the right of the rear seat – its location having been decided by the engineer, sitting in the seat and making a trial installation. What we did not know was that although the engineer and I were about the same height, my back was much longer, and my legs and functional reach were much shorter. The latter affected my downwards reach when strapped in tightly (see the **Note** above, on a measurement missing from Table 4–3).

TABLE 4-3 (Ref. 4.5)

A Bideltoid breadth   (7)
B Biacromial breadth   (-)
C Hip breadth, sitting   (8)
D Stool height   (4)
E Thigh clearance height   (-)
F Acromial height, sitting   (-)

G Shoulder height, sitting   (-)
H Sitting height, eye   (6)
J Sitting height   (5)
K Vertical functional reach, sitting   (-)
L Knee height, sitting   (12)
M Functional reach   (9)

N Cervical height, sitting   (-)
O Elbow rest height, sitting   (-)
P Stomach depth   (-)
Q Buttock - knee length   (11)
R Buttock - heel length   (10)

continued

TABLE 4–3 (Ref. 4.5) – continued

(1) Age		(2) Weight Standing on spring scale			(3) Stature			(4) Stool height D		
Percentile values %	Year	Percentile values %	kg	lb	Percentile values %	mm	in	Percentile values %	mm	in
1	19.89	1	55.75	122.90	1	1638.0	64.49	1	366.3	14.42
2	20.25	2	58.38	128.71	2	1651.6	65.02	2	371.6	14.63
3	20.57	3	59.46	131.08	3	1660.5	65.37	3	376.3	14.82
5	21.15	5	61.41	135.38	5	1672.7	65.85	5	382.6	15.06
10	22.52	10	63.90	140.88	10	1693.7	66.68	10	391.6	15.42
15	23.61	15	65.87	145.21	15	1708.6	67.27	15	398.9	15.71
20	24.55	20	67.43	148.65	20	1721.1	67.76	20	402.6	15.85
25	25.60	25	68.80	151.68	25	1732.4	68.20	25	407.2	16.03
30	26.41	30	69.96	154.23	30	1741.3	68.56	30	410.4	16.16
35	27.07	35	70.99	156.51	35	1750.8	68.93	35	414.2	16.31
40	27.82	40	71.99	158.72	40	1758.4	69.23	40	417.4	16.43
45	28.70	45	73.34	161.68	45	1766.0	69.53	45	420.0	16.54
50	29.79	50	74.46	164.15	50	1774.8	69.87	50	423.3	16.66
55	31.06	55	75.75	166.99	55	1781.6	70.14	55	427.1	16.81
60	32.35	60	76.90	169.54	60	1789.2	70.44	60	429.9	16.92
65	33.68	65	78.27	172.56	65	1796.4	70.72	65	433.2	17.06
70	35.09	70	79.44	175.13	70	1805.6	71.09	70	436.5	17.19
75	36.30	75	80.94	178.45	75	1814.2	71.43	75	439.9	17.32
80	37.36	80	82.44	181.75	80	1824.4	71.83	80	444.2	17.49
85	38.31	85	84.32	185.88	85	1838.3	72.37	85	449.2	17.68
90	39.75	90	86.37	190.41	90	1854.3	73.01	90	456.2	17.96
95	41.59	95	90.01	198.45	95	1879.3	73.99	95	466.8	18.38
97	42.90	97	92.42	203.75	97	1892.8	74.52	97	469.0	18.46
98	43.70	98	93.88	206.97	98	1905.0	75.00	98	469.9	18.50
99	44.63	99	96.50	212.75	99	1924.0	75.75	99	470.9	18.54

Mean: 30.76 yr
Standard deviation: 6.49 yr
Coefficient of variation: 21.09%
Range: 18.67–45.92 yr
Number of subjects: 1999

Mean: 75.04 kg; 165.43 lb
Standard deviation: 8.81 kg; 19.42 lb
Coefficient of variation: 11.74%
Range: 51.00–109.00 kg; 112.44–240.30 lb
Number of subjects: 1998

Mean: 1774.4 mm; 69.86 in
Standard deviation: 62.3 mm; 2.45 in
Coefficient of variation: 3.51%
Range: 1514.0–2009.0 mm; 59.61–79.09 in
Number of subjects: 2000
Check measure deviation: 2.3 mm; 0.1%

Mean: 423.9 mm; 16.69 in
Standard deviation: 24.3 mm; 0.96 in
Coefficient of variation: 5.74%
Range: 333.0–473.0 mm; 13.11–18.62 in
Number of subjects: 2000

continued

TABLE 4–3 (Ref. 4.5) – *continued*

(5) Sitting height J			(6) Sitting height, eye H			(7) Bideltoid breadth A			(8) Hip breadth, sitting C		
Percentile values			Percentile values			Percentile values			Percentile values		
%	mm	in	%	mm	in	%	mm	in	%	mm	in
1	864.7	34.04	1	750.8	29.56	1	418.7	16.49	1	323.7	12.74
2	871.4	34.31	2	759.4	29.90	2	423.5	16.67	2	328.3	12.93
3	876.2	34.50	3	764.8	30.11	3	426.8	16.80	3	332.1	13.08
5	883.4	34.78	5	772.1	30.40	5	431.7	17.00	5	337.1	13.27
10	895.3	35.25	10	783.0	30.83	10	439.3	17.30	10	343.5	13.52
15	903.0	35.55	15	790.1	31.11	15	443.8	17.47	15	348.0	13.70
20	909.1	35.79	20	796.4	31.35	20	447.8	17.63	20	351.4	13.83
25	914.6	36.01	25	802.9	31.61	25	451.4	17.77	25	354.3	13.95
30	919.6	36.20	30	806.8	31.76	30	453.8	17.87	30	356.8	14.05
35	924.7	36.41	35	811.3	31.94	35	457.1	18.00	35	359.5	14.15
40	928.8	36.57	40	815.7	32.11	40	459.6	18.10	40	362.5	14.27
45	932.6	36.72	45	819.5	32.27	45	461.9	18.19	45	364.8	14.36
50	936.2	36.86	50	823.4	32.42	50	464.8	18.30	50	367.0	14.45
55	939.5	36.99	55	827.4	32.58	55	467.5	18.40	55	369.6	14.55
60	943.7	37.15	60	831.2	32.72	60	470.2	18.51	60	371.8	14.64
65	947.4	37.30	65	834.7	32.86	65	473.7	18.65	65	374.5	14.74
70	951.7	37.47	70	839.0	33.03	70	476.3	18.75	70	377.6	14.87
75	957.3	37.69	75	844.1	33.23	75	479.3	18.87	75	381.3	15.01
80	962.2	37.88	80	849.3	33.44	80	482.8	19.01	80	383.9	15.11
85	967.8	38.10	85	854.9	33.66	85	486.7	19.16	85	388.1	15.28
90	973.9	38.34	90	862.1	33.94	90	491.9	19.37	90	393.2	15.48
95	986.1	38.82	95	872.4	34.35	95	499.7	19.67	95	400.3	15.76
97	992.3	39.07	97	881.0	34.69	97	505.4	19.90	97	406.1	15.99
98	998.0	39.29	98	887.8	34.95	98	509.8	20.07	98	409.5	16.12
99	1007.0	39.65	99	895.5	35.26	99	513.7	20.22	99	414.7	16.33

Mean: 936.0 mm; 36.85 in
Standard deviation: 31.0 mm; 1.22 in
Coefficient of variation: 3.31%
Range: 824.0–1026.0 mm; 32.44–40.39 in
Number of subjects: 2000
Check measure deviation: 6.4 mm; 0.7%

Mean: 823.5 mm; 32.42 in
Standard deviation: 30.8 mm; 1.21 in
Coefficient of variation: 3.74%
Range: 727.0–919.0 mm; 28.62–36.18 in
Number of subjects: 2000

Mean: 465.8 mm; 18.34 in
Standard deviation: 20.8 mm; 0.82 in
Coefficient of variation: 4.47%
Range: 396.0–547.0 mm; 15.59–21.54 in
Number of subjects: 1993
Check measure deviation: 6.6 mm; 1.4%

Mean: 368.3 mm; 14.50 in
Standard deviation: 19.5 mm; 0.77 in.
Coefficient of variation: 5.29%
Range: 310.0–436.0 mm; 12.20–17.17 in
Number of subjects: 2000
Check measure deviation: 1.9 mm; 0.5%

continued

TABLE 4–3 (Ref. 4.5) – *continued*

	(9) Functional reach M		(10) Buttock–Heel length R		(11) Buttock–Knee length Q		(12) Knee-height, sitting L	
Percentile values %	mm	in	mm	in	mm	in	mm	in
1	722.0	28.42	974.5	38.37	549.6	21.64	505.3	19.89
2	729.7	28.73	989.6	38.96	554.6	21.83	511.3	20.13
3	735.6	28.96	997.9	39.29	557.9	21.96	513.9	20.23
5	744.6	29.31	1007.1	39.65	563.9	22.20	518.5	20.41
10	756.6	29.79	1021.9	40.23	573.4	22.58	525.9	20.71
15	763.5	30.06	1035.0	40.75	579.2	22.80	531.5	20.92
20	770.2	30.32	1045.2	41.15	583.5	22.97	536.0	21.10
25	776.6	30.58	1054.1	41.50	588.0	23.15	541.1	21.30
30	782.0	30.79	1062.1	41.82	592.4	23.32	545.3	21.47
35	787.3	31.00	1069.5	42.11	596.6	23.49	549.0	21.61
40	791.5	31.16	1075.8	42.36	599.9	23.62	551.9	21.73
45	795.7	31.33	1082.5	42.62	603.3	23.75	555.2	21.86
50	800.1	31.50	1087.9	42.83	606.5	23.88	557.5	21.95
55	804.7	31.68	1094.8	43.10	609.7	24.00	560.2	22.05
60	810.0	31.89	1100.5	43.33	613.6	24.16	563.6	22.19
65	815.2	32.09	1109.1	43.66	617.4	24.31	567.2	22.33
70	820.1	32.29	1115.8	43.93	621.3	24.46	571.1	22.48
75	825.1	32.48	1125.3	44.30	625.7	24.63	574.9	22.64
80	830.5	32.70	1134.9	44.68	629.9	24.80	578.9	22.79
85	837.5	32.97	1143.8	45.03	634.7	24.99	583.8	22.98
90	845.9	33.30	1155.2	45.48	641.4	25.25	590.6	23.25
95	859.2	33.83	1173.1	46.19	652.4	25.69	602.0	23.70
97	871.1	34.30	1188.1	46.78	658.5	25.93	610.0	24.02
98	878.0	34.57	1201.1	47.29	663.8	26.13	615.7	24.24
99	889.4	35.01	1210.7	47.66	671.7	26.45	622.5	24.51

(9) Functional reach M
Mean: 801.7 mm; 31.56 in.
Standard deviation: 35.8 mm; 1.41 in
Coefficient of variation: 4.46%
Range: 678.0–946.0 mm; 26.69–37.24 in
Number of subjects: 1996
Check measure deviation: 8.7 mm; 1.1%

(10) Buttock–Heel length R
Mean: 1089.9 mm; 42.91 in
Standard deviation: 51.4 mm; 2.02 in
Coefficient of variation: 4.71%
Range: 889.0–1276.0 mm; 35.00–50.24 in
Number of subjects: 1993
Check measure deviation: 12.0 mm; 1.1%

(11) Buttock–Knee length Q
Mean: 607.6 mm; 23.92 in
Standard deviation: 26.9 mm; 1.06 in
Coefficient of variation: 4.42%
Range: 515.0–693.0 mm; 20.28–27.28 in
Number of subjects: 2000
Check measure deviation: 7.4 mm; 1.2%

(12) Knee-height, sitting L
Mean: 558.9 mm; 22.00 in
Standard deviation: 25.4 mm; 1.00 in
Coefficient of variation: 4.54%
Range: 453.0–662.0 mm; 17.83–26.06 in
Number of subjects: 2000
Check measure deviation: 3.5 mm; 0.6%

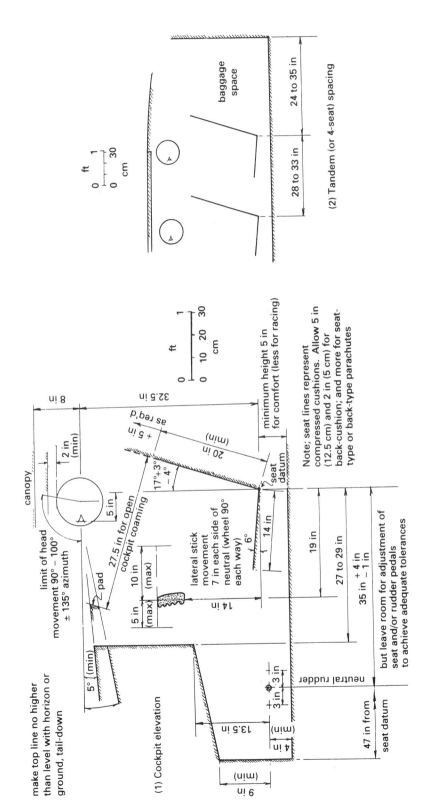

Fig. 4.4a Cockpit spacing for comfort (Ref. 1.5).

Surrounding cockpit structure should be strong enough to resist penetrating the enclosed volume shown below. The floor and supporting structure should be able to resist deformation which, in turn, distorts the seats, causing injury

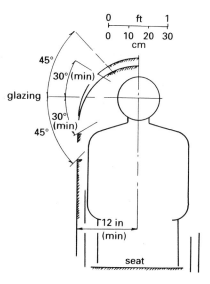

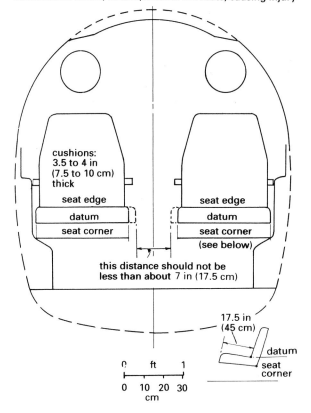

(4) Extent of glazing and cockpit sills to enable the pilot to look out sideways in 30 and 40 degree banked turns. It is essential to have an unobstructed view inwards, above the plane of the turn with an angle of bank of at least 30 degrees

(3) Generalized inboard profile of cabin or cockpit section of light business-executive aeroplane

Fig. 4.4b Cockpit sections for minimum comfort.

The gear-selector had a range of unrestricted movement, with final UP and DOWN-lock accomplished by pulling and pushing the lever beyond a spring-loaded ball which acted as a detent, at each end of the range. Because every two-seat Spitfire on the British Register had previously landed wheels-up, the engineer had sensibly provided each ball with a stiffer spring.

The rear cockpit had not the later mark of blister canopy, but only a neat, bulged, standard one. This precluded any attempt to land the aeroplane from the back seat, which I occupied for insurance purposes. The gear-selector could be reached before strapping in, by leaning over.

Overshooting from a test balk, the final downwind leg was in bright sunlight, from directly behind. This had the effect of illuminating by reflection the green gear-DOWN lights in both cockpits. I had pulled the harness tight for the balked landing, before the P1 suggested that I might like to carry out the downwind checks. Tightening the harness *first* was a wrong move. Pushing the gear selector DOWN and feeling firm resistance against stretched finger tips, I assumed I had reached the end of the range for the lever, saw the

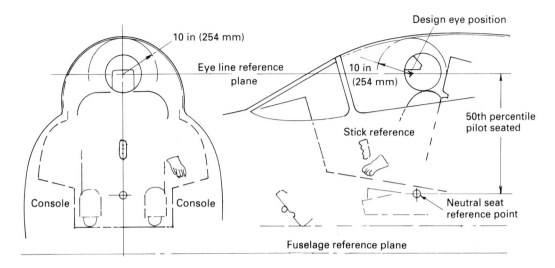

Fig. 4.5a NATO cockpit proportions.

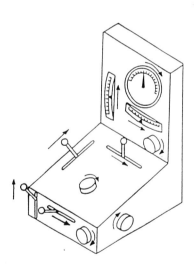

Fig. 4.5b Expected display changes for given movements of controls. (Courtesy of RAF Institute of Medicine)

spurious '*two-greens*', and handed over to the P1, who then concentrated upon his final turn. Result: ... a Wheels-Up!

Subsequently the P2 gear-selector lever was cranked upwards slightly to ensure that no-one else with 28 percentile arms came up against a stiff ball-detent with reflected sunlight giving a false impression of gear-DOWN.

Weights of aircrew

Table 4–3 shows the percentile weight range of British military aircrew in 1973. They are

Plate 4–2 Refurbished and modified Spitfire Mk IX with locally incorporated rear cockpit to resemble the trainer variant provided for the Rusians in World War II (Ref.: Example 4–1). The Rolls-Royce Merlin engine needed right rudder on take-off. (Author)

typical of the occidental Caucasian male. In developing countries weights and stature are increasing slowly. Flying clothing and a parachute, or a flight bag, can increase weight by a further 14 kg, say 30 lb.

Note: It is easy to underestimate civilian crew and passenger weights. The average used for design purposes of 77 kg (170 lb) is still applied to young and old, oriental and occidental male and female crew and passengers alike. Today the average occidental weight is probably nearer 82 kg (180 lb) clothed with deviations of ±14%, and ethnic variations can make a considerable difference to calculations.

I have found gross errors in weights of light aircraft prepared for test because the 77 kg assumption was made. Better to resort to a set of good bathroom scales when in doubt. Many engineering organizations keep them to hand for the purpose.

Personal Preparations

No test flight should ever be carried out without preparation. This involves certain basic items of personal test equipment and flying kit. A golden rule is:

NEVER CARRY PASSENGERS ON TEST FLIGHTS – only *essential* crew.

Personal test equipment

☐ A *knee-pad* with paper or cards that can be secured and turned over without them blowing everywhere. Some pilots will say at this point: '*But I have a pocket voice-recorder*'. That is a bonus, electric gadgets break down, or batteries fail, or a plug can come adrift at the wrong moment. A knee-pad strapped to the leg, with a sharp pencil (and a sharpener) tied to the pad are essential 'belt-and-braces' items of equipment. I have found that a metal pad (bought years ago in Oshkosh) which is around the size of A4 paper when folded in half, with a clip for holding either paper or a stenographer's notebook, is a convenient size for use in an enclosed cockpit (i.e. when there is no provision for baling out and no turbulent propwash).

An open, draughty, cockpit is another matter. For that I have a home-made box pad of aluminium alloy, with internal rollers each end. It involves tediously winding into it a prepared roll of paper before flight, which then curls itself into an annoyingly tight roll when finally removed. But, such a pad has the advantage that, in the event of the paper being ripped across accidentally, or when baling out, the ends spring back inside the protecting box. Whereas the stenographers notebook would almost certainly be wrenched out of the clip and lost or shredded.

There is no ideal or perfect pad. Figure 4.6 shows in-flight examples of recorded results. The narrower paper with spin hieroglyphics is from the roller-pad. (Incidentally, should any reader stumble across a Challenge notebook, with a red cover and the number (14) in the top right hand corner, with my name, and dated mid-1985, please send it back to claim the reward.)

☐ A *device for estimating stick position* can be most useful. A cross was made out of stiff but soft wire and fastened to the instrument panel for tests to find the possible cause of the break-up of a Gloster Javelin. The wire was stiff enough to resist bending under applied g, but not so stiff and strong as to restrict the pilot in an emergency. The arms and trunk of the cross intersected over the stick with elevator and ailerons neutral. All was painted black, with white markings every quarter to full-scale deflection in each direction. It enabled a crude grid reference to be used in manoeuvres.

☐ *Parachutes* should be carried by each member of a test crew, especially when tests involving spinning and aerobatics are being carried out. It is the duty of the pilot to make sure that he and any crew member know how to get out of the aeroplane in a hurry, and how to use the parachutes (i.e. know exactly where the rip-cord is and work out how your fingers will find it if you cannot see it). This involves selecting a decision height at which, without heroics, the aeroplane must be abandoned. A crew member should be got out of the aircraft *before* the pilot – in any case, some aeroplanes have recovered from spins when that has been done, and it would be a pity to abandon a good aeroplane and crew member, surrendering an opportunity to maybe regain control. Owners and manufacturers can be very unforgiving.

Note: There are many thin and neat back-pack type parachutes designed for use in gliders which present few difficulties for the wearer – and what you prepare for tends not to happen.

☐ *Life-jacket(s)* when tests are to be carried out over (or sufficiently near) substantial areas of water. The wind can blow a parachutist from an area of safety to one of hazard.

☐ A '*meat-hook' spring balance* for measuring stick-forces. It is not a safe lump of metal to tuck into a seat pocket. It tends to be heavy and is best kept in a tough pocket in one's flying-overalls – and tied to the pilot, NOT to the aircraft.

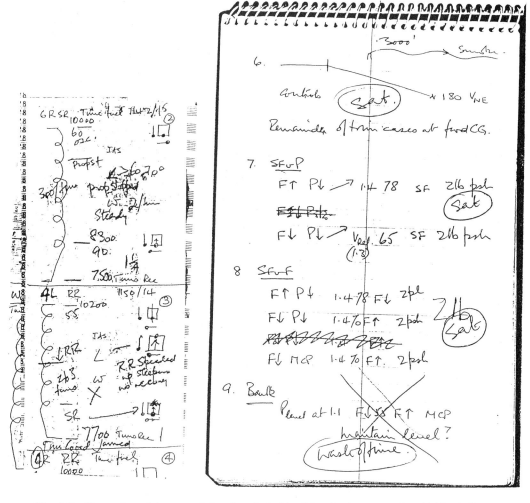

Fig. 4.6 Examples of the use of shorthand: (left) with spinning notes (see Table 4–5) using a roller-type knee-pad which protects paper when torn; (right) results of tests 6, 7, 8, and 9 in Table 4–6 using a clipboard and more vulnerable, but convenient, stenographer's notebook.

☐ A small *accelerometer* for measuring applied 'g', which can be taped in the cockpit when the aircraft is static. It is not an essential instrument, but it can be extremely useful when testing agile and highly manoeuvrable aeroplanes.

Note: A steady balanced turn with 60° bank *applies* 1.0 g.

☐ A *stop-watch*, preferably non-digital with an easily read second-sweep hand, because it reduces the load of mental gymnastics when the brain is needed for other things.

☐ A good quality, small, *voice-recorder with throat microphone* – but still take the knee-pad.

☐ A *day book* or *large diary* (I have used more than 20 stenographer's notepads for a succession of flight tests, including preparatory calculations and after-flight notes). One must never delay making such notes, even for a few hours, never mind days, because the

Plate 4–3   A device for 'eyeballing' stick position with reasonable accuracy when using a Gloster Javelin to find out why a similar aircraft broke up in flight. The wire was strong enough to resist *g* but soft enough to yield in the event of an ejection. It is marked in $\frac{1}{4}$ divisions either side of neutral. (Author, courtesy of RAE Farnborough)

longer they are left the more faulty is recollection and no matter how well intentioned one may be, the less truthful and valuable they become.

Note: One's notes form the basis of the flight test report. ALWAYS keep your notes. If there is a subsequent accident and you have to justify your reported findings in court because you are in the firing-line, it is much better to produce authentic notes in evidence, no matter how oiled and scruffy, than an unverifiable report, immaculately presented.

(I speak with feeling, having scrapped notes when the formal report had been written, only to be faced long after with an inquiry into a near fatal accident in another aircraft of the same type.)

Personal flying kit

The reason for the following advice is that one can never be certain how a test flight might end.

☐ *Warm and comfortable clothing*, in case of an emergency. Body heat is lost with shock, and a test crew may take some finding.

Note: NEVER wear socks and clothing with nylon in them – when exposed to fire it melts into the flesh, which may later pull away when the clothing is removed!

- ☐ *Heavy duty lace-up shoes*, NOT 'trainers', 'flip-flops' or slip-on fashion sandals (which is what a young lady arrived wearing once for a check-out in a Tiger Moth). Leather footwear which supports and protects the ankles is best.
- ☐ *Flying-overalls, cape-leather gloves* (for a sensitive touch) *helmet* and *goggles* or *visor*, as a protection against flash-fire and air blast.
- ☐ A *sharp knife*, easily reached with either hand and kept spotlessly clean (in case you cut yourself or someone else), and never used for anything other than in emergency. Tie it to you with a practical length of cord, enough for it to be used with either hand (I tie mine to the knee-pad – not ideal but convenient).

Note: It is possible that you could be in water, and/or upside down, your weight hanging on the shoulder-harness, causing the quick-release buckle to jam. You will then need to cut yourself free. Cramped, you might have to kick your way out through a cockpit

TABLE 4–4

Configuration shorthand

Control	Symbol	Configuration	
		Setting	Change
Flap	F	F ↑ UP F_{TO} take-off F_{50} 50 degrees F ↓ DOWN	↓ , ↘ select DOWN ↑ , ↗ select UP
Slats	S	(code as required)	
Power	P	P ↑ full, max. P_{75}, P_{50}	↓ , ↘ CLOSE throttle (75% and 50%, etc.)
Reheat	R	(code as required)	
RPM	N	e.g., (2100)	(2300) ↑ , ↗ increase ↓ , ↘ (1800) decrease
Torque	Q	ft lb/psi/	↑ , ↗ increase ↓ , ↘ decrease
(RPM/boost) piston	(/)	e.g., (2100/2.5)	↗ (/) ↘ (/)
Undercarriage (U/C, or landing gear)	G	G ↑ UP; G ↓ DOWN	↓ select DOWN ↑ select UP
Airbrake	A/B, or B	B ↑ IN B ↓ OUT (deployed)	↓ extend ↑ retract
Trim (e.g., elevator) Other controls	FNUT/N/FND RR/LR, RA/LA FBS, etc., etc.	full nose-UP/NEUTRAL/-DOWN RIGHT/LEFT rudder, aileron FULL-BACK stick (elevator), etc., etc.	

TABLE 4–5

Manoeuvre shorthand

Manoeuvre	Symbol	Application	
Straight	→	Stall → P↓ F↓ (60)	straight stall, throttle closed full flap (looking for)
Climb	↗		
Descent	↘	V_{MO} ↘ V_{NE}	max operating speed followed by dive to never-exceed speed
Oscillation/buffet	∿∿∿	⌇	phugoid
Roll	(L R)	−30 ⌒ +30	roll from 30°L to 30°R
Turn	left / right (loops)	(P₇₅ F↓)	Turning stall to left, 75% power, full flap
Bank	<	<60° L, R	60° bank L, R or (+), (−)
Spin	L (left) / R (right)	L	entry altitude (number of turns) (seconds/turn) altitude recovery initiated (turns taken) altitude level
Stick	L R		full-back stick (FBS) full left aileron (FLA) full right rudder (FRR)
Rudder	L ─o─ R	─o	

transparency (it can be tough and resilient), or walk out of a fuel fire. All of these things have happened and, if you are unfortunate enough to experience anything similar, then you could need every item listed here.

Again: *'What you prepare for rarely happens'*.

Easing the workload – shorthand

Writing on a knee-pad is not easy, especially in a light aeroplane which is dancing about. Pilots resort to their own shorthand squiggles, interspersed with numbers and maybe only a single word. Tables 4–4 and 4–5 show examples. The tables are not exhaustive and may not suit every taste. However, your test card or pad, prepared beforehand, is a reminder of what you intend to do, a record of what you actually did, and of what happened next. As mentioned already, it is made more useful by adding or noting within it after landing a short narrative summary, conclusions and recommendations.

Harben shorthand (Ref. 4.6) (also: *Stone spin shorthand* (Ref. 4.7)) could suit some people, because it helps one to remember control positions during a particular exercise by simply marking them on a prepared knee-pad for later debriefing. A shorthand is as open as any other to adaptation according to needs. An example is shown at the bottom of Table 4–5 showing stick (elevator and aileron) and rudder positions, required or actual, at some point in a test.

A copy of listed tests to be carried out must be left behind when you fly, as a pre-flight record of intentions. It is of no use if your shorthand is not intelligible enough for others to guess *accurately* and *easily* what it means, should you be unable to explain it yourself at the time. You should also leave a note of the area in which the test flight will be carried out, and the Air Traffic Control (Centre or Airfield), and radio frequencies you have arranged to use.

Table 4–6 lists, for example, a package of basic flight tests used for many years in the UK to probe basic handling qualities and performance climb rates (measured at maximum weight and forward CG, for least excess lift) of single-and twin-engined light aeroplanes with both engines operating. The table also shows how the shorthand is used. The tests assume re-tractable landing gear 'up' in the clean configuration ('down' when gear is fixed).

Two CG positions are shown, and such a package would take two and probably three sorties. A third could be needed to investigate at least at take-off and landing at a light weight and the most forward CG defined by the manufacturer, as a check of elevator author-ity at low speeds. In the table a landing without elevator (14), using trimmer alone, is shown as being carried out at aft CG (to check excessive sensitivity, while reducing the chance of dropping the aeroplane onto the ground, nosewheel first, during the tests). If the landing on trimmer alone is too easy, it is then worth checking an approach at a forward CG, to simu-late a solo pilot with fuel almost gone, faced with an emergency.

Figure 4.6 shows the usefulness of shorthand when space is cramped.

Plate 4–4   Scale replica World War II Junkers Ju 87. Ref.: Example 4–2. (Author)

TABLE 4–6

Checklist: Basic* (non-aerobatic) flight tests

Serial No	Test	Configuration	Fwd CG	Aft CG
1	Takeoff V_1, V_R, V_2	normal, mistrimmed, minimum lift-off	✓✓	✓
2	Climb initial / 5 minute	$P\uparrow$ F_{TO} $G\uparrow$ / $P\uparrow$ $F\uparrow$ $G\uparrow$	✓ / ✓	— / —
3	Stickforce/g (L R) (SF/g)	$\angle 60°$ V_A and V_{MO}	—	✓
4	Stalls, trim $1.5\,V_s$ $P\downarrow$ $P_{75\%}$	$F\uparrow G\uparrow$ (check $F\uparrow G\downarrow$) / $F_{TO}\,G\downarrow$ / $F\downarrow\,G\downarrow$	✓ / ✓ / ✓	✓ / ✓ / ✓
5	Stalls (L,R. $P\downarrow$ trim $1.5\,V_s$ $P_{75\%}$	$F\uparrow G\uparrow$ (check $F\uparrow G\downarrow$) / $F_{TO}\,G\downarrow$ / $F\downarrow\,G\downarrow$	✓ / ✓ / ✓	✓ / ✓ / ✓
6	Change of trim with V and P and control jerks	$P\uparrow$ $F\uparrow$ $G\uparrow$ V_x or $1.4\,V_{s_1} \rightarrow V_{MO}$ / $P5\%DG$ $F\downarrow G\downarrow$ $1.3\,V_{s0}$ (3° glideslope)	✓ / ✓	—
7	Change of stick force with P	$1.4\,V_{s_1}$ $P\downarrow F\uparrow G\uparrow$ ↗$P\uparrow$ / $V_{app}(1.3\,V_{s0})$ $P\downarrow F\downarrow G\downarrow$ ↗$P\uparrow$	— / —	✓ / ✓
8	Change of stickforce with F	$1.3\,V_{s_1}$ P_{FL} $G\downarrow$ $F\uparrow \rightarrow F\downarrow$ $1.3\,V_{s0}$ / V_F $P\downarrow$ $G\downarrow$ $F\downarrow$↗$F\uparrow$ / $1.2\,V_{s0}$ $P\uparrow$ $G\downarrow$ $F\downarrow$↗$F\uparrow$	✓ / ✓ / ✓	✓ / ✓ / ✓
9	Balked landing $1.2\,V_{s0}$ to $1.3\,V_{s0}$	$P5\%DG$ $F\downarrow G\downarrow$ → $P\uparrow$ $F\uparrow$ $G\uparrow$	✓	—
10	Longitudinal Stability	$1.4\,V_{s_1}$ ↗$1.5\,V_{s_1}$ ↘$1.3\,V_{s_1}$ $P\uparrow$ $F\uparrow$ $G\uparrow$ ($G\downarrow$) / $1.4\,V_{s0}$ ↗$1.5\,V_{s0}$ ↘$1.3\,V_{s0}$ $P5\%DG$ $F\downarrow G\downarrow$ / $1.3\,V_{s_1} \rightarrow V_{MO}$ $P_{75\%}$ $F\uparrow$ $G\uparrow$($G\downarrow$)	— / — / ✓	✓ / ✓ / ✓
11	Lateral and Directional Stability	$1.2\,V_{s_1}$ $P\downarrow F\uparrow G\uparrow$ ($G\downarrow$) / $1.2\,V_{s0}$ $P_{75\%}$ $F\downarrow G\downarrow$ (use full power for directional) / $1.2\,V_{s0}$ $P\downarrow$ $F\downarrow G\downarrow$	— / — / —	✓ / ✓ / ✓
12	Controllability sideslips	$1.2\,V_{s_1}$ $P\downarrow$ $F\uparrow G\uparrow$ ($G\downarrow$) / $1.2\,V_{s0}$ $P\downarrow$ $F\downarrow G\downarrow$ / $1.2\,V_{s0}$ $P5\%DG$ $F\downarrow G\downarrow$	— / — / —	✓ / ✓ / ✓
13	Minimum trim airspeed and stickforce 10lbf = 50N	✻V_2+10lbf $P\uparrow$ F_{TO} $G\uparrow$($G\downarrow$) / ✻$1.4\,V_{s_1}+10$lbf $P\downarrow$ $F\uparrow G\downarrow$ / ✻$1.4\,V_{s_1}$ $P5\%DG$ $F\uparrow G\downarrow$ / ✻$V_{app}+10$lbf $P\downarrow$ $F\downarrow$ $G\downarrow$ / ✻V_{app} $P5\%DG$ $F\downarrow G\downarrow$	✓ / ✓ / ✓ / ✓ / ✓	— / — / — / — / —
14	Landing	$V_{app}-5$knot $P5\%DG$ $F\downarrow G\downarrow$ / V_{REF} or $V_{AT}-5$knot $P5\%DG$ $F\downarrow G\downarrow$	✓ / ✓	—
	Without elevator (trimmer alone)	achieve zero rate of descent for touchdown $P5\%DG$ $F\downarrow G\downarrow$	—	✓
NOTE	✱ No spinning or aerobatics.	($G\downarrow$) fixed landing gear		

Example 4–2: Smaller Scale Replica – Junkers Ju 87

An awkward situation was caused by insurance regulations, when attempting to carry out a test flight which relied upon the generosity and goodwill of an owner. In this case a knee-pad and shorthand saved the situation. Without the test flight a valuable portion of the expensive visit would otherwise have been wasted.

It occurred at Oshkosh some years ago. My first visit, in 1970, led to test flying some 17 completely different and unusual types of single and two-seat homebuilt aeroplanes. By the mid-1980s insurance regulations reduced the opportunities for flying to two-seaters with a nominated and insured pilot in the front seat. There were times when the second seat lacked the necessary flying controls, making a proper test assessment impossible.

The opportunity to sample a scale Junkers Ju 87 involved having a retired USAF Colonel at the controls in the front cockpit, while I was in the back with none.

Fortunately there was enough room to wriggle forward under the transparent canopy linking both cockpits and lean over the left shoulder of the pilot, who had my notepad and used my shorthand, while operating the rudder pedals, brakes and flaps. It had the advantage that I could withdraw out of the way without restricting his retaking full control if need be. At a safe altitude and held in a prone position by the canopy, it was possible to reach over his left shoulder to the stick and throttle. It worked because it had been pre-planned, using the shorthand as an *aide-mémoire*. A short test programme was comfortably completed for an aeroplane with fixed gear and a fixed pitch propeller (see Table 4–6), through close Anglo-American cooperation. The parts which relied entirely upon the Colonel's assessment were the take-off and landing.

Having got out of the way some of the difficulties, and the basic equipment needed to record results which tell it as it was, the next step is the disciplined approach, using pre- and in-flight rating systems, the calculation of weight, balance, and airspeeds needed for the tests.

References and Bibliography

4.1 Stinton, D., Instrument flight – a human problem. *Air Clues*, May–September 1959, Assistant Chief of Air Staff (Training), Royal Air Force

4.2 Stinton, D., Mistake-making in man–machine systems, Vol. 15, *Technology Common to Aero and Marine Engineering.* Proceedings of an International Conference organized by the Society for Underwater Technology, London, UK, 26–28 January 1988

4.3 Learmount, D., Human factors, *Flight International*, 22–28 *July 1992*

4.4 Crossfield, A. Scott as quoted in 'Test flying at the speed of sound', *The Aeroplane, 10,* September 1954

4.5 Bolton, C. B., Kenward, M., Simpson, R. E. and Turner, G. M., FPRC/1327 *An Anthropometric Survey of 2000 Royal Air Force Aircrew 1970/71*. London: Ministry of Defence (Air Force Department), 1974

4.6 Harben, N. R., *The Complete Flying Course – A Handbook for Instructors and Pupils.* London: C. Arthur Pearson, 1939

4.7 Stone, R. R., *Engineering Flight Test Report: Stone Spin Shorthand*, including *Spin Data Card and Sample Criteria Spin Set*. Wichita: Beech Aircraft Corporation, Code Ident No. 70898, 1974

4.8 Alkov, R., *Life Changes and Accident Behaviour*, Approach Reprint. US Naval Publication, 1975

4.9 Rahe, R. H. *Life Crisis and Health Change*. Report No. 67–4, Navy Medical Neuropsychiatric Research Unit, San Diego, California

CHAPTER 5

The Disciplined Approach

'This is real test-flying, when the pilot notes everything that is happening and is able to render a story, not only coherent but constructive, on landing.'
Said of Michael Daunt, OBE (1909–91) by John Grierson, also of the Gloster's team, after Daunt had identified three serious faults and carried out an experiment during a maiden flight of just three-and-a-half minutes in a Gloster Meteor. (*Times* obituary, 30 July 1991)

Before buying or testing a particular aeroplane it is useful to assess, as far as possible, what to expect of it compared with others which are broadly similar. This chapter presents two rating systems, the first of which is aimed at making just such a comparison. The second is for use in flight. The chapter also deals with initial calculation of weight and balance, and then re-calculation involving changes of ballast and fuel-load. Finally there are ways of sorting airspeeds needed for the tests.

Pre-Flight Parametric Rating of Aircraft

In the absence of a full set of authoritative information about an aircraft it was useful to devise a personal parametric method when working in Operational Requirements of the Ministry of Defence, many years ago. It has been a flexible tool for comparing other aircraft since (Refs 5.1 and 5.2). It has the advantage that anyone may construct whatever numerical parameters they wish for a particular task. But they must be kept simple. Other systems have been published by Chevalier (Ref. 5.3) and Wild (Refs 5.4 and 5.5). Parameters have the advantage of enabling one to judge whether or not salesmen and representatives of the manufacturer are being evasive, or sometimes misleading, long before the aircraft starts telling you the truth in its own way.

Method

The method is broadly similar to that published in Refs. 3.8 and 4.8, in that at least two aeroplanes are needed one (let us say $\mathbf{A}$) acts as the yardstick for $\mathbf{B}$. Working out the values of the same parameters for each we then ratio the specific values of $(\mathbf{B}/\mathbf{A})$ in each case. If the $\mathbf{B}$ value is better than $\mathbf{A}$, then the ratio $(\mathbf{B}/\mathbf{A}) > 1.0$. If worse, then it is less than 1.0. A calculator and a pad of graph paper is needed, as are manufacturer's data sheets. The small print on the back of a manufacturer's brochure generally warns that all specifications and performance data are subject to change without notice. Let your arithmetic be coarse-cut.

Two places of decimals are sufficient. Avoid comparing very large and very small numbers. It is better to write:

$$1\ 787\ 000 \text{ as } 178.7 \times 10^4$$

or,

$$0.0103 \text{ as } 10.3 \times 10^{-3}$$

as long as comparisons of the same parameter between aircraft are in terms of 10^4 or 10^{-3}.

Choose ratios which reflect *improvement* in terms of numbers which grow *larger*, and vice versa. The value of using ratios is that they can be turned upside down, so that one with a small number in the numerator and a large in the denominator, can be changed from a fraction into a whole number by inversion. Finally, use compatible units. Avoid mixing dollars with francs, or kilogrammes with pounds-weight.

Examples of parameters have appeared already, in Chapter 3 for example:

$$\text{geometric aspect ratio} = b/\bar{c} \tag{3-11a}$$

$$\text{wing loading}, w = W_0/S \tag{2-1}$$

and,

$$\text{power loading} = W_0/P \tag{2-1}$$

The following list is not exhaustive. It is a guide. The examples are taken from one particular study. A different class of aircraft could need changes:

☐ *Seating/Occupants, N,* is an indication of likely payload. Many accidents are caused by overloading, with CGs which are outside the approved limits. With baggage limited to 44 lb (20 kg) per economy class passenger one should calculate (payload/seat) near to:

$$220N \text{ lb } (100N \text{ kg}) \tag{5-1}$$

The maximum number of occupied seats results in minimum range because, as we saw in Equation (3–82), disposable load is made up of exchangeable items. The heavier the payload, the lighter the weight of useable fuel. Payload and useable fuel load are often nearly equal in small public transport aeroplanes.

☐ *Power unit, type and number* enable us to find total rated power, P, or rated thrust F. Knowing the design gross weight, W_0, we may then calculate:

$$\text{power/weight}, \quad P/W_0 \tag{5-2}$$

and,

$$\text{thrust/weight}, \quad F/W_0 \tag{5-3}$$

which are indicators of ability to accelerate, manoeuvre and climb high. Reciprocals are the power and thrust loadings:

$$W_0/F \text{ and } W_0/P \tag{5-4) and (5-5}$$

the last of which appeared in a general form in Equation (2–1). Both are measures of the physical effort needed to move the weight (i.e. mass) of the aircraft. When combined with speed ratio, V_c/V_{s0}, (see Equation (5–40)) they reveal improvement (or otherwise) of both propulsive and aerodynamic efficiency.

☐ *Wing span, b*: the longer the span of the wings across the flight path the smaller the rate of change of momentum imparted to the air when it is pushed downwards to produce lift.

The lower the lift-dependent drag generated in the process, the less fuel that is burned developing the thrust to overcome it (see Equation (3–21). It follows that the longer the span for a given aircraft weight, the more efficient is the aeroplane in terms of flying for range, endurance and in its ability to fly high.

Lift-dependent (or induced) drag is inversely proportional to:

span loading, W_0/b (5–6)

which term gives a feel for the drag of one aeroplane compared with another. Its reciprocal

b/W_0 (5–7)

enables comparison of cruise performance at a given altitude to be made.

When cruising altitude, H, is known (we use the published altitude for maximum level speed) it is possible to deduce the relative density and, hence, $(\sigma^{0.5})$, from Fig. 3.1, or Fig. 5.1 if preferred). As height is increased the relative density of the air decreases, and the performance at altitude is compared by means of an improved version of Equation (5–7):

Hb/W_0 (5–8)

or,

$(b/W_0)/(\sigma^{0.5})$ (5–9)

☐ *Wing area, S*: too little area and the aeroplane stalls faster than it should, needing longer runways and/or elaborate and expensive high lift devices. Too much wing area and there is more 'wetted area' of surface exposed to the relative airflow and more skin friction drag than there should be. That spoils cruise lift/drag ratio and economics, shortening endurance and range, R, for a given design gross weight, W_0. In short, there are economic penalties either way. Wing area is a direct measure of 'wetted area' of skin, A_w. Typically:

A_w/S = 4.5 to 5.0 (multi-engined)   (5–10)
        = 3.5 to 4.5 (single-engined)   (5–11)

Thus, it is possible to gain some idea of power or thrust expended per unit of aircraft surface area direct from the wing area. If the aircraft being compared are sufficiently close in configuration as to share similar constants in Equations (5–10) and (5–11) then simply use:

P/S and F/S (5–12)

Work done to overcome drag results in the cruising speed achieved in level flight, V_C, so:

$V_C/(P/S)$ and $V_C/(F/S)$ (5–13) and (5–14)

are indicators of aerodynamic and propulsive efficiency. They are measured, e.g., in knots per horsepower per unit wing area, or knots power kW per unit of wing area (i.e. per unit 'wetted area' of skin, as implied by transposition of either Equations (5–10) or (5–11)).

☐ *Everling number*, E, is an old but useful indicator of efficiency named after Professor Everling, its originator (1926), in the period between both World Wars (Refs. 1.5, 5.6 and 5.7). In the project stage we may use design cruising speed, V_C, or if it is known, the maximum operating speed, V_{MO}. Here we shall use V_C:

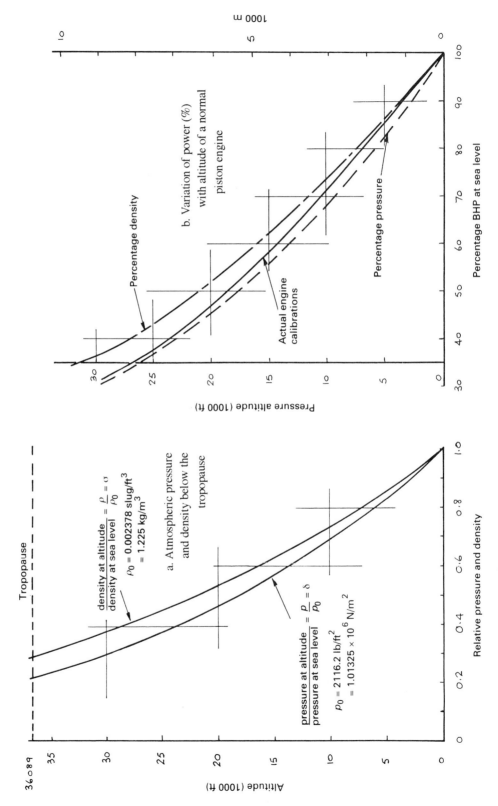

Fig. 5.1 Atmospheric pressure and density and their effect upon % power delivered by a normally aspirated piston engine (Ref. 1.5).

$$E = [V_C^3/96 \times 10^3 \, (\sigma^{0.5})] \, (W_0/P) \, / \, (W_0/S) \qquad (5\text{–}15)$$
$$= \eta/C_D \qquad (5\text{–}16)$$

in which η is the propeller efficiency, C_D the total drag coefficient (Equation (3–28)), and the last two terms appeared originally in the generalized form of Equation (2–1).

☐ *Price $*, can be used in various ways, as long as accurate values can be found. The first is in terms of occupants (seats) per $:

$$N/\$ \qquad (5\text{–}18)$$

But, as occupants and fuel must be traded (Equation (3–82)) a large number of seats per $ often reveals a shorter practical range, or a lighter and maybe weaker structure. Other ratios follow from the breakdown of weight (Equation (3–78) *et seq.*):

gross weight = operating empty weight + disposable load

Empty weight, W_E, is a measure of the mass of structure and machinery needed to hold the aircraft in the air at the right speed for the price paid, so that value for money is shown by:

$$W_E/\$ \text{ and } W_0/\$ \qquad (5\text{–}19) \text{ and } (5\text{–}20)$$

If disposable load is not quoted (and often it is not), an approximation can be found from Equation (3–82). The following Equations (5–21) to (5–23) can be used either alone, or expressed per $.

By using:

disposable load = payload + useable fuel (+ any necessary ballast)

and then inserting the ratio of empty to gross weights, W_E/W_0, we have:

$$\text{disposable load/gross weight} = 1 - (W_E/W_0) \qquad (5\text{–}21)$$

Or, as structural adequacy is indicated by the disposable load per unit of empty weight:

$$\text{disposable load/empty weight} = (W_0/W_E) - 1 \qquad (5\text{–}22)$$

If the ratio is too large, say in excess of 0.85, then the structure weight might be inadequate, either in strength or in stiffness, in which case be wary of Equation (5–22).

But if the reciprocal:

$$W_E/(W_0 - W_E) \qquad (5\text{–}23)$$

is too large (more than about 1.2) then the structure, engines and systems may occupy an uneconomic proportion of the gross weight. Again, be wary. When Equations (5–22) and (5–23) are expressed per $ and exceed these limits, they then give unreal impressions of cost-effectiveness.

☐ *Stall speeds*, V_{S1} and V_{S0}, are the building blocks of many airworthiness requirements. Stall quality is the keystone. For example, in JAR-23.49(c) (Ref. 5.8) the highest stall speed in the landing configuration, V_{S0} (measured at forward CG and scheduled), must not exceed 61 KCAS for single-engined aeroplanes; or those twins of 6000 lb or less maximum weight (2730 kg) which are unable to meet the minimum gradient of climb of 1.5% at a pressure altitude of 5000 ft, in a specified configuration, with the critical engine inoperative. This requirement is comparable with FAR 23.49(b) (Ref. 5.9).

Most small aircraft with take-off flap selected and landing gear up have stall speeds around 10% faster than V_{S0}, i.e. gear and flap down. So, as take-off safety speed, V_2, is determined by the manufacturer to include a safety margin on the minimum control speed), we find:

$$V_2 = 1.2V_{S1} \tag{5-24}$$

With flaps in the take-off position and gear up this makes V_{S1} about 1.1 V_{S0}, so that:

$$V_2 = 1.2 \times 1.1 \ V_{S0} = \text{approx. } 1.3 \ V_{S0} \tag{5-25}$$

As kinetic energy is proportional to (speed)^2 on take-off:

$$(V_2)^2 = \text{approx. } 1.7 \ (V_{S0})^2 \tag{5-26}$$

The approach speed in the landing configuration, V_{app}, has a safety margin of 30% added, making:

$$V_{app} = \text{not less than } 1.3V_{S0} \tag{5-27}$$

for which the kinetic energy at the same weight is comparable with that for V_2.

☐ *Take-off distance, TOD, to 35 ft screen*, is achieved in a short distance when the stall speed, (hence V_2) is low, and in the case of a propeller-driven aeroplane, both power loading and wing loading, (W_0/P) and (W_0/S), are low. Thus:

$$\text{TOD to 50 ft varies as } (W_0/P) \ (V_2^2/1000) \tag{5-28}$$

in which V_{S0}^2 may be used instead of V_2^2, and the reciprocal taken:

$$1000P/W_0 \ (V_{S0})^2 \tag{5-29}$$

which grows larger with improving (shortening) field lengths.

Similarly, for jet aircraft, in terms of wing loading times thrust loading:

$$\text{TOD to 35 ft varies as } (W_0/S) \ (W_0/F) = W_0^2/FS \tag{5-30}$$

while, again, the reciprocal:

$$FS/W_0^2 \tag{5-31}$$

grows larger as field lengths shorten.

☐ *Landing distance, LD, from 50 ft varies more or less as:*

$$W_0(V_{app})^2 \tag{5-32}$$

but as V_{app} is related to V_{S0} (see Equation (5–27)), if one reciprocal value of:

$$1/W_0 \ (V_{S0})^2 \tag{5-33}$$

is larger than another, then the corresponding landing distance is shorter.

☐ *Field factor* has no precise definition, but it arises from 'gut-feeling' that, as take-off and landing distances are of comparable magnitude, they can be averaged and used as a measure of the mean distance needed in which to lift a given weight of aircraft into the air, and land again safely:

$$\text{average distance} = (\text{TOD} + \text{LD})/2 \tag{5-34}$$

When gross weight, W_0, is divided by this distance we have a measure of merit:

$$2W_0/(\text{TOD} + \text{LD}) \tag{5-35}$$

the magnitude of which tells us which aircraft has the better combination of propulsion-plus-aerodynamic high lift devices. At this point cost could usefully be introduced into the denominator. The number 2, which is a constant, may be omitted if so wished.

☐ *Sea level rate of climb*, v_c or v_y (USA) can usually be expressed in terms of rated power or thrust/weight ratio (see Equation (5–30)):

$$P/W_0 \text{ and } F/W_0 \tag{5–36}$$

Roughly 2.6 lb jet thrust, F, is equivalent to 1 TEHP (thrust equivalent horsepower), so that, for example, a turbopropeller engine which produces shaft horsepower (SHP) plus F lb residual jet thrust, has an approximate output of:

$$\text{power produced} = (\text{SHP} + \text{TEHP})$$

i.e.

$$\text{ESHP} = \text{approx. SHP} + (F/2.6) \tag{5–37}$$

For practical (non-sales-talk) purposes, the engineer counts only SHP.

Note: It is worth adding at this point that a propeller produces the ultimate thrust horsepower, THP and TEHP, and it is necessary to multiply the SHP, or *brake horsepower* (BHP) in the case of a piston-propeller unit by the propeller efficiency, η. A wooden propeller has an efficiency around 70%, while a metal propeller is about 80% efficient, so that, roughly:

$$\begin{aligned} \text{THP} &= \text{propeller efficiency}, \eta \times \text{SHP (or BHP)} \\ &= 0.7 \text{ to } 0.8 \text{ of the rated power the engine} \\ &\quad + \text{any TEHP from the exhaust} \end{aligned} \tag{5–38}$$

under the actual conditions occurring in flight. Exactly the same equation can be used when working in kW, as in FPSR units.

☐ *Maximum level speed*, V_{M0} or V_C (if V_{M0} has not yet been determined): the larger the speed ratio:

$$V_{M0}/V_{S0}, \text{ or } V_C/V_{S0} \tag{5–39) and (5–40}$$

the better is the high speed performance of the aircraft.

☐ *Aerodynamic (and propulsive) efficiency* of jet aeroplanes was dealt with in Equations (2–10), (2–11) and (2–11a), being given in the latter equation as:

$$W_0 M/F \qquad \text{(see Equation (2–11a))}$$

☐ *Range*, R, must compare like with like, i.e. using either maximum useable fuel, or when leaving 45 minutes reserves, then it may be treated as:

$$R/\$ \tag{5–41}$$

or as a weight ratio:

$$R/W_0 \text{ or } R/W_F \tag{5–42) and (5–43}$$

in which W_F is the weight of useable fuel. If W_{ZF} is the zero-fuel weight (i.e. useable fuel gone):

$$W_F = W_0 - W_{ZF} \tag{5–44}$$

so that fuel weight ratio:

$$(W_F/W_0) = (1 - W_{ZF}/W_0) \tag{5–45}$$

which is an indication of the volumetric efficiency of the structure as a container of fuel. But beware, it could also mean a much smaller payload than volumetric efficiency, so make sure that another parameter, like Equation (5–22), is paired with this as a cross-check.

Other relationships may also be devised using these equations, for example, the Breguet range equation which we encountered in Chapter 2 and saw that it is the product of the following three parts, aerodynamics, propulsion and structure (see Equation (2–7)). Restating it in its essential parts:

☐ *Breguet range equation*:

$$\text{(lift/drag)} \times \text{(propeller } \eta/\text{specific fuel consumption)} \times (W_F/W_0) \qquad (5\text{--}46)$$

For a given cruising speed and altitude:

$$(\eta/\text{sfc}) \ (L/D) \text{ may be re-written as } (\eta/D) \ (W_0/\text{sfc})$$

enabling us to say that, as cruise power setting and sfc may be assumed constant:

$$\text{range, } R, \text{ varies as } (\eta/C_D) \ (W_F/W_0) \qquad (5\text{--}47)$$

which is a measure of range flying efficiency in terms of the Everling number, (η/C_D) in Equations (5–15) and (5–16), and fuel weight ratio, (W_F/W_0), of Equation (5–45), enabling us to write:

$$\text{E } (W_F/W_0) \qquad (5\text{--}48)$$

Alternatively, using the disposable weight ratio Equation (5–21), when the weight of fuel carried is unknown, Equation (5–48) may be replaced by:

$$\text{E } [1 - (W_E/W_0)] \qquad (5\text{--}49)$$

But if the zero-fuel weight is given, then Equation (5–48) can be replaced by:

$$\text{E } [1 - (W_{ZF}/W_0)] \qquad (5\text{--}50)$$

☐ *Service ceiling, H_S,* is, as we saw the foot of Table 2–2, the altitude at which the rate of climb has dropped to 100 ft/min (0.51 m/s). It is not much use as a pure number, but the higher the service ceiling the better able is the aircraft to fly above the worst weather. It is also a fair confirmation of adequate climb characteristics on the way to altitude. But, if the cruising altitude, H, is given then it is better to use that instead of H_S.

Aircraft built to fly high are costly (they need longer wings, oxygen systems, pressurization, turbocharged engines, special navigational and communications equipment. Also, because of the requirement to make a safe emergency landing by night or day in IMC (*Instrument Meterological Conditions*), they are required to *demonstrate* safety levels equivalent to multi-engined aircraft, this results in more than one engine as the easier and cheapest solution). Having said that:

$$H/\$ \text{ or } H_S/\$ \qquad (5\text{--}51\text{a}) \text{ and } (5\text{--}51\text{b})$$

can show relative merit between competitors, as long as one is comparing like with like, i.e. at either the service ceiling or cruising altitude.

Figures 5.2 and 5.3 (see Ref. 1.5) enable us to couple with Equation (5–51b) to give a more comprehensive formula, in terms of service ceiling and price of aircraft:

$$H_S = 10^3 \ n \ (Pb/W_0) \qquad (5\text{--}52\text{a})$$

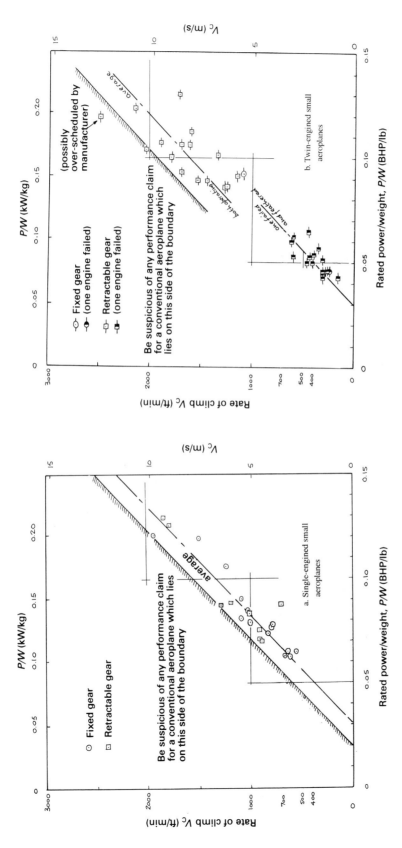

Fig. 5.2   State-of-the-art rates of climb for propeller-driven light and small aeroplanes (Ref. 1.5).

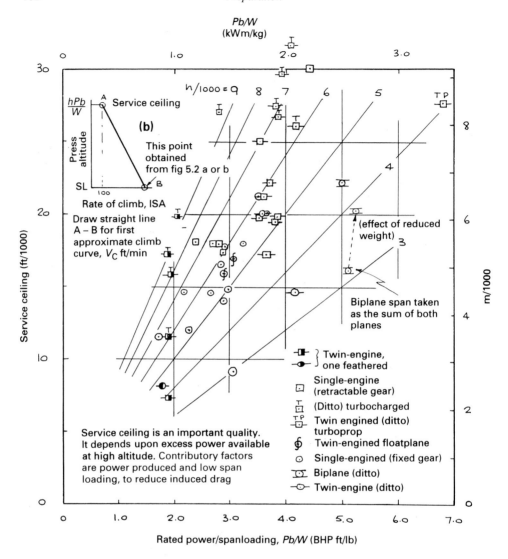

The way to use this information is to first calculate P/W for the aeroplane. Insert its value at the base of fig. 5.2a or b to find the sea level rate of climb. Then use the legend above to 'eyeball' a value of $n/1000$ for the aeroplane. Use the wing-span, b, to find $n(P/W)b$ and construct a straight climb line, shown as inset b.

Fig. 5.3 Construction of 'state-of-the-art' climb rate using estimate of service ceiling and Fig. 5.2.

such that:

$$H_s/\$ = n(Pb/W_0)/(\$/10^3) \tag{5–52b}$$

in which n is a 'configuration factor'. The larger the value of n, the higher the service ceiling and the better in general is the climb performance. The term, which is proportional to span loading and inversely to rated power, can be constructed using Equations (5–6) and (5–52a):

$$\text{'configuration factor'}, \; n = (H_s / 10^3 \, P) \, (W_0 / b) \tag{5–53}$$

which is another way of comparing relative merit.

☐ *Powerplant effectiveness*, $\bar{p}_\eta$, suggested itself as a result of its original form being published in a general paper by Waldemar Voigt on the Messerschmitt Me-262, Schwalbe (Ref. 5.15). The formula provided a key to power and thrust required for a given level of 'good' performance in subsonic and, by extension, transonic flight. From this the size of engine might be determined. Here it enables us to check the reasonableness of that choice, using four basic terms: maximum lift coefficient in the landing configuration, C_{Lmax}, total drag at maximum level speed (lift-dependent + parasite), C_{Dmin}, powerplant weight, W_p, and all-up weight, W_0. That is not all. Aerodynamics also enter the otherwise simple picture, and the proportion of the total drag occupied by its parasitic and lift-dependent elements:

$$\bar{p}_\eta = (C_{Lmax} / C_{Dmin}) \, (W_p / W_0) \tag{5–54}$$
$$= 43.5 \pm 0.03 \text{ for a 'good' subsonic aeroplane}$$
$$= \text{say, } 28 \text{ to } 29 \text{ at transonic speeds, owing to compressibility effects}$$

This is a useful statement, because, as we saw in Chapter 3, the performance engineer regards total drag as:

$$D = D_p + D_i \tag{3–23}$$

which, by means of drag coefficients and Equation (3–28), and cancelling the qS terms on each side translates into the form:

$$C_D = (C_{DP} + C_{Di}) \tag{3–28a}$$

C_{Di} depends primarily upon lift coefficient squared, $C_L{}^2$, together with the 'induced drag' factor, K', geometric aspect ratio of the wing, A, and π:

$$C_D = C_{Dp} + (K' C_L{}^2 / \pi A) \tag{5–55}$$

Note 1: The balance between both parts of this equation is important. In subsonic (subcritical) flight, where the compressibility effects are small enough to be ignored, C_{DP} is almost constant. The lift-dependent drag coefficient in brackets, C_{Di}, is not constant. Angle of attack and C_L decrease with increasing airspeed as $(1/V^2)$ at constant lift and weight. Therefore the total drag coefficient must also decrease, being the sum of one almost constant term, and another which grows smaller with increasing airspeed. This means that C_{DP} increasingly dominates the total drag equation as shown in Fig. 5.4. At maximum speed, which with retractable gear and variable pitch propeller may be more than three or four times the stall speed, the parasitic drag of the aircraft exceeds 9/10 of the total. With fixed gear and fixed pitch the speed ratio, V/V_{s0}, is nearer 2.0 to 2.5, so that parasitic drag is 3/4 to about 4/5 of the total.

Note 2: K' is inversely proportional to 'Oswald's (planform) efficiency factor, e, which appears in standard works on aerodynamics, and in Ref. 1.5. It is also a useful dump for a number of other sources of drag losses, Δ, which arise from the configuration and condition of an aircraft: interactions between lifting airflows at junctions, fairings, scuffed surfaces, 'hangar-rash', mixing of wakes caused by gaps and leaks, and propulsion effects, such as damaged propeller leading edges:

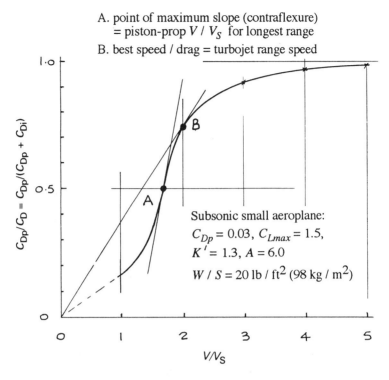

A. point of maximum slope (contraflexure)
   = piston-prop V / V_S for longest range

B. best speed / drag = turbojet range speed

Subsonic small aeroplane:
$C_{Dp} = 0.03$, $C_{Lmax} = 1.5$,
$K' = 1.3$, $A = 6.0$
$W / S = 20$ lb / ft^2 (98 kg / m^2)

Fig. 5.4 Although the curve is for a light aircraft, nevertheless it is typical, showing how the parasite drag coefficient, C_{Dp}, approaches in magnitude the total drag coefficient, C_D, with increasing airspeed, expressed here in terms of the stall speed in the relevant configuration.

$$K' = (1/e) + \Delta \qquad\qquad (5\text{--}55\text{a})$$

Figure 5.5 shows a range of values of K'. The term Δ is hard to determine. For practical purposes first assume $K' = (1/e)$, and then examine the aeroplane to see how dirty and superficially damaged is the airframe and the propeller. With fixed gear DON'T over-look the fact that the spats might be missing. If one cannot find an equipment list, the AFM or POH generally includes a three-view drawing which confirms whether or not they are fitted and, by deduction, one can usually decide whether or not the incorpo-rated data refers to spats-on, or spats-off. Having found the answer add $\Delta = 5\%$ or 10% to $(1/e)$ as a rough estimate of K'. For a well cared for aeroplane, beautifully kept, it could be fair to let $\Delta = 0$. In reality, wear and tear does not allow it to remain zero for long.

Note 3: K' is also affected by wing-tip form (and the same is true of tail surfaces, which make their contributions to total drag). This is discussed further in Chapter 11. For the present, suffice it to say that wings and their tips should be shaped in such a way as to keep the trailing edge as long as possible. The aerodynamic span between the tip vortex cores is then greatest, the aerodynamic aspect ratio which is directly proportional to its square, approaches more nearly the geometric aspect ratio, and lift-dependent drag is closest to the theoretical minimum for the planform.

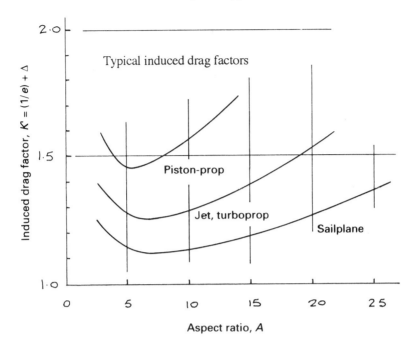

Fig. 5.5 The induced drag factor, K', changes with aspect ratio for different classes of aircraft. The real indicator of induced drag of an aircraft at a given angle of attack ($\alpha : C_L$) is the value of K'/A.

Note 4: Powerplant weight is a heavy item. It depends upon the amount of power required and, this in turn determines the size and weight of the engine together with its mounting, systems forward of the firewall, propeller and cowlings. The fractional weight ratio, W_P/W_0 (Ref. 1.5) varies from about 0.23 for a light military trainer or cabin monoplane, to 0.20 for piston twin or a jet fighter, to 0.18 for a turboprop twin and a subsonic jet transport. But an aerobatic sport aeroplane may have a fractional weight ratio as high as 0.28, because it is not designed to travel far and carries a proportionally lighter disposable load.

☐ *Pazmany 'efficiency' factor*, η_K, was devised by Ladislao Pazmany, a professional aeronautical engineer and a designer of homebuilt aircraft, for an Experimental Aircraft Association contest at Oshkosh in Wisconsin (Ref. 5.10). It takes the form:

$$\eta_K = (V_c/V_{SO})\,(S/P)^{1/3}\,(W_0/S)^{0.5} \qquad (5\text{–}56)$$

In this equation the term (S/P) is another way of writing $(W_0/P)/(W_0/S)$, which appears at the tail-end of the expression for Everling number (Equation (5–15)). Had we used it in (S/P) form earlier, its origin would have been hidden.

When inverted its reciprocal gives another physical insight, i.e.

$$P/S = 1/(S/P) \qquad (5\text{–}56a)$$

which has been called the *'propwash factor'* (rated power/wing area), an indicator of the rate of change of energy in the wash from the propeller(s) diffused per unit of area of aircraft skin, A_w (see Equations (5–10) and (5–11)). The lower the rated power per unit (wing) area to achieve a given cruising speed, V_C, the more efficient is the aeroplane. Thus both Everling and Pazmany cross-relate with:

$$V_C/(P/S) = V_C (S/P) \qquad (5\text{--}13)$$

to give alternative indicatorsof aerodynamic and propulsive efficiency.

☐ *Figure of merit, $\bar{M}$,* is vintage (Ref. 5.17), here we shall use it in the form:

$$\bar{M} = \text{payload} \times \text{distance flown/weight of fuel used}$$
$$= (W_{Pay} \times R)/W_F \qquad (5\text{--}57)$$

Other figures of merit may be created to suit one's particular purpose. For example, taking the concept of *Transport Efficiency* devised by G. Gabrielli and T. von Kármán (which appeared in their paper *What Price Speed?*, in *Mechanical Engineering*, Vol. 72, 1950, of the American Society of Mechanical Engineers) and calling their term *TE*, we have:

$$TE = \text{horsepower/ton} \qquad (5\text{--}57a)$$

in which the tonnage represents the gross weight, but might just as easily be adapted to payload, W_{Pay}, and be further modified into a 'dimensionless' power coefficient:

$$P/(WV_C), \text{ or } P/(W_{Pay}V_C) \qquad (5\text{--}57b \text{ and } 5\text{--}57c)$$

representing the installed power divided by the average sector, or block, or cruising speed and weight, in whatever units one wishes to use.

Another measure of merit uses the average sector, or block, time as a divisor in Equation (5–57), turning it into (payload × average or cruising speed/weight of fuel used):

$$(W_{Pay}V_C)/W_F = \text{productivity/weight of fuel used} \qquad (5\text{--}57d)$$

Taking the idea a step further, in straight and level cruising flight where $L/W_0 = 1$, on multiplying the payload fraction by the lift/drag ratio, we have:

$$(W_{Pay}/W_0)(L/D) = (W_{Pay}/D) \qquad (5\text{--}57e)$$

Now divide by speed to give:

$$W_{Pay}/(DV_C) = W_{Pay}/P \qquad (5\text{--}57f)$$

the ratio of (payload carried/installed power). Finally, multiplying this equation by speed, V_C, turns it back to Kármán–Gabrielli form – only this time as a reciprocal of Equation (5–57c):

$$(W_{Pay}/P)V_C = 1/(P/W_{Pay}V_C) \qquad (5\text{--}57g)$$

This last equation is the same as Equation (5–57f), but without the speed terms cancelled.

☐ *Rating of condition or quality* is a subjective scale which depends upon past experience and having inspected an aircraft, but it can be as misleading as judging a used car. If you like the idea, then choose a set of even numbers, such as 1 to 10, to avoid taking the easy way by picking the odd one in the middle.

Note: To be of use, one must have all of the aircraft being compared inspected or test flown by the same person.

Application

It is unnecessary to employ every suggested parameter, but those chosen should be relevant to one's needs. Usually not more than ten will suffice. One of the least reliable items, because it is so hard to find, is an accurately comparable price for each aircraft. If one cannot compile a complete set of comparative prices – or of any other item for that matter – then reject it. Replace it with one for which a full set of values can be assembled.

Example 5–1: Comparison of Two Aircraft for Same Role

The method can be applied to any aircraft, either new or used. The most accurate figures are those provided by flight testing by the same qualified test pilot. Table 5–1 compares two real aeroplanes in terms of data published by the manufacturer. The prices are 'ballpark' but fictitious.

Table 5–2 shows the result of manipulating ten of the simplest parameters, including a subjective rating by the same test pilot for both aircraft.

Normalization

Selecting aeroplane **A** as the yardstick (**B** would also have served well enough), Table 5–3 has been constructed by dividing both **A** and **B** by the relevant value of **A** in Table 5–2. The columns are then totalled, giving **A** as 10.00 and **B** as 9.24, which shows that *on the basis of the chosen parameters* **A** is marginally better than **B**. Subjectively, though, the pilot gave **B** a higher assessment than **A**, which aircraft also had a slightly better field performance than **A**.

It may suffice, as here, to compare the totals, item by item, when the sample of aircraft is small. When the sample is larger, pillar charts ease visual comparison of results, by drawing lines which link the segments representing each item.

TABLE 5–1

The brochure
(Selected brochure items listed for two real aircraft but with fictitious prices)

Brochure item	Symbol	Aircraft A	Aircraft B
Occupants	N	4	4
Power	P (BHP)	235	285
Wing area	S (ft^2)	174	181
Gross weight	W_0 (lb)	3100	3400
Empty weight	W_E (lb)	1794	2094
Stall speed	V_{SO} (KEAS)	50	51
Maximum speed	V_C (KEAS)	187	182
Range	R (nm)	1010	889
Take-off distance to 50 ft	TOD (ft)	1570	1769
Landing distance from 50 ft	LD (ft)	1320	1324
Price	$	72×10^3	85×10^3

TABLE 5–2

The evaluation, using ten parameters

Ratios	Parameter	Aircraft A	Aircraft B	Equation
Weight *v* power	W_0/P	13.2	11.9	(5–5)
Speed *v* power and wing area	$V_c/(P/S)$	138.46	115.58	(5–13)
Disposable load *v* gross weight	$1 - (W_E/W_0)$	0.42	0.38	(5–21)
Disposable *v* empty weight	$(W_0/W_E) - 1$	0.73	0.62	(5–22)
Weight *v* field length	$2\,W_0/(\text{TOD} + \text{LD})$	2.15	2.20	(5–35)
Top speed *v* stall speed	V_c/V_{SO}	3.74	3.57	(5–40)
Range per dollar price	$R/\$$	14.02×10^{-3}	10.45×10^{-3}	(5–41)
Range per pound gross weight	R/W_0	0.33	0.26	(5–42)
Occupants per dollar	$N/\$$	5.55×10^{-5}	4.7×10^{-5}	(5–18)
Subjective rating		5	7	

TABLE 5–3

The comparison

Aircraft A is used as the yardstick. Each number in columns A and B of Table 5–2 is divided by the relevant number in column A (which has the effect of making every number in A equal to 1.0). The larger the total in either column, the better is the aircraft. In this case – *on the basis of these parameters* – aeroplane A appears superior to B.

	Parameter	Aircraft A	Aircraft B
Weight *v* power	W_0/P	1.0	0.90
Speed *v* power and wing area	$V_c/(P/S)$	1.0	0.83
Disposable load *v* gross weight	$1 - (W_E/W_0)$	1.0	0.90
Disposable load *v* empty weight	$(W_0/W_E) - 1$	1.0	0.85
Weight *v* field length	$2\,W_0/(\text{TOD} + \text{LD})$	1.0	1.02
Top speed *v* stall speed	V_c/V_{SO}	1.0	0.95
Ranger per dollar price	$R/\$$	1.0	0.75
Range per pound gross weight	R/W_0	1.0	0.79
Occupants per dollar price	$N/\$$	1.0	0.85
Subjective rating		1.0	1.40
TOTAL		10.0	9.24

In-Flight Rating of Handling Qualities

The subjective nature of the pilot's task, coupled with changes in life events which affect him adversely from time to time, make it essential to equip test pilots with a disciplined means of accurately assessing the handling qualities of aircraft in flight, under a wide range of operating conditions. Further, as we saw in Chapter 2, successful flight started with small aircraft which, like the Wright Flyer, were nearly all control, with little or no stability. During and after World War I scout pilots preferred aircraft which were '*as quick as a squirrel*' in response to control. They rejected longitudinal stability which, among other things, reduced controllability, increasing stick-forces, making an aeroplane harder to land.

In certain respects the flying qualities liked by pilots did not appear to accord with those sought by the scientist and the engineer who, aware of the accident statistics, sympathized with the operator. Thus, there were advantages to be gained by all concerned in establishing a common base and language so that each knew exactly what the other was talking about. Once a common understanding is reached one tends to find that differences between opponents in technical arguments are more apparent than real, being caused by semantics more often than not.

Various rudimentary rating systems have been tried, one I have found useful when looking back at flight test notes in the control-force ratio A : E : R quoted in Equation (2–5). A pencilled (2 : 2 : 2) or (3 : 2 : 5) in the margin beside a particular phases of flight, serves to jog the memory of heavy ailerons, or an over-light, or over-heavy, rudder and a lack of control harmony. But it tells a reader nothing of how heavy or light were the controls, or how effective they were compared with those of another aircraft.

Cooper–Harper rating scale

Mooij, discussing criteria for low-speed longitudinal handling qualities (Ref. 5.11), dealt at length with an effective scale, which is known among professional test pilots as the *Cooper–Harper rating*. The scale originated in work of G. E. Cooper in 1957. The Cornell Aeronautical Laboratory (CAL), aware of early shortcomings, attempted to clarify the scale in certain respects, revising it in an improved form which is sometimes referred to as the Harper scale.

A subsequent paper by both Cooper and Harper led to a request by the Advisory Group of Aerospace Research and Development of the North Atlantic Treaty Organisation (AGARD) to produce a standard pilot's handling rating scale. This was published as the Cooper–Harper *Handling Qualities Rating Scale* in 1969 (Ref. 5.12).

Table 5–4 (Ref. 5.11) defines the Cooper–Harper terms and shows the rating scale. It is largely self-explanatory. Its use in practice rests upon the pilot being caused to make certain clear decisions while handling the aircraft in different configurations throughout the flight envelope. Examples of areas and configurations where handling qualities are of critical importance are those ticked at aft CG in Table 4–6. Many others between items 1 and 14 at forward CG have handling aspects which are proper subjects for evaluation. One would not only wish to know about control forces and their effectiveness, but also about the ease or difficulty and the level of skill needed by the pilot (see Compensation in Table 5–4) while gathering each numerical result.

Masefield (Ref. 5.13) used a mixture of civil and military requirements to size the ailerons of a military trainer. The reason for the military requirements, in the form of US Military Specifications MIL-F-8785B and 8785C, was that the aeroplane had to reproduce the handling qualities of a fighter as far as possible. It should be added that, compared with Mil Specs the civil requirements for roll performance, in the form of FAR 23 and especially the then BCAR Section K, were sparse as far as roll performance was concerned. JAR-23 has now aligned with FAR 23, but even so the requirements still concentrate on minimum, weight-dependent, rates of roll on take-off and landing, '*using a favourable combination of controls*'. Mil Specs, on the other hand, specify a wide range of detailed roll requirements, as one would expect, for aeroplanes in the fighter trainer category.

Matching the rating scale to requirements and findings

Tables 5–5 and 5–6 are derived from Refs 5.13, 5.14 and Table 5–4. They show the route for finding a military requirement and correlating Mil Spec flying quality levels with Cooper–Harper ratings.

Rating for civil airworthiness purposes

Unsuccessful attempts have been made to improve the *Cooper–Harper* system, by ordering the assessments linearly, making them a basis for calculation. The original system continues to be widely used and remains (in the language of airworthiness) as '*proven by useage*'.

Although the *Cooper–Harper* system is applicable to tasks in which demands are made upon the pilot, it cannot be used for assessing compliance of an aircraft, or a device, with airworthiness requirements. Strictly, there is either compliance, or non-compliance. Pilot compensation is not an element to be considered – except in so far as one must not expect skill beyond that of the average pilot. Even so there is tolerance and mitigating factors are taken into account. For a long time I used the scale shown in Table 5–7 for assessing compliance with airworthiness requirements.

Preparing the Aircraft: Weight and Balance

The preparation of an aircraft for a test flight requires accurate knowledge of the weight and CG position relative to datum. By means of such a datum one is able to calculate the CG location on the mean chord.

There are many and varied ways of presenting weight and balance information in the Aircraft Flight Manual. One may show graphically the moment against weight, leaving the pilot or engineer to work out the sum of the moments. Another may show the weight-CG envelope fitted with a cursor, with slots, calibrated for each element, into which the point of a pencil is inserted, one slot at a time, to build up the loading diagram item by item, to avoid exceeding the CG limits. Yet another may present a table with the moment arm of each element listed down the side. The pilot then has to insert the relevant item-weights and work out the sum of their products.

TABLE 5–4

Definitions of terms used when rating handling qualities

Aircraft role: the primary purpose for which the aircraft is used.

Compensation: the measure of additional pilot effort and attention needed to achieve a continuing level of man–machine performance when aircraft qualitieis and characteristics are insufficient for the role intended.

Performance: the efficiency with which the pilot is able to use the controls to achieve required precision in the performance of a task.

Task: is the work contributed by the pilot towards the aircraft, as a man–machine system, to accomplish a defined segment of a flight.
   Control: is that part of the task which requires direct and continuing operation of the primary flying controls and system selectors.
   Auxiliary: is the remainder of the task which does not involve *control*.

Workload: the combination of mental and physical effort needed to perform a piloting task.
   Physical: the effort expended by the pilot in operating the controls to carry out a specific task.
   Mental: the intellectual effort (alluded to in Figs 4.1 and 4.2) which, although it cannot be quantified, can be evaluated either by a trained pilot, or by measurement of the physical input needed to achieve measurable output in the performance of a specific task.

continued

Table 5–4 – *continued*

Revised Cooper–Harper handling qualities rating scale

Aircraft characteristics	Demands on the pilot in selected task or required operation*	Pilot rating
Excellent / Highly desirable	Pilot compensation not a factor for desired performance	1
Good / Negligible deficiencies	Pilot compensation not a factor for desired performance	2
Fair – some mildly unpleasant deficiencies	Minimal pilot compensation required for desired performance	3
Minor but annoying deficiencies	Desired performance requires moderate pilot compensation	4
Moderately objectionable deficiencies	Adequate performance requires considerable pilot compensation	5
Very objectionable but tolerable deficiencies	Adequate performance requires extensive pilot compensation	6
Major deficiencies	Adequate performance not attainable with maximum tolerable pilot compensation. Controllability not in question	7
Major deficiencies	Considerable pilot compensation is required for control	8
Major deficiencies	Intense pilot compensation is required to retain control	9
Major deficiencies	Control will be lost during some portion of required operation	10

Adequacy for selected task or required operation*

Pilot decisions:

- Is it satisfactory without improvement? — NO → Deficiencies warrant improvement — YES → (ratings 1–3)
- Is adequate performance attainable with a tolerable pilot workload? — NO → Deficiencies require improvement — YES → (ratings 4–6)
- Is it controllable? — NO → Improvement mandatory — YES → (ratings 7–9)

*Definition of required operation involves designation of flight phase and/or subphases with accompanying conditions

TABLE 5–5

Route for finding military requirements in Mil Specs
(based upon Ref. 5.13)

Definition	Mil Spec
Type of aircraft	Classification
Flying task	Flight phase
Rating of flying qualities	Level

TABLE 5–6

Correlation between flying quality level and Cooper–Harper rating
(see Table 5–4 and Refs 5.13 and 5.14)

Level	Flying quality	Cooper–Harper rating
1	Clearly adequate for mission	<3.5*
2	Increased pilot workload	3.5 to 5
3	Pilot workload excessive	5 to 7

* Rating 3.5 is interpreted as borderline between 3 and 4.

TABLE 5–7

Classification of civil airworthiness items

Unacceptable: an item which fails to comply with the required standard.

Reservation: an item for which compliance with the required standard is not clearly proven and the issue is reserved for later decision. Resolution will require either:
1. A policy ruling by the regulatory Authority, or
2. Additional substantiation.

Concession: an item which does not comply with the required standard, but which can be accepted on the basis that:
1. An equivalent level of safety is achieved, or
2. The specific requirement is inappropriate.

Temporary derogation: an item which, while failing to comply with the required standard, involves an additional risk small enough to be acceptable for a defined limited period of time.

Major recommendation for improvement: an item which meets the required standard, but where considerable improvement is recommended.

Recommendation for improvement: an item which marginally meets the required standard, but further improvement is recommended.

Unserviceability: a device which is temporarily inoperative or performing below its nominal level.

Comment: self-explanatory.

Figures 5.6 and 5.7 are typical of the information presented to the pilot in a Flight Manual/ Pilot's Information Manual, in this case that of an unspecified American light aeroplane. Such information tends to follow a standard pattern.

Example 5–2: Finding Centres of Gravity

To find the location of the CG it does not matter what units are used, as long as they remain consistent: pounds or kilogrammes, inches, feet or metres throughout. This actual example, using weight in pounds and the moment arms in inches aft of datum, with a test crew of two at design gross weight is continued through Examples 5–3 and 5–4. Working, in Table 5–8, lists the various items of weight and each moment arm. P1 and P2 are first and second pilots respectively. *JP4* is turbine fuel (see Jet 'A' calculations in Fig. 3.26, showing turbine fuel as weighing 6.7 lb per US gal). The CG in this case is found to lie at 136.1 inches aft of the datum specified by the manufacturer for such calculations.

Example 5–3: Adjustment of CG by Movement of Ballast

Now assume that the CG in the previous example is too far aft, by 1.5 in (3.81 cm), and it must be moved forwards to 134.5 in AOD. The gross weight for test must be maintained, but ballast can be moved. To find out how many inches forward 340 lb of ballast should be moved, we resort to this equation:

$$\pm \Delta \bar{x} = \pm (W_n / W_0) \Delta x_n \qquad\qquad (3\text{--}77)$$

Forward movement is in the negative direction so, transposing for Δx_n:

$$-\Delta x_n = -\Delta \bar{x} / (W_n / W_0) = -\Delta \bar{x} \ (W_0 / W_n)$$
$$= -1.5 \times (4789 / 340)$$
$$= -21.13 \text{ in } (54.1 \text{ cm}) \text{ (forward) shift of ballast}$$

This means that the ballast must now be located at:

$$145 \text{ in AOD} - (21.13 \text{ in}) = 123.87 \text{ in } (0.315 \text{ m}) \text{ AOD}$$

Example 5–4: Adding Ballast and Off-Loading Fuel

But what if it is found that the CG in Example 5–2 is too far forward by 0.88 in (2.24 cm)? The gross test weight cannot be changed, nor the main ballast increased at that station, but there is room for more ballast further aft, 60 in (1.524 m) nearer the tail, i.e. 145 in + 60 in = 205 in AOD. Fuel weight is on the top limit (the tanks are full) and so the whole problem can only be resolved by off-loading fuel to compensate for the additional weight of extra ballast needed at 60 in AOD.

As the reduction in fuel weight and the increase in ballast aft are equal, we resort to the following equation to find what change in both weights is required:

$$\pm \Delta W_1 = \pm W_0 \Delta \bar{x} / (x_1 \mp x_2) \qquad\qquad (3\text{--}76)$$

noting carefully the (+) and (–) signs:

$$+\Delta W_1 = (4789 \times 0.88) / [(145 + 60) - (154.72)]$$
$$= + 83.8 \text{ lb } (38.1 \text{ kg})$$

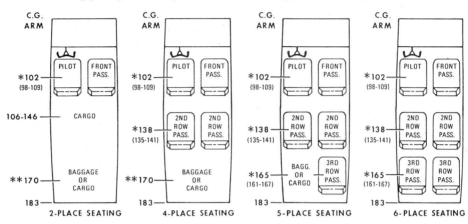

Fig. 5.6 Typical presentation of weight and balance information in the Aircraft Flight Manual/Pilot's Information Manual.

Total ballast is now (340 + 83.8) = 424 lb (193 kg), rounded up to the nearest whole unit. While the fuel load is reduced by an equal amount to (1120 – 83.8) = 1036 lb, or 155 US gallons (see Fig. 3.25).

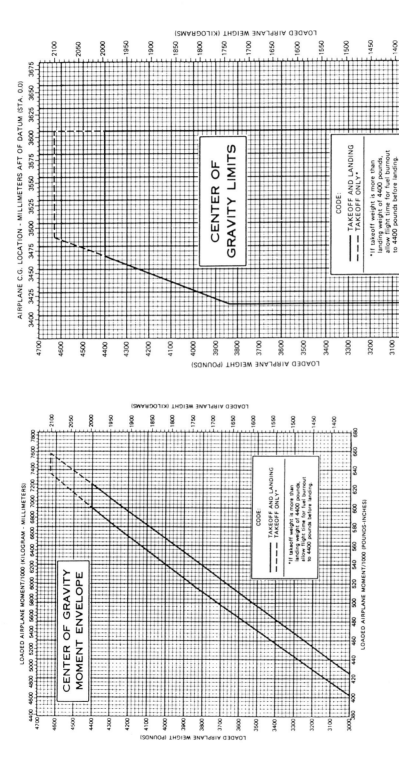

Fig. 5.7 Typical centre of gravity moment and limit information presented in the weight and balance section of an Aircraft Flight Manual/Pilot's Information Manual.

TABLE 5–8

Weight and balance with two crew at gross weight, W_0

Item	Weight (lb)	Arm in AOD	Moment (lb in)
Basic empty (Equation (3–80))	2954	130.04	384 162
P1	205	120	24 600
P2	170	120	20 400
Full fuel (167 US gal of JP4)	1120	154.72	173 286
Ballast	340	145	49 300
TOTALS	4789	136.1	651 758

The CG is located at $(651\ 758/4789) = 136.1$ in AOD (3.46 m, aft of datum).

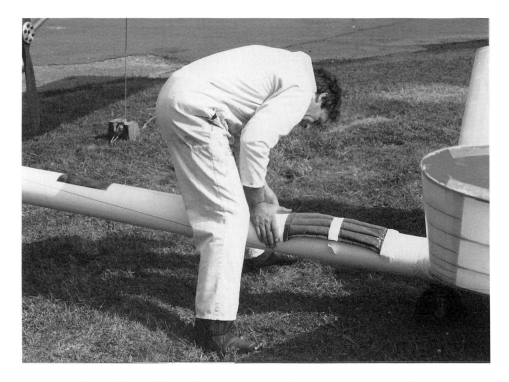

Plate 5–1   Adding ballast, taped to the tail of a three-axis microlight, to achieve aft CG (Pipistrelle) pusher, with a vee-tail. Ref.: Example 5–3. (Author)

Airspeed calculations

In Chapter 3 airspeeds were defined at length. Their measurement is subject to errors:

☐ *Instrument error* is inaccuracy which is largely due to mechanical losses. The error is unique to each instrument, is subject to wear, and must be calibrated by bench test

Plate 5–2 This plate, along with Plate 5–1, shows that it is often hard to find a convenient and secure place to put required ballast for aft CG tests. Ref.: Examples 5–2 and 5–3. (Author)

before an instrument is used in flight. A non-mechanical contribution arises from (see Chapter 7):

☐ *Pressure* or *Position error* which arises from differences between the far-field values of pitot and static pressures and those in the near-field, which is disturbed by the presence of the aircraft and by compressibility effects at high speeds. The pressure error must be known accurately for operational and flight safety reasons.

☐ *Lag error* caused by rapid changes in airspeed and altitude by the aircraft, which are too fast for the pressure-sensing sources.

At this point we assume that some sources of airspeed information exist which can be used by the pilot to construct a package of comprehensive flight tests. They might be the result of a quick look at a prototype on its first flight. On the other hand they might be culled from a published and approved AFM, POH or other document used for certification.

To use the information to greatest effect remember that the ambient air temperature and barometric pressure profoundly affect performance results, particularly those measured in the climb. The climb lines plotted in Fig. 2.5 are a case in point, in that their validity is 'not proven'. They show marked differences in climb performance between the Mercedes and BMW-engined and one other Fokker DVII. Comparisons with the Bristol F2B are also open to doubt, simply because we know nothing of the ambient conditions: time of day, time of year, weather (was a warm or cold front approaching, or had one just gone through, and for which aircraft?), or of the flying experience and standards of the pilot(s).

Take, for instance, Fig. 5.8 which shows an aeroplane flying through a column of air at some 'tape-measure' height above mean sea level (amsl). The height of the column of air to the notional 'top of the atmosphere' varies constantly as one might gather from satellite pictures taken from space, showing cloud formations covering vast tracts of each hemisphere with linked vortex motions of differing intensity. For a given barometric pressure at

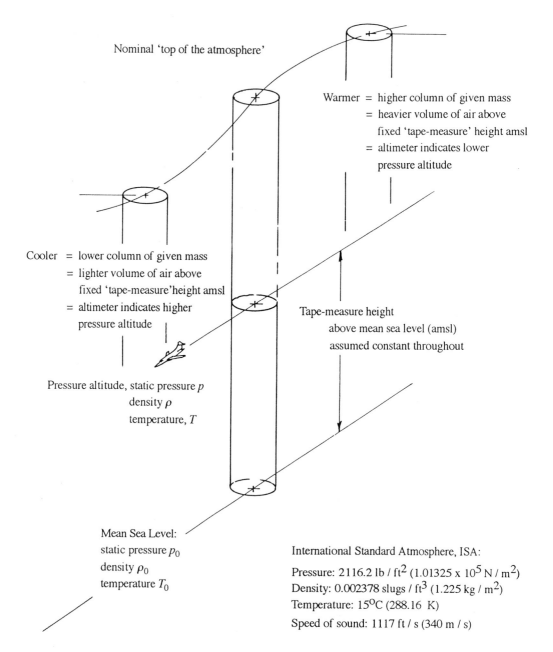

Nominal 'top of the atmosphere'

Warmer = higher column of given mass
       = heavier volume of air above
         fixed 'tape-measure' height amsl
       = altimeter indicates lower
         pressure altitude

Cooler = lower column of given mass
       = lighter volume of air above
         fixed 'tape-measure' height amsl
       = altimeter indicates higher
         pressure altitude

Tape-measure height
    above mean sea level (amsl)
    assumed constant throughout

Pressure altitude, static pressure p
         density ρ
         temperature, T

Mean Sea Level:
    static pressure p_0
    density ρ_0
    temperature T_0

International Standard Atmosphere, ISA:

Pressure: 2116.2 lb / ft^2 (1.01325 x 10^5 N / m^2)
Density: 0.002378 slugs / ft^3 (1.225 kg / m^2)
Temperature: 15°C (288.16 K)
Speed of sound: 1117 ft / s (340 m / s)

Fig. 5.8 The reason why pressure altitude indicated at a given 'tape-measure' height amsl changes with ambient conditions.

mean sea level (weight of the column of air pressing down on unit area of MSL surface), a warm column is taller than one of the same weight which is cooler. What this means is that an aircraft flying on the radar altimeter at a fixed 'tape-measure' height over the sea, or at a fixed altitude to clear a mountain top, transits through air masses of varying temperature and vertical extent. If we represent these as columns of equal mass, each exerts the same sea level static pressure. But the aneroid altimeter in the aircraft registers apparent changes in altitude as the flight progresses, and as the instrument is a barometer it shows changing pressure-altitudes on encountering each different air mass, not 'tape-measure' height.

For example, a column of ISA air presses down on the surface at MSL with a pressure of:

$1.013\ 25 \times 10^5$ N/m² (2116.2 lb/ft²)

If the column has a SL temperature of (ISA + 10°C), and the lapse of temperature with altitude corresponds with the theoretical ISA law, then an aircraft flying at 10 000 ft on the radar altimeter would have a pressure altitude indicated of 9650 ft, because the altimeter capsule would be measuring the ambient pressure due to the weight of the column of air above it. But if the SL temperature is (ISA–10°C), then the same aircraft at 10 000 ft on radar altimeter would have an indicated altitude around 10 400 ft.

Looked at the other way around, when flying at a constant indicated pressure altitude through hot and cold air masses, then with all else equal, the pilot would see his radar altitude changing as he climbs and descends through air masses of differing temperatures.

Changing air temperature and pressure are associated with changing air density, ρ, which appeared in Equations (3–26) and (3–63) to (3–66). The dynamic pressure, q, of Equation (3–26) has other useful forms, for example, when making calculations for flight at high Mach numbers, which flow from the equation for the speed of sound in ft/s:

$$a = (\gamma p / \rho)^{0.5} \tag{5–58}$$

where $\gamma =$ the ratio of the specific heats for air, approx. 1.4; and a, p and ρ have their earlier meanings. From this we may show that:

$$q = 1/2\rho V^2 = 1/2\gamma p M^2 \tag{5–59}$$
$$= 0.7\ p\ M^2 \tag{5–60}$$

As an aircraft pushes through the air it adds kinetic energy to the molecules, which is measurable as an adiabatic temperature rise in terms of the true airspeed in mph:

$$\Delta T = (V\ \text{MTAS}/100)^2\ °C \tag{5–61}$$

to within three parts in 1000. Raising the temperature of the air reduces its density. This is a source of error in the measurement of dynamic pressure. Compressibility is another source of such error. The magnitude of the various errors at any given altitude, airspeed and Mach number, depend upon the locations of the sources measuring the dynamic and static pressures on the airframe.

Figure 5.9 summarizes the various factors involved in the calibration of an airspeed indicator. During early test flights it is a fairly simple matter to arrange for a static pressure-measuring cone, towed point first, within which is the static sensor (a hole) pointed aft. The cone shown is out of scale, as is the hose. A thin hose connecting the static source to the instruments should take no strain and so it is clipped to a thin but strong line. Both are then wound on to a drum inside the aircraft, from which the device can be trailed and recovered in flight. The pitot head measures (dynamic + ambient static pressure), the cone measures the ambient static pressure. When the pressures are fed to chambers on opposite sides of a diaphragm (one chamber is formed by the aneroid capsule inside the airspeed indicator),

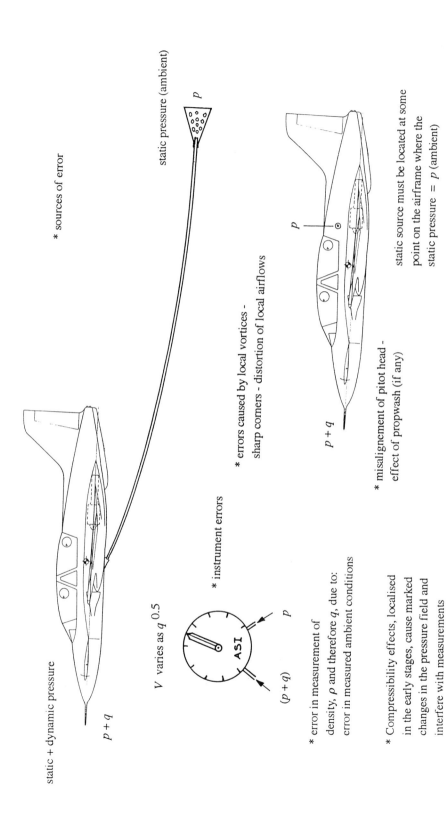

static + dynamic pressure

$p + q$

* sources of error

static pressure (ambient)

p

V varies as $q^{0.5}$

* instrument errors

$(p + q)$ p

ASI

* error in measurement of
  density, ρ and therefore q, due to:
  error in measured ambient conditions

* Compressibility effects, localised
  in the early stages, cause marked
  changes in the pressure field and
  interfere with measurements

* errors caused by local vortices -
  sharp corners - distortion of local airflows

$p + q$

* misalignment of pitot head -
  effect of propwash (if any)

p

static source must be located at some
point on the airframe where the
static pressure = p (ambient)

Fig. 5.9a   Measurement of pressure difference $q = (p + q) - p$, and causes of error between pitot head and static source. In this case the test aircraft (top) is towing a trailing static cone.

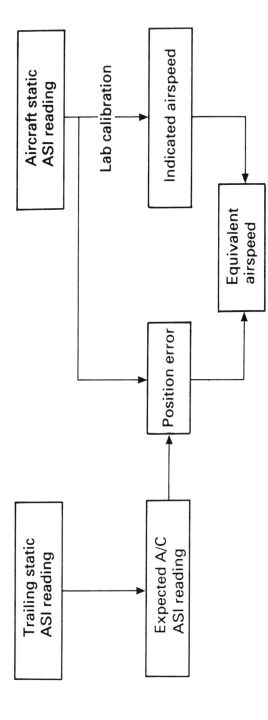

Fig. 5.9b The process of airspeed calibration. (Courtesy of the former Norman Aeroplane Company Ltd)

the static pressures cancel and the diaphragm is deflected by the dynamic pressure alone. The deflection is transmitted by means of a suitable linkage to the ASI display. The device is trailed below and behind, well clear of the near-field of the aircraft.

For higher speeds a pressure-sensing 'bomb' is often used. This is trailed clear of the aircraft and is equipped with its own pitot head and separate static source.

The true airspeed of the aeroplane is measured using radar 'fly-by' techniques, or over kinetheodolites spaced at a known distance apart on the ground. These involve corrections for windspeed, on top of which they are expensive. Another method is to fly in formation with a second chase aircraft which has calibrated instruments. A reliable check is to measure the average times taken to fly reciprocal tracks between surveyed points on the ground, or around a circuit of known length, as used for air or yacht and power-boat racing.

The same dynamic and static sources are used to measure airspeed, altitude and Mach number. Figure 5.10 is an actual example in which CAS and IAS were plotted from manufacturer's data for use during a test flight to validate the type-certification of a foreign-built aircraft. When an ASI has been calibrated instrument errors are recorded on a card mounted close to the instrument, so that the relevant corrections might be applied in flight.

Example 5–5: Using the Aircraft Flight Manual and Pilot's Operating Handbook

When carrying out a test flight to renew or validate an existing C of A one has to interpolate from IAS to CAS and vice versa, using information like that ringed in Fig. 5.11 (extracted from an unspecified Pilot's Operating Handbook). For normal, non-test, operations it is enough to use tabulated data. During certification flight tests factored calibrated and indicated airspeeds are needed ($1.2V_S$, $1.4V_{S0}$, etc.) and they are not easy to derive direct from tabulated data. Therefore Fig. 5.12 shows a way of plotting tabulated forward CG data, given in Fig. 5.11, coupled with a grid for quick conversion of stall speeds into factored CAS and IAS values. Its construction needs explanation.

The coordinates run from the lowest indicated speed: 50 KIAS in the bottom LH corner of the top LH (Normal Static Source) calibration table in Fig. 5.11. Both coordinates are graduated equally. A line at 45° is drawn through the origin to represent 1.0 V_S. Further lines, for 1.1 to 1.5 V_S are then constructed as a fan-like grid. The ordinate is labelled in the relevant units CAS (i.e. KCAS for this aircraft). The abscissa is labelled both KIAS and KCAS, because the graph may be used for both.

Plot the flaps-UP and flaps-FULL calibration curves, one with a solid line the other broken, using the data for the Normal Static Source (top LH calibration table). The major portions of these two curves are with power required for level flight, or maximum power during a descent and annotated as such by the manufacturer.

Factored speeds start with point A, which is the stall speed in the landing configuration, V_{S0} = 64 KCAS (bottom RH table, FULL flap), for which the corresponding indicated stall speed is 57 KIAS. Mark that as point A'. Erect perpendiculars from A and A'. Where that from A intercepts the 1.0 V_S line, mark this as B and draw a line parallel with the abscissa through B to intercept the perpendicular from A' at G'. We do not know how this stall speed was obtained – was it throttle closed (UK) technique, or with RPM set for zero thrust (US)? Either way, little will be lost by marking the stall speeds as being power-OFF.

Now do the same for the power-OFF stall speed in the clean configuration V_{S1} = 72 KCAS (bottom RH table), marking it as point F. Erect a perpendicular from F to meet the 1.0 V_S line at G. A perpendicular from V_{S1} = 66 KIAS meets a line drawn parallel with the abscissa through G at G'.

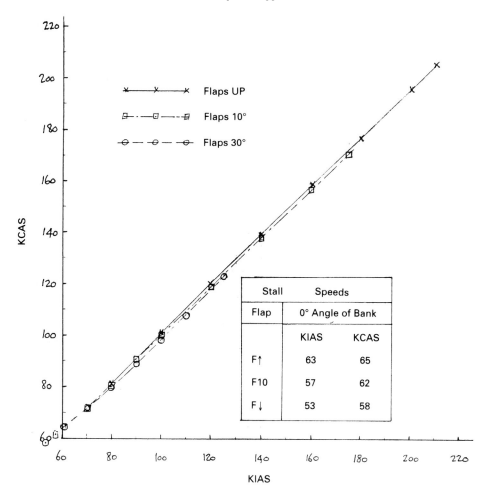

Airspeed data prepared from manufacturer's information for validation tests of a
twin-engined light aeroplane at gross weight (MTOW). Such a presentation is cramped
compared with fig. 5.12. Stall KIAS values are approximate because of unpredictable
errors which defy accurate measurement and intrude at speeds less than 1.2 V_S.
The AFM notes that stall speeds are approximate

Fig. 5.10   Conversion of tabulated airspeed data to graphical form to assist interpolation
during subsequent flight tests.

To obtain, for example, $1.3V_{S0}$ and $1.2 V_{S1}$ as KIAS, extend the perpendiculars from A
and F (both of which are CAS) to meet the respective $1.3 V_S$ and $1.2 V_S$ lines at C and H.
By drawing lines parallel with the abscissa through C and H we obtain $1.3V_{S0}$ KCAS and
$1.2 V_{S1}$ KCAS where they intercept the ordinate. In the other direction, where they intercept
the calibration curve at the points D and I, drop perpendiculars to E and J to obtain
$1.3 V_{S0}$ KIAS and $1.2 V_{S1}$ KIAS.

STALL SPEEDS

CONDITIONS:
Power Off
Gear Up or Down

NOTES:
1. Altitude loss during a stall recovery may be as much as 450 feet.
2. KIAS values are approximate.

REARWARD CENTER OF GRAVITY (FUSELAGE STATION 142)

WEIGHT LBS	FLAP DEFLECTION	ANGLE OF BANK							
		0°		30°		45°		60°	
		KIAS	KCAS	KIAS	KCAS	KIAS	KCAS	KIAS	KCAS
4630	UP	63	70	68	78	75	83	89	99
	1/3	58	66	62	71	69	78	82	93
	FULL	51	61	55	66	61	73	72	86
4300	UP	61	68	66	73	73	81	86	96
	1/3	56	64	60	69	67	76	79	91
	FULL	49	59	53	63	58	70	69	83
4000	UP	59	65	63	70	70	77	83	92
	1/3	54	61	58	66	64	73	76	86
	FULL	47	57	51	61	56	68	66	81

FORWARD CENTER OF GRAVITY (FUSELAGE STATION 137.4)

WEIGHT LBS	FLAP DEFLECTION	ANGLE OF BANK							
		0°		30°		45°		60°	
		KIAS	KCAS	KIAS	KCAS	KIAS	KCAS	KIAS	KCAS
4630	UP	66	72	71	77	78	86	93	102
	1/3	62	68	67	73	74	81	88	96
	FULL	57	64	61	69	68	76	81	91
4300	UP	64	69	69	74	76	82	91	98
	1/3	60	66	64	71	71	78	88	93
	FULL	55	62	59	67	65	74	78	88
4000	UP	61	67	66	72	73	80	86	95
	1/3	58	63	62	68	69	75	82	89
	FULL	53	60	57	64	63	71	75	85

AIRSPEED CALIBRATION

CONDITIONS:
1. Power required for level flight or maximum power during descent.
2. 4630 pounds.

NORMAL STATIC SOURCE

FLAPS UP								
KIAS	60	80	100	120	140	160	180	200
KCAS	67	81	101	121	141	161	181	202

FLAPS 1/3								
KIAS	60	70	80	90	100	120	140	160
KCAS	64	73	81	91	100	120	140	161

FLAPS FULL							
KIAS	50	60	70	80	90	100	110
KCAS	59	64	72	81	91	102	112

ALTERNATE STATIC SOURCE

FLAPS UP								
NORMAL KIAS	60	80	100	120	140	160	180	200
ALTERNATE KIAS	63	85	108	130	152	173	195	217

FLAPS 1/3								
NORMAL KIAS	60	70	80	90	100	120	140	160
ALTERNATE KIAS	64	75	86	97	107	129	150	171

FLAPS FULL							
NORMAL KIAS	60	70	80	90	100	110	- - -
ALTERNATE KIAS	74	83	94	105	117	128	- - -

Fig. 5.11 Typical tabular presentation of airspeed calibration and stall speeds in AFM.

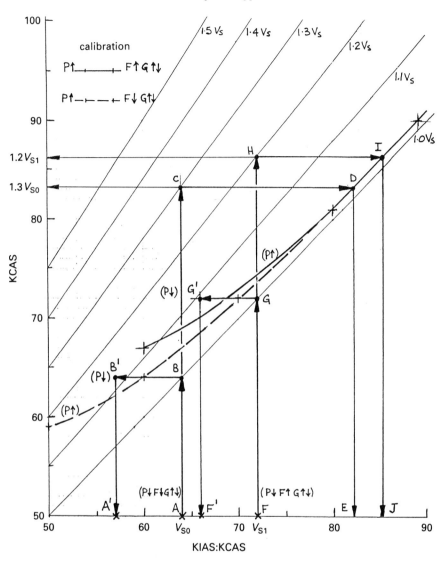

Fig. 5.12 Plot of calibrated speeds in Fig. 5.11 in a form amenable to interpolation and factoring between CAS and IAS.

When making conversions from AFM and POH data to graphical form before factoring as shown, make sure that:

☐ *Correct data* is used to determine the test CG.

☐ *Configuration* (*Conditions* in Fig. 5.11) to which the data applies are noted. In Fig. 5.12 the various critical points have been annotated Power-ON, or Power-OFF, using the shorthand suggested in Table 4–4.

When an ASI Fails

Early test flying should be carried out VMC and in sight of the ground, in case of the loss of primary flight instruments before systems are sorted out (it can happen). The failure of an ASI can be particularly disconcerting when the pilot is unsure of stall speeds, power settings and corresponding airspeeds and attitudes. Extreme changes of attitude and airspeed must be avoided. Roger Vee's comment on flying the Fokker DVII, in Example 2–2, is relevant, in that the pitot tube and other instruments had been removed, leaving him with only the wind on his face. So, he, '*intended to take good care to get up sufficient speed before taking the Fokker off the ground'*.

If caught out on instruments and the *turn and slip indicator* is working, first use it to level the wings, while keeping an eye on the altimeter. If altitude is increasing the nose is above the horizon, if decreasing, the nose is below. By checking forward or backward on the stick, the moment the altimeter begins to move in the other direction, reverse and halve the stick movement, keeping the wings level ... and so on until again in steady level flight.

The *vertical speed indicator* ceases to register a rate of climb or descent as the nose crosses the horizon. It is more sensitive than the altimeter, so use the altimeter for a coarse indication of where the nose is relative to the horizon. Then gradually, transfer attention to the VSI.

When back in more or less level flight, use power changes and the Vertical Speed Indicator, VSI, to keep changes in height and airspeed small. An *artificial horizon* makes everything much easier. Otherwise, keep rates of turn slow, to turn on to and hold a compass heading.

The situation may be easier in an open cockpit, when the main thing is to follow the advice of a flying instructor, H. Barber, in 1916:

'Keep the aeroplane in such an attitude that the air pressure is always directly in the pilot's face.'   (Ref. 5.15)

Example: 5–6 Flight with a Failed ASI

Early in the 1970s I was given the opportunity to test a homebuilt Breezy. The parasol monoplane had a bare *Warren-girder* fuselage, no windscreen, seats out in front of the strut-braced wing. The pusher engine and tail-surfaces followed behind, and all sat firmly on a nose-gear undercarriage.

When airborne in VMC, concentrating on attitude, the ASI reading between my feet dropped to zero. There was no turn and slip indicator and the only other flight instrument was an altimeter.

Swaying at the narrow end of a travelling crane, the buzz of the engine assuring that the aeroplane was still attached somewhere behind, things began to fall into place:

☐ *Mouth closed*, sideslip flattened the upwind nostril.
☐ *Mouth open*, sideslip inflated the downwind cheek.
☐ *Increasing airspeed* bent eyelashes downwards behind the sunglasses – so here was V_{NE}.

Plate 5–3 The face is an instrument, when one can feel the wind. Ref.: Example 5–6. Breezy at Oshkosh. (Courtesy of the late Harold Best-Devereux)

☐ *Throttled back* and easing the nose up slowly, an increasing 'thump – thump – thump –... of aileron cables inside the wing gave comforting audio low-speed warning.

Barber was right – when you can feel the wind the face is an instrument panel!

References and Bibliography

5.1 Stinton, D., *A consumer's approach to the brochure* Supplement to Vol. 39–7. Geneva: Interavia, 1984

5.2 Stinton, D., Number-crunching nine company aircraft. *Business Aviation 1986* Supplement to Vol. 41–7. Geneva: Interavia, 1986

5.3 Chevalier, H. L., Comparison of loading parameters for various single-engined light airplanes. Oshkosh (Wisconsin): *Sport Aviation*, November 1969

5.4 Wild, R. H., Format for flight, *Flight International*, 11 March 1971

5.5 Wild, R. H., *The State of the Art of Light Aircraft Design*. Geneva: *Interavia*, 1972

5.6 Everling, E., Comparative qualities in aircraft statistics, *Flight*, Aircraft Engineer Supplement, 1926

5.7 Mettam, H. A., Comparison of aircraft performance, *Flight*, Aircraft Engineer Supplement, 1927

5.8 JAR-23, Cheltenham: Civil Aviation Authority, Printing and Publications Services, 1993

5.9 FAR Part 23, Washington: Federal Aviation Administration, The Office of the Federal Register, National Archives and Records Administration, 1990

5.10 Pazmany, L., A method devised for the *Experimental Aircraft Association* of Oshkosh, Wisconsin, c1980

5.11 Mooij, H. A., NLR TR 83037U *Criteria for Low-Speed Longitudinal Handling Qualities of Transport Aircraft with Closed-Loop Flight Control Systems*. Amsterdam: Martinus Nijhof Publishers, 1984

5.12 Cooper, G. E. and Harper, R. P. *The Use of Pilot Rating in the Evaluation of Aircraft Handling Qualities*. AGARD-R-567, 1969; and NASA TN-D 5153, 1969

5.13 Masefield, O. L. P., Doctoral Thesis, *Design of a Manual Roll Control for a Trainer Aircraft*. Loughborough University of Technology, Leicestershire (Department of Transport Technology), 1990

5.14 Moorhouse, D. J. and Woodcock, R. J., AFWAC-TR-81-3109, *Background Information and User Guide for MIL-F-8785C, Military Specification for Flying Qualities of Piloted Airplanes*, Wright-Patterson Air Force Base: Flight Dynamics Laboratory, Ohio 45433, 1982

5.15 Voigt, W., Gestation of the Swallow, *Air International*, March 1976

5.16 Gibbs-Smith, C. H., *The Aeroplane, An Historical Survey*. Her Majesty's Stationery Office, 1960

5.17 Penrose, H., *Architect of wings. A Biography of Roy Chadwick – Designer of the Lancaster Bomber*. Shrewsbury: Airlife Publishing, 1985

5.18 Dommasch, D. O., Sherby, S. S., Connolly, T. F., *Airplane Aerodynamics*, New York: Pitman Publishing Corporation, 1959

SECTION 4

PERFORMANCE

CHAPTER 6

Airfield Performance

'... that TriStar was lucky to get home with all but one hydraulic pipe cut by that errant centre-engine fan. Except that such a "10^{-7}" was postulated long ago by a British airworthiness engineer, who said at some meeting "*we'd better have one of those pipes over there, just in case*" .... We owe a lot to airworthiness engineers – and to the airworthiness test pilots. Thank them, and of course many other carnally knowledgeable test pilots, for the crashes that didn't and won't happen ... a test pilot's job is to produce an aircraft that doesn't need a test pilot to fly it.'
Roger Bacon, Straight and level, *Flight International*, 9 January 1982

Performance is the flying quality which has the most influence upon commercial sales and success. If one aeroplane can take off and land shorter from a rougher airfield, climb higher, cruise faster and fly farther than another with the same payload, for the same price, then it will be the one which sells better. The purpose of this chapter is to see how performance information is collected, treated and presented for use by the operator, not how complicated performance tests are carried out.

The development of compact and powerful gas-turbines driving propellers wrought a revolution in the development of small passenger and freight-carrying aeroplanes. Turbo-prop engines are tiny, light, reliable and highly expensive jewels, and with their combination of high power and economy they led to the introduction of the important commuter class of aeroplane. The commuter, we saw in Chapter 3 (UK Group B/JAR-OPS Class A), has no more than 19 seats and a MTOW not exceeding 19 000 lb (8600 kg). Because it is potent it is the main type of aircraft to bear in mind in this chapter on airfield performance (i.e. the ability of an aeroplane to take off and land with a publicly acceptable level of assurance of safety on airfields of declared lengths at safe weights, in defined operational conditions). In size it might not be much larger than many small twin-engined aeroplanes in JAR-OPS Class B (i.e. less than nine seats and an MTOW of not more than 12 500 lb (5700 kg)). Yet the commuter has capabilities enough to warrant a level of scheduled performance similar to much larger transport category aeroplanes. It effectively straddles and links the light and the heavy ends of airworthiness certification and operation.

Twin and single-engined small aeroplanes are discussed separately later in the chapter.

Performance and probability

The purpose of performance regulation is to ensure to an acceptably high probability that the space required for the flight should not exceed the space available. Numerically, the required probability of success must be 1 000 000 to 1 (in shorthand, 10^6), and of failing to do so, only 1 in 1 000 000 (or, 10^{-6}).

Performance operating rules are essentially flight safety rules. They ensure that an aeroplane used for public transport is despatched at an acceptably safe weight. The rules are concerned only with the despatch weight, called the *Regulated Take-Off Weight* (RTOW), because when this has been determined and the aircraft despatched at this weight (mass) the pilot will be called upon to make critical performance decisions under pressure, only in the event of unlikely failures (e.g., engine failure, dragging brake).

Unlikely is emphasized because performance requirements are concerned with probabilities: like the probable risks shown in Table 6–1. Operators sell reliability of their operations to passengers, emphasizing safety, the risk-free assurance of getting there. Thus, risk and reliability are complementary, like opposite sides of a coin. One is the head and the other the tail.

To see how probability theory works, let us apply it to the simplest take-off case of one type of aeroplane with all engines operating. On take-off the aeroplane is heavy with fuel and payload. Although the engines are at take-off power, acceleration is slowest, and in the event of an engine failing, the asymmetric yawing moments are higher than with power reduced for landing. But, exactly the same probability methods apply to both take-off and landing.

Probability Theory

If the reliability of a system is 70%, then risk of failure is 30%, and vice versa.

TABLE 6–1

Scale of probabilities

Description	Probability	Examples
Frequent: Likely to occur often in the life of the aircraft		Popping of a circuit-breaker Failure of a light
	10^{-3}	
Reasonably probable: Unlikely to occur often but may occur several times during the life of each aircraft		Engine failure
	10^{-5}	
Remote: Unlikely to occur to each aircraft during its life but may occur several times during the life of a number of aircraft of the type		Minor accident damage Possible passenger injuries
	10^{-7}	
Extremely remote: Possible but unlikely to occur in the total life of a number of aeroplanes of the same type		Ditching, double engine failure Extensive accident damage Possible loss of life
	10^{-9}	

Therefore:

reliability % = 100 – risk %

or, dividing through by 100:

$$\text{reliability} = 1 - \text{risk} \tag{6–1}$$

The reduction of risk involves the spending of time and effort, which cost money. The problem always comes down to how much can one afford to spend? When there is competition, how much dare one not afford?.

If we toss a coin with heads representing reliability and tails the risk, then in one throw:

probability of reliability (head) in one throw = 1/2
probability of risk (tail) in one throw = 1/2

The *certainty* of getting either a head or a tail in one throw with a coin which cannot land on its edge is 100%, because then there are no other choices, and this can be expressed as:

certainty = 100 %

which is:

$$= (1/2) + (1/2) = 1 \tag{6–2}$$

While the probability of getting neither (as when the coin lands and stays on edge) is so remote as to be virtually impossible:

$$= 1 - (1/2 + 1/2) = 0 \tag{6–3}$$

Hence, the chance of getting a head in only a few throws can vary from 0 to 100% in theory. We might have a run of heads, or a run of tails which could make us think that the chance of one was better than the other. But after a very large number of throws, say 100 written as 10^2, or 1000 (10^3), or 10 000 000, i.e. 10^7, we would come closer and closer to the *true probability* of the system, which in the case of the coin toss is:

$1/2 = 0.5 = 5 \times 10^{-1}$

The probability, P_A, of an event like failure to lift-off in a given distance along a runway in a number of attempts is given by the ratio of the number of failures, A, to the total number of test runs carried out, $(A + B)$, in which B is the number of times lift-offs were accomplished within the required distance:

$$P_A = A/(A + B) \tag{6–4}$$

In reality the true probability is an unknown and its determination would need time-consuming and expensive testing, far in excess of what would be practicable. Fewer (more cost-effective) tests produce a band of values from which a mean value can be deduced as an approximation to the true probability. Even though it cannot be found with the certainty of a single number, a unique point, as long as we know the mean value and shape of the distribution of results about the *mean*, defined by what is called their *standard deviation*, then we have enough to go on.

The Airfield and Take-Off Performance Rules

Figure 6.1 is a sketch of a runway on a typical airfield. It has a taxiway, which is commonly part of the perimeter track, giving access to the point A on the runway at which the pilot

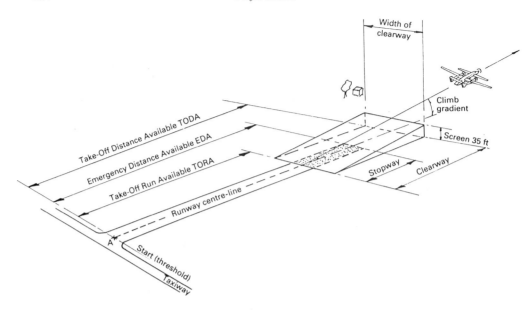

Fig. 6.1   Typical declared distances on take-off.

completes his line-up and stops to complete final checks – like *"CONTROLS FULL AND FREE"* – before opening up and starting his take-off run. There is room for lining up earlier or later, so that each aeroplane does not start its run at exactly the same spot. One pilot may turn sharply from the taxiway, so gaining slightly in distance available over another who takes a wider, lazier, turn on to the runway and then stops further down. Yet another may turn wide without stopping, gently easing on power in the turn, building up forward speed early, saving a few seconds, but wasting distance since the early part of the take-off suffers from slow acceleration due to the low thrust while the engines 'spool-up'.

The figure shows three *declared distances* from the defined start of the take-off run. In order they are *take-off run available, emergency distance available* and *take-off distance available*:

☐ TORA = paved length for acceleration on the ground   (6–5)
☐ EDA = TORA + *stopway* (room to stop safely in an emergency)   (6–6)
☐ TODA = TORA + *clearway* (room to fly over obstacles – including
            ditches – but which cannot exceed the first upstanding obstacle)   (6–7)

The defined start is assumed to be the runway threshold. This may be either the outer edge of the taxiway, at the grass, or it may be displaced inwards as shown, although this latter case normally applies only to landing rather than take-off.

The rules governing take-off performance specify a number of indicated airspeeds, normally obtained by factoring, for example, the stall speed in the take-off configuration, V_{S1}. At the start of the take-off run the pilot calls for *'full power'* (not *'take-off power'*, which has eye-watering consequences if the throttle is chopped instead (as has happened)). He releases the brakes and the aeroplane accelerates. At a predetermined *rotation speed*, V_R, the pilot pulls on the stick or control-yoke to lift the nose-wheel off the ground. The aeroplane then continues the take-off run on the mainwheels at an increased angle of attack. To ensure that the pilot will not attempt to haul the aeroplane into the air before it is ready:

$V_R > V_{EF}$ (6–8a) which is

$$\geq 1.10\ V_{S1} \qquad\qquad (6\text{–}8b)$$

where V_{EF} is the speed at which an engine is assumed to fail in the calculation of the decision speed, V_1.

Acceleration continues until lift-off speed, V_{LOF}, is reached. Ideally, the aeroplane should leave the ground cleanly at this speed. But, as it is possible to lift-off earlier in ground effect (ground cushion under the wings and fuselage) by tweaking back on the stick prematurely, a *minimum unstick speed*, V_{MU}, is defined to afford protection, by making:

$$V_{LOF} \geq 1.10V_{MU} \qquad\qquad (6\text{–}9)$$

Lift-off must be achieved in the length of take-off run available, TORA.

At lift-off the aeroplane is climbed straight ahead to penetrate an imaginary screen (defined in the rules as 35 ft unless otherwise stated), which is located at the end of the TODA. A speed of V_2 must have been achieved within the TODA, such that:

$$V_2 \geq 1.10V_{MC} \text{ or } 1.2\ V_{S1} \qquad\qquad (6\text{–}10)$$

The term $1.10\ V_{MC}$ is a multiple of the minimum control speed with one engine inoperative, to provide lateral and directional controllability protection in the possible event of engine failure.

Scatter in measured results

Not every aeroplane of the same type and weight, each flown by a different pilot, leaves the ground at exactly the same point on the runway. Nor will they penetrate the screen in the same place at the end of the TODA, even if all start from precisely the same point. There is scatter like that shown in Fig. 6.2, raising the question: '*What is the probability of achieving, say, TORA (or TODA and V_2) with an average pilot in an average example of this type of aeroplane?*' Engineers may seek an answer by using '*engineering judgement*', i.e. looking at the scatter, using intuition and past experience, while being well aware of the consequences of over-optimism in one direction, and excessive caution in another. But engineering judgement alone is not reliable when attempting to weigh what is achievable and likely to satisfy the level of safety which the public considers acceptable.

Scatter in the measurements is caused by two groups of variables:

☐ *Operational variables*, which can be measured or 'forecast', such as weight, altitude, temperature and runway slope.
☐ *Statistical variables*, which are subject to everyday variations and include shortfall in thrust, increased drag, braking force coefficients between wheels and dry or wet (said to be 'contaminated') runway surfaces, initial climb speeds and gradients. These are accounted for using analytical techniques.

To cope with scatter mathematically we use the concepts of *gross* and *net performance*.

Gross and net performance

The following definitions are based upon those of the CAA:

☐ *Gross performance* is the *average* performance which a fleet of aeroplanes of the same type is as likely to achieve as not, if satisfactorily maintained and flown in accordance

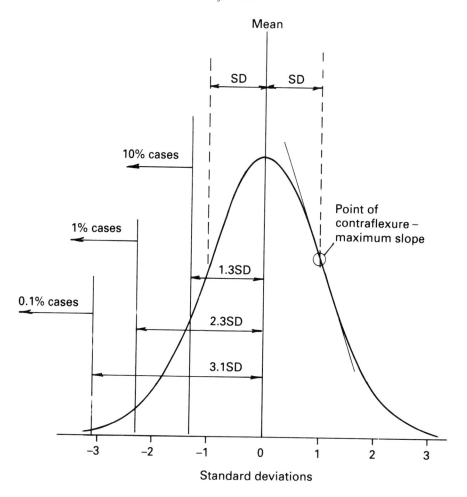

Fig. 6.2 Properties of the normal Gaussian distribution. 68.3% of all cases are contained within ±1 SDs (standard deviation). Within ±2 SDs lie 95.4% of the cases, and 99.7% within ±3 SDs. The diagram shows % of cases lying outside it.

with the techniques prescribed in the Flight Manual. Although we call it gross perform-ance it embraces many things, e.g., *average* distances for take-off or landing, airspeed after lift-off. Aiming to take off in a gross distance, the pilot has a 50:50 chance of succeeding.

☐ *Net performance* is the average performance *adjusted* by an amount necessary to cater for various contingencies which, while causing scatter, cannot be accounted for operationally.

Where the gross take-off distance is the average in a normal distribution, the net take-off distance is longer, by an increment equal approximately to 5 standard deviations (see below). Thus the probability of failing to achieve the net performance is REMOTE. Net distances are always longer than gross:

$$\text{net distance} = \text{gross distance} \times (1 + \text{a gross-to-net factor}) \qquad (6\text{–}11)$$

Determination of realistic gross-to-net factors is complicated. There are two sets of contributions. The first, like turbulence, variations in piloting technique, and temporary below-average performance, cannot be accounted for individually although there are ways of allowing for their effects collectively. The second set can be measured and include:

☐ *Weight* of the aeroplane at the start of the take-off run.
☐ *Altitude* of the airfield.
☐ *Air temperature* at the airfield, and
☐ *Relative humidity* at the airfield, which has an adverse effect upon engine performance and is usually taken into account during certification. In some cases correction factors are provided in the Operating or Flight Manual.
☐ *Condition of the surface* of the runway including coefficients of friction, wet and dry, (which split as follows):
   (1)   rolling coefficient of friction, which affects TORR and TODR; and
   (2)   braking coefficient of friction, which affects the *Emergency Distance Required*, EDR;
   and adverse contamination by snow, slush, ice and standing water, which variously affect drag (due to displacement and impingement of water, snow, or slush) and available braking friction.
☐ *Slope(s)* of the surface(s) in the direction of take-off along the TORA, TODA and EDA. An uphill slope increases the take-off ground-run, and a downhill slope increases the landing distance, and the deceleration portion of the EDR.

☐ *Wind accountability*: the headwind used for despatch calculations must not be more than 50%, the tailwind not less than 150% of the reported wind component along the runway.

Note: If there is 90° crosswind there is no beneficial headwind component, and handling of the aircraft may dominate any other consideration. Crosswind limitations included in the aircraft flight manual have been *demonstrated*, and thus may appear to be surprisingly low. They are to be respected because a test pilot has been there and done it.

Example 6–1: Take-Off Performance in Rain – Motor-Glider

Rain drops (insects and ice) contaminate the leading edges of aerofoil surfaces, including propellers, degrading dangerously in some cases the efficiency of laminar flow surfaces. Stall speeds and all distances are increased and climb gradients decreased.

A motor-glider with rain drops on its highly polished laminar wings, was still firmly on the ground, 14 knots above its speed for lift-off and much further along its take-off run than normal, because of the breakdown of the wing boundary layer from laminar to turbulent flow.

Handling Scatter

The key to handling scatter lies in the mathematical properties of an elegant curve, shown in Fig. 6.2, which approximates to a *normal*, or *Gaussian, distribution* of results, named after the German mathematician K. F. Gauss (1777–1855). The curve is used widely in statistics, because it can be calculated. Its form describes numerically the distribution of a variable, clustering (more or less) symmetrically about some central mean value, with a dispersion which is random (like sets of aircraft climb, cruise or airfield performance test points adjusted to standard conditions). The form of the curve is given by:

$$y = A\,e^{-bx^2} \tag{6-12}$$

in which e is the exponential number, 2.718, while A and b are constants.

Before this distribution can be used with a set of data points, all sources of systematic variability must first be corrected out of the results, as far as possible. In the context of aircraft performance this usually means that the data must first be corrected to standard conditions or expressed as a difference between measured and calculated (or estimated) values (Ref. 6.1).

The Gaussian distribution is completely defined when we know the mean value, together with the standard deviation. Relevant definitions are:

☐ *Median* of a sample of n results contained within the curve is the value that is exceeded by one half of the sample, and not exceeded by the other half.

☐ *Mean* (arithmetic) value of a sample of size n, each member of which has some unique value, x, is given by:

$$\bar{x} = \Sigma_1^n (x)/n \tag{6-13}$$

☐ The *standard deviation*, SD of a sample, σ_x (σ for a population, s for a small sample), which is the distance between the mean value and the point where the greatest slope of the curve occurs (in Fig. 6.2 – the point of contraflexure), is given by:

$$\sigma_x = [\Sigma_1^n (x - \bar{x})^2/n]^{0.5} \tag{6-14}$$

☐ *Coefficient of variation* is the standard deviation of a set of data, expressed as a percentage of the mean

$$(\sigma_x/\bar{x}) \times 100 \tag{6-14a}$$

☐ *Range* of a set of data is the difference between the largest observation, x_{max}, and the smallest, x_{min}.

Example 6–2: Standard Deviation and Probability

The significance of standard deviation is that it enables us to handle probabilities in a more useful way. For example, the space between ±1 SDs, in Fig. 6.2 contains 68.3% of the measured results; ± 2.0 SDs bracket 95.4%, while 99.7% lie between ± 3.0 SDs, Knowing this gives us a feel for the odds involved in achieving a desired result. As 68% lies between $\frac{2}{3}$ and $\frac{7}{10}$, then about 2 out of 3, or 7 out of 10 average pilots and aeroplanes in a fleet can be expected to produce results contained within plus or minus one standard deviation of the mean. Further, 95.4%, near enough 19 out of every 20 pilots will produce results within plus or minus two SDs.

Put another way, Fig. 6.2 shows that 1 in 10 (10%, or 10^{-1}) of the pilots will fail to achieve a standard within 1.3 SDs of the mean (i.e. 9 out of 10 will do better than that). Only 1 in 100 (10^{-2}) will fail to achieve a standard 2.3 SDs away from the mean. While 1 in 1000 (10^{-3}) will fail to get within 3.1 SDs.

The curve of normal distribution can be manipulated into the form of Fig. 6.3. This tells us that 10% (10^{-1}) of the cases occur at or beyond the mean plus 1.3 SDs (i.e. 90% or 9 out of 10 will do better than this, 10^{-2} (or 1%) are at or beyond the mean plus 2.3 SDs, 10^{-3} or 1 in 1000 are at or beyond the mean plus 3.1 SDs, while 10^{-7} or 1 in 10 000 000 are between 5 and 6 SDs beyond the mean, and so on. Table 6–1 takes that information a step further by putting meat on the bones in real terms of things which can be expected to happen to an aeroplane during its operating life.

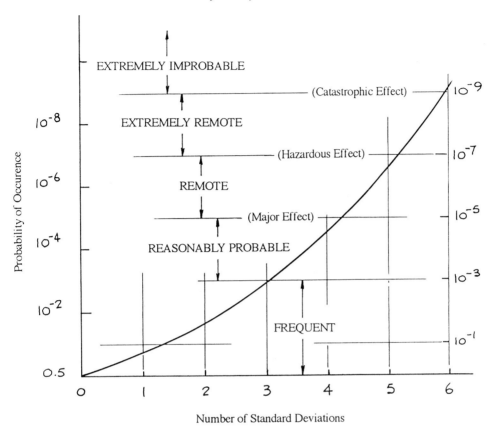

Fig. 6.3 Relationship between probability of an occurrence and a number of standard deviations of scatter away from the mean in Fig. 6.2. Definitions of effects in terms of probability are also shown.

Table 6.2 shows four factors which have a fundamental effect upon performance, and the percentage variation one might expect to find among 68% of the measured results lying within ±1.0 SDs of the mean.

□ *Combined standard deviations* of a number of samples must be calculated because many factors affecting performance are parameters, variables of which other variables are functions. An example is the thrust produced by a turbo-prop unit. This depends upon the pressure ratio within the engine, the fuel flow rate and pressure, the efficiency of combustion of fuel and air, the rotational speed of the engine, tailpipe thrust, the size, rotational speed, and efficiency of the propeller. Each is subject in turn to variation caused by, among other things, ambient atmospheric conditions (temperature, pressure, density), wear and tear, as well as the propeller pitch, and power lever settings by the pilot.

When it is necessary to find the overall SD of a parameter from K constituent variables at $n_1, n_2, n_3, ..., n_K$ data points, with standard deviations, $\sigma_1, \sigma_2, \sigma_3, ..., \sigma_K$, it is found by taking the square root of the sum of the squares (RSS).

Table 6–3 is a root sum square example, derived from the SDs of the effects on take-off run/distance of all of the various factors affecting take-off run/distance, with all engines

Performance

TABLE 6–2

Typical values of 1.0 standard deviation

Item	1.0 SD
Thrust*	2.25%
Drag**	2.00 to 2.50%
Braking force coefficient	13.00%
Initial climb speed	2.40%

Note: This means that 68.3% of the cases in each sample do not deviate from the mean by more than ± the amounts shown.
* With engine set accurately to the nominal thrust setting parameter (e.g., low pressure rotor speed, N_1, engine pressure ratio, EPR, torque, Q, and with no allowance for thrust-selling errors by the pilot).
** Depending upon configuration (i.e. clean or flaps extended).

TABLE 6–3

Root sum square standard deviation (see Equations (6-15))

Manoeuvre	1.0 SD
Take-off run/distance (all engines operating)	3.00%

operating. To carry out this calculation we need to know not only the SD of each parameter, but also the rates of exchange (influence factors, denoted ξ) between each of the parameters and take-off distance (e.g., a 1.0% change in thrust will cause approximately a 1.0% change in take-off distance, all else being equal, thus, the influence factor of thrust on TOD is approximately 1.0). The effect of a 1.0% change in engine low-pressure rotor speed, N_1, on thrust is about 5.0% for a conventional turbojet, so the influence factor, ξ, of N_1 on TOD is about 5. Therefore the root sum square of the deviations is written:

$$\text{RSS } \sigma_x = (\sigma_1^2 + \sigma_2^2 + \sigma_3^2 + \dots + \sigma_K^2)^{0.5} \qquad (6\text{–}15a)$$

$$= (\sum_0^n (\xi \times \sigma)^2)^{0.5} \qquad (6\text{–}15b)$$

where n is the number of parameters taken into account.

Products of probabilities

When it is necessary to know the probability of an event which is dependent upon the probabilities of occurrence of a number of other independent events, these are multiplied together:

$$P_E = P_1 \times P_2 \times P_3 \dots P_n \qquad (6\text{--}16)$$

by treating them mathematically in the same way as indices:

$$P_E = 10^{-2} \times 10^{-3} \times 10^{-1} \times 10^2 = 10^{(-2-3-1+2)} = 10^{-4} \qquad (6\text{--}16a)$$

Probability applied to take-off run and distance: TORR and TODR

Airfields have declared available distances: TORA, EDA, TODA, which we saw in Fig. 6.1. The next step is to ensure that the required take-off run, TORR, required emergency stopping distance, EDR, and required take-off distance, TODR, fit TORA, EDA and TODA.

As it is impossible to have absolute certainty that they will fit, we say instead that there should only be a remote chance of failing to take off in the declared distances available – indeed, so remote as to be verging on the extremely remote. This is where Fig. 6.3 is useful, because REMOTE/EXTREMELY REMOTE corresponds with a probability of 10^{-7} which is about 5 SDs away from the mean. A gross distance is the average distance taken by average pilots flying average examples of the type of aeroplane. A pilot loading his aeroplane to a despatch weight for take-off in a gross distance which fits exactly the declared distance for the runway in use, has only an even chance of achieving it. But, if he reduces the take-off weight to shorten his take-off distance by an amount corresponding with 5 SDs, he gives himself a $10^7 : 1$ chance of success, i.e. only a remote 10^{-7} probability of failing to take off in the distance available.

Example 6–3: Standard Deviation – Gross and Net Take-Off Distances

To understand what 5 SDs represent, look at Table 6–3, which shows the root sum square of the standard deviations of the effects on take-off distance on those factors which affect take-off performance. The RSS single overall SD is 3.0%, so that 5 SDs $= 5 \times 3\%$, or 15% of the gross take-off run or distance. To ensure that 10 000 000 aeroplanes of a type take-off in the available distances (or that one aeroplane of a type manages to take-off safely on 10 000 000 occasions from the same airfield), we use Equation (6–11) to produce the net take-off run and take-off distances *required* with all engines operating:

$$\text{net distance} = \text{gross (pilot's average) distance} \times (1 + 15/100)$$
$$= 1.15 \times \text{gross distance} \qquad (6\text{--}11a)$$

Figure 6.4 shows the application of the method to required take-off run, TORR, using the RSS value in Table 6–3. When the gross take-off run and take-off distances are multiplied by $(100\% + 15\%) = 1.15$, they become the net *Take-Off Run Required* (TORR) and net *Take-Off Distance Required* (TODR) respectively. They must not exceed TORA and TODA if a risk of failing to take off higher than 10^{-7} is to be avoided. Similarly there must be only a remote probability that the emergency distance required, EDR, exceeds EDA, the emergency distance available, but in this case the EDR is scheduled as the gross (unfactored) accclerate-stop distance and the achievement of an acceptably low rate of fatal over-run accidents is based on a combination of the very low probability of an engine failing during the critical period close to V_1 and the low probability of a low-speed over-run being fatal.

Catering for Engine Failure

Engine failure on take-off is rare. The operational rules are framed in such a way as to

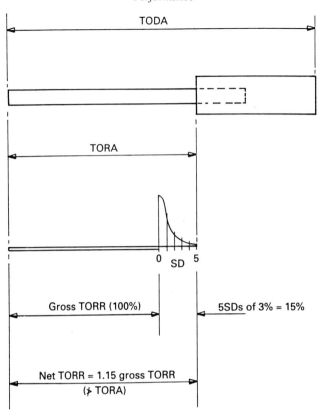

Fig. 6.4   Conversion of gross to net distance.

ensure that there is always the distance available *either* to stop or to go if the most critical engine fails. There is a decision speed on each take-off below which, if the critical engine fails, there will be enough *runway-plus-stopway* left to reject (abandon) the take-off and stop within the available distance. This speed, referred to as V_1, is illustrated in Fig. 6.5. Should the critical engine fail above that speed, then it will be equally safe to continue the take-off. V_1 is presented to the pilot as indicated airspeed. Sizing of the engines provides enough power or thrust to match the aircraft to the types of airfields it is expected to operate from, and to achieve acceptable climb gradients after take-off with the critical engine inoperative.

Before reaching V_1 there is a moveable point at which the most critical engine is assumed to fail, which is related to V_1 and other significant speeds as shown in Equations (6–17a) and (6–17b). The point corresponds with the *engine-failure speed*, V_{EF}. If the pilot elects to abandon the take-off the accumulated kinetic energy (which is a function of the take-off weight and ground speed squared) must be dissipated in the remaining length of the declared emergency distance available, EDA. As we saw earlier, speeds spelt out in the requirements are expressed as ratios of particular 'bench-mark speeds', like the stall speed, V_{S1}, in the take-off configuration. To define V_{EF} we use V_{MC} or V_{MCG}, the minimum control speed, or minimum control speed on or near the ground, and the decision speed, V_1, such that the minimum permitted value of:

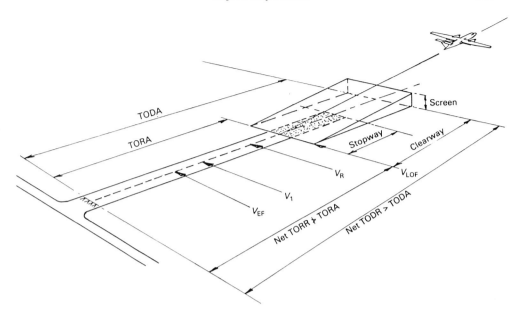

Fig. 6.5 Elements considered on take-off with lift-off at the point where TORR = TORA.

$$V_{EF} \text{ is not } < 1.05\ V_{MC} \text{ or } \geq V_{MCG} \tag{6–17a}$$

but it is:

$$= V_1 - \Delta V \tag{6–17b}$$

Here ΔV is the speed gained accelerating on the remaining engine(s) between the speed at which the engine failure actually happens, V_{EF}, and the speed at which the pilot recognizes the failure at V_1. Thus, ΔV is the speed gained during the pilot's interval of 'recognition time'. As an engine might fail at almost any point during the ground run, the pilot must first correct yaw and then check which engine has failed (this is not easy, especially if a third centre-line mounted engine, out of sight at the tail, has quietly failed without trace of yaw). It takes a short time to realize what has happened, during which the aeroplane continues to accelerate at a reduced rate to V_1, the speed at which the pilot has decided what he must do next.

The reaction of anyone to engine failure is never instantaneous. Therefore the requirements are framed *so as to allow the average pilot two seconds to react after* V_1, as shown in Fig. 6.6. Here it is assumed that he elects to abandon the take-off, doing so by reducing power to idle and selecting the first means of retardation, which may be lift-dumpers, air-brakes, reverse-thrust or pitch (provided that an asymmetric yawing moment is not made unacceptably worse), drag-chute, or simply wheel-brakes. Having said that, with some older military aircraft the application of wheel brakes had to be delayed to a lower speed after closing the throttles, to avoid burning out the brakes.

The despatch rules are written in such a way as to ensure that if an engine failure is recognized at an airspeed less than V_1, then EDA is long enough to stop, and the operational procedure for the pilot is to reject (abandon) the take-off.

If an engine fails at or after V_1, the pilot must continue the take-off, because there is then insufficient distance remaining in which to stop. V_1 is specified in terms of the minimum control speed with asymmetric power, or the speed at which the aeroplane is rotated into the attitude for take-off:

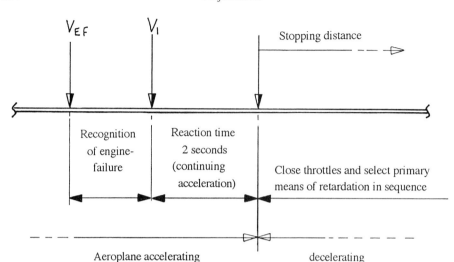

Fig. 6.6 Sequence of events following engine failure at V_{EF} and rejection of take-off at decision speed V_1.

$$V_{\mathrm{EF}} < V_1 \leq V_{\mathrm{R}} \qquad\qquad (6\text{--}18\mathrm{a})$$
$$\geq 1.05\, V_{\mathrm{MC}} \qquad\qquad (6\text{--}18\mathrm{b})$$
$$\geq V_{\mathrm{MCG}} \qquad\qquad (6\text{--}18\mathrm{c})$$

On failure of one engine, an aeroplane with four engines loses only one quarter of the thrust available. A twin loses half, but the engines of a twin are sized to provide ample thrust for take-off on the remaining one if failure of the other occurs at or beyond V_1. The thrust of an aeroplane with three engines lies in between.

Performance Margins and Abandoned Take-Off

Although engine failure has been treated so far as the reason for rejecting a take-off, it is not the only one. If, for example, a red warning light came on at a speed less than V_1, or there was hydraulic failure, or a major bird-strike, then take-off would be rejected. A glancing impact with a swan on take-off wrecked the starboard gun bay of a North American F-86 Sabre and almost wrecked the aeroplane. The risk involved in attempting to continue to fly after such an emergency could have more disastrous consequences than one following engine failure close to the runway. Hydraulic failure of a flying control, which has no effect upon the power delivered by the engines, may necessitate rejection at a much higher speed than V_1. For emergencies of this kind, airworthiness requirements set publicly-acceptable, low, target levels for the probability of fatal accidents. This is done by defining scenarios which might lead to a fatal accident, assessing the probability of such an occurence, P_{O}, and the probability of the resulting accident being fatal, P_{F}.

To achieve a desired (low) risk airworthiness target, P_{T}, we use an equation such as Equation (6–16). But as despatch rules produce scheduled (net) performance a further term is introduced, to account for the probability of the factor which converts gross to net performance in Equation (6–11) being exceeded. By definition, the probability of achieving better than, equalling or achieving worse than net performance is one, since there are only three possibilities. Hence, if P_{G-N} is the probability of bettering or equalling the net performance, the probability of achieving worse than net performance must be $(1 - P_{G-N})$. Consequently, from Equation (6–16):

$$P_T = P_O \times P_F \times (1 - P_{G-N}) \tag{6–19}$$

from which the magnitude of P_{G-N} is sized to achieve the desired risk target, P_T:

$$P_{G-N} = 1 - P_T / (P_O \times P_F) \tag{6–20}$$

Example 6–4: Probability of Fatality in the Accelerate-Stop Case

The probability of a turbine twin-engined aeroplane suffering engine failure is approximately (1×10^{-6}) per engine second at take-off power. This means that the probability of engine failure within one second after passing V_{EF} is about 2×10^{-6}, or two in one million, and the probability of this happening on a distance-critical runway is about one in ten, 10^{-1}. Thus, in Equation (6-19):

$$P_O = 2 \times 10^{-6} \times 10^{-1} = 2 \times 10^{-7} \tag{6–21}$$

The probability of a low-speed over-run accident involving fatalities, P_F, is about one in ten, 10^{-1}. So that if the target fatal over-run rate is to be as low as one in 100 000 000, or 10^{-8}, then substituting in Equation (6–20):

$$\begin{aligned} P_{G-N} &= 1 - (10^{-8}/[(2 \times 10^{-7}) \times 10^{-1}]) \\ &= 1 - \tfrac{1}{2} = 0.5 \end{aligned} \tag{6–22}$$

From Fig. 6.3 this corresponds with a factor of zero standard deviations. In other words the net to gross margin is zero and it is acceptably safe to schedule the mean performance for the accelerate-stop case. Therefore, in the case of engine failure at such a critical point, gross distances are used without any factor being applied to convert them to net.

  This example is, of course, a grossly over-simplified form of the rigorous method of calculating the risks in the accelerate-stop case. However, it is useful as an example of the basic principle applied in rigorous calculation, in that it illustrates the combination of independent probabilities. In a rigorous calculation, instead of calculating the peak risk arising from stopping just beyond V_{EF}, it would be necessary to integrate the risks arising from trying to stop after engine failure at all possible speeds throughout the acceleration, to lift-off and beyond. It would also be necessary, for every speed, to allocate probabilities to the pilot faced with deciding whether to stop, or to continue the take-off. Such calculations are complicated but, when carried out, the answer is that the probable risk of a fatal accident appears to be nearer 2×10^{-8} than 1×10^{-8}. The current target rate is 1.6×10^{-8}. Large transport aeroplane statistics for 1962 to 1985 indicate an actual fatal accident rate for this cause around 7×10^{-8} (per flight hour).

Weight, Altitude and Temperature (WAT) Limits

Take-off performance is dependent upon the weight (mass) of the aeroplane, the pressure altitude and ambient air temperature of the take-off surface, and the declared distances, TORA, EDA and TODA.

  Both altitude and temperature are intimately related by the gas laws, which affect the thrust and power output of the engine, and the true airspeed for a given EAS/CAS. The former are named after Robert Boyle (1627–1691), an Irish chemist, and Jacques Alexandre

Cesar Charles (1746–1823), a French physicist. Charles' law is sometimes attributed to Louis Gay-Lussac (1778–1850), who discovered it simultaneously.

☐ *Boyle's law* states that when the temperature is constant:

$$(\text{pressure})_1 \times (\text{volume})_1 = (\text{pressure})_2 \times (\text{volume})_2 \quad (6\text{--}23)$$

☐ *Charles' law* states that at constant pressure (at absolute temperature K):

$$(\text{temperature})_1 / (\text{temperature})_2 = (\text{volume})_1 / (\text{volume})_2 \quad (6\text{--}24)$$

showing pressure and volume are proportional to temperature.

Both laws are usually combined, with p, v and t substituted for pressure, volume and temperature respectively.

As $(\text{mass} / \text{volume}) = \text{density}$, then:

$$p_1 v_1 / t_1 = p_2 v_2 / t_2 \quad (6\text{--}25)$$

and:

$$p_1 / \rho_1 t_1 = p_2 / \rho_2 t_2 \quad (6\text{--}26)$$

or, generally:

$$p = \rho R t \quad (6\text{--}27)$$

in which R is the constant for the particular gas concerned.

The hot, high airfield

Imagine two identical airfields, one at sea level in a temperate latitude, the other at a higher altitude in a hotter latitude, where the air is thinner and less dense. Both have the same declared distances. If the pilot attempts to take off at the same weight at the hot and high airfield as he calculated for TORA, TODA and EDA at sea level, it takes the aeroplane longer to accelerate to the required decision, lift-off and safety speeds IAS/EAS, which are reached farther down the runway, because there is less thrust to accelerate the mass of the aeroplane to the correspondingly higher true airspeeds. Thus, if the pilot rejects the take-off for any reason, the aeroplane has a higher kinetic energy (proportional to TAS squared) than it would have at a lower altitude and temperature at the same airspeed indicated by the ASI. Therefore more runway is needed in which to dissipate the kinetic energy and come to a stop. To match EDR to the declared EDA for the same aeroplane on both airfields, the kinetic energies (proportional to WV^2), must be equal. This can only be made so at the higher TAS *by reducing take-off (despatch) weight inversely with true airspeed squared* at the hotter, higher airfield. Weight (mass) will also need to be reduced to achieve the same acceleration with the lower power/thrust available in hotter, less dense air.

Note 1: The reverse argument applies to lower altitude and colder airfields.

Note 2: The WAT limit minimum gross gradients of climb must be used when calculating despatch weight, even though the airfield may have no obstacles.

Note 3: Zero wind is assumed but not stated.

Balanced Field Lengths

Originally, a *balanced field* was a runway which had no declared stopway or clearway, so that total runway length included EDA, which was one and the same as the TODA and TORA. As a result some early Aircraft Flight Manuals (AFM) and Operating Data Manuals (ODM) contained information which related maximum take-off weight, MTOW (see Equation (3–78)), to the balanced field and, hence, the runway length. Today a balanced field has EDA = TODA, which means that the clearway and stopway (where they exist) are assumed to be equal in length, and:

$$\text{TODA} = \text{EDA} = \text{TORA} + (\text{clearway} = \text{stopway}) \tag{6–28}$$

This eases matters for the operator, because now the longer TODA/EDA (instead of available runway length TORA) becomes the determining factor, enabling operations to be carried out at heavier take-off weights.

When neither a stopway nor a clearway exist, such a field may still be described as balanced, because (clearway + stopway) = (0 + 0) = 0, and:

$$\text{TODA} = \text{EDA} = \text{TORA} \tag{6–29}$$

Unbalanced Field Lengths

When the stopway and the clearway are of differing lengths we call this:

$$\text{an } \textit{unbalanced field} \text{ because EDA is not equal to TODA} \tag{6–30}$$

The heaviest possible take-off weight is then achieved by utilizing fully the different lengths of stopway and clearway.

Decision Speed, V_1

Clearly, the most important speed on take-off is the decision speed, V_1, identified by means of the pilot's airspeed indicator. The higher the decision speed, the higher the true airspeed and kinetic energy to be destroyed in the event of abandoning the take-off. The maximum permitted speed for V_1 is either V_R, the speed at which the pilot initiates nose-up rotation into the take-off attitude, or the maximum brake energy speed, V_{MBE}. The lowest permitted V_1 is the minimum speed on the ground, V_{MCG}, at which the aircraft can be controlled within an acceptable lateral deviation in the event of failure of the critical engine.

It is possible to calculate a range of values for V_1, which depend upon the circumstances. The ground run to lift-off is in two segments. First is the all-engines acceleration distance. Second is the distance with one engine inoperative and a slower acceleration from the failure speed to the lift-off speed. The optimum weight at which take-off can be achieved is limited by the balanced field, in which case there is only one value of V_1 corresponding with:

$$\text{TODR} = \text{TODA} \tag{6–31}$$

$$\text{EDR} = \text{EDA} \tag{6–32}$$

Figure 6.7 shows that:

☐ As the *emergency distance*, EDA, is the runway length in which work is done by the engines to accelerate to V_1 and then by the brakes in bringing the aircraft to a halt by absorbing the kinetic energy $(W/g)V_1^2$, then as despatch weight (mass, W/g) varies inversely with V_1^2 for a given EDA/EDR (as we saw for the hot, high airfield), again V_1 varies inversely as the square root of the weight (mass) ($V_1 \propto W^{-0.5}$).

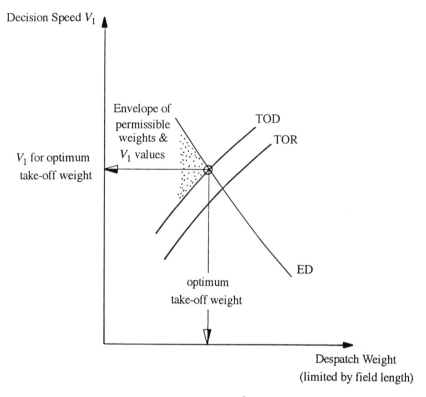

Decision Speed V_1

Envelope of permissible weights & V_1 values

TOD

TOR

V_1 for optimum take-off weight

ED

optimum take-off weight

Despatch Weight (limited by field length)

- Emergency Distances, ED, vary as $W V_1^2$ (kinetic energy to be dissipated) i.e., for a given EDA $V_1 \alpha W^{-0.5}$

- Take-off Runs, TOR, vary as V_1^2 for a given weight, and

- Take-off Distances, TOD, also vary as V_1^2 for a given weight i.e., $V_1 \alpha W^{0.5}$ for a given TORA or TODA

- Take-off Distances vary as W^2 for a given V_1, because: TOD α $(W/S)(W/F)$, or, $(W/S)(W/P)$, the wing, thrust or power loadings

Fig. 6.7   Determination of decision speed, V_1, for a given despatch weight, declared available and required runway distances.

□ For a given weight (mass) the take-off run depends upon how quickly the weight of the aircraft can be lifted. It follows from Equations (3–26) and (3–27b) that weight (mass) W varies as V_1^2, so that for a given TORA/TORR, V_1 varies as $W^{0.5}$.

□ Similarly, the take-off distance varies as V_1^2, and for a given TODA/TODR, $V_1 \propto$ to $W^{0.5}$.

If the pilot is not constrained to take-off at maximum weight, then a higher V_1 may be selected closer to V_R (and lift-off speed). While the aeroplane is on the ground longer with all engines up to that point, the remaining ground run and time to lift-off is shorter if one engine then fails. The ground run with all engines is proportional to the square of the engine failure speed, but this is more than cancelled out by the shorter distance needed to lift-off at the higher speed. Therefore there is a modest reduction in TODR with increasing V_1. If TODR must be kept short because of obstacles restricting the extent of the clearway, it is safest to select a higher V_1 – and if EDR must be shortened to account for, say, a cliff or a mountain rising from the end of the stopway, a lower V_1 can be selected.

Usually a V_1 value can be found which makes TODR = TODA and EDR = EDA, this is called the 'balanced V_1'.

Climb Gradients After Take-Off: The Net (Segmented) Flight Path

Probability methods are used in much the same way as on take-off when dealing with the climb gradients needed for safe departure beyond the imaginary screen, located at the up-wind end of the TODA.

All public transport aeroplanes, except those in UK Performance Group E and US FAR 23, are checked during certification for the ability to clear obstacles after take-off. For this a segmented flight path is defined, as shown in Fig. 6.8. Net climb performance is calculated for each segment, based upon *true gross gradients* (i.e. measured *geometrically* and not as a change of pressure altitude with distance flown). These ensure that there is only a remote possibility of the climb path not being achieved. To enable the operator to construct the basic gross flight path, considerable flight testing and performance reduction is needed to provide the information which the Flight Manual contains. The gross gradients required in each segment, with one engine inoperative, are shown in Table 6–4.

The UK Performance Group A (JAR-OPS Class A) multi-engined aeroplanes have their one-engine-inoperative true gross climb gradients reduced to net values by adding the following percentage margins, in exactly the same way as for net factored distances in Equation (6–11):

$$\text{twin-engined} = -0.8\% \tag{6–33a}$$
$$\text{three-engined} = -0.9\% \tag{6–33b}$$
$$\text{four-engined} = -1.0\% \tag{6–33c}$$

These gross to net margins correspond with about 1.5 SDs, a risk of 10^{-1}, i.e. 1 in 10. The reason for such a small number of SDs is that the probability of a multi-engined aeroplane suffering an engine failure close to V_1 on a distance-critical runway is of the order of 10^{-6}, one in a million (see Equation (6–21)). The combination of both, their product, is a 10^{-7} remote probability that the event will occur and that the net performance level will not be achieved.

The net flight path extends from a point 35 ft (50 ft in the case of some older aeroplanes) above the end of the TODA for a wet or dry runway, as appropriate, to a height of at least 1500 ft above the surface of the aerodrome. Manual propeller feathering is allowed at an altitude not less\than 400 ft. The scheduled flight paths are synthesized from free-air climb gradient measurements, supplemented by a small number of first segment gradient measurements made in ground effect, usually in the form of continued take-offs with engine failure earlier than V_R is reached. The climb gradient is continued to a height at which the aircraft is out of ground-effect (i.e. >2 wingspans).

The configuration conditions, show in knee-pad code (as suggested in Table 4–6), are broadly as follows. The first, second and fourth segments are subject to the WAT minima:

☐ The *first segment*: starting at the screen, in the take-off configuration, climbing at V_2 (not less than 1.1 V_{MC} and 1.2 V_{S1}).

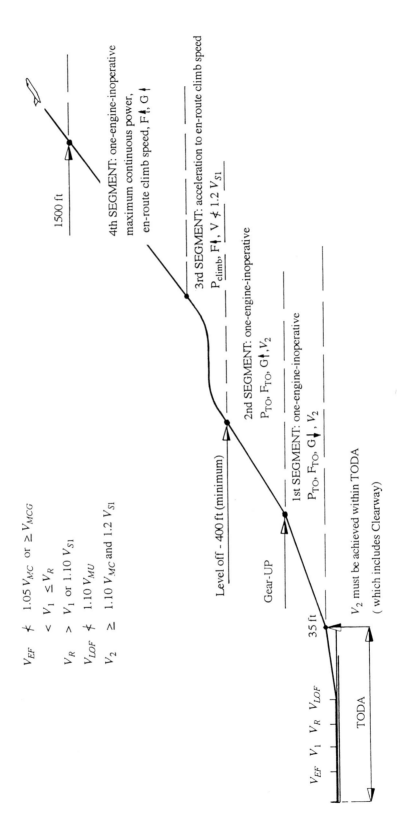

V_{EF} $\not<$ $1.05\,V_{MC}$ or $\geq V_{MCG}$

V_R $>$ V_1 or $1.10\,V_{S1}$

V_{LOF} $\not<$ $1.10\,V_{MU}$

V_2 $\geq$ $1.10\,V_{MC}$ and $1.2\,V_{S1}$

V_{EF} V_1 V_R V_{LOF}

V_{EF} V_1 V_R V_{LOF}

TODA

35 ft

V_2 must be achieved within TODA
(which includes Clearway)

Gear-UP

Level off - 400 ft (minimum)

1500 ft

1st SEGMENT: one-engine-inoperative
P_{TO}, F_{TO}, G↓, V_2

2nd SEGMENT: one-engine-inoperative
P_{TO}, F_{TO}, G↑, V_2

3rd SEGMENT: acceleration to en-route climb speed
P_{climb}, F↓, V ≮ $1.2\,V_{S1}$

4th SEGMENT: one-engine-inoperative
maximum continuous power,
en-route climb speed, F↓, G↑

Fig. 6.8 Take-off flight path from screen height for JAR-OPS Class A aeroplanes (see Table 3–3).

TABLE 6–4

Segmented gross climb gradients
Group A (JAR-OPS Class A) aeroplanes

| Configuration | Number of engines | | | Segment |
	2	3	4	
Landing gear retracting	0%	0.3%	0.5%	1st
Landing gear UP /flaps TAKE-OFF	2.4%	2.7%	3.0%	2nd
Flap retract and acceleration	level	level	level	3rd
Final take-off	1.2%	1.5%	1.7%	4th

- The *second segment*: starting at the point at which the landing gear is retracted. The climb is maintained without changes of take-off power or flap, at V_2 to at least 400 ft above the take-off surface.
- The *third segment*: starting when flap retraction is initiated at a minimum height of 400 ft. The aeroplane is then accelerated in level flight from not less than 1.2 V_{S1} to the *en-route* climb speed.
- The *fourth segment*: (final take-off) starting when the *en-route* climb speed is achieved (not less than the greater of 1.1 V_{MC} and 1.2 V_{S1}). With maximum continuous power set on the live engine, the climb continues to 1500ft above the take-off surface.

Often, the limiting case for Group A aeroplanes is the second segment climb requirement, with gear up and take-off flap selected, so that the second segment governs the size of wing needed to compensate for any inadequacy on the part of the flaps. These are limited in size to what remains of the wing trailing edge on which to hang them, between aileron root and nacelles or fuselage, so that they might be less effective than the designer hoped. This is the reason for elaborate and costly flap combinations and configurations in what are called *soft wings*. More than one WAT-limit may have to be determined, to take into account multiple flap settings and/or power-augmentation by means of, for example, water-methanol injection.

Note: Some smaller turboprop and other multis are limited by the first segment climb, during which the drag of the gear-plus-take-off flap can be critical, because the smaller the aeroplane the higher the ratio of gear drag/total drag. Also the additional drag of the gear doors opening to permit gear retraction can lead to the first segment becoming limiting.

Landing distances

Measuring and scheduling landing distances

When measuring landing distances to provide net information for use in Flight Manuals, the same statistical methods are used as for measuring take-off performance. In other words pilot, aircraft and runway friction variability produce scatter, and there are factors such as obstacles in the surrounding area which determine the descent path, and others which degrade braking and lengthen stopping distances.

Tests are carried out to produce data which enable the required landing distances at destination and alternate airfields to be calculated. These must never exceed the landing distances available. Landing distance is dependent upon the kinetic energy to be dissipated by the braking system(s). The estimated landing weight, and airspeeds at critical points on the flight path are subject to careful regulation. For example, the distance required to land from a height of 50 ft must not exceed 70% of the landing distance available (LDA) on the most suitable runway for use in still air.

The landing distances shown in Fig. 6.9 are scheduled in the Flight Manual, and the factors which affect them are as follows:

- *Landing weight*: the heavier the weight the higher the stall speed, and the higher the kinetic energy to be destroyed in coming to a stop, increasing the distance required.
- *Altitude* of the aerodrome: the lower the ambient pressure and density the higher the true airspeed and the higher the kinetic energy to be dissipated in stopping, again leading to increased distance.
- *Temperature* at the landing surface at altitude affects the TAS equivalent of the target threshold speed, V_{REF}, and landing distance. At high altitude or high ambient temperature, both of which reduce density, the TAS equivalent of CAS is increased, and with it the kinetic energy to be dissipated. Thus, the TAS equivalent of V_{REF} and landing distance increase with altitude and rising temperature, and fall as both are reduced. High temperature affects adversely power (and thrust) available in the balk (see Example 6–5 below).
- *Runway gradients*: a downhill slope lengthens float before touchdown, and ground roll afterwards. An uphill slope has the opposite effect.
- *Wind accountability*: a headwind shortens the landing distance required, while a tailwind can lengthen it dramatically. But, as we saw when considering wind effects on take-off, a crosswind can cause handling difficulties which outweigh anything else (see Example 6–6 below).
- *Delay* in selecting reverse thrust, airbrakes and wheel brakes.
- *Friction coefficient* of the runway, and the consequent:
- *Ground-roll* from touchdown to stop.

The airspeed scheduled is normally $V_{REF} \geq 1.3\ V_{S0}$, which must be achieved by the threshold. For Commuter Category aeroplanes V_{REF} is not to be less than 1.05 V_{MC} (the minimum control speed with the wing flaps in the landing position).

Specialized Transport, Normal, Utility and Aerobatic Category aeroplanes may also have demonstrated by flight tests that a maximum steady approach gradient, steeper than 5.2% (3°), can safely be maintained down to the 50 ft screen. It follows that an instrument available to the pilot is needed for steeper than normal approaches.

Note: From the cockpit, a 3° approach looks about right to the pilot, which is why it is adopted universally. It is easy to over-estimate the steepness of nose-down angle of descent/approach/dive/spin when looking ahead through the windscreen. Many test pilots mark angles in china-graph on the side-screens, which help to estimate angle relative to the horizon.

When runway surfaces are other than hard paving, ways of adjusting the distances scheduled in the AFM, based upon tests, must be provided for operators. When it is impracticable for the manufacturer to carry out the necessary tests, seek the advice of the certificating authority.

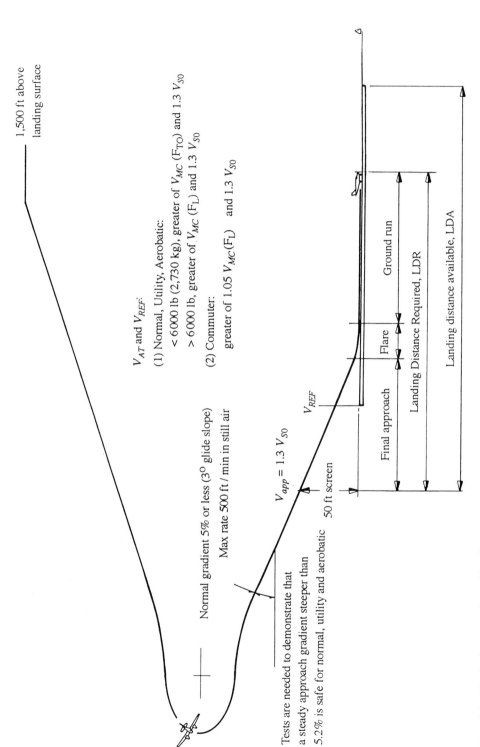

1,500 ft above landing surface

V_{AT} and V_{REF}:

(1) Normal, Utility, Aerobatic:
< 6000 lb (2,730 kg), greater of V_{MC} (F_{TO}) and 1.3 V_{S0}
> 6000 lb, greater of V_{MC} (F_L) and 1.3 V_{S0}

(2) Commuter:
greater of 1.05 $V_{MC}(F_L)$ and 1.3 V_{S0}

Normal gradient 5% or less (3° glide slope)
Max rate 500 ft / min in still air

V_{app} = 1.3 V_{S0}

Tests are needed to demonstrate that
a steady approach gradient steeper than
5.2% is safe for normal, utility and aerobatic

50 ft screen

V_{REF}

Final approach

Flare

Ground run

Landing Distance Required, LDR

Landing distance available, LDA

Fig. 6.9 Typical approach and landing.

Example 6–5: Effect of Ambient Temperature on Overshoot (Go-Around)

The long nosed, twin-engined Gloster Meteor NF 12 and 14 (now classed as 'warbirds') are privately owned and appear in air displays.

   When they were operational, back in the 1950s, tests demonstrated that they were unable to overshoot when balked while carrying full fuel, warload and with one engine inoperative, after 09.00 in summer heat at Nicosia, Cyprus, although the lighter single-seat Mk 8 was well able to do so.

Example 6–6: Crosswind Limitation

The earliest mark of one foreign light twin had a demonstrated crosswind limitation of about 13 mph. On a test flight, when landing in a slight turbulence from curl-down over buildings upwind of the threshold of the runway, full aileron was needed, and a long pause for them to take effect, even though the crosswind was not on the limit.

Balked Landings

Regulations governing landing weight ensure that the Commuter Category aeroplane, like

Plate 6–1 The ponderously heavy Gloster Meteor NF14 night fighter was unable to overshoot on one engine when balked with full fuel and warload, after 0900 in the summer heat in Cyprus, unlike the lighter single-seat MK8. Ref.: Example 6–5. (Courtesy of the then Air Ministry)

its heavier sisters, must be able to discontinue a one-engine-inoperative approach if balked. Tests are to demonstrate that there is enough remaining power in hand [(thrust − drag)/ weight] to achieve both a minimum gradient of climb and a further circuit with one engine inoperative.

Performance to be demonstrated in the balk

☐ *Commuter Category* aeroplanes, in the landing configuration, with all engines operating at power no higher than that which can be achieved in 8 seconds from flight idle (throttle closed), must achieve a gross climb gradient not less than 3.2% at a climb speed equal to V_{REF} with flaps in the landing configuration (i.e. not less than the greater of V_{MC} and 1.3 V_{SO}).

☐ *Normal*, *Utility* and *Aerobatic Category* aeroplanes:

  (1) *weighing not more than 6000 lb (2730 kg) maximum weight*, with reciprocating engines, must be able to maintain a steady climb at sea level in standard temperature of at least 3.33% (1 in 30, i.e. 2°), at a climb speed not less than 1.2 V_{S1}. The configuration is design maximum weight with take-off power on each engine, and flaps in the landing configuration (flaps may be retracted if they can be raised in 2 seconds or less without loss of altitude, sudden change of attitude or exceptional piloting skill).

  Note: At a hot and high airfield the 'climb' gradient might be negative. There is a move afoot in JARs to make all small aeroplanes meet the >6000 lb piston/all turbine WAT limits for take-off and landing, applied to those:

  (2) *weighing more than 6000 lb (2730 kg) maximum weight*, reciprocating engined aeroplanes (and those with turbine engines, regardless of weight), must be able to achieve a climb gradient of not less than 2.5% (which is 1 in 40, or 1.5°) at a speed of V_{REF} which is the greater of V_{MC} with the wing flaps in the landing position and 1.3 V_{SO}.

Turbine-engined aeroplanes have slower engine acceleration times from flight idle than reciprocating engines and no concession is made for aeroplanes weighing 6000 lb or less.

Maximum despatch weights

The despatch weight of an aeroplane is determined by considerations of the maximum for take-off and also for landing at its destination, plus high ground clearance *en route* in the event of engine failure. When dealing with weight and balance in Chapter 3 (see Equation (3–78)) we looked at the way in which maximum take-off weight, MTOW, is constructed, and also saw that it is the maximum weight at which all flight requirements must be met. In fact there is more to it than that:

☐ *Maximum regulated take-off weight* is the lowest weight derived from the following:

  (1) Maximum authorized take-off weight (structural limit).
  (2) Maximum take-off weight for altitude and temperature (WAT limits).
  (3) Take-off field lengths.
  (4) Obstacle clearance (take-off flight path).
  (5) *En-route* climb.
  (6) Maximum authorized landing weight.
  (7) Maximum landing weight for altitude and temperature (landing WAT limits).

(8)   Landing field length.

(9)   Other airworthiness limitations, e.g., tyre speed limits, brake energy limits (WV^2), etc.

Items (1) to (9) require specific calculations to be made using various parameters.

Often the despatch weight of an aeroplane for flight over a short sector is limited by the *maximum authorized landing weight*. This is a structural limitation, specified in the Flight Manual. The maximum landing weight may be limited to less than the MALW by field-lengths and WAT conditions, in exactly the same way as those for take-off.

☐ *Maximum landing weight for altitude and temperature*: tests are carried out to determine go-around climb performance as a function of weight, altitude and expected air temperature at the time of landing. The combinations of weight, altitude and temperature at which the relevant climb gradient minima are just met are scheduled as Landing WAT Curves.

The Flight Manual is the approved source of data for all such calculations and its contents are backed by law. To ignore the advice given in the Flight Manual could, if such action is then shown to have led to an accident, invalidate the insurance.

Operating Limitations on Light Twins

Operational weather minima (cloud base and visibility) for light twins are stiffer than those for commuters. The situation is summarized in Table 6–5 and comes about because the potent commuter has performance scheduled which enables it to take off into murky weather, by day or night, without restriction – like much heavier transport category aeroplanes in JAR-OPS Class A. The light twin on the other hand (JAR-OPS Class B) is not allowed such freedom. For example, the second segment climb of the commuter aeroplane has a minimum gradient not less than 1.2%, while that of the light or small, twin is unspecified numerically but, by implication, need be no better than zero. Operational limitations on minimum visiblity and cloud base are imposed, and the safety of the take-off and landing

TABLE 6–5

Differences between light twin and commuter aeroplanes

Flight phase	Light twins
Engine failure on take-off	No performance scheduled
Net flight path	None scheduled as there is no verification of obstacle clearance. The pilot must see and avoid obstacles, hence the need for a regulated cloud base.
Factored take-off distance	Likely to be introduced to provide for the abandoned take-off
Weight, altitude, temperature advice	Sparse, making loss of an engine tricky on take-off or landing (unless flaps are UP)
One-engine-inoperative (2nd segment) climb gradient	Numerical minimum not specified but, by implication, not less than zero

relies upon the pilot keeping a visual look out instead of relying entirely on internal instrument and other aids.

Table 3–3, showed the increasing severity of operational regulation imposed upon such aeroplanes. It has to be so because they are small, and they would otherwise incur high certification costs, out of proportion to their capability. Tough operational rules enable costly airworthiness regulation to be less severe.

Operating Limitations on Single-Engined Aeroplanes

We have already seen that single-engined aeroplanes in the UK are prohibited from operating for the purpose of Public Transport at night or when the cloud or visibility prevailing at the aerodrome of departure, or forecast for the destination or an alternate, are less than certain minima. The reason is that departure, destination and alternate airfield weather limits do nothing to help a pilot make a safe forced landing following single-engine failure *en route*. So, along the planned route and in the planning of diversions, minimum visibility of 1 nautical mile (1.6 km) and a cloud base no closer than 1000 ft (305 m) to the surface are as specified in Ref. 3.7, Sect. 3, 10(2). These UK limits allow flight above a cloud layer, while giving a sporting chance of a safe forced landing in visual meterological conditions (VMC) if the engine stops.

The CAA in the UK is not entirely alone in specifying such limits, although there is vacillation between different nations in Europe. France and Germany, for example, do not permit single-engine passenger operations at night or when Instrument Flight Rules (IFR) are in force, although cargo and freight operations are allowed. Canada, as we saw, is reported to be leading the way to permitting operations in IFR with paying passengers in turboprop singles. Australia is expected to follow suit, followed by the USA. This will increase pressure on the JAA and, within it, the UK CAA.

The subject is expected to remain a source of contention for some time. It can be argued technically that the probability of an in-flight engine shut down increases directly with the number of engines (i.e. one is less likely to fail than when four are fitted) but the fact is that long range, large, multi-engine aeroplane maintenance and professional piloting standards are necessarily higher, and there is a lower probability of an accident. Also, the loss of one engine in four involves a proportionally smaller loss of thrust than one out of two. On the other hand a twin is relatively overpowered on both engines, having a lower power-loading than one with three or four engines. Thus the loss of one out of two (well-spaced) engines still leaves adequate residual thrust from the other to meet the performance requirements at maximum weight – more than is possible with a four-engined machine with three shut down.

Coupled-Engines and Contra-Props versus Single-Engines

Asymmetry when a wing-mounted engine fails brings with it the possibility of mishandling. The opposite solution is to save money in almost every respect by using instead one highly reliable, nose-mounted, single turboprop engine. Operators in remote areas (like the Australian Flying Doctor Service) are said to favour single turboprop-engined aeroplanes for economic reasons. The flaw in any single-engine solution is that of inability to demonstrate equivalent twin-engined safety following engine failure in the same flight conditions as the twin.

Setting aside the matter of costs, one proposal for getting rid of asymmetry, while cleaning up the airframe and maintaining a semblance of twin-engined safety, is to couple two engines in a low-drag, nose-mounted, 'twin-pack', driving contra-props. The Soloy Corporation, Olympia, USA is developing a coupled 1300 SHP PWC PT6A-114A with a Soloy Dual Pac gear-box, for a 17-seat Cessna 208 Caravan 1, designed originally as a 13-passenger single-engined aeroplane. But, with an arrangement of this kind there arises the question of reliability of the concentric drive shafts and hubs in a single gearbox.

Coupling engines and using counter-rotating propellers are not new ideas – they go back to the 1930s Schneider Trophy races at least, and to the Italian Macchi M 72 racing seaplane. Increasingly powerful engines brought with them increased torque. Not only did this threaten control authority in roll and yaw, it also meant overloading the buoyancy of one float more than the other, causing added resistance and yaw, which degraded further the effectiveness of directional control (see Appendix).

The civil certification problem with coupled centre-line mounted engines lies in providing and demonstrating equivalent twin-engined reliablity and safety with a pair of closely nesting engines and a complex gear train. Engines mounted side by side, with intakes and rotating parts close together, are more sensitive to sympathetic aerodynamic and mechanical failures than when twin engines are mounted well away from one another, on each wing, or on opposite sides of a fuselage. Always there is too the problem of non-containment of bits and pieces of a disintegrating rotor (as has happened).

If the probability of losing one's single engine is 10^{-6} per second, then the probability of losing one out of two in a twin with separate engines is 2×10^{-6} per second. But with coupled engines, the mutual risk of sympathetic failure is increased by the complexity of their

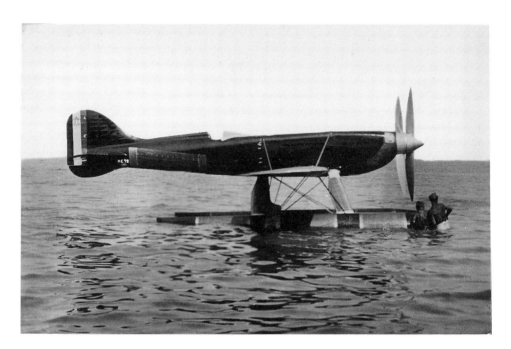

Plate 6–2   Italian Schneider Trophy Macchi MC 72 of 1931 which had engines in tandem driving separate, counter-rotating propellers through concentric shafts. Ref.: Example 8–2. (Courtesy of the Philip Jarrett Collection)

Plate 6–3 The Fairey Gannet AEW Mk3 of the Fleet Air Arm with Bristol Siddeley double Mamba propeller turbines. Each drove its own counter-rotating propeller with the intention of shutting down one for economical cruise and loiter. Note the additional fins on the tailplane to improve directional stability. (Courtesy of Cyril Peckham via Philip Jarrett Collection)

combination. For example, whether two engines drive one propeller through a gearbox, or they drive counter-rotating propellers, each on separate concentric shafts. Coupled, independent engines and propellers have been used in the past – the Fairey Gannet of the Royal Navy (design work started in 1945) had a Double-Mamba installation driving separate propellers on concentric shafts, either one of which could be shut down and the propeller feathered. While there are no Double-Mamba failure statistics to hand, military levels of risk cannot be compared with those which are acceptable for civil operations, because the risks in relation to operational use are different. From the naval point of view, such turboprop powerplants were 'full-throttle' engines which operated efficiently only over a narrow band of gas-generator RPM. Shutting down one engine when full power is not needed not only extended the range, it also reduced maintenance costs. Sometimes, though, it was hard to relight the shut-down engine. There is no comparable equivalent in passenger-carrying civil operations.

Third-party risk and population density

Attempting to provide equivalent twin-engined safety with a single-engine, or with coupled engines, is not the end of the argument. Just as with the military, there are operational considerations. Third-party risk and population density are major factors in the safety equation. Thus, one may expect national (social-geographic) variations between states. The Canadians and others who would lead the way in the certification of singles have more open, underpopulated areas than many parts of the UK and Western Europe. Thus, the JAA, which normally resorts to US regulations as the basis for its rule changes, has decided in the case of JAR-OPS to follow tougher ICAO guidelines. These limit operations as described earlier, to risking forced landings in single-engined aeroplanes *only in VMC*.

Net Performance Data for Light and Small Aeroplanes

Light and small aeroplanes used for public transport operations also need advice for the guidance of pilots. Information is less elaborate than that provided by operators of commuter and larger transport aeroplanes. It is published in the Owner's Manuals, or Pilot's Operating Handbooks of light aircraft certificated to JAR-23 or FAR 23. The information is often presented in tabular form, as a function of weight, altitude, temperature, head and tailwind components, and runway slope, and includes:

(1)   Factored all-engines take-off distance.
(2)   Take-off WAT limits.
(3)   *En-route* gross climb gradient (including engine-out for twins).
(4)   Glide gradient (singles).
(5)   Factored landing distance.
(6)   Landing WAT limits.

Sometimes one must fall back upon more general information, like that shown in Table 6–6 (Ref. 6.2). This provides conservative CAA guidance to General Aviation pilots on factors which increase the all-engines take-off distances to 50 ft. The FAA publishes a body of similar general advice for the benefit of owner-pilots.

   The factors shown in Table 6–6 are cumulative and must be multiplied when one or more occur together. Further, their product is then multiplied by the Public Transport Factor, 1.33, (which applies to all single-engined and multi-engined aeroplanes in Performance Group E) to produce the required take-off distance, TODR.

Contamination of Runways and Strips (Rolling and Braking Friction)

Coefficients of rolling and braking friction affect field performance on take-off and landing, and are of critical importance in their effect upon the performance of small and light aircraft on different surfaces. Generally, paved surfaces have coefficients of friction, μ, close to 0.02 (see Table 6–7).

   The thrust needed to accelerate an aeroplane on take-off is wasted in part by having to overcome rolling resistance. Frictional resistance, R, is given by:

$$R = \mu(W - L) \tag{6-34}$$

in which W is the weight of the aeroplane, L the lift and μ the coefficient of friction. If a runway is wet but not flooded it usually has a lower braking μ than when dry, typically:

$$\mu_{wet} = \text{about } 0.5\ \mu_{dry} \tag{6-34a}$$

Runways with standing water, slush and snow, are considered to be *contaminated*. In many ways grass, turf, sand or clay surfaces may also be thought of as contaminated, when μ is changed and take-off or landing runs and distances are altered adversely.

TABLE 6–6

Performance factors for light aeroplanes in UK Performance Group E (Ref. 6.2)
FACTORS ARE CUMULATIVE AND MUST BE MULTIPLIED

Condition	Take-off		Landing	
	Increase in distance to height 50 feet	Factor	Increase in landing distance from 50 feet	Factor
A 10% increase in acroplane weight, e.g., another passenger	20%	1.2	10%	1.1
An increase of 1000 ft in aerodrome elevation	10%	1.1	5%	1.05
An increase of 10°C in ambient temperature	10%	1.1	5%	1.05
Dry grass* – Short, 5″ (13 cm)	20%	1.2	20%**	1.2
– Long, between 5″ and 10″ (13–25 cm)	25%	1.25	30%**	1.3
Wet grass* – Short	25%	1.25	30%	1.3
– Long	30%	1.3	40%	1.4
A 2% slope*	uphill 10%	1.1	downhill 10%	1.1
A tailwind component of 10% of lift-off speed	20%	1.2	20%	1.2
Soft ground or snow*	25% or more	1.25 **	25% or more	1.25 **
NOW USE ADDITIONAL SAFETY FACTORS (if data is unfactored)		1.33		1.43

Notes: * Effect on Ground Run/Roll will be greater.

** For a few types of aeroplane, e.g., those without brakes, grass surfaces may decrease the landing roll. However, for safety, assume the *INCREASE* shown until you are thoroughly conversant with the aeroplane type.

Any deviation from normal operating techniques is likely to result in an increased distance.

So, if the distances required exceed the distances available, changes will HAVE to be made.

Many years ago extensive tests were carried out at the then RAE Farnborough on the rolling friction and aquaplaning characteristics of a variety of aeroplanes, in standing water and slush. For tests in slush we waited for snow. Aquaplaning tests were carried out using shallow pools built on the runway, with plasticine walls of differing heights for different depths of standing water. These were then filled by the fire services. During the test runs under carefully controlled conditions linear accelerations (and decelerations) were measured using sensitive accelerometers.

T<small>ABLE</small> 6–7

Rolling coefficients of friction
(Ref. 6.3)

Runway surface	Coefficient, μ
Hard paved	0.02
Short dry grass (sod)	0.05
Short wet or long dry grass	0.08
Long wet grass	0.13

It will be recalled that, many years ago, there was an accident on the snow-contaminated runway at Munich Airport, when an Airspeed Ambassador crashed fatally while carrying 'Matt Busby's babes' the Manchester United football team.

The subject is a minefield, and there is much scope for research.

Hard paved wet versus dry runways

Acceleration and deceleration are usually slower on wet runways than when the same runway is dry. The decision speed, $V_{1\,wet}$, is around 5 (twin-engine) to 10 knots (four-engine) less than $V_{1\,dry}$ with the decision point moved closer to the start of the take-off run, because the aircraft takes a longer distance to stop on a wet surface. Changes are also made to take-off run and distances required, TORR and TODR. The take-off weight associated with the reduced V_1 would be severely limited were the aeroplane required to clear a normal 35 ft screen at the end of the TODR. Because of this and because a marked improvement in accelerate-stop safety may be exchanged for a relatively small loss of safety in the continued by take-off, the definitions of TORR and TODR allow revision:

☐ A reduction in screen height to not less than 15 ft (say 5 m) at the end of the required take-off distance.
☐ Lift-off at the end of the required take-off run.

The problem is how to obtain the highest despatch weight of an aeroplane, while revising the decision speed, V_1, for operation from a wet instead of a dry runway, knowing the declared runway distances EDA, TORA and TODA. Regulations list the operational variables which affect field performance, and these were discussed earlier. There we saw that the limiting despatch weight of the aeroplane depends upon: the aerodrome pressure altitude and ambient temperature; condition of the runway surface (which affects directly the friction characteristics to be contended with); headwind or tailwind component along the runway (a reported headwind must be factored by 0.5, a reported tailwind by 1.5).

Consider first the dry runway case. For each of the three declared distances the all-engine and critical engine inoperative cases must be satisfied. The all-engines net take-off run and distances are derived by applying the same 1.15 factor to their gross values, as we saw in Equation (6–11a). This ensures that on a normal take-off without failure of an engine the probability of exceeding the declared distances is less than 10^{-7} (the standard deviation being 3% (Table 6–3)). The one-engine-inoperative distances are unfactored because the full distance is required *only when engine failure occurs at the most critical engine failure speed*, and the probability of this happening lies in the 'remote' range of Fig. 6.3.

With the ambient and airfield factors unchanged, the all-engines TORR and TODR are constant. But the one-engine-inoperative distances are a function of the friction characteristics of the runway surface, the speed at which engine failure is assumed to occur, and the associated decision speed, V_1. These are shown diagramatically in Fig. 6.10, which is an extension of Fig. 6.7 and relates V_1 and despatch weight for wet and dry runways.

If a runway is dry, the any combination of weight and V_1 which lies to the right of point X exceeds at least one of the available distances. Only a weight/V_1 combination which lies within the acute angle AXB complies with the regulation requiring EDR, TORR and TODR to be less than EDA, TORA and TODA respectively. The optimum despatch weight, W_{max}, and $V_{1\ dry}$ for a balanced field length are those corresponding with point X. When the actual take-off weight is less than W, then a range of decision speeds exist, for example, those between A and B.

When the runway is wet then only a (weight/$V_{1\ wet}$) combination corresponding with point T in Fig. 6.10 satisfies the despatch weight requirement, where the broken TOD_{wet} line DT cuts the broken ED_{wet} line. In practice the curves in Fig. 6.10 vary between individual aircraft and the runway distances available, however, $W_{max\ wet}$ is not allowed to exceed $W_{max\ dry}$. A twin-engined aeroplane is usually limited by the one-engine-inoperative case.

Note: Manufacturers must schedule both wet and dry runway distances, and it follows that tests are carried out for that purpose on dry and wet runway surfaces.

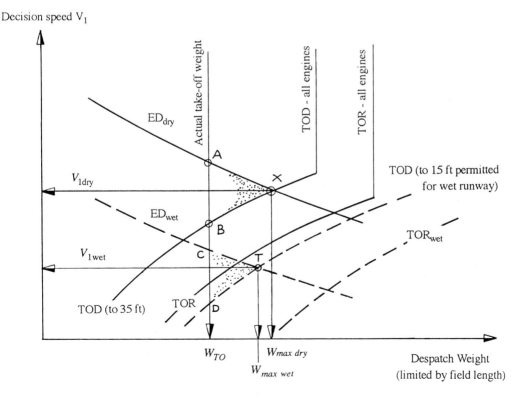

Fig. 6.10 Determination of decision speed/despatch weight relationship for dry and wet paved runway for a particular aeroplane.

Softer surfaces

Just as there are differences between wet and dry take-off run, take-off distance and emergency distance, there are differences caused by softer runway surfaces. There is nothing to be gained by exploring the problem here. Instead we shall look at it from a different angle, arising out of a flying accident.

Example 6–7: Accident on Take-Off – Piper PA-46 Malibu

Some years ago there was an investigation into the effects of altitude and the rolling coefficient of friction of turf on an airstrip as likely contributions to a fatal accident with a piston-propeller Piper Malibu. It was impossible to discover values for the different terms in the performance equation for the crashed aircraft at the time of the accident, and the problem had to be solved another way. Values of μ shown in Table 6–7, had their origins in Australia (Ref. 6.3), the UK and in Canada. It was possible to show that if the required take-off distance, s_1, was known for a rolling coefficient, μ_1, then the distance, s_2, due to a higher coefficient of rolling friction, μ_2, appeared to relate to s_1 in this way:

$$s_2/s_1 \text{ varies as } [(1 - \mu_1)/(1 - \mu_2)] \tag{6–35}$$

When compared with conservative information adopted and published by the CAA (see Ref. 6.3 and Table 6–6), Equation (6–35) was somewhat optimistic. Using the CAA information, published for use by General Aviation pilots, a better approximation was made by adding a constant, 0.19:

$$\text{TORR (and TODR) varied as } s_2/s_1 = k[(1 - \mu_1)/(1 - \mu_2)] + 0.19 \tag{6–35a}$$

To use these results bear in mind that:

- TODR = (TORR + airborne distance), and the airborne distance is unaffected by the rolling coefficient of friction. They were lumped together for conservative convenience. We have dropped TODR from the equation.
- The factor $k*$ is there to account for changing engine characteristics with altitude. A turbo-charged or a flat-rated engine, like a turboprop, would maintain $k = 1.0$ to a much higher altitude. A normally aspirated engine, losing power with altitude, might be expected to have k exceed 1.0 by several percent on airfields located well above sea level. Table 6–6, for example, shows TOD increased by 10% for a 1000 ft increase in airfield elevation.
- The values of μ used in Fig. 6.11 are *effective rolling coefficients*, not actual values, because of the conservative factors injected by the CAA. It is felt that they are an overkill*.

Note*: There is ample scope here for research.

Table 6–8 shows the increments obtained using Equation (6–35) at sea level, ISA. The above factors must be applied in addition to any factors called for by the performance operating rules.

Figure 6.12 shows a classification of grass used in Table 6–7 (see Ref. 1.5). Figure 6.13 from the same reference shows the way in which weight must be reduced in order to take off from a softer surface than one that is hard paved, to obtain the same factored take-off distance.

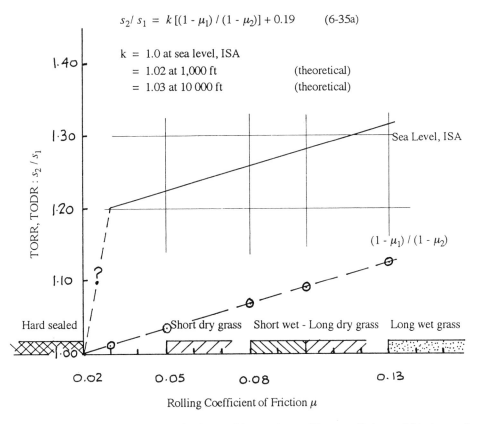

$$s_2/s_1 = k[(1 - \mu_1)/(1 - \mu_2)] + 0.19 \qquad (6\text{-}35\text{a})$$

$k = 1.0$ at sea level, ISA

$\quad = 1.02$ at 1,000 ft (theoretical)

$\quad = 1.03$ at 10 000 ft (theoretical)

Fig. 6.11 Effect upon TORR and TODR of increasing rolling coefficient of friction and altitude when $\mu_1 = 0.02$, using information published in Ref. 6.2. The results are 'ball-park' and should be used with caution pending further research.

TABLE 6–8

Factored take-off distance* with μ at sea level, ISA
(Equation (6–35), $k = 1.0$)

Runway surface	Increment	Factor
Short dry grass (sod)	+ 21 %	1.21
Short wet / long dry grass	+ 25 %	1.25
Long wet grass	+ 31 %	1.31

* See fair correspondence with Table 6–6.

Field performance measurements using simple equipment

The camera grid in Fig. 6.14 is one of the simplest devices, based upon similar triangles, for gathering gross data on take-off and landing distances, to and from screen heights. It must

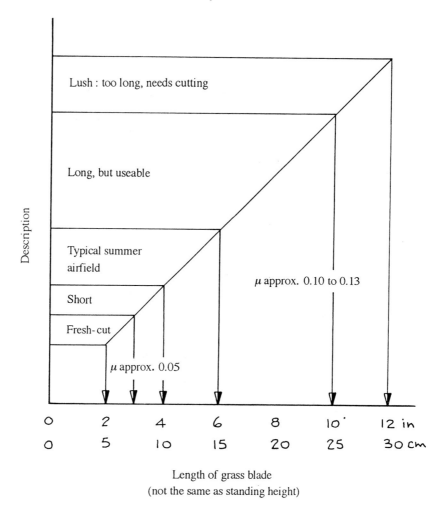

Fig. 6.12   Grass length and approximate friction coefficients.

be mounted firmly in the ground. An accurately surveyed site is needed, at a known distance from the runway. Ensure that the wire grid is within the focal range and that the camera is free to traverse in azimuth. The grid is scaled to measure imaginary 35 ft (or other desired) screens, representing notional obstacles, at any point between the thresholds at each end of the runway. There is no need to calibrate the camera and grid as long as the geometry and trigonometry are correct.

A sensitive anemometer is needed to measure both wind velocity and direction. A properly shaded thermometer is used to take the true air temperature in bright sunlight; and an altimeter for the pressure altitude.

Note: Performance testing is carried out at the forward CG limit. The scheduled speeds are functions of stall speed, also determined at the forward CG limit to yield the highest values. Wind speeds, measured close (but not too close) to the runway should not be more than about 10 knots from any direction, say 12% of V_{S1} is reasonable, because of the large wind gradient effects associated with higher wind speeds.

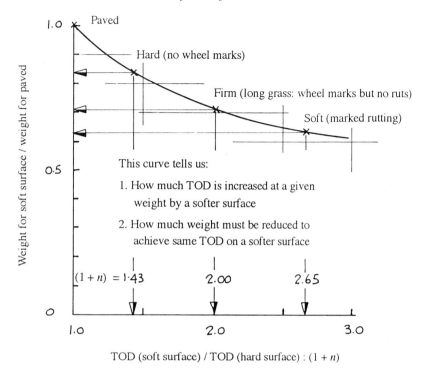

Fig. 6.13   Take-off weight for constant TOD and take-off distance at constant weight, both on the same soft surface (Ref. 1.5).

Rule of Thumb for a (First) Flight From a Short Field or Airstrip

When faced with carrying out a flight from a field or airstrip which you think might not be long enough, then this method is worth trying. It is based upon that of Dip. Ing. Hans-Werner Lerche (Ref. 6.4) a test pilot at Rechlin (the German equivalent of RAE Farnborough) during World War II, who employed it when flying captured allied aircraft for the first time, often from unprepared fields in which they had forced landed during operations.

☐ Calculate the weight and stall speed of the aeroplane in the take-off configuration. Equations (3–9) and (3–26) enable us to find stall speeds, knowing the lift coefficient. If this is not known accurately a value of $C_L = 1.0$ is conservative. Alternatively, Fig. 6.15 is a carpet which, by relating V_{S1} in KEAS, wing loading and lift coefficient, simplifies calculations. Now, using the airspeed legend in Fig. 6.8 one may estimate the airspeed at which to rotate into the take-off attitude. This should be not less than 10% faster than the calculated stall speed, i.e.:

$$V_R \text{ not} < 1.10\ V_{S1} \qquad \text{(from Equation (6–8))}$$

It is wise to aim to lift off faster than V_R at, say:

$$V_{LOF} = 1.10\ V_R, \text{ about } 1.20\ V_{S1} \qquad (6\text{–}36)$$

Note: If the weight differs from that for which the stall speed is quoted (and it is always prudent to carry out a first flight at a lighter weight) then from Equations (3–9) and (3–26):

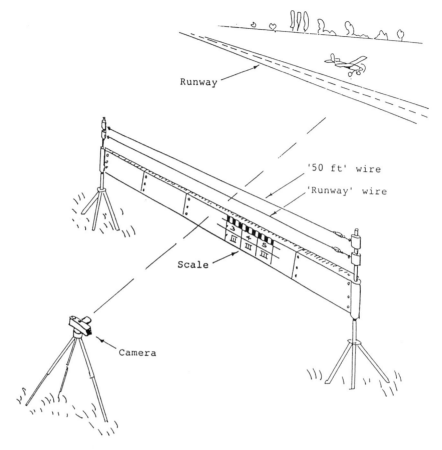

Runway

'50 ft' wire

'Runway' wire

Scale

Camera

Triangulation Device for Take-off and Landing Measurements:

The scale is arbitrary and is converted into distance along the
runway, using similar triangles. The gaps between the scale and
the wires are set to coincide with the runway centre-line and the
35 ft or 50 ft screen at the mid point of the runway. The device
must be erected level and parallel with the runway. Each section
of the scale is 1.5 to 2.0 m in length, in light alloy channel, bolted
together for rigidity while remaining portable.

Fig. 6.14   Take-off and landing measurement. (Courtesy of Terence Boughton)

new stall speed V_{S1} = original V_{S1} × *(new weight/original weight)*$^{0.5}$ (6–37)

☐ Measure the distance available (TORA) in which to lift off, taking into account the need
to clear any obstacles by about 35 ft (not less than 10 m) on a climb path of 3° (3/60 or
1 in 20 – see Fig. 3.23).

☐ Make a clear mark at half the distance available (TORA/2) to the point selected for lift-
off.

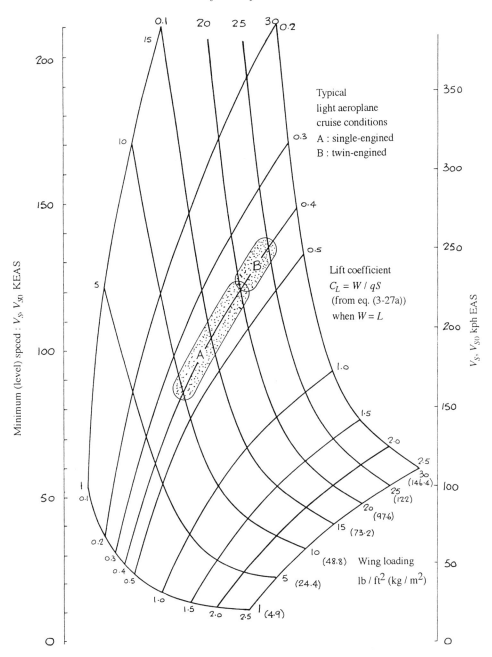

Fig. 6.15 Minimum level speed in terms of lift coefficient and wing loading (Ref. 1.5).

☐ Check airspeed during the take-off run to ensure that at least $(\frac{1}{2})^{0.5}$, or 71% of the calculated V_{LOF} is achieved by the time the half-way mark is reached. This comes from the general equation linking distance, s, acceleration, f, initial and final velocities, u and v:

$$v^2 = u^2 + 2fs \qquad (6\text{--}38)$$

in which distance: $(\text{TORA}/2)$ varies as $[V_{\text{LOF}}/(2^{0.5})]^2$, or $(0.71\,V_{\text{LOF}})^2$ (6–39)
Use this as the decision speed, V_1.

☐ If $V_1 = 0.71\,V_{\text{LOF}}$ is not reached by the half-way point, STOP.

Comparative Safety of Jet, Turboprop and Piston-Engined Aeroplanes

A rough comparison between the safety of jet, turboprop and piston-engined aeroplanes may be deduced from Table 6–9, which is the result of a separate study. The aim is to continue to close the gap between piston and turboprop aeroplanes on the one hand, and big jets on the other.

Table 6–9

Airworthiness catastrophic accident rates ($\times 10^{-6}$ per flight hour)

Class of aeroplane	Rate	Factor
JAR Part 25 large turbojet	0.07	1.0
JAR Part 23 commuter (mainly turboprop)	0.77	11.0
JAR Part 23 light twin (mainly piston)	1.58	23.0

Example: 6–8: Commuter/Outback Transport – Tri-Motor

A recent design study at Loughborough University was that of a requirement for a multi-role transport aeroplane to fit a number of international markets. Roles included commuter, light transport, air-ambulance and disaster-relief. It was argued that the aeroplane must have some STOL capability, and be easy to repair and maintain on out-stations, away from fixed-base operators, especially in areas such as those among nations of the Pacific Rim, Africa and South America.

Operators want turboprop engines. With organizations such as the Australian Flying Doctor Service and wide open spaces in mind one requirement would be for an aircraft with only one engine, because of the importance of keeping costs down. On the other hand, my own experience of operations in the (then) British Far East Air Force, was that of operation out of primitive short soft strips, with climb-out or balked landing paths among and over 200 ft (61 m) trees in primary jungle. For this STOL aerodynamics, power enough to cope with engine-failure, and easy loading (with wide CG range), repairability and maintainability were essential.

An aeroplane which attempts to provide the salient features listed is sketched in Fig. 6.16, showing side and rear-loading, room for 18 passengers, or six stretcher-cases, a doctor and two attendants as an air-ambulance, and two flight-crew. The configuration lends itself to fitting tanks and spray-booms for treating oil-spills. A tri-motor arrangement still has 2/3 power remaining with one engine inoperative and asymmetric handling could be eased; while cruise performance might be improved by shutting down the centre-engine. Further, in the event of irreparable damage to one outer engine in the field, there is the possibility of exchanging it with the centre-engine, and flying the aeroplane back to a fixed base as a symmetrical twin.

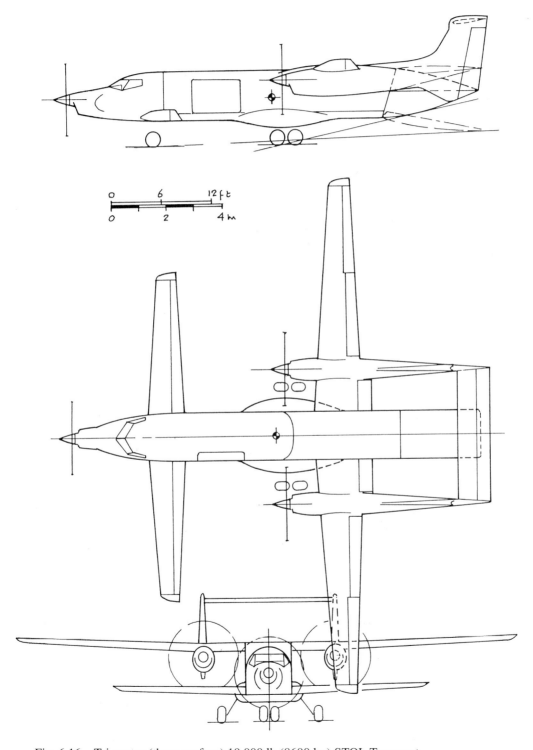

Fig. 6.16 Tri-motor (three-surface) 19 000 lb (8600 kg) STOL Transport.

A three-surface configuration was chosen because tandem fore and mainplanes lift in the same sense and provide the required CG-range, greater flexibility in loading and STOL. The tail surface offers improved control. The stabilizer between the booms is optimized as the dominant pitch and trimming control, because it allows the foreplane to get on with the main job of lifting. With flaps-DOWN on fore and rear planes, the tailplane may also be trimmed to provide positive lift, using the elevator for control. Although the configuration suffers increased drag from additional wetted-area and junctions, fuel may be pumped between tanks in the foreplane and mainplane, to put the cruise-CG in a position where trim-drag is least.

Kendall of *Gates Learjet Corporation* compares in Ref. 6.6 the performance trades of two-surface and three-surface configurations, concluding that:

- Two-surface airplanes cannot have minimum induced drag at all CG locations.
- Pure canard airplanes cannot attain the minimum induced drag trimmed condition and be inherently stable.
- Three-surface airplanes can have minimum induced drag at all CG locations and be inherently stable.
- Tests are needed to determine the potential drag benefits of three-surface airplanes.

The same author summarizes these points, in Ref. 6.7, when observing that the three-surface aeroplane can have: 'better cruise-efficiency in a stable trimmed condition over a practical range of CG locations than the conventional tail-aft design, which in turn has much better cruise efficiency than the two-surface canard configuration'.

Special flight test considerations

The three-surface aeroplane falls in the commuter class (19 000 lb (8600 kg)) and would be assessed against JAR-23 and FAR-23. It has novel features which increase its complexity, and require special attention in a number of critical areas bearing upon field performance:

- *Three engines* require tests with one engine inoperative on take-off, and two engines inoperative on landing.
- *Three surfaces and the wider CG-range* which is proposed must complicate investigation of controllability and stability at forward and aft CG.

Note 1: There is a risk of wake-interference between the three surfaces at large angles of attack, which are in line if the aeroplane is over-rotated with the tail-bumper on the ground. Interference could severely affect stall quality. Added to which is a likelihood of cross-coupling between rolling moment with sideslip and directional stability in cross-winds and when yawed, affecting adversely handling on take-off and in the flare.

Note 2: The chisel tail incorporating clamshell doors which form the end of the afterbody and are intended to open in flight, have an included angle in the side elevation close to 40°. Taper in excess of 25° is likely to cause turbulent separation and buffet. For example, the military version of the Armstrong Whitworth Argosy (1958) had a chisel-tail which could be opened and closed in flight. The civil version had an egg-shaped fairing. The aeroplane had suffered a low-frequency vibration in the tailplane, accompanied by a lateral vibration in the rear fuselage. Vortex-generators were fitted to the steep reverse slope where the flight deck joined the fuselage, which was the source of flow separation and buffet, discovered during flight tests. These had followed some 2320 wind-tunnel tests, occupying more than 1370 hours of running time (Ref. 6.5).

Plate 6–4 Civil version of Armstrong Whitworth Argosy, showing vortex generators to reduce effects of separation on reverse slope of cockpit hump. Ref.: Example 6–9, Note 2. (Courtesy of RAF Museum)

Wind-tunnels are eminently useful, but there is no substitute for carnal knowledge of an aeroplane at full-scale Reynolds numbers.

Fixes would include reshaping of the rear fuselage profile; and fitting vortex-generators and strakes (see Chapter 8). Even so, they never did succeed in fixing anti-symmetric nodding of the tail-booms of the Argosy.

☐ *Effectiveness of arrangement and size of flying-controls, and flaps*, which involve interlinking in pitch, would need close attention. There might be a case for some power-assistance in that respect. If so, then such technical complication would prejudice success as an aeroplane for use in the Australian outback and similar remote areas.

☐ *Field performance tests* must involve measurements of the effects of a wide range of paved, contaminated and softer surfaces, their hardness (California Bearing Ratio, CBR), and effects of grass of different types and lengths.

☐ *Ground handling tests* would determine the suitability of the undercarriage: the effect of CG position on steering, rotation on take-off, stability on the ground, wheel arrangement, size and type of tyre upon quality of ride, limiting surfaces, and upon take-off and landing distances.

Probability methods shape the philosophy behind field performance requirements of commuter (and heavier) transport aeroplanes. Considerable airworthiness testing of an aeroplane like the Trimotor is needed for CAA Group B and JAR-OPS Class A certification, as we saw in Fig. 3.24 and Table 3–3.

References and Bibliography

6.1 ESDU DATA ITEM No. 91017, *Statistical Methods Applicable to Analysis of Aircraft Performance Data*. London: ESDU International plc, 27 Corsham Street, London N1 6UA, or P O Box 1633, Manassas, VA 22110, USA

6.2 General Aviation Safety Sense 7A, *Aeroplane Performance*. Cheltenham: Civil Aviation Authority

6.3 *Australian Air Navigation Order, Section 101.22*. Canberra: P O Box 367, ACT 2601, Australia

6.4 Lerche, H-W., *Luftwaffe Test Pilot*. London: Jane's Publishing Company, 1977

6.5 Tapper, O., *Armstrong Whitworth Aircraft since 1913*. London: Putnam, 1973

6.6 Kendall, E. R., *Performance Trades of Two-Surface and Three-Surface Configurations*. Paper 84-2221 of the American Institute of Aeronautics and Astronautics, 1984

6.7 Kendall, E. R., *The Minimum Induced Drag, Longitudinal Trim and Static Longitudinal Stability of Two-Surface and Three-Surface Airplanes*. Paper 84-2164 of the American Institute of Aeronautics and Astronautics, August 1984

CHAPTER 7

General Performance
(Climb, Glide and High-Speed)

'Tactics apart, the vital question is that of performance. A machine with a better speed and climbing power must always have the advantage.'
Cecil Lewis, on single-seater combat in 1917. (Ref. 1.8)

'When, soon afterwards, I got the contract ... I made an attempt to equip the smallest possible light aircraft with a powerful engine in order to produce a fighter which exceeded in its performance anything that had been seen before.'
Willy Messerschmitt on the Bf 109 (1942), quoted by Frank Vann (Ref. 7.1)

Often in test flying one has to make rapid initial assessments of the general quality of performance of an aeroplane, without resort to specialized equipment. A timed climb and descent, with a run to the never exceed speed or Mach number, V_{NE}/M_{NE} reveals much. In the case of a new, uncertificated aeroplane, V_{NE} might not yet have been established, in which case one would look at V_D/M_D (or V_{DF}/M_{DF}) the design maximum, or demonstrated diving speed or Mach numbers (but NOT on a first flight!). A glide with throttle closed gives a fair indication of best emergency lift/drag ratio with propeller(s) or fans windmilling. The rate or gradient of climb is an indication of the amount of power left in hand to haul the mass of the aircraft upwards at a given true airspeed. If there is insufficient excess power, too much drag, or the aeroplane is too heavy, then a previously scheduled rate or gradient of climb will not be achieved.

Even if the calculation of weight and balance before the flight is correct, the weight schedule can be wrong. This is often found to be the case when testing older aeroplanes, the weights of which have changed over the years, leaving scope for small errors to creep into the records. All aeroplanes grow heavier with age, and their centres of gravity migrate aft – because there is usually more airframe to repair and repaint behind the wing than lies ahead of it.

Power, it will be recalled, is the rate of doing work, i.e. force times velocity (force × distance/time), in appropriate units. Figure 7.1 is a reminder of the force and speed vector diagrams in unaccelerated flight, during climb, level run and descent. Figure 7.2 shows rates of climb and descent, measured as follows, and plotted in relation to power.

Measurement of climb and descent performance

Table 7–1 Shows a typical presentation of light aeroplane climb performance. To measure

a. Glide: resultant drag, D = residual thrust - actual drag

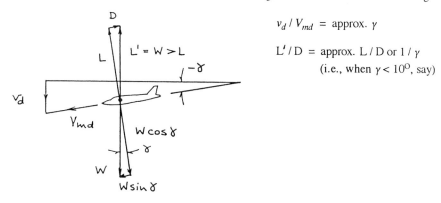

v_d / V_{md} = approx. γ

L'/D = approx. L/D or $1/\gamma$
(i.e., when $\gamma < 10°$, say)

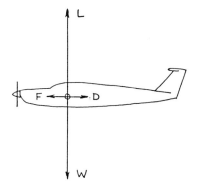

b. Steady level flight: $L = W$ and $F = D$

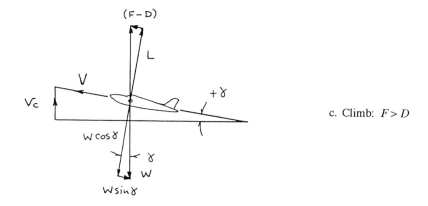

c. Climb: $F > D$

Fig. 7.1   Speed and force diagrams in unaccelerated flight.

climb and descent performance the following points should be considered before take-off. It may be that the climb and glide speeds have been scheduled already, in which case they should be used. If not, then a set of timed *partial climbs* and *partial descents*

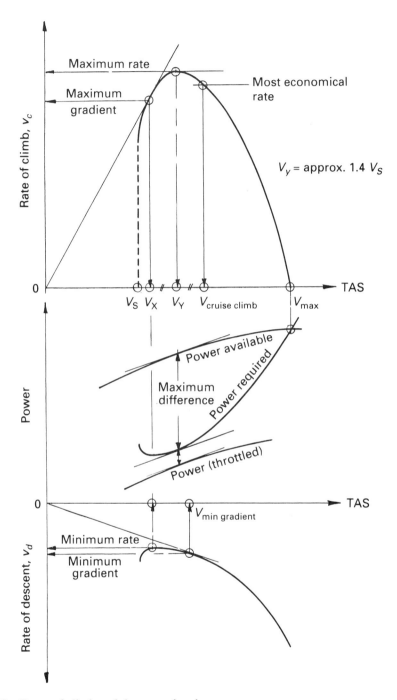

Fig. 7.2   Rates of climb and descent related to power.

are carried out at different airspeeds through the same altitude bands, working out the fuel consumption, quantity used and, hence, the mean weight of the aircraft for each partial.

Performance

TABLE 7–1

Typical layout of light aircraft climb performance

NORMAL ENROUTE RATE OF CLIMB
FLAPS AND GEAR UP – 110 KIAS

CONDITIONS:
2600 RPM
31 inches Hg
Mixtures Set at 90 PPH
Cowl Flaps Open

Weight	Press Alt (FT)	Rate of climb (FPM)			
		–20°C	0°C	20°C	40°C
4630	S.L.	1185	1020	860	690
	4000	1055	890	720	545
	8000	920	745	575	400
	12 000	780	605	425	–
	16 000	625	450	275	–
	20 000	465	295	–	–
4300	S.L.	1325	1150	980	805
	4000	1190	1015	835	655
	8000	1050	865	685	500
	12 000	900	715	530	–
	16 000	740	555	380	–
	20 000	580	400	–	–
4000	S.L.	1465	1285	1105	920
	4000	1325	1140	950	760
	8000	1180	985	795	600
	12 000	1025	830	635	–
	16 000	860	665	480	–
	20 000	690	505	–	–

Pre-test (with emphasis on the climb)

☐ Work out the weight and balance of the aeroplane – and it is prudent to check the CG position at a number of test weights as fuel is consumed. By doing so we may then construct a graph of test weight and CG against indicated fuel remaining in the tank(s). The latter is important on two counts. First, we then know in which direction the CG migrates will fuel-burn and whether it is likely to go beyond the test limits. Second is that the rate of climb increases with fuel consumed, which enables us to plot rate of climb against fuel state and weight at a given altitude.

Note: CG movement is significant. Climb results obtained with it in the position which degrades excess power most adversely are those used as the basis of climb performance scheduled in the Flight Manual. A conventional aeroplane with its tail at the back needs more up-elevator to hold attitude as the CG moves forward. This increases the tail download, detracting from the total lift of the airframe, while increasing tail lift-dependent drag. A tail-first (canard) aeroplane needs more foreplane lift as the CG moves forward which, added to wing lift, is theoretically advantageous. But it also introduces lift-dependent trim-drag which detracts from the excess power available.

Example 7–1: Tractor Versus Pusher Propellers

A number of pusher-propeller configurations have appeared, especially among tail-first configurations and microlights. They raise the question of relative 'efficiency' – how much better or worse is the pusher compared with the tractor? The answer is that the pusher propeller, working on the air as an actuator disc in the wake of the various parts of the airframe ahead of it, is not as efficient as a tractor, which turns in free and largely undisturbed air. Where the pusher gains is that the wash through the disc is discharged to atmosphere without suffering drag losses caused by friction behind the propeller, as propwash slides and tumbles past the skin on its way aft. As a thrust-producer the pusher is usually superior to the tractor.

Note: The Rutan Voyager has its front propeller feathered in long distance cruise configuration (Ref. 2.10).

But that is not the end of the story. Lift generated by a tractor propeller under power, acting ahead of the CG at an angle of attack, is destabilizing. Lift of a pusher at the same angle of attack is behind the CG. This time it acts in a stabilizing sense.

When a pusher is aft of the fin and rudder surfaces there is no asymmetric wash to strike them, so that there is little or no directional trim change with change of power and need to trim out footloads before settling into the climb. Therefore, one might change from a RH rotating engine and propeller to one that is LH, without noticing much difference in directional handling – quite unlike the state of affairs in Example 8–1.

A disadvantage of the pusher which can affect climb rate adversely is that it tends to swallow more damaging stones and grit thrown up by the wheels. Even so, it is my experience that push–pull configurations, like the Cessna 336 and 337, and the Rutan Defiant, have better rates of climb on the rear engine, with the front engine stopped and feathered, than the reverse.

Note 1: Always ensure with such aircraft that during run-up the idling speed of the rear engine is comfortably (but not excessively) faster than that of the front. Although both front and rear engines of a Defiant were idling at the same RPM before take-off, I found that during subsequent stalls with both throttles closed, the rear engine quietly stopped more than once in the much slower wake.

On the last occasion this occurred over-the-hedge, with throttles closed at the end of its very flat and long approach to touch-down (a consequence of it having a lot of lifting surface area and, because of the position of the CG between fore and aft planes, no flaps are fitted). Had it been necessary then to carry out a go-around in an emergency it might have been impossible to climb away on the front engine alone, out of ground-effect, or to restart the rear one in time.

Note 2: In Chapter 13 we see that an important advantage of the tractor over the pusher is in ground-handling, and the benefit of propwash over the tail surfaces.

Test Conditions

☐ The climb(s) must be carried out crosswind, so that one may spiral down again at the end of each, if necessary, to repeat the climb on a reciprocal heading, through the same slice of air. The air will be drifting downwind, so, plan the heading and route before take-off. For example, when carrying out reciprocal single-engined climbs crosswind in a Britten-Norman Islander, we drifted off-shore. Over the land the first climb produced a positive rate of climb. The second resulted in sink (a negative rate) even though the aeroplane was lighter. We were being affected by subsidence over the cooler sea in the normal diurnal cycle of a sea breeze. Over the warmer land the air ascended. Over the sea it descended.

☐ *Choose your weather with care*: climbs must be started at a comfortable pressure altitude, out of ground turbulence. The clean gradient wind, above the immediate boundary layer of the earth, is found at an altitude which varies with surface conditions. Elsewhere, over a flat plain, or water, in smooth air a timed climb might be started as low as 1000 ft (300 m), without introducing unreasonable error. If any turbulence from natural and other features is present then 2000 ft at least should be the rule rather than the

Plate 7–1   Pilatus Britten-Norman Turbine Defender is a much modified variant of the piston-engined Islander. Here, increased keel surface area forward has been balanced by a much enlarged dorsal fin aft. All major modifications affect flying qualities and extensive additional testing in flight is needed to clear them for certification or service release. (Courtesy of Pilatus Britten-Norman)

exception. One has to balance inaccuracy caused by attempting to fly in a level of turbulence which the pilot might argue is reasonable, to get the job done, against weight errors through fuel-burn which creep in while spending time searching for smoother conditions. It is not only downright stupid, but professionally unforgivable to penalize an aeroplane unfairly, wasting an owner's money.

Note: For operational safety reasons in the UK single-engined climb tests of twins are not started below 3000 ft above terrain.

☐ Aim to carry out straight climbs. Bank reduces the lift component and a larger angle of attack is needed to compensate for such loss of lift. Turning wastes power by increasing the lift-dependent drag in direct proportion to the square of the lift coefficient and, hence, the square of the increased angle of attack, within the working range of the wing (see Chapter 5, powerplant effectiveness, $\bar{p}_\eta$, and the preamble to Equation (5–55)).

☐ Ensure that the propeller is in good condition. Equations (5–37) and (5–38) give the power relationships, showing that the thrust and therefore spare horsepower for a climb is directly proportional to propeller efficiency. A propeller with a scuffed and chipped leading edge suffers a rapid loss in propulsive efficiency (Fig. 7.3).

☐ Carry out a run-up before take-off to check the ground-static RPM. As long as there is no danger of overheating the engine, or damage to the flying controls, this should be done

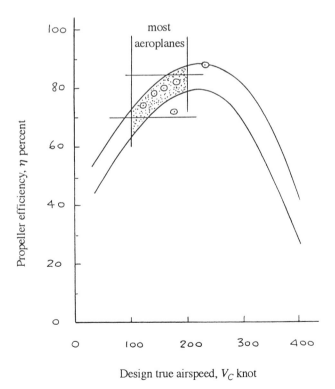

Fig. 7.3 Practical propeller efficiencies. Contamination, erosion and chewed leading-edges rapidly reduce values to the bottom of the band – and even lower.

crosswind, to avoid windmilling effects. Power generated is proportional to RPM cubed, which means that a false rise in revolutions of, say, 1% gives the impression of:

$$(101/100)^3 = 1.03 \text{ (an increase of 3\% in power)} \tag{7–1}$$

Normal wear and tear on an engine increases power lost to many causes, such as increased internal friction from deteriorating bearings, losses of working pressures and torques, chipping and erosion of propeller surfaces which increase their rotational drag. Thus, an accurate ground static RPM is either a sure sign that all is well, or a first sign that something is wrong.

Power Required in the Climb

The power required for a steady climb is in two parts. The first is expended kinetically, pushing the aeroplane through the air at its climb speed (even tractor propellers push – think of the way in which they are bolted on to the propeller shaft). The second is used to haul its mass upwards against gravity. As long as the engine is not suffering cooling problems, the best climb speed is not far from that at which minimum power is required for level flight. This is the best endurance speed with a propeller, so that minimum power for flight is found to lie somewhere between 1.3 and 1.4 V_{S1} (see point C in Fig. 7.4). The scheduled climb speed is often set closer to the latter, for improved engine cooling. The power equation is generally stated in the form:

$$FV = DV + Wv_c + (W/g)a \tag{7–2}$$

in which:

$$DV = C_D \left(\tfrac{1}{2}\rho \, SV^2\right)V \tag{7–2a}$$

using terms already familiar, with the exception of a, the linear acceleration (dV/dt). In steady flight $a = 0$, so that the performance equation may be written in terms of rated horsepower, P, and the corresponding FPSR conversion factor:

$$FV = 550\eta P = DV + Wv_c \tag{7–2b}$$

Power varies as (TAS)^3. For simplicity let us assume that the stall speeds clean and in the landing configuration are sufficiently close as to be replaced by a single term, V_S, making the speed ratios V_C/V_{S0} and V_C/V_{S1} simply equal to V_C/V_S. So, the maximum power available from the engine(s) is proportional to V_C^3, the maximum speed in level flight. The minimum power to fly at the best climb speed is proportional to $(1.4V_S)^3$. Therefore from Equation (7–2), and with acceleration $a = 0$:

maximum power available = minimum power for level flight + power for climb + 0

Assuming that the drag coefficient remains constant at both 1.4 V_S and V_C, we may transpose this statement, to say in terms of the total power available:

$$\text{power for climb is proportional to } [(V_C/V_C)^3 - (1.4 \, V_S/V_C)^3]$$
$$= 1 - [1.4/(V_C/V_S)]^3 \tag{7–3}$$

Example 7–2: Loss of Climb Rate due to a Drop in Ground-Static RPM

Table 7–2 lists numerous common causes of inadequate piston-propeller climb performance, all of which result from reductions in excess power. For example, if the ground

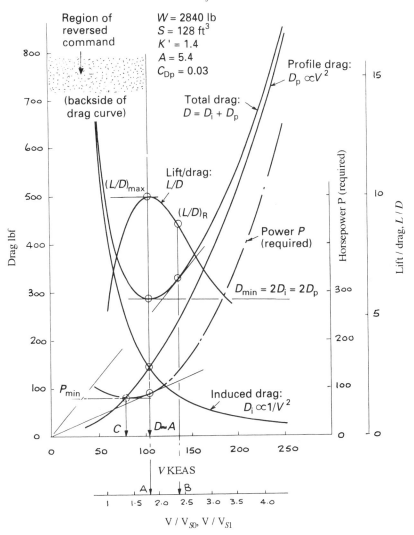

Fig. 7.4 Typical drag and power curves for a piston-engined trainer.

A = Speed for minimum total drag and, hence, best lift / drag.
  = $(L/D)_{max}$ best endurance speed for a jet aeroplane, and speed for flattest glide.

B = Best range speed, corresponding with best $(L/D)_R$ for range and speed / drag, V/D.
  = approx' 0.94 $(L/D)_{max}$ for a jet (here, $(L/D)_R$ = approx' 0.9 $(L/D)_{max}$, piston-prop').
  = 1.32 V_{md}, the EAS for minimum drag.

C = Speed for minimum power, the best piston-propeller endurance speed, V_{mp} (which
  is near enough 0.76 V_{md}, i.e., about 1.3 to 1.4 V_S, the 'ball-park' initial climb speed).

D = Best range speed for piston-propeller corresponding with (speed / power)$_{max}$, $(V/P)_{max}$.
  As fuel consumption of a piston engine is roughly proportional to power, maximum
  range results from (fastest speed / unit fuel consumed). This is close to point A.

static RPM of a piston engine drops by 1.0%, from 2450 to 2400, the reduction in
power is $(1/100)^3$ or 3%. The overall effect of this in the case of a relatively clean single
engined monoplane, with retractable gear, for which $V_c/V_S = 2.7$ say, may be seen by first

TABLE 7–2

Common causes of inadequate climb performance
of a piston-propeller aeroplane
(Civil Aviation Authority: Ref. 7.2)

(a) General
  (i)   Pilot out of practice.
  (ii)   Weather: turbulence, waves, temperature inversion, high humidity.
  (iii) Incorrect reading of IAS (it is easy to confuse, or to substitute, CAS for IAS, or knots for mph).
  (iv)   Faulty ASI (e.g., leaks, blockages, (including water), instrument unserviceable).
  (v)   Failure to bank towards the operating engine.
  (vi)   Faulty OAT indicator.

(b) Airframe (Sources of Drag)
  (i)   Aircraft flown without wheel fairings or spats.
  (ii)   Oversize wheels and tyres.
  (iii) Badly fitting, or leaking, doors and panels.
  (iv)   Dented leading surfaces.
  (v)   Incorrect rigging of wing-flaps (trailing, edge up or down a fraction can make a large difference either way, because it can also adversely affect trim drag from the taiplane/stabilizer).
  (vi)   Cowl flaps left open on the inoperative engine.
  (vii) Additional external equipment: de-icing boots, aerials, beacons, etc.

(c) Engine
  (i)   Air fuel ratio: a too-rich mixture setting.
  (ii)   Pre-heating of induction air through wrong setting of carbair lever or hot/cold flap not seating properly.
  (iii) Inability to achieve maximum RPM through fault in throttle or governor linkage.
  (iv)   Incorrect fuel delivery pressure, causing too rich a mixture.
  (v)   Lack of adequate cylinder compression (e.g., spark plug seating).
  (vi)   Incorrectly fitted exhaust system.
  (vii) Spark plug gaps and condition.
  (viii) Ignition timing ($5°$ error has been known to account for 125 ft/min).
  (ix)   High engine temperatures (e.g., incorrect setting of cowl flaps).
  (x)   Carburettor ice accumulated during operation at part-throttle, failing to clear before operation at full throttle.
  (xi)   Turbochargers inoperative.

(d) Propeller
  (i)   Correct type of propeller not fitted.
  (ii)   Excessive propeller cropping.
  (iii) Incorrect fine pitch stop setting.

  Note: Sometimes simply changing the propeller for another of the same type has remedied matters, as in the case of one or two Cessna 150s.

(e) Weight
  (i)   Unrecorded growth of weight empty.
  (ii)   Incorrect weight scheduled.
  (iii) Miscalculation of test weight.

  Note: When comparing a climb result with one obtained previously, CG position can make a difference. A forward CG requires more nose-up elevator, which causes the aeroplane to fly at a larger angle of attack with higher associated drag, than when the CG is aft. Climbs carried out at the same weight, but with differing CG positions can produce different results.

calculating the total excess power available to climb at the scheduled airspeed, using Equation (7–34):

$$\text{total power available for climb} = 1 - (1.4/2.7)^3 = 1 - 0.14 = 0.86$$

i.e. about 86% of the engine output.

The drop of 3% in power output caused by a 1% loss of ground-static RPM reduces this amount at the scheduled speed by:

$$0.03 \times 86/100 = 2.6\% \text{ of the rate of climb} \tag{7–3a}$$

which is nearly 3%. With one engine inoperative at the same speed (assuming no additional drag caused by the failed engine – which is impossible), the loss of 1% in ground-static RPM would cause the aeroplane to lose more than 5% in rate of climb.

Frequently one finds that ground-static RPM of a piston-propeller aeroplane are, say, 2200 instead of 2300 (or more). This represents a drop of nearly 4.5% in RPM, representing a power loss between 12% and 13% i.e. 125 ft/min in a rate of 1000 ft/min; and a reduction in climb gradient of 1.3% at $1.4V_s = 96$ KCAS. Such a loss in gradient exceeds substantially the difference between gross and net gradients of –0.8%, corresponding with 1.5 standard deviations, in Equation (6–33a). Not only is such a figure too large to ignore, it illustrates well that deterioration in climb performance is one of the most sensitive and significant indicators of the health of an aeroplane; hence the emphasis upon carrying out run-up check *crosswind*.

Determination of Best Climb and Glide Speeds

Basic test equipment

The following must be accurate and calibrated where applicable:

- Airspeed indicator.
- Sensitive altimeter.
- Thermometer for measuring ambient temperature (OAT).
- Engine power instruments (tachometer, torque gauge, thermometer, etc.).
- Stop watch.

Test technique

The test techniques for partial (*sawtooth*, USA) climbs and descents are almost identical, except that a climb is carried out with throttle(s) open, the descent with them closed, or with power set for zero thrust. Tests must be carried out in smooth air, clear of cloud and away from hills which may produce wave action. Recording of data should not start until the aeroplane is established in the climb at the correct speed. In the case of a piston-propeller aircraft the change in weight during the time elapsed from take-off may be estimated approximately, knowing the rated BHP:

$$\text{weight of fuel consumed to start of climb} = 0.5 \text{ lb/BHP/hour} \tag{7–4}$$

At constant weight the rate of climb at a given pressure altitude varies with airspeed, first increasing to a maximum at the best climb speed, before again decreasing. At low altitudes the curve is almost flat topped, sharpening at higher altitudes. Up to full throttle height the rate of climb is fairly insensitive to small changes in airspeed. Engine cooling problems, sloppiness in control or a tendency to instability can make it necessary to recommend an increase in airspeed beyond the optimum, even though some of the climb rate may be lost.

Partial climbs are made by timing the aeroplane over small changes of pressure altitude (setting 1013.2 mb, 29.92 inHg) in straight and steady flight, at around six different forward speeds, through different altitudes. Indicated airspeed must be maintained within ±2 knots. The height intervals should be as small as is compatible with timing by means of a stop watch: 500 ft for heavies, up to 2000 ft for agile aerobatic and fighter aircraft. One should aim to note the indicated altitude every half minute, including the times of entering and leaving the selected band. This should be arranged to bracket the test altitude equally. Record the fuel state during each partial, engine conditions (RPM, torque, temperatures and pressures), and outside air temperature (OAT). Leave space on the test pad for comment. Finally, to achieve steady conditions through each height band, a partial must start and end comfortably outside it.

If time allows, the airspeeds should be worked through twice, in ascending and descending order. A level speed run with the aeroplane in the same configuration may usefully be interposed. For the tests differences in airspeed of about 10 knots are usually sufficient.

By changing the engine setting(s) to either throttle-CLOSED, or power for ZERO THRUST to descend back through the height band through which one has just climbed, it is possible to make measurements of glide performance and, hence, rate and angle of descent – and from that the lift/drag ratio of the aeroplane.

We have assumed that in these tests the pilot is taking a look at performance for some initial purpose, like a quick check, and not gathering data from which Flight Manual information will be scheduled. Therefore, time would not be spent on precise reduction of rates of climb, descent and airspeed to their absolute values. Having said that, care is needed and results should be as accurate as possible, otherwise they are devalued.

Figure 7.5a is a typical plot of partial climb results. A similar curve would be drawn showing descent rates, only this time they would appear beneath the base line, rather like negative climb rates. The best rate of climb at each altitude is then plotted as shown in Fig. 7.5b (which may usefully be compared with Figs 2.4 and 2.5, both of which were crude attempts to compare the climb performance of a Fokker DVII and Bristol F2B).

Absolute and service ceilings

To complete the picture in Fig. 7.5b, a straight climb line should be extrapolated through the results, up to the service ceiling (at which the rate of climb is 100 ft/min (about 0.51 m/s)). If the service ceiling is not yet known it may be estimated, using Equation (5–52a) and Figs 5.3a and 5.3b. A sea level rate of climb may be also be estimated by means of Fig. 5.2a or 5.2b.

Gathering and presenting climb data

Figure 7.6 shows a typical test card layout, of a size to fit a knee-pad. When completed it is small enough to be transferred to the safety of a flight case, or a leg pocket of flying overalls. The card is clear and neat, but the final results should not be presented in this form. They must be plotted graphically, by the test pilot or the flight test engineer (I am a great believer in making test pilots work, by plotting their own results).

It is rare to find final climb information presented in the form of Fig. 7.5a or 7.5b. Light and small aeroplane manufacturers usually provide 'user-friendly' tables of rates of climb

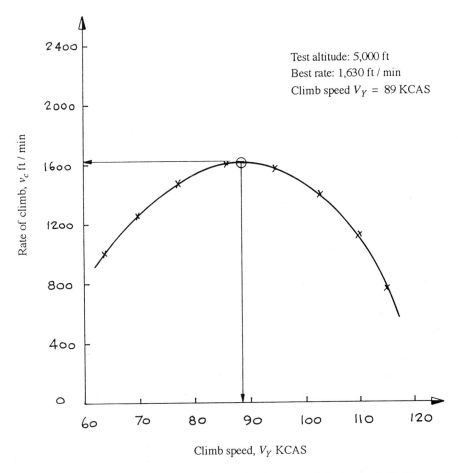

Fig. 7.5a   Plot of partial climb results to determine best rate of climb speed, V_Y. The method can be used for best gradient-of-climb speed, V_X.

with guidance on the configuration to which it applies, as given in '*Conditions*' at the top left-hand of Table 7–1 (Cessna T337H). Manufacturers of larger public transport aircraft schedule the information graphically (as gradients for each segment, where appropriate) like the final *en-route* climb data in Fig. 7.7. This includes precise guidance on the configuration to which the information applies, with airspeeds scheduled separately, e.g., Fig. 7.8.

Note: All information for pilots must be complete and unambiguous. There are times when it may be necessary to change a flight plan in the air and revise climb data, under stressful conditions. Therefore, when considering the best and clearest way of presenting information in the cockpit, always bear in mind that the crew may be under high pressure at the time.

Manipulation of results

Having obtained climb results at different airspeeds, identities given in Equations (6–23) to (6–37) are useful for manipulating it into useable forms involving pressure, volume and absolute temperature.

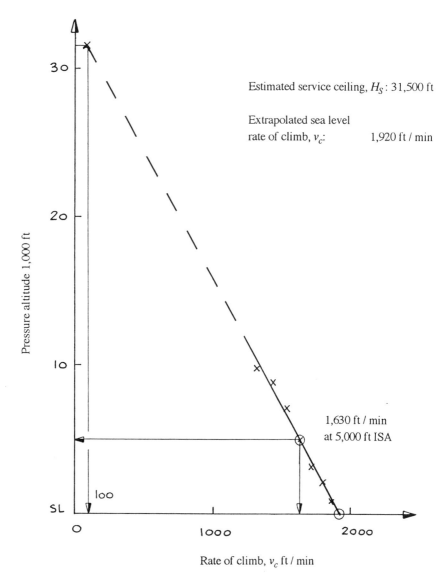

Fig. 7.5b   Plot of ISA climb performance using results of partial climbs, service ceiling estimated from Fig. 5.3 and extrapolation to find sea level value. Compare this figure with Fig. 2.5 which, although crude, employed the same method using different sources for the Fokker DVII and Bristol F2B.

Knowing the pressure and instrument errors, once a rate of climb line has been constructed it may then be adjusted for any other pressure, density and temperature lapse rate by means of Figs 5.1a and 5.1b.

Subsequent Climb Checks

Subsequent climb checks are carried out to:

Performance Climb

Clear of cloud, out of turbulence.

Altimeter set 1013 mb/29.92"

Climb power RPM MP

Cowl flaps OPEN/CLOSED/POSITION

Climb speed (V_Y)

(Check Hot air/Alternate air OFF)

Establish a steady climb in trim. Record:

TIME	ALTITUDE	IAS KIAS/MPH	OAT °C/°F	ALTITUDE	IAS KIAS/MPH	OAT °C/°F
0						
½						
1						
1½						
2						
2½						
3						
3½						
4						
4½						
5						

Near/at TOC. Record

MP RPM OIL P OIL T °C/°F

CHT °C/°F Fuel contents

Fuel pressure

Trims Elevator Aileron Rudder

Sheet 6

Fig 7.6

Fig. 7.6 Example of climb test card used for certification. (Courtesy of Bob Cole (CAA))

- Establish whether or not the climb performance of a new aeroplane has been realistically scheduled.
- Check a new aeroplane of an established type for the effect of changes introduced to update, otherwise improve, or simply refashion the marque over the years.
- Examine an individual aeroplane in service, or a fleet, for deterioration.

Plate 7–2   Twin engined push-pull Cessna FT337GP. Such aeroplanes with one engine inoperable suffer less resistance with the front engine shut down, and so perform better on the rear engine, producing higher single-engine rates of climb. Ref.: Examples 2–3 and 7–1. (Author)

☐ Complete an airborne inspection following maintenance, or repair after an accident or incident.
☐ Assure the condition of a renovated or used aeroplane before making an offer to buy.

To carry out such a test involves using information in the AFM using a standardized test card for the results, like the one already suggested.

Dealing with Sparse Information

Sometimes difficulties arise when checking very simple piston-propeller aeroplanes – the Piper Pawnee is one example – for which scheduled information is only listed as follows:

$$\left.\begin{array}{l}\text{rate of climb (ft/min)}\\ \text{service ceiling (ft)}\end{array}\right\}\ \text{for a given maximum weight and best climb speed}$$

One has to know that the rate of climb stated is the sea level value in the standard atmosphere and the service ceiling is that altitude (ISA) at which the rate of climb has fallen linearly to 100 ft/min (0.51 m/s). This enables a graph to be reproduced which is similar to Fig. 7.5b. Sometimes the service ceiling referred to might be an earlier 150 ft/**min**(0.76 m/s) but that does not matter. The service ceiling is usually high enough for a difference of 50 ft/min to make an error which is too small to be worried about. When completed, we may then use the graph to interpolate a practical rate of climb (ISA) for the test altitude, against which the rate of climb obtained during the tests is then compared.

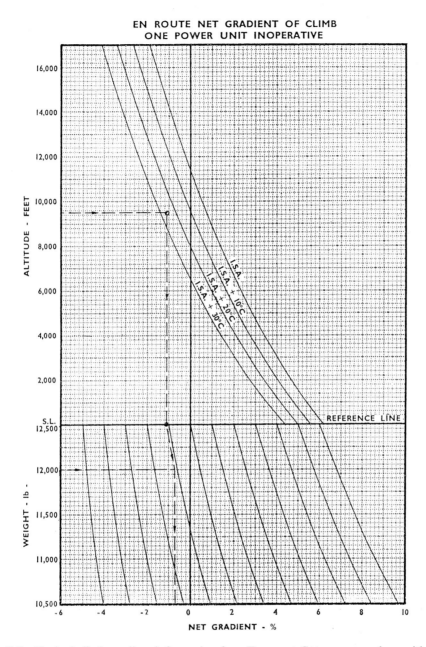

Fig. 7.7 Typical climb gradient information for a Transport Category aeroplane with one power unit inoperative. (Courtesy of CAA)

However, another source of error for such simple aircraft can be a lack of information in the Manual on the variation of scheduled climb performance with temperature. The following corrections can be useful:

☐ *Observed outside air temperature below ISA*: then the scheduled rate of climb, or the rate of climb interpolated as above, should be *increased* by 4 ft/min/°C (0.025 m/s/°C)

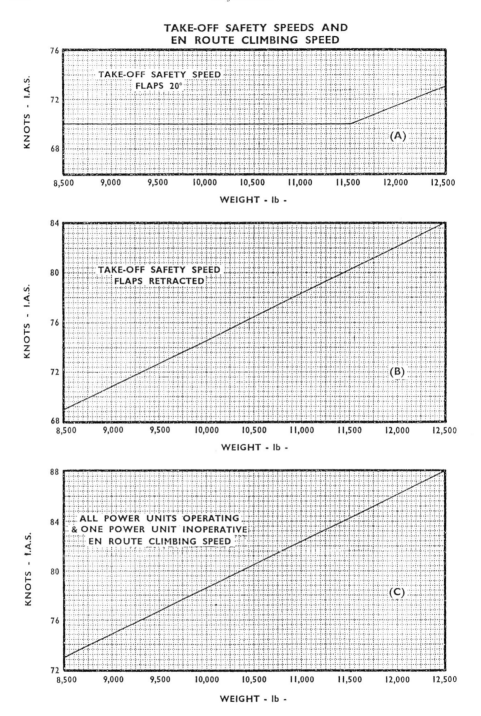

Fig. 7.8 Presentation of airspeed information for Transport Category aeroplane in Fig. 7.7 – for which graph (C) applies. (Courtesy of CAA)

for every observed degree centigrade *below* the ISA temperature for the mean climb altitude.

☐ *Observed outside air temperature above ISA*: involves the opposite action to that above, in that the scheduled rate of climb should be *decreased* by 4 ft/min/°C (0.025 m/s/°C) for every observed degree centigrade *above* the ISA temperature for the mean climb altitude.

Such corrections would be applied to results like those shown in Fig. 7.9 had the recorded mean OAT deviated sufficiently from ISA at the mean test altitude (see the plot on the broken line at the right-hand side of the graph). In this case no correction was thought to be necessary, as the 'eyeballed' mean value lay on the line, even though the temperature above and below showed evidence of lapsing faster than the standard rate.

Common Causes of Inadequate Piston-Propeller Climb Performance

Piston-propeller aeroplanes represent the largest proportion of those on national registers in the developed world: around three quarters. Many carry passengers for pleasure if not hire and reward and it is useful for owners and engineers to know the common causes of climb shortfall which aeroplanes suffer. These are listed in Table 7–2 (Ref. 7.2).

Tolerances

Any shortfall in climb performance between the test results and that scheduled is treated in exactly the same way as scatter in field performance (discussed at some length in Chapter 6). If the shortfall is not more than 20 ft/min (0.1 m/s) below that scheduled, then it is reasonable to accept it as being caused by the pilot being somewhat out of practice on type, or weather conditions, and other small but cumulative errors in the normal cockpit instrumentation. But if the shortfall exceeds 20 ft/min, the aeroplane should be checked for causes against Table 7–2.

If the shortfall exceeds 80 ft/min (0.41 m/s), then advice should be sought from the certificating authority. In the case of the UK CAA, Flight Department staff are always helpful, and it is probably still true that such help costs nothing.

Example 7–3: Standard Deviations and Light Aeroplane Climb Shortfall

Assuming a climb speed of 84 KCAS at a pressure altitude of 5000 ft (ISA), where the square root of the relative density is about 0.93 (see Fig. 5.1b), what is the reduction in gradient of climb represented by a shortfall in climb rate of 80 ft/min? Assume zero instrument and pressure errors.

From Equation (3–66)

$$TAS = EAS \text{ (or } CAS)/(\sigma^{0.5})$$
$$V \text{ KTAS} = 84 \text{ KCAS}/(\sigma^{0.5})$$

which in ft/min (see Table 3–9):

$$= 84 \times 1.688 \times 60 \times 0.93$$
$$= 7912 \text{ ft/min}$$
the gradient for 80 ft/min shortfall $= -(80/7912) = -0.01$, or -1.0% \hfill (7–5)

which is slightly more than the –0.8% applied to the gross gradient of a twin to obtain the

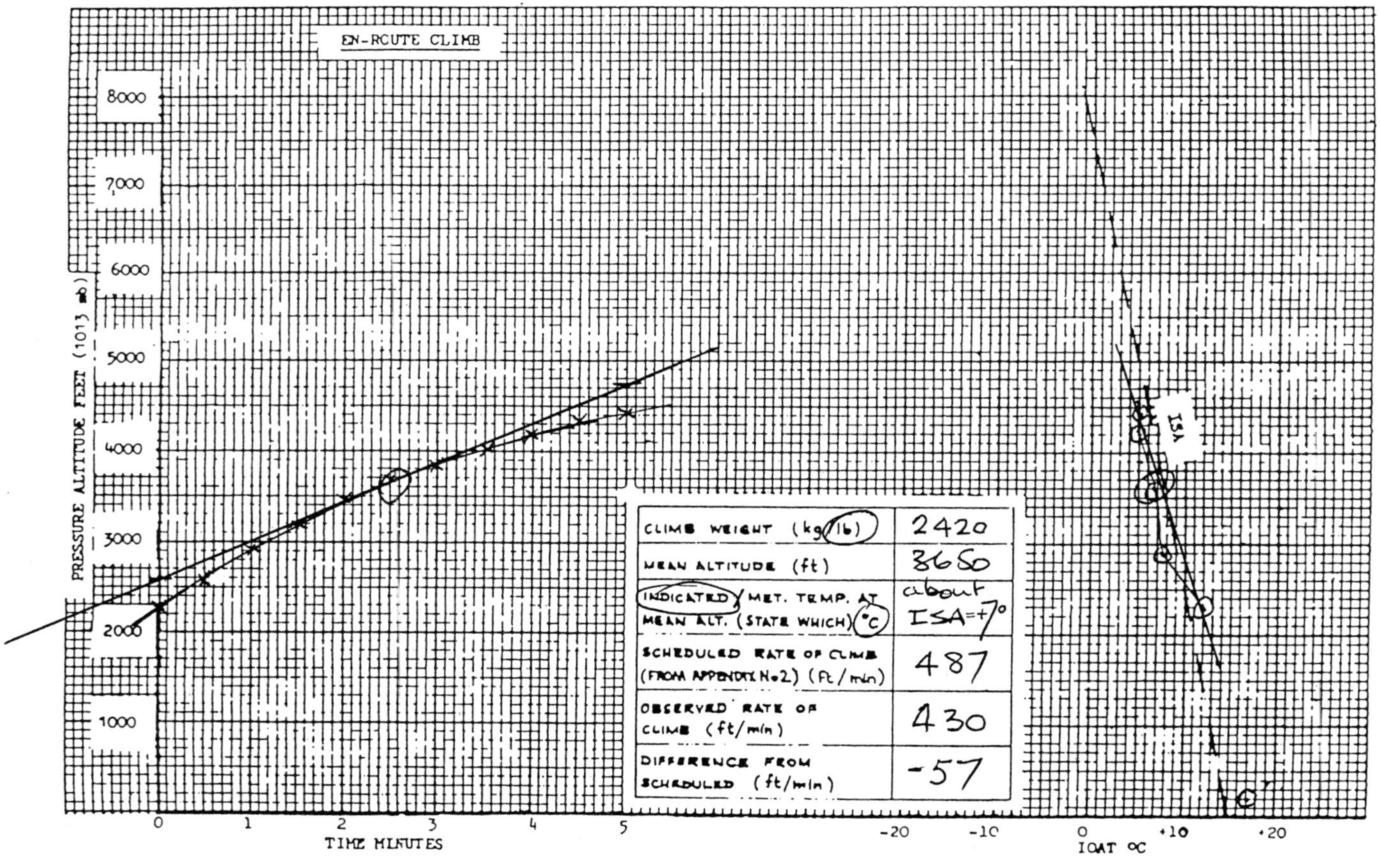
EN-ROUTE CLIMB
PRESSURE ALTITUDE FEET (1013 mb)
8000
7000
6000
5000
4000
3000
2000
1000
0
1
2
3
4
5
TIME MINUTES
-20
-10
0
+10
+20
IOAT °C
ISA
CLIMB WEIGHT (kg/lb)
2420
MEAN ALTITUDE (ft)
3650
INDICATED/MET. TEMP. AT MEAN ALT. (STATE WHICH) (°C)
about ISA=+7°
SCHEDULED RATE OF CLIMB (FROM APPENDIX No2) (ft/min)
487
OBSERVED RATE OF CLIMB (ft/min)
430
DIFFERENCE FROM SCHEDULED (ft/min)
-57

Fig. 7.9 Actual working example of climb results gathered during an air test, showing the rate of climb to be around 57 ft/min less than scheduled for the test weight and ambient conditions. The engineers were asked to find the cause, of which the propeller was the prime suspect.

net gradient in Equation (6–33a) and corresponds with around 2.0 standard deviations. This means that only 1 in 100 aeroplanes should be as much as 80 ft/min down. Thus, the same safety techniques are applied to light small aeroplanes as to the heaviest of public transports in JAR-OPS Class A, as implied by the graph in Table 3–3.

High Speed and Level Cruise Performance: Everling Number, E

Although airworthiness requirements applicable to high speed flight are mainly concerned with handling – i.e. flutter-free and no undesirable features – nevertheless speed is said to be the aristocrat of movement, so Everling number, E, is a useful tool for a making a crude assessment of drag coefficient. With this purpose in mind it was suggested that a level run at climb power should be interspersed with the partial climb tests. All that we need assume at this stage is knowledge of the theoretical stall speed of the aeroplane in the clean configuration, or one actually measured at aft CG (which is the CG position giving the highest airspeed).

We saw in Chapter 5, when dealing with parameters, and Equations (5–15) and (5–16), that Everling number relates maximum level speed, relative density, weight, rate power and wing area in an expression which is equivalent to the ratio (η/C_D):

$$E = \eta/C_D = [V_C^3/(96 \times 10^3 (\sigma^{0.5}))] (W_0/P)/(W_0/S) \qquad \text{(5–15) and (5–16)}$$

Propeller efficiency, η, if unknown, may be estimated by inspection of overall condition (especially of the blade tips and leading edges), and from Fig. 7.3, knowing the design true airspeed. Aeroplanes designed to cruise efficiently will have propellers tailored for cruise conditions.

The final two terms form the reciprocal of the *'propwash factor'* in Equations (5–56) and (5–57).

Example 7–4: Estimation of Drag Coefficients, Using Everling Number

A light aeroplane weighing 2600 lb (1180 kg) with an engine of 150 BHP (112 kW), has a wing area of 130 ft^2 (12 m^2) and a cruising speed at sea level, ISA, of 143 KTAS. Using Fig. 7.3 to find the efficiency of the new propeller which was fitted for the test, calculate a 'ballpark' overall drag coefficient, and then estimate the parasite drag coefficient. The stall speed, V_{S0}, is 57 KCAS.

Figure 7.3 suggests a value of $\eta = 0.8$. Being sea level, the square root of the relative density of the air is 1.0. Inserting the various numerical values in the combination of Equations (5–15) and (5–16) above, we have:

$$E - 0.8/C_D = [143^3/96\,000 \times 1.0] (130/150) = 26.4 \qquad \text{(7–6)}$$

therefore:

$$C_D = 0.8/26.4 = 0.03 \qquad \text{(7–7)}$$

From Fig. 5.4a we may derive the 'ballpark' ratio (C_{Dp}/C_D), knowing the ratio of (V_C/V_{S0}):

$$V_C/V_{S0} = 143/57 = 2.51 \qquad \text{(7–8)}$$

such that:

$$C_{Dp}/C_D = \text{about } 0.88$$

making:

$$C_{Dp} = \text{approximately } 0.88 \times 0.03 = 0.026 \qquad\qquad (7\text{–}9)$$

Thus, a relatively simple package of partial climb (and glide) tests, coupled with a straight and level high speed run, can produce a considerable amount of basic information for designers, engineers and aerodynamicists. The minimum rate of descent, as a ratio of the true airspeed, subtends an angle which is near enough $1/(\text{lift}/\text{drag})$ ratio of the aeroplane.

Knowing the aspect ratio, A, it is possible to extend the method further to estimate other characteristics, like the induced drag factor, K', which is shown in Fig. 5.4b. Conversely, an aspect ratio and, hence, lift slope, a_A, may be found (Fig. 3.10c), by selecting a desired value of K' from Fig. 5.4c, and so on. The pilot is in the useful position of being able to contribute an opinion on the validity of a number of estimates.

Sources of Performance Shortfall

We might now list major reasons why an aeroplane fails to achieve its rate of climb, or glide distance, ceiling, level speed, take-off or landing distances:

- Excessive weight.
- CG too far forward.
- Instrument errors.
- Stall speeds too high.
- Insufficient excess power.
- Insufficient lift to spare.
- Excessive drag.

Taking them in order:

Excessive weight

There are four common reasons why an aeroplane might be too heavy:

- An *unrecorded growth of empty weight* (poor weight control and monitoring). It often arises as a result of repainting; or a change of equipment without a record being made.
- *Incorrect weight scheduled* for the test. This can happen with ease, especially when an aeroplane is ageing and various modifications have been made. The weight schedule has been altered by calculation and those making the calculated record are only human. Errors creep into the schedule, remain undiscovered, and carry through.
- *Miscalculation of test weight.* This too is easily done. The pilot should check both weight and balance personally before accepting an aeroplane for test.
- *Blocked drain holes*, causing an aeroplane standing in the rain to take in water. The CG then migrates aft. We found a large nosewheel aeroplane sitting on its tail in dispersal, with tail surfaces full of water, and holes blocked by birds nests and droppings. Tail-heaviness with nose-gear makes it too easy to rotate on take-off – and much harder to do so with a tail-dragger.

CG too far forward

See foot of Table 7–2, where it is explained that as the CG moves forward, more up-elevator is needed to trim and this introduces trim-drag, which spoils performance.

It follows, of course, that as the CG moves aft trim-drag is reduced. Indeed, as is well known to racing pilots, if the aeroplane is flown with the CG at the point at which no balancing tail-load is needed, trim-drag is eliminated and it is then possible to attain the highest airspeeds.

Instrument errors

Common causes of instrument error are:

- *Position error* of the static pressure source.
- *Misalignement of the pitot head* with the local airflow (usually not critical if less than ±15° in any direction). Light and slow aeroplanes can exceed this angle with ease.
- *Propwash* causing change of angle of the local airflow at the pitot-head.
- A *local vortex* affecting the pitot head, being shed by a sharp corner, or a gap through which there is a leak of air into the local flow.
- *Error in measurement* of the ambient temperature.
- *Mechanical errors*: common with damaged second–hand and used instruments.

Stall speeds too high

Field performance suffers when stall speeds are too high, as we saw in Chapter 6. Increased stall speeds increase the kinetic energy to be gained on take-off and dissipated on landing. Both require longer runways.

Stalling is discussed in the next chapter.

Insufficient excess power

As mentioned earlier, a worn propeller is a common source of reduced power. A propeller with an efficiency of 85% when new can rapidly deteriorate to 70% in a short time through wear and tear.

Table 14–1 is a piston engine fault-finding table (Ref. 14.1) introduced for good measure, because defective running reduces power. The information is as useful today as it was in 1937.

Insufficient lift to spare

Common causes of insufficient lift are:

- Poor slat and flap rigging.
- Poor aerofoil surface profiles.
- Poor sealing of gaps.
- Excessive angles of deflection of control surfaces.
- Lack of elevator control authority, preventing the wing reaching the stall.

Note: This common deficiency is most noticeable power-OFF, with the stick pulled fully aft. There is no 'g'-break, or nose-down pitch. The aeroplane simply nods its way earthwards, usually without any trace of a wing-drop.

- Laminar flow wing sections can now be constructed using modern glass and composite GRP technology, but they are vulnerable to the flow breaking down and becoming turbulent. The result is a loss of lift which can be gross – and asymmetric, depending upon the cause. Rain, bug-splat and adhesive speed-tape have all been known to destroy laminar flow and degrade the efficiency of lifting surfaces. Small-scale homebuilt aeroplanes are particularly vulnerable, having low Reynolds numbers, which as we saw in Chapter 3 is:

$$\text{R}_x = \frac{\text{airspeed} \times \text{size (chord length)} \times \text{density of air}}{\text{viscosity (treacliness) of air}}$$

(from Equations (3–30a) and (3–30b))

Wing, control surface and flap chords which operate at relatively low Reynolds numbers can suffer sharp stalls, excessive drag and reduced lift/drag ratios.

Reynolds number effects, when coupled with small moments of inertia, make very light and small aeroplanes over-quick in response and sometimes vicious.

Loss in Range, Endurance and Agility

In Chapter 2 the Breguet range equation was discussed in relation to Example 2–3, the Rutan Voyager. Equation (2–8) showed that for:

☐ Piston and turboprop aeroplanes range varies as:

$$(\eta/c') \, (L/D)_{max} \tag{2–8a}$$

☐ Turbojet and turbofan aeroplanes range varies as:

$$(V/\mu') \, (L/D)_R \tag{2–10a}$$

and:

$$(M/\mu') \, (L/D)_R \tag{2–11a}$$

It follows that any reduction in propeller efficiency, η, leads to an increase in power setting to maintain the scheduled cruising speed (the reason for which the pilot is usually unaware, having nothing to tell him that it is the propeller which is less efficient), and so more fuel is consumed.

Any increase in drag reduces airspeed, Mach number and the ratio of lift/drag. Again, the pilot increases the power to compensate, and the fuel consumption rises.

Endurance, the time one may remain airborne on a given quantity of fuel, is shortened by an increase in fuel consumption. If $(L/D)_{max}$ is decreased for any reason, then the glide angle steepens and more power and fuel are needed to slow the rate of descent while maintaining the required airspeed.

Agility depends upon the excess power available at a given speed. When an aeroplane is manoeuvred, angle of attack, applied g and lift increase – and with it lift-dependent drag (in proportion to C_L^2). If excess power, [(thrust – drag) × speed], is degraded by reduced thrust and/or increased drag, speed is reduced and with it agility. A reduction in airspeed through an insufficiency of excess power leaves less scope for manoeuvring, and takes one closer to the stall.

In each of these items there is an insufficiency of power. The pilot on discovering this can always run a check on the rate of climb. If this is below schedule, then the first check should be the propeller.

In the next section we look at handling qualities, starting with the stall – the most vital of all tests that a test pilot is asked to carry out.

References and Bibliography

7.1 Vann, F., *Willy Messerschmitt First Full Biography of an Aeronautical Genius*, Sparkford: Patrick Stephens, 1993

7.2 CAP 520, *Light Aircraft Maintenance*. London: Civil Aviation. Authority. c1990

SECTION 5

HANDLING QUALITIES

CHAPTER 8

Stalls

A stall is the lowest practicable speed at which an aircraft can be flown. It is marked by noticeable, uncommanded changes in pitch, yaw or roll and/or by a marked increase in buffet. In principle the stall must be detectable; the aircraft must pitch nose-down when it occurs; and up to the moment at which the nose goes down uncontrollably until speed builds up, the pilot must be able to use the ailerons and rudder effectively, in the correct sense. Thus, there are four reasons for carrying out stall tests:

□ *Effective operations*, for which good, safe, stall qualities are paramount. Investigation occupies a major portion of the time spent on civil certification flight tests. Although in military flying stalling is a phase of flight which was hard to associate with a particular task (unlike civil scheduling of performance), reliable stall quality is essential for VSTOL operations. The ability of advanced combat aircraft to manoeuvre at angles of attack beyond those achieved by previous generations calls for accurate assessment of stall and post-stall qualities. Handling at the stall is essential for exercising pilots to cope with their aircraft off-design, around the edges of the low speed, high normal acceleration and high altitude portions of the flight envelope. Pilots need information and protection as follows.

□ *Stall speed*: measurement of the minimum in-flight airspeeds and their variation with weight, CG position, altitude, aerodynamic and power configurations (symmetric and asymmetric).

□ *Handling qualities*: to enable an average pilot first to identify the stall then to achieve a safe recovery from the highest angle of attack attainable in normal flight, including when asymmetry is present.

□ *Stall warning*: either aerodynamic or artificial, which gives an *average* pilot time to recover from any achievable high angle of attack, without stalling the aeroplane.

If we imagine safety in flight as an arch, then safe stall characteristics are the keystone but, paradoxically, there is no satisfyingly formal, legal, definition of a stall to be found any-where – certainly not on either side of the Atlantic. A French word, *decrocher*, to 'unhook' comes closest to defining what happens when an aeroplane flies too slowly and so *departs* from totally controlled flight. The words 'departs' and 'totally' are deliberate. 'Totally', because although the airflow over the aerofoil and body surfaces, which provides the lift-sustaining static pressure field, rarely suffers a complete breakdown, remaining control may be significantly less than total. Usually, as in the case of Fig. 8.1, plenty of lift is still being generated at angles of attack beyond the maximum, even though a rapid increase in drag and insufficient thrust make it unuseable. A stall, more often than not, is the conse-quential response of the aircraft to a partial breakdown of the lifting pressure system. The most dangerous stalls are those in which this occurs asymmetrically.

The word 'departure' needs explanation.

Departures from Fully-Controlled Flight

A departure occurs when an aircraft takes it into its head to leave the flight path intended by the pilot. Although it may be swift, it is fortunately rare for a departure to be so violent that the pilot cannot continue to manipulate the controls. The greatest danger lies in being taken unawares by the sly departure, what some like to call the 'killer stall', which sneaks up unawares and catches those who are inexperienced, or sluggish, or whose cutting edge has been temporarily blunted (see Chapter 4, Tables 4–1 and 4–2, *et seq.*).

The 'safest' stalls are those in which loss of lift is equal on either side of the plane of symmetry of an aeroplane, accompanied by a lady-like pitching down of the nose, with little or no tendency to drop a wing. All controls remain effective in the normal sense. Then the departure has occurred in the pitching plane alone, and recovery is effected by gently but firmly easing forward on the stick to lower the nose, decrease the angle of attack and increase airspeed, while simultaneously applying power.

The moment at which a sedate stall occurs is sometimes hard to detect. With a sensitive accelerometer it may show as a '*g-break*', i.e. a reduction in the normal acceleration below $+1.0\,g$ as the lift decreases, the aircraft drops slightly, the nose goes down and there is an initial reduction in the angle of attack, as during the first eleven seconds in Figs 8.2a, b and c.

When a stall is asymmetric about the plane of symmetry, i.e. when one wing loses some of its lift before the other, a departure in roll occurs. This is followed closely by a departure in yaw, because the downgoing wing suffers an increase in angle of attack with increased drag, which pulls more on one side of the CG than the other, yawing the nose and causing the flight path to curve in the same direction. The speed of the relative airflow over the wing on the inside of the curve is further reduced, steepening the local angle of attack while the outer wing, sweeping through the air faster, has a local angle of attack which becomes shallower. Asymmetric moments grow, yaw and roll rates and angles increase, and the nose drops rapidly below the horizon.

What happens next is largely a matter of speed and correctness of reaction by the pilot – ESPECIALLY in the use of his feet to prevent yaw. The sequence of positive actions is crucial if he is to slow (even if he cannot prevent entirely) the rate of build up of aerody-namic and inertia moments in pitch roll and yaw.

Complicating everything for the pilot are the yaw-roll coupling characteristics of the aeroplane, which behaves rather like a number of flywheels, each rotating independently about its own inertia axis, all of which are concentrated at the CG (see Fig. 12.2a). It takes time and effort of aerodynamic moments, and/or moments from the propulsion system (e.g., loss of an engine, either by failure, or actually falling off – which has happened more than once) to wind-up the mass of an aircraft about its axes. When wound-up it takes at least

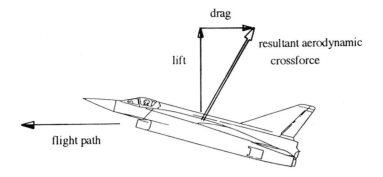

The aerodynamic crossforce is the resultant of all of
the pressures acting upon (and within) the aircraft. It
is resolved into lift and drag components normal and
tangential to the flight path

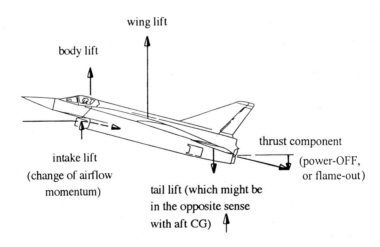

Wing and non-wing lift components

Fig. 8.1 Lift generated by the airframe is due to wing + body + nacelles + tail + propulsion
system. Here thrust is negative – as with a 'flame-out'.

an equal and opposite effort on the part of the pilot, using his flying controls primarily
(sometimes assisted by engine power) to then destroy the rotary motions and bring the
machine back to undisturbed flight.

It always takes TIME for flying controls to bite and to have the required effect. Even
though the correct actions have been carried out, failure to recover again to fully-controlled
flight may be caused by the pilot fearing that his actions are not working, simply because

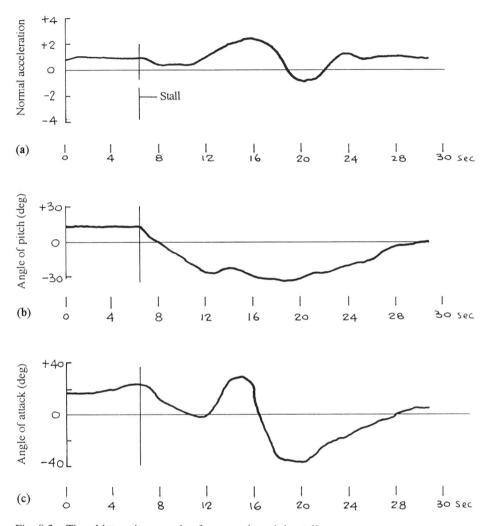

Fig. 8.2   Time history in seconds of a normal straight stall.

the aeroplane is taking longer than he thinks it should to respond. This is especially true when spinning (a simultaneous combination of departures in pitch, roll and yaw). Then doubting himself he tries something else (as we saw in Chapter 4, *Symptoms of pilot stress under test conditions*). Anxiety makes him expect things to happen faster than the inertias allow, especially when he is out of spinning practice and is caught by an inadvertent spin.

All propeller-driven aeroplanes fly in an asymmetric airflow to an extent, and the asymmetry is worst when tractor (nose or front-mounted) propeller(s) are both the same hand, i.e. RH tractor (rotating clockwise as seen by the pilot), or LH tractor (anticlockwise). Then the aeroplane and its control surfaces are bathed in a swirling spiral of air. The higher the power absorbed by a propeller, the more energetic is the spiral of propwash and the greater the asymmetry. Counter-rotating propellers, and 'handed-engines' (the propeller of an engine on one wing rotating in opposition to that on the other) are engineering tricks to reduce the adverse effects of such aerodynamic asymmetry. The reason why they are not commoner is that they are heavy and expensive to build, certificate, to buy and maintain.

'P'-Effects in Relation to the Stall

The lower the airspeed and the closer to the stall, the less able are the flying controls to counter the *'P'-effects*, also called the *'P' factor* (see Figs 10.4 and 10.5):

☐ *Asymmetric blade effect*, when the plane of the propeller is not at right angles to the flight path, so that blades on one half of the disc have a component of advance into the wind, while those on the other have a component of retreat (easier to see if you imagine the extreme case of the propeller rotating in the plane of the airflow). The result is an uneven lift and thrust distribution across the propeller disc. Together they produce a yawing couple when the propeller is inclined in the pitching plane, and a pitching couple when the propeller is yawed. Pitching couples, accompanied by small but potentially misleading changes in indicated airspeed, are noticeable when sideslipping and maintaining attitude visually, by the position of the nose relative to the horizon.

☐ *Rigidity*, in that the propeller, acting as a gyroscope, tries to remain fixed in space, which gives rise to:

☐ *Precession*, in that if a couple is applied to tilt the plane of rotation, the resulting motion is precessed 90° onwards in the direction of rotation. For example, an aeroplane with a RH tractor propeller will, when pitched nose-up, yaw to the right, and vice versa.

☐ *Pitching moment* is introduced when a propeller, which is invariably mounted forward or aft of the centre of gravity, is inclined to the relative airflow, which causes a component of thrust to act in the pitching plane. This affects longitudinal stability and response to elevator, aggravating the other effects.

☐ *Propwash* moves in a helix spiralling backwards around the aeroplane. The wash strikes body, wing and tail surfaces asymmetrically, with especially powerful effect in yaw (which, by imposing an angle of attack on the fin and rudder, causes the rudder to be deflected by the pilot to make the aeroplane fly straight.). A RH tractor causes the helical flow to attack the fin and rudder predominantly from the left, attempting to swing the nose to the left, so that the pilot has to counter with right rudder, and vice versa.

Note: Often large and powerful historic and vintage propeller-driven aeroplanes had fixed fin-offsets built into them for the purpose. This is a point to be watched when using a different engine in an old aeroplane – CHECK which way the propeller rotates!

☐ *Torque* is the opposite reaction to the engine turning the propeller, which tries to stand still and cause the aeroplane to rotate in the opposite direction instead. A RH tractor propeller and propwash cause an aeroplane to roll left, and yaw left. This the pilot counters with stick or wheel to the right (left aileron down). The left aileron then causes more drag than the right, yawing the nose to the left, adding to the adverse effect of the propwash.

Example 8–1: Rolls-Royce Merlin and Griffon-engined Spitfires

Later marks of Spitfire, which originally had a Rolls-Royce Merlin engine (RH tractor), were fitted with the more powerful Rolls-Royce Griffon, which turned LH. Early conversion accidents were caused by the reflex action of pilots to counteract swing on take-off by leading with the right (Merlin) foot, when the left foot was needed for the Griffon.

Example 8–2: The Piston-Propeller Torque-Roll

Torque-rolls caused many accidents, especially when opening up to go around again,

Plate 8–1   Supermarine Seafire 47 of the Fleet Air Arm with contra-props to counter torque-roll at low speed when increasing power. It also had more fin and rudder area. This was the fastest of the Spitfires with a Rolls-Royce Griffon, descended from the Schneider Trophy-winning R engine. Ref.: Example 8–2. (Courtesy of the late Charles Brown via Philip Jarrett Collection)

close to the stall, after wave-off from the relatively limited flight decks of aircraft carriers. Aeroplanes such as the Supermarine Seafire 47, with a 2350 BHP Rolls-Royce Griffon (almost double that of the original Spitfire), and the Supermarine Seagull in Fig. 8.13, needed counter-rotating propellers to deal with the problem.

Example 8–3: Stall Speed Estimates – Système 'D' (Example 1–3)

Designer's first calculations

☐ *Erect and inverted flight*: While the aeroplane has a triplane form the angles of the upper and lower planes reduce the likelihood of normal biplane/triplane interference. It is arguable to take average values of lift slopes for monoplane and biplane with aspect ratio, shown in Fig. 3.10c. The projected span of the upper and lower wings is about 14 ft, while that of the main (middle) wing is about 22.5 ft. Wing chords are 6 ft.

Aspect ratios: mainplane, $A = 22.5^2/(6 \times 22.5) = 3.8$
and lift slope (from Fig. 3.10c) = 0.07/degree (monoplane) and 0.055 (biplane)

average:

$$= 0.06, \text{ say}$$

For simplicity, assume the same lift slope for the projected areas (maybe it is not, but no matter at this early stage – there are too many other imponderables to worry about accuracy). Assuming that the mainplane stalls, power-OFF, around $10°$, then:

maximum (unflapped) $C_L = 10 \times 0.06 = 0.6$

and if the ailerons are drooped $10°$ to form plain flaps, we might anticipate 10% more lift coefficient:

$C_L = 1.1 \times 0.6 = 0.7$, near enough

The inclined upper and lower planes have angles of attack which decrease as the cosine of their $55°$ dihedral and anhedral angles, i.e. 0.57 of the mainplane angle of attack. When the mainplane is stalled they might not be, simply because their angles of attack are 0.57×10. So it may be impossible to stall the aeroplane because the projected wing areas, erect or inverted, are:

$(6 \times 22.5) + (2 \times 0.57 \times 6 \times 14) = 135 + 95.8 = 231 \text{ ft}^2 (21.5 \text{ m}^2)$

of which (95.8/231), i.e. about 41%, could be unstalled.

The lift coefficients at the maximum angle of attack of the mainplane are approximately 0.7 for the mainplane, and $0.7 \times 0.57 = 0.4$ for the upper and lower planes, giving a rough average:

$(0.7 \times 135 + 0.4 \times 95.8)/231 = 0.57$

while:

wing loading $= 4700/231 = 20.35 \text{ lb/ft}^2$

Thus, using the carpet plotted in Fig. 6.15, both values intersect at a minimum level speed, power-OFF around 100 KEAS, in a $10°$ nose-up attitude and the mainplane at the point of the stall. That angle corresponds, near enough, with the ground attitude. The implication is that touch-down, three-point and power-OFF, will be about 100 KEAS.

☐ *Knife flight (90 degrees bank)*: Using the same method as before, this time the mainplane makes no contribution to lift and the aeroplane is, in effect, an X-wing biplane. Their total projected wing area when inclined at $35°$ is 225 ft². Their average lift coefficient at $10°$ angle of attack is roughly 0.57, at which the minimum level speed of 100 KEAS is almost unaltered. What is not known is the angle of attack at which the X-wing configuration will stall.

Project test pilot's first comments

The stall calculations point to an unsatisfactory situation. If minimum level speeds, $10°$ nose-UP are roughly 100 KEAS, then an engine failure during a manoeuvre will give the pilot much to think about. The probability of an engine failure is higher with such an expensive one-off experimental machine than the once in a million discussed in Chapter 6. The pilot workload will be high, both mechanically, aerodynamically, when handling engine, rotor and flying controls in a complicated sequence of manoeuvres.

The counter-rotating rigid rotor-propellers are around 17.00 ft diameter (5.2 m). Turning at 630 RPM (to avoid the tip speed exceeding a Mach number of M 0.5), they will bathe flying surfaces in propwash, so it is possible that with such large rotors, intended for hovering, the wings will never be stalled, power-ON. With calculations showing that the upper and lower (X-wing) planes will probably not reach their stalling angles, it will be impossible to spin, or carry out flick/snap manoeuvres. If so, then of what value is the aeroplane for 'unlimited' manoeuvres?

Further, if the pilot rolls the aeroplane sharply, or experiences a rate of rolling departure entering an attempted spin of 360°/s (which is not unusual), that represents 60 RPM added to one rotor and subtracted from the other, rotating in the opposite direction. Surely a Δ RPM increment of ±9% transmitted, without slip, through gearboxes direct to the engines must overspeed the gas-generator of one and underspeed the other? That cannot be good.

The calculated stall speeds are based upon gross assumptions. What is the effect upon angle of attack of dihedral nearing 60°? Can one really use the projected areas of the upper and lower planes (whichever they may be) and work out a reasonable lift slope and wing loading?

Power failure could occur in an unusual position at an airspeed less than the theoretical power-OFF stall speed, making it impossible to push the nose down and glide. Then the only way out for the pilot might be to rotate the aeroplane nose-up, before kinetic energy in the rotors decayed, converting to an autorotational mode by turning the aeroplane into a tail-sitting autogyro. A successful landing would involve a tail-first arrival – an inevitable crash, even if the rotor β-control enabled one to flare and touch down with zero speed. A further criticism is that if failure occurred at too low an airspeed, there would be insufficient elevator and rudder authority to pitch or yaw the aeroplane into a nose-up attitude and become an autogyro.

Resolution of the problem might lie in fitting longer wings with more area and lift slope, but this would wreck any supposed advantage. An alternative would be to lighten the aeroplane. Even if it were possible to halve the gross weight, this would only reduce the minimum level speed to $(\frac{1}{2})^{0.5}$, i.e. 0.71, of the calculated values. However, it would appear to be impracticable to reduce weight, which is already a minimum for the engines, rotors, gearbox and airframe structure.

In any case, even if the minimum level speed could be further reduced to a more practical and useable value, the handling problem in an emergency, caused by reduced authority of the control surfaces, would be made worse.

The aeroplane appears to have far more technical and handling problems than advantages.

Effects of Wing Shape and Tail Location

Within the working range of angles of attack of a subsonic wing with conventional aspect ratios around 6.0 the lift slope, a_A (Figs 3.9 and 3.10), is approximately a straight line. The lower the aspect ratio, the more the lift slope decreases with increasing α, as shown in Fig. 8.3, and this markedly affects stall and post-stall behaviour. With high aspect ratio wings the loss of lift at the stall tends to be much sharper and more clearly defined than it is with a wing of low aspect ratio. The higher the aspect ratio the more taper is needed to keep structure weight within reasonable bounds. Taper makes it easier to stall the tip before the root and drop a wing.

A low aspect ratio wing has altogether 'softer' stall characteristics than one of higher aspect ratio. It reaches a larger angle of attack before anything resembling a stall occurs, while suffering a considerable increase in profile drag caused by separation of the boundary layer. A sharply-swept slender wing of low aspect ratio (rather like the upper sketch of the experimental Handley Page HP 115, in Fig. 8.3a(1)) behaves more like two wing tips joined than a short-span wing with two swept tips. Powerful lifting vortices – part of the trailing (horseshoe) vortex system – provide a non-linear boost to their normal lift, introducing additional lift-dependent vortex drag. The net result is that the lift/drag ratio of an aeroplane with a wing of low aspect ratio decreases faster than it does with one of higher aspect ratio. The rate of descent increases rapidly and the pilot can find himself on the backside of

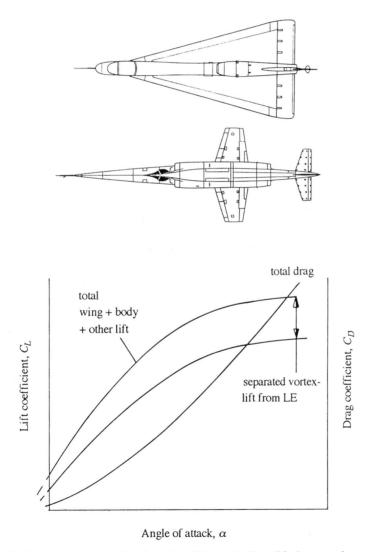

Fig. 8.3a(1)   Low aspect ratio (slender) aircraft have shallow lift slopes and steep drag rise with increasing angle of attack. Handley Page HP 115 (upper), Douglas X-3 (lower); but not drawn to the same scale.

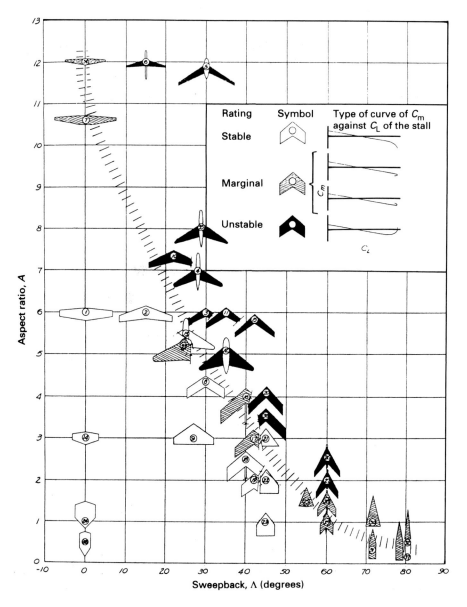

Fig. 8.3a(2)   Low aspect ratio may be combined with considerable sweep-back and still avoid pitch-up into the stall. High aspect ratio coupled with sweep may be a troublesome combination, forcing a deeper stall (Ref. 1.5, from NACA TN, No. 1093, Washington, May 1946).

the drag curve (on the left-hand side of Fig. 8.4) mushing rapidly groundwards with insufficient thrust to recover the situation.

Swept wings

Swept-back wings have their own particular problems. While high aspect ratio wings stall at shallower angles of attack than wings of lower aspect ratio, sweepback causes peaking of

Aircraft at high speed and small angle of attack

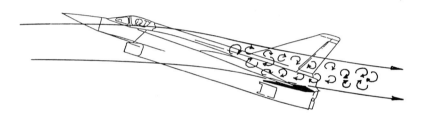

Same aircraft at large angle of attack

Fig. 8.3b   The advantage of a low-mounted tail stabilizer.

the local lift coefficient towards the tips (Fig. 8.5e). When combined with high aspect ratio, as shown in Fig. 8.3a(2) it is possible to find oneself on the '*backside of the drag curve*' (see Fig. 8.4a), combined with irrecoverable *pitch-up* – and a tip-stall just around the corner. This has led to numerous accidents in air displays with swept-wing aircraft.

Pitch-up is a pitching departure, exacerbated by the juxtaposition of foreplane and rearplane. Although Fig. 8.3d shows differing locations of a conventional tail, the situation applies equally to canard configurations, for which an aft-mounted wing is affected by the wake of a foreplane.

Pitch-up caused by a canard foreplane may be less severe than the effect of a wing upon a tail following behind, but a pilot should expect to encounter complications when yaw is present at or near the stall. The wake and tip vortices of a canard then sweep asymmetrically across a wing when yaw is present, and can precipitate a rolling departure.

Forward sweep avoids the tip-stall. The highest lift coefficients occur towards the root (see Fig. 8.5f). Considerable stiffness must be built into such wings if the tips are not to suffer an aero-elastic distortion – a tendency to rotate and diverge leading-edge up.

STOL-Kit Modifications

In recent years special STOL-kits have been marketed to improve the short take-off and landing field performance of light aeroplanes. My experience is that such kits simply

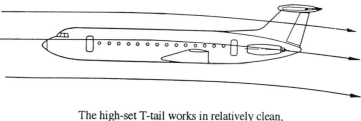

The high-set T-tail works in relatively clean,
undisturbed air at small angles of attack

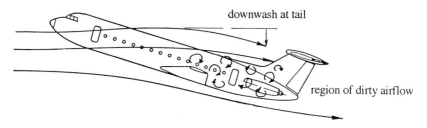

Danger of T-tail causing pitch-up at large angle of attack

Fig. 8.3c Flow effects upon T-tails and rear-mounted engines. Care is needed to tailor the
aerodynamics for optimum results.

replace an old shortcoming with a new one. The stall speed might be lower, but this makes
it easier to misjudge a steep approach at low airspeed, believing that the reduction in stall
speed makes a slower approach safer. But, in the event of a balk, unless an early decision is
made at a safe altitude to throw the approach away and try again, a pilot can find himself on
the backside of the drag curve and dropping into a bucket from which he has no escape.

First, the power/weight ratio will almost certainly be insufficient to overcome the increased drag. One simply hits the ground in a flatter attitude, with the risk of exceeding the
structural limits of the landing gear, if not one's own.

Second, pushing forward on the control column, applying full power to gain airspeed,
merely increases the steepness of the subsequent impact. As the NACA test pilot, Scott
Crossfield, is said to have said: '*It's not the drop it's the sudden stop*'.

In matters of type-certification by the UK Authority, never in my time was any performance credit given to any light aeroplane with STOL-kit fitted, and I have no reason for
believing that this policy has changed.

Airflow Separation and Side-Effects

Wing sections exhibit more than one type of flow breakdown and separation at the stall.

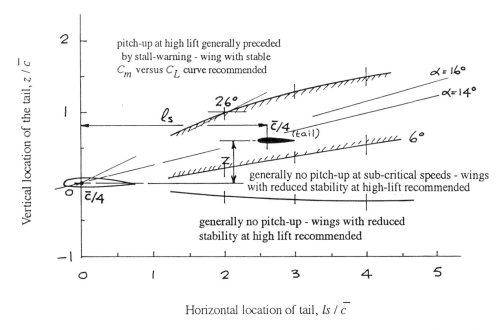

Fig. 8.3d Boundaries of horizontal and vertical tail or rearplane locations (Ref. 1.5, after NASA TMX-26, Washington, August 1959). A canard foreplane has the potential for causing combined roll, yaw and pitch-up.

Although the term *separation* is commonly used to describe the way in which the relative airflow behaves, the nature of the concept depends upon the way in which one looks at what is happening. The 'wind-tunnel' view is that of the eye travelling with the wing section. Then a breakdown in flow appears as a separation of the moving stream away from the surface, as the boundary layer of air between it and the airframe thickens and stagnates.

The view seen by the eye when fixed relative to the air is that of an aircraft coming out of nowhere and punching its way through the previously static mass of molecules. A lump of now highly agitated air is gathered up and displaced some distance by the aircraft before being dumped back in its wake. That lump of air is constantly being replenished and left behind, so that it is in a sense separated from the originally undisturbed mass to which it belongs.

Turbulence is the result of displacement, shear and vorticity imparted to the air, all of which absorb energy from the store of internal energy carried in the fuel. It often occurs asymmetrically and, when large in scale, leads to loss of lift and control. If the pilot attempts to replace a loss of lift by increasing the angle of attack, this only increases drag, making matters worse. The remedy is to reduce the angle of attack and attempt to regain airspeed.

Separation of the airflow into turbulent, vorticular, lumps is not always gross, except at the stall. Poor junctions and fillets, bulky and bluff sections of airframe, propellers which are driven like windmills and beginning to disc in a dive, slow the airflows in their wakes, reducing velocity and dynamic pressure recovery at the tail (see Equations (3–50) and (3–50a)). If the aerodynamicists have miscalculated the drag coefficient, or wing, body and nacelle junctions have not been faired as well as they might have been, then the local

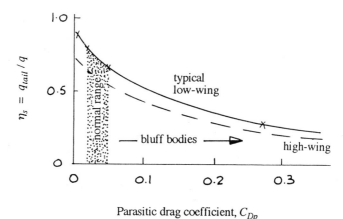

Total drag $= D_i + D_p$ eq. (3-23)

Parasitic drag coefficient, C_{Dp}

Efficiency of stabiliser surfaces:

$$\eta_s = q_{tail}/q = (V_{tail}/V)^2$$
(*see* eqs (3-50) and (3-50a))

A sluggish wake reduces the forces generated by the tail surfaces, decreasing stability by shifting the Neutral Point (NP) forward, shortening the static and manoeuvre margins.

In practice the size of the stabiliser surfaces must be increased as $1/\eta_s$, which depends upon the parasite drag coefficient of the real aircraft. A high-winged aircraft sheds a wake which reduces η_s more than one with a low wing. Tail surface areas (and volumes) must be made larger than for the same aircraft with a low wing.

Fig. 8.3e Thickness, sluggishness of wake and dynamic pressure recovery downstream depend upon drag coefficient and geometry. Here the curves refer to an aeroplane of normally average proportions (derived from Ref. 1.5).

parasitic and other drag coefficients have an effect upon dynamic pressure recovery at the tail as shown in Figs 8.3d and 8.3e. The Supermarine Seagull grew a third, central, fin during flight tests. Other aeroplanes since, with high wings, have grown additional fin surfaces, or had dorsal and ventral fin surfaces added, or were designed with larger fin surfaces from the start.

Figure 8.3e goes beyond the Appendix, Fig. A24b. Not only does a slowed wake reduce fin effectiveness, it reduces the component of lift generated by a tail stabilizer. Thus, we can see on looking back at Fig. 3.15 that when this occurs the neutral point, NP, must move

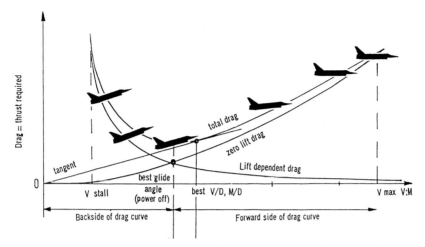

a. Variation of drag components with airspeed (see also figs. 5.4 and 7.4)

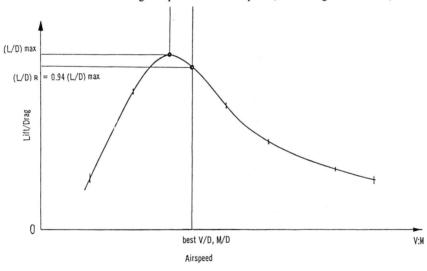

b. Variation of lift / drag with airspeed (see also fig. 7.4)

Fig. 8.4   Variation of lift/drag with airspeed at constant weight (Ref. 1.4).

forward, closer to – or even ahead of – the CG, diminishing the static margin and longitudinal stability.

In the approach to and at the stall wake effects are magnified and may grossly degrade tail authority.

Rules for 'Good' Stall Quality

☐ There must be WARNING of an approaching stall, IDENTIFIABLE by the pilot.

☐ The aeroplane must PITCH NOSE-DOWN at the stall. This is to reduce the angle of attack, enabling it to adopt an attitude which leads naturally to a build-up of airspeed, without assistance from the pilot.

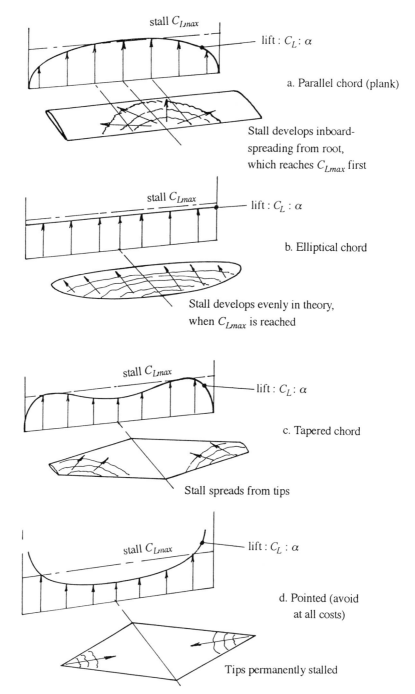

Fig. 8.5a to f Comparison between planform and development of stall. Local C_L is proportional to (local aerodynamic load/local chord). So, an increase in wing chord (as in f. which shows a root fillet) will reduce the local lift coefficient, causing it to retreat from C_{Lmax}. The stall C_{Lmax} varies with Reynolds number and, hence, local geometric chord.

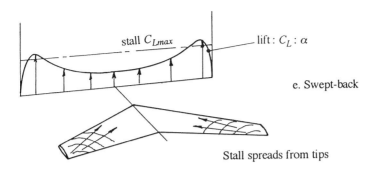

e. Swept-back

Stall spreads from tips

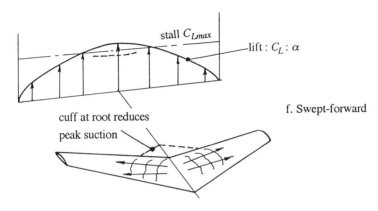

f. Swept-forward

cuff at root reduces
peak suction

Fig. 8.5a to f – Continued.

☐ Up to the moment at which the nose goes down uncontrollably (until speed builds up), it must be possible to CONTROL ROLL AND YAW BY NORMAL USE of the ailerons and rudder. To this add that it shall be possible to control any wing drop within ± 15°.

Were it possible to incorporate these rules, like some Napoleonic Code, then aeroplanes would give us much less trouble than they do. In reality, when they are in early stages of their development, they often drop wings more than 15°, pitch up at the stall, suffer degraded control, have rudder reach the stop, lose tail authority and try to change ends, have an idling engine slow and then stop, sometimes long before reaching the stalling angle of attack.

Factors Affecting Stall Quality

Stall quality is affected by:

☐ *Shape of the lift curve* with angle of attack. A sharp peak is undesireable. It can catch a pilot unawares.
☐ *Wing aspect ratio and twist* (wash-in and wash-out), taper ratio, sweep and tip form, aerofoil section profile. The aim is to avoid a sharp, asymmetric wing drop.

☐ *Rolling moments* are functions of aerodynamic asymmetry (slip, skid and propwash), aerodynamic roll damping by wings and tail surfaces acting about the CG, and moments of inertia, mainly in roll, but also in pitch and yaw.

☐ *Lateral rolling moment with sideslip* (dihedral effect of the wings and, to a lesser extent, contributions from other surfaces).

☐ *Authority of the lateral controls* (which may be by weight shift and not by ailerons and spoilers).

☐ *Adverse yaw* caused by higher drag of the aileron which is deflected downwards.

☐ *Directional stability* (weathercock stability) which depends upon the effectiveness of the fin and other keel surfaces in providing favourable yawing moments.

☐ *Yaw damping*, which is a function in the main of aerodynamic damping of all keel surface areas acting about the CG in the yawing plane. However, the moment of inertia in yaw helps initially with a contribution but, once the aeroplane has started to yaw, the same moment of inertia opposes aerodynamic damping.

☐ *Authority of the yaw control(s)*, i.e. rudder control power, which is affected by rudder throw (range of movement) and proneness to tip stalling – which is a function of taper-ratio, tip chord and local Reynolds number.

☐ *Longitudinal stability* (in pitch) provided by surfaces fore and aft of the CG.

☐ *Longitudinal aerodynamic damping* in the pitching plane.

☐ *Authority of the longitudinal pitch control* (which may be by weight-shift, and not by elevator or stabilator).

Types of stall

The type of stall depends upon where the breakdown in lifting airflow occurs, i.e. whether the stall starts at the leading or the trailing edge, the tip or the root. The latter depends upon planform. Figure 8.5a to 8.5f show, in simplest terms, the way in which planform (taper and sweep) affect the regions in which a stall develops on a wing. Taper is employed for lightness, to decrease wing area and mass outboard, so reducing aerodynamic roll damping and the structural bending moment at the root. Excessive taper is dangerous because it forces an increase in local lift coefficient (local wing (pressure) loading/dynamic pressure (Equation (3–27a))). Although in these sketches stall C_{Lmax} is shown as a constant across the span in all cases, it is not. In reality it varies with local Reynolds number and, hence, with local wing chord. The lower the Reynolds number the lower the maximum attainable lift coefficient (see Fig. 8.6). The narrower the chord the sharper and earlier the stall.

Pointed and highly tapered aerofoil tips should be avoided at all costs. Optimum taper ratio is not less than about $\frac{1}{2}$ if one is to achieve a 'good' compromise between docile aerodynamics and structural lightness.

As far as aerofoil sections are concerned, the 'classic' stall we have in mind is one accompanied by warning, in the form of aerodynamic buffet, and a marked (not too sharp) nose-drop. This is the stall associated with early subsonic wing sections, which had their deepest camber forward, near the leading edge. However, stalls are of three basic types with combinations, as shown in Figs 8.7 and 8.8. Camber close to the leading edge has a significant part to play in the flow breakdown which appears. The types of stall described below may modify considerably what is shown in Fig. 8.5:

☐ *Thin aerofoil stall.* Relatively thin wing sections often have marked camber of the mean chord line near the leading edge, which might be a reason for Ref. 8.2 introducing an aerofoil upper-surface ordinate at 1.25% of the chord, as shown in Fig. 8.7. A steep adverse pressure gradient is established downstream, even at shallow angles of attack.

R = Reynolds number (chord x air density / viscosity) from eqs. (3-30a and b)

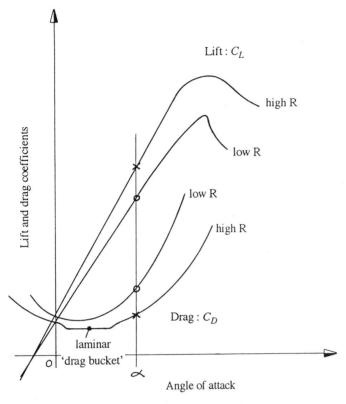

Fig. 8.6 The lower the Reynolds number then, in general, the sharper the stall, the higher the drag and the lower the ratio of lift/drag at a given angle of attack.

The airflow, which is initially laminar, cannot negotiate the curvature of the upper surface without separating. As it does so it forms a small *separation bubble* containing a mini-circulation around a low pressure vortex core which, unless conditions change, remains attached to the aircraft. The flow outside the bubble, if energetic enough, becomes turbulent and reattaches to the surface on the downstream side. On a very thin wing a separation bubble may eventually spread aft to the trailing edge with increasing angle of attack, until the wing stalls.

☐ *Leading-edge stall* (Figs 8.7 and 8.8c). On a thicker wing which has plenty of curvature well forward so that a separation bubble forms, an increase in angle of attack rapidly steepens the pressure rise downstream, causing a chordwise shortening of the bubble. Eventually the bubble collapses and an abrupt leading-edge stall follows.

☐ *Rear or trailing-edge stall.* Modern aerofoil sections, like those in the NACA 63- and 64-series with thickness ratios of 12% or more, gentler slopes near the leading edge and camber and maximum thickness further aft, have less 'peaky' stalls, because primary, turbulent separation of the airflow starts at the trailing edge. A separation bubble with turbulent re-attachment might still form just behind the leading-edge with increasing angle of attack, but the dominant feature is a trailing-edge vortex like that sketched in

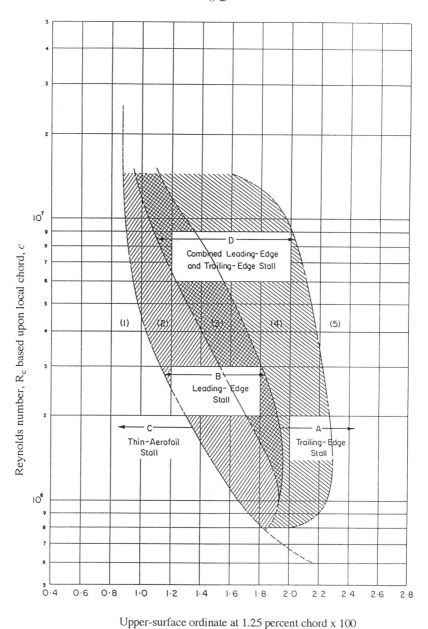

Upper-surface ordinate at 1.25 percent chord x 100

Fig. 8.7   Stall characteristics based upon aerofoil section upper-surface ordinate measured above the chord line at a point 1.25% of the chord behind the leading-edge (Ref. 8.2, sheet 66034).

Fig. 8.9. The start of such a vortex could be seen clearly during handling tests with the Norman Aeroplane Company Firecracker, which had the starboard wing and aileron tufted for the purpose with short strands of wool. The *slightest* touch of right aileron DOWN caused tufts along the aileron trailing-edge to reverse immediately, and point forwards.

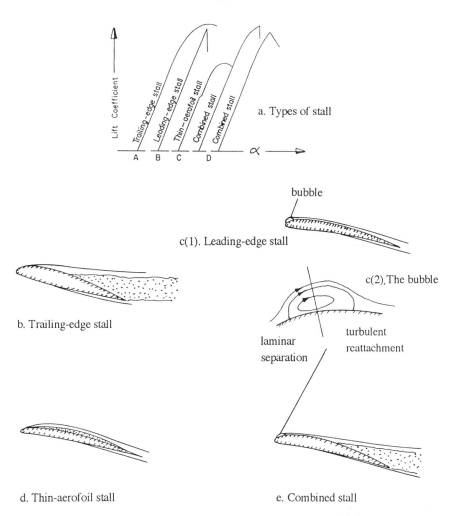

Fig. 8.8   Types of aerofoil section stalls (refer to Fig. 8.7 and Ref. 8.2).

The primary vortex at the trailing-edge induces a secondary vortex, with an opposite direction of rotation, further forwards, aft of the aerofoil section crest. With increasing angle of attack the secondary vortex strengthens and its presence is sensed by the air ahead which becomes agitated. Air ahead of the wing begins to vibrate until the secondary vortex has become so dominating that the boundary layer upstream separates, undercut by the reversed flow adjacent to the skin, which is moving forwards. With full-back stick the aeroplane begins to nod in the approach to the stall, because the boundary between the stalled region at the trailing edge and the region over the forward portion of the wing (ahead of the CG) is still lifting and the stall does not develop completely. As lift is lost the aeroplane sinks and the tail (ideally) causes nose-down pitch. There follows an increase in speed and a regaining of lift, which causes the nose to rise and leads again to forward migration of the stalled region. The cycle of gentle nodding continues as long as the pitch control is held fully nose-UP – or until coordination of the pilot's feet is relaxed, causing yaw and a wing to drop.

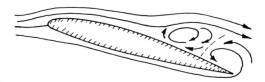

a. Mechanics of the trailing-edge stall

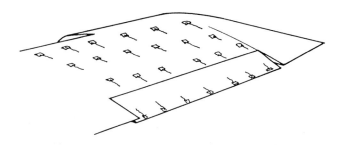

b. Observed reversal of wool tufts at the trailing-
edge when the aileron was slightly deflected
(*NDN Firecracker* trainer)

Fig. 8.9 The commonest form of stall encountered with modern subsonic aerofoil sections is that occurring at the trailing-edge of the wing, which tends to provoke a gentle nodding motion, unlike the sharp leading-edge stall.

Note 1: A complete rear stall is rare (as long as slip and skid are absent). Instead one tends to find a full-back stick minimum flying speed, which may be hard to determine accurately, because of oscillation of the ASI needle. In this condition a light aeroplane adopts what in French has been described as *une descente parachutale*, which is steep and usually slow, around 300 to 500 ft/min (5 to 8 ft/s or 1.5 to 2.5 m/s).

Note 2: Low rates of descent of this order suggest that if one is badly caught out and faced with a forced landing in poor visibility in a single-engined light aeroplane, then a full-back stick descent might offer the least destructive way of getting down. There is a fighting chance of walking away from the wreckage – *as long as no yaw is allowed to develop* on the way down, because that will provoke a spin.

Danger of yaw

Even slight yaw (rotation of the aeroplane in azimuth about its normal axis through the CG) is dangerous in the approach to and at the stall. Usually it results from inadequate coordination of hands and feet by a pilot whose backside is not sensitively tuned to the onset of slip and skid. If allowed to develop then the wing which is advancing and moving faster relative to the air than the other (which has a component of retreat in its motion) generates most lift. Asymmetry in the lift distribution then leads to departures in roll and yaw.

Wing-dropping is no indication that a pilot has two left feet. An asymmetric stall might be caused by errors in manufacture or in rigging of surfaces. A new Piper Cherokee derivative, during type-testing for certification many years ago, fell over more than 90 degrees on to its right ear in a straight stall, every time, no matter how carefully this was approached. Later we were told that it had a production wing on one side and a pre-production wing on the other, with small constructional differences in rigging between them. It was enough to cause wing-dropping which could not be checked, even with coarse use of opposite rudder in an effort to reduce any trace of yaw. The pre-production wing was changed and the wing-drop was cured.

The way for the pilot to prevent inadvertent yaw is to keep one eye looking outside the cockpit, aiming the aeroplane at a distant object (for example, the edge of a cloud is useful because, being in the same air mass, the aeroplane and the cloud have no residual relative motion across the line of flight). Any tendency to yaw can then be countered immediately.

Handed propellers

The spiral propwash from propeller(s) (a *'P'-effect*) is a primary cause of asymmetry and yaw. This can be alleviated to an extent with multi-engines, and more than a little expense, by fitting handed engines. It might help matters to mount the engine with the LH tractor propeller to the left (port) wing, and the RH tractor on the right (starboard) wing, to increase the angles of attack inboard provoking the wing to stall first near the root.

But if the opposite-handed engines are mounted the other way around, the angles of attack of the longest portions of the wings, outboard of the nacelles, are increased by the propwashes. This, in theory at least, improves the lift at low airspeeds. The choice of mounting depends upon the designer's priority and what the test pilot has found: a bad tip-stall, or a stalling speed that is too high.

A single-engined aeroplane has a more extreme problem, in that the local angle of attack of the wing behind the blades on the up-going blade half of the propeller disc is increased by upwash. The angle of attack of the wing behind the opposite half is reduced by a component of downwash. This means that with, e.g., a RH tractor propeller the local upwash increases the local angle of attack of the port wing, increasing its lift and making it more likely to stall before the starboard wing. For this reason leading-edge stall breaker-strips (which we shall look at in due course) often (but not always) have to be mounted asymmetrically between port and starboard wings.

Turning Flight and Accelerated Stalls

The difference between normal turning stalls and accelerated stalls lies in the rate of approach. Turning stalls are entered using not more than 30° of bank, but whereas speed is reduced at −1.0 knot/s in a turning stall, the accelerated stall is approached with an airspeed reduction of up to −5 knot/s. In the latter case one may either attempt to hold a constant acceleration, while using the elevator control to reduce speed, or use the 'wind-up turn' technique – which is closer to that used in operations – while maintaining constant airspeed or Mach number. During accelerated stall tests at constant applied '*g*', airspeed reduces rapidly. In such tests the pilot is looking primarily at the adequacy of stall warning, particularly at aft CG and heavy weight in the take-off, *en-route* and approach/landing configurations.

Yaw, slip, skid and turning flight are related motions in that asymmetric lift is generated between opposing wings which, if uncorrected, causes roll. At speeds close to the stall yaw leads inevitably to roll which then translates into a spiral or spin. The spiral dive was the original maverick motion, before the '*spinning nose dive*' (as a spin was first called) and the recovery were explored. In 1918 these departures were noted, but not described as such:

> 'a pupil will get into a spinning nose dive through making a faulty spiral on certain types of machine on which the area of the stabilizing tail fin is too small....'   (Ref. 8.4)

A low speed spiral at or close to the stall in a conventionally proportioned (wing span longer than the body) aeroplane rapidly becomes a spin under the influence of the aerodynamic and inertia 'driving moments' of the wings. But if the stall does not develop and instead the angle of attack reduces, then airspeed builds rapidly and delayed recovery can result in the pilot overstressing the aeroplane. Structural failure occurs if V_A, the design manoeuvring speed, is exceeded by more than the safety margin provided for in the design.

Figure 8.10 shows an aeroplane in a steady level turn with angle of bank, ϕ. Equilibrium is maintained by lift, L, and weight, W, being equal (were they not then the aeroplane would either climb or descend). The applied normal acceleration, angle of bank and centripetal acceleration are given by Equations (3–31a) to (3–31c). Expressing the airspeed in terms of a factored (true) stall speed, mV_{S1}, such that lift:

$$L = nW, \text{ and varies as } (mV_{S1})^2 \qquad (8–1)$$

we see that:

$$n \text{ varies as } (m)^2 \qquad (8–2)$$

which enables us to plot the curve shown in Fig. 8.11, in which the angle of bank is included.

In a balanced turn the outer wing travels faster than the inner. Rudder is used as the primary balance control to prevent slip or skid and is applied in the direction of indicated imbalance. Because the apparent weight of the aeroplane, nW, is heavier one must increase the angle of attack to maintain a level turn, which leaves a smaller margin before the stalling angle is reached. But the slower inner wing is generating less lift, and it is necessary to apply aileron out of the turn to prevent the aeroplane rolling further in. Application of down-aileron on the inner wing increases geometrically the *apparent* angle of attack (measured tangentially with a straight-edge from deflected trailing edge of the aileron to touch the under-surface of the wing close to the leading edge) as shown in Fig. 8.12.

It is argued that an increase in the apparent angle of attack through aileron deflection causes an earlier stall. I am not sure of this in practice. It can be demonstrated that AS LONG AS NO YAW IS ALLOWED TO DEVELOP one can apply full aileron, in either direction, almost at the point of the normal stall without provoking a premature asymmetric stall. If you then try to hold the aileron at full deflection for any length of time you may find that a wing drops, because it is hard to continue to fly straight at the point of the stall, with the tail working inefficiently in a dirty wake. This leaves the pilot with less control in yaw.

Aileron deflection is invariably accompanied by some aileron drag (see Fig. 9.11), and, unless one is quick-footed, adverse yaw is unavoidable. Of course, the consequence is a stall of the inner wing, the one with down-aileron. With modern sections this usually takes

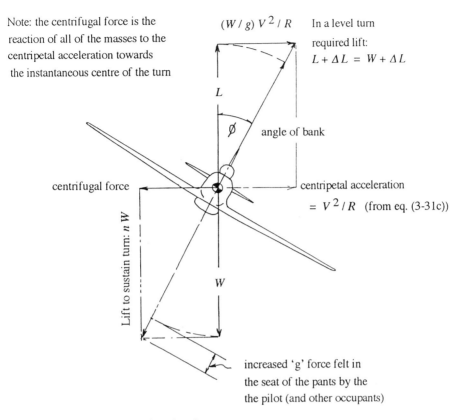

Note: the centrifugal force is the reaction of all of the masses to the centripetal acceleration towards the instantaneous centre of the turn

$(W / g) V^2 / R$ In a level turn

required lift:

$L + \Delta L = W + \Delta L$

$\varnothing$ / angle of bank

centrifugal force

Lift to sustain turn: $n W$

centripetal acceleration

$= V^2 / R$ (from eq. (3-31c))

L

W

increased 'g' force felt in the seat of the pants by the the pilot (and other occupants)

Fig. 8.10   Balance of forces in a level turn.

the form of a rear stall, which is gentler than a leading-edge stall. Often the pilot may be unaware that the inner wing has stalled, because there may be little or no warning buffet. As aileron drag can be marked the dominant impression is of increasing rates of roll-in, yaw and nose dropping.

Slip, skid and the stall

If the angle of bank is excessive for the amount of rudder and elevator applied, then the rate of turn will be insufficient and the aeroplane will sideslip in the direction of turn. The weathercock authority of the tail surfaces may often be inadequate at low airspeeds because of a reduction in tail efficiency as it dips into the thickening slow wake, in which tail efficiency is reduced, as we saw in Chapter 3.

$$\eta_s = q_s / q \text{ (or } q_{ws} / q)$$ (3–50)

Figure 8.3e shows that tail efficiency is reduced by the magnitude of the drag (wake-dirtiness) coefficient of surfaces lying ahead of it, and by location of the tail surfaces relative to the wake. A slowed airflow over areas of the tail surfaces has the effect of reducing their authority in direct proportion. Somewhere in the flight envelope most if not all aircraft have insufficient tail area to a degree. The solution, if weight and CG permit, is to enlarge the tail area in the ratio $(1 / \eta_s)$ as shown in Equation (3–50b). Extra fin area, either as a third central fin added to an aircraft with twin fins and rudders, or in the form of smaller auxiliary

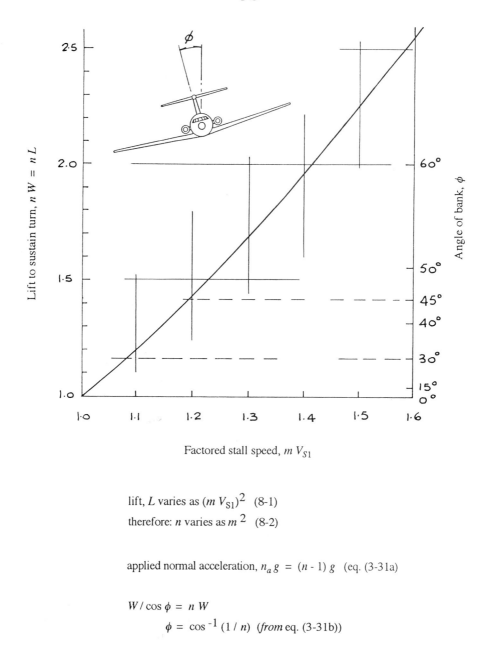

Factored stall speed, $m \, V_{S1}$

lift, L varies as $(m \, V_{S1})^2$ (8-1)

therefore: n varies as m^2 (8-2)

applied normal acceleration, $n_a \, g = (n - 1) \, g$ (eq. (3-31a))

$W / \cos \phi = n \, W$

$\phi = \cos^{-1} (1 / n)$ (*from* eq. (3-31b))

Fig. 8.11   The relationship between (factored) stall speed and lift to sustain a level turn.

fins fitted to the tailplane surfaces of one with a single fin and rudder, were common when design was still largely a matter of non-computerized trial and error (even so, computers have not solved the basic problem). Historic World War II examples were the box-like, high-winged, Avro York transport which was developed from the mid-winged Lancaster bomber; and the Supermarine Seagull amphibian, fitted with a deep variable incidence wing, parasol-mounted on the high engine nacelle (Appendix A).

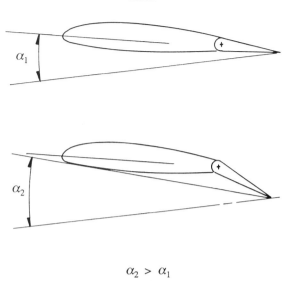

$$\alpha_2 > \alpha_1$$

Fig. 8.12   Increase in effective angle of attack by deflection of a simple flap (e.g., an aileron).

The response of an aeroplane to defective fin and rudder area is to sideslip with the same ease as when the pilot is under-ruddering in a turn. Sideslip also causes a yawed wake from the forward fuselage, coupled with any propwash, to trail across the inboard sections of the trailing (downwind) wing, resulting in a loss of lifting efficiency inboard. The sideslip angle, if it continues to increase, reaches a point at which the leading (upwind) wing rises and the outer (downwind) wing, now descending and steepening the local angle of attack, then stalls. The aeroplane rolls out of the turn and may carry on to enter a spin in the opposite direction.

Note 1: When a back-swept wing sideslips the roll described above does not necessarily lead to stalling and spinning. Instead it is pre-empted by oscillatory tail-wagging, 'Dutch roll', as each wing is caused to sideslip alternately. This is discussed later when considering the interactions between over-powerful lateral stability and weak directional stability. For the present bear in mind that flying qualities are a melange of mutually dependent interactions.

Note 2: A canard aeroplane, if allowed to stall with yaw, may suffer a more dramatic combination of departures than one with a conventional, tail-last, layout. The wing, being located aft, suffers more interference from yawed wakes ahead of it. Lying behind the CG, loss of lift causes pitch-up, which exacerbates matters. Further, the moment arm(s), of fin(s) and rudder(s) are short, so that it is also much easier to cause a fin stall. The result may be a combined roll, pitch-up and loss of directional stability, causing the aeroplane to attempt to perform a rolling backflip and changing ends (see Fig. 8.13).

Skidding, the opposite of sideslipping, is caused by over-ruddering, or by a rudder control which is too powerful for the lateral roll control and stability, both of which are weak. This causes a rate of turn that is too fast for the amount of bank. The outer wing then travels faster and generates the more lift. Ailerons must then be deflected against rudder in the direction of skid to counter the roll. This causes the pilot to apply crossed controls, for

Plate 8–2   The prototype Avro York was an early 1940s development of the Lancaster bomber. The Lancaster's midwing was raised
and a box-like fuselage fitted. Such was the loss of effectiveness of the
tail due to reduced dynamic pressure recovery in the wing-body wake
that ....
(Courtesy of A&AEE via Philip Jarrett Collection)

example, skidding to the right with too much left rudder, and aileron to the right to keep the
left wing up. As the aeroplane rolls under the influence of the rudder the nose drops, and the
rudder acts increasingly as an elevator – which depresses the nose even more. The drag of
the downwards-deflected aileron causes increasingly adverse yaw, slowing the trailing wing
and increasing its angle of attack more than the outer. Finally the trailing wing stalls first
and the aeroplane enters a spin in the direction of the applied rudder.

Stalling in climbing and descending turns

Operationally, the handling technique used in a climbing turn is the same as for one that is
level, except that the usual purpose of a climbing turn is to gain height as fast and with as
much economy of effort as possible. Speed is adjusted by means of the elevators, and the
angle of bank must not be too steep, to avoid loss of rate of climb from an excess of drag and
an insufficiency of the lift in hand to overcome the weight (Fig. 8.14).

During a *climbing turn* the aeroplane is moving about all three axes, while *rolling away
from the direction of turn*. At first sight this is unexpected. However, simulating an exaggerated climbing turn with the hand, or with a model, it is readily apparent that this is so, as

YORK C MK I
MERLIN
JUNE 1945

Plate 8–3   ... the production Avro York grew a third central fin. It was not the only aeroplane to suffer loss of tail aerodynamic efficiency from the same cause.
(Courtesy of the former Ministry of Aircraft Production via Philip Jarrett Collection)

shown in Fig. 8.14a in which the aeroplane is climbing in a left-hand spiral out of the page. Figure 8.14b shows the opposite. Here the aeroplane is *descending* in a left-hand spiral and so *rolling in the direction of the turn.*

In the climb case the outer (starboard) wing has the larger angle of attack, while in the descent it is the inner (port) wing which has the larger angle of attack. Thus in a climb at too low an airspeed the outer wing will tend to stall before the inner, causing the aeroplane to roll out of the turn. Whereas in the descending turn the reverse happens, and the inner wing, stalling first, causes the aeroplane to roll inwards.

When carrying out turning stall tests great care is needed to ensure that such differences in behaviour are clearly noted, and the correct inferences drawn.

Note: If a descending spiral dive starts it will tighten and tend to get out of hand, unless the pilot is aware of what is happening. Later, when we come to stability, we will see how important it is that roll with sideslip should be no worse than neutral. When a pilot is flying in, say, 'goldfish-bowl' conditions (no external horizon), is out of practice on instruments, and allows yaw to develop by failing to pay sufficient attention to maintaining balance with the rudder, a dangerous spiral dive can result if lateral control and dihedral effect are weak. Attempting to pull out without rolling the wings level, killing the turn and any slip or skid with rudder, can cause a high speed stall and a flick (snap, sharp wing drop), possibly with structural damage.

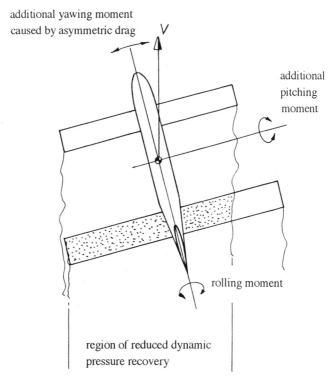

Fig. 8.13 Effect of yaw upon motions in pitch, roll and yaw with a canard/tandem configuration, at an angle of attack which brings the rearplane into the wake of the foreplane.

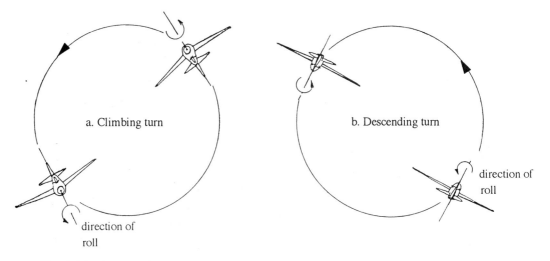

Fig. 8.14 An aeroplane rolls out of a climbing turn, increasing the angle of attack of the outer wing while decreasing that of the inner. The outer wing tends to stall first in a balanced turn. In a descending turn the situation is reversed: the aeroplane rolls inwards, increasing the angle of attack of the inner wing which then stalls before the outer, as long as the turn is balanced.

The North American AT-6 (known as the Texan or Harvard) is a valuable piston-engined trainer, not least because of the alacrity with which it will flick if mishandled.

Effects of Change of CG Position on Stall Speed

Stalls carried out at forward CG produce the highest stall speeds, for which reason they are called 'performance stalls' and are used for performance calculations (see Chapter 6). Stall speeds at aft CG are lowest, and the lower dynamic pressures reduce the authority of tail surfaces, making handling more difficult. For this reason aft CG stalls are called 'handling stalls'.

Fore and aft position of the CG affects stall speeds. If the stall speed at the forward CG is known, then the overall lift coefficient of the aeroplane, C_{Lfwd}, which is the sum of wing (plus body and nacelle) and tail stabilizer lift may be calculated in terms of dynamic pressure, q, and the wing area, S. To do this we use Equation (3–9), letting lift equal weight for the purpose.

If the test CG is moved aft by an amount $(\Delta \bar{c}/\bar{c})$ which may be written as $(CG_{test} - CG_{fwd})$, the wing will still provide the same C_L at the stalling angle of attack. But the tail now has to bear more of the weight in normal flight and nose-down elevator is needed to increase stabilizer lift by an increment ΔC_{Ls}. Taking moments about the new test CG position it may be shown that when the distance through which the CG is shifted is small in relation to the

Plate 8–4   A wonderful trainer with a sharp bite for the unwary or the casual. The North American AT-6 Texan or Harvard (RAF), which will flick to the right if mishandled when pulling *g* or turning too steeply at low speed. Ref: Example 8–4. (Courtesy of owner Dr Paul Franceschi)

tail arm, l_s, then the new lift coefficient, C_{Ltest}, is balanced by an increment in stabilizer lift, ΔC_{Ls}, such that:

$$C_{Ltest} = \text{approx. } C_{Lfwd} \, (1 + (Cg_{test} - Cg_{fwd})) / \bar{V}_s \, \eta_s \tag{8–3}$$

and as the CG is moved aft for the test:

$$C_{Ltest} > C_{Lfwd}$$

so that, as C_L varies as $1/V^2$:

$$V_{S1test} < V_{S1fwd} \tag{8–4}$$

In Equation (8–3) the terms in the denominator are the stabilizer (tail) volume coefficient (Equation (3–53)) and stabilizer efficiency (Equation (3–50)) respectively. For practical purposes the range of stabilizer volume coefficients is as shown in Table 8–1 (Ref. 1.5). Tail arms are around 2.5 to 3.0 times the mean chord of the wing for many conventional, subsonic aeroplanes. However, in the drive for cruise efficiency and long range, modern materials and methods of construction make practicable wing aspect ratios nearer 10.0 than 6.0. The consequence is that for a given gross weight and wing area (now a combination of longer span with narrower mean chord) the trend is towards tail arm ratios around 4.0 to 5.5 times the mean chord.

Note: In the foregoing no distinction is made between the standard mean chord and the aerodynamic mean chord. It is always the latter, unless otherwise stated (see Equations (3–4) to (3–8)). Both are usually close enough in length to interchange as required.

TABLE 8–1

(Ref. 1.5)
Stabilizer volume coefficient, $\bar{V}_S$

Power, flap	Sport, competition and agricultural aeroplanes*	Transport	All-moving stabilizer
Propeller-driven simple flaps	0.3–0.65	0.5–0.8	0.35–0.55
Propeller-driven high-lift flaps	rare	0.85–1.2	0.5–0.65
Jet, simple flaps	0.2–0.55	0.4–0.65	0.2–0.4
Jet, high lift flaps	0.3–0.4	0.5–0.68	about 0.4
Sailplane/SMG*	0.57–0.6 (0.64, on total area of a Vee-tail)	–	0.53–0.59
Manpowered	–	–	say, 0.2–0.25

Note: Canard designs are not commonplace enough for there to be much evidence, but the tail-volume coefficient with an economically-sized foreplane appears to be about 0.8 to 0.85 times $\bar{V}_S$ with a tailplane.
* Too much stability makes agricultural aeroplanes exhausting to handle and pilots want stick-forces to be light.

The other side of the coin, as we have seen, is that the narrower the chord of a surface the lower the Reynolds number (Equation (3–30b)). This can mean a sharper stall, higher drag, a lower ratio of lift/drag at a given angle of attack (Fig. 8.6) and, therefore, the need for subtle tailoring of aerodynamic surfaces and controls, high lift devices and 'fixes'. Such tailoring increases the interest but complicates the task of the test pilot, who has to explore every 'fix'.

Test Procedures and Techniques

Test requirements involve checking two essential features. That the aircraft can be:

- ☐ *controlled* into the stall and recovered with ease; and that
- ☐ *controls* are used normally throughout.

Thorough planning is essential, with a slow build-up to the tests themselves. It is sensible to research every reliable source that one can find on the subject of stalling and the prediction of stall characteristics. The aeroplane and its systems should be known thoroughly. The most basic instrumentation is an accurately calibrated test ASI, a tape recorder and knee-pad, a visual normal accelerometer, a fuel-consumption indicator and stop watch. If available an automatic trace recorder should be fitted, or a camera for photographing the instruments.

Unfortunately most pitot-static systems are not accurate enough for the measurement of stall speeds. Properly calibrated instruments are essential, and are often in the form of a separate airspeed-measuring system, using a trailing 'bomb' or cone, or a calibrated nose or wing boom. Finally, ensure that you are in current practice – never carry out stalls, or any tests, 'cold'.

Note: If the stall characteristics are untested, then a parachute should be carried (and worn if possible) – with one for the flight test engineer. Look back to the advice to wear a parachute in Chapter 4 before groaning, you might have cause to be grateful.

A set of basic straight and turning stall tests were suggested in Table 4-6 (Serial Nos 4 and 5) at forward and then aft CG positions. Although the ultimate object is to carry out a full package of stalls at the maximum weight for which evidence is sought, for initial tests it is prudent to use a mid CG-range and a test weight based upon the minimum amount of fuel, plus a reserve. Of course, the aeroplane will then reach lower stall or minimum full-back-stick flying speeds than at heavier weights, and allowance must be made for this fact.

What to look for

It is convenient to split the test programme into four parts: (1) the *approach*, (2) the *developed stall*, (3) *post-stall behaviour*, (4) the *recovery*, during each of which the following are noted:

- ☐ *Exact behaviour about all axes* together with control positions and forces associated with each characteristic.
- ☐ *Airspeed and rate of descent* at which the various characteristics occur.
- ☐ Any *tendency to spin or spiral* at the stall, or in the post-stall.
- ☐ The most *effective method of recovery without altering the power settings*.

From past experience one aims to complete recovery from any previously unexplored or unproven stall by:

Jet:	15 000 ft (4570 m)
Commuter:	8000 ft (2440 m)
Light piston:	5000 ft (1525 m)

The point at which an aeroplane is considered to be stalled varies and depends largely upon its configuration (and pilot opinion). The effects of altitude are explored when credit is sought for variations in speed and stall characteristics which depend upon Reynolds and Mach numbers (scale-effect and compressibility). When stall speeds are not thought to be sensitive to Reynolds and Mach number effects, AND PROVEN BEHAVIOUR IS BE-NIGN, subsequent stall tests are carried out down to a minimum altitude of 1500 ft (about 460 m) above the altitude of the highest take-off and landing surface from which the aeroplane is intended to operate.

Importance of a 1 knot/s Speed Reduction Rate

Mean stalls, the sort that creep up unawares, or intrude without warning, can happen at almost any rate. The first tend to be slow and are hardly noticeable. The second can be fast. Because the possibilities are legion, test stalls are the subject of a precisely-defined procedure which enables test pilots, engineers, designers and accident investigators to talk with accuracy, anywhere in the world. Significantly, there are marked differences between test and 'training' stalls.

The difference between test and training stall techniques is that whereas the aim of training stalls is to teach handling – so as to recover with the minimum loss of height – test stalls are to gather numbers, to explore characteristics and discover any dangerous features. Training stalls are carried out in level flight, the aim being to avoid loss of altitude. In a throttle-closed test stall the aircraft is in a descent. During stalls with power-ON the aircraft is in climb when thrust is greater than drag – which is not always the case on the backside of the drag curve (see Fig. 8.4a).

Test stalls are highly controlled and scientifically disciplined exercises. One must always bear in mind that mission accomplishment is the object, and poor or inadequate stall quality may prevent the execution of some tasks. Therefore, *if the characteristics permit*, the stage beyond the normal tests described here must be a controlled investigation of simulated, inadvertent, stalls in mission-related manoeuvres, using representative operational procedures.

One starts from a comfortably trimmed speed – not so far away from the expected stall speed as to waste time in the approach – not too slow and close to the stall that everything then happens too quickly for the pilot to note what is happening. A comfortable speed is between 1.3 V_S and 1.5 V_S, with engine(s) idling and the aeroplane in the configuration prescribed for the particular test. However, tests are included to check that stall speeds are not appreciably lower with engine(s) idling than with zero-thrust. Negative thrust at the stall, caused by propellers windmilling and which slightly increases the stall speeds is acceptable, because this is what a pilot will experience on chopping the throttles. If the difference between the stall speeds at idle-thrust and zero-thrust is not more than 0.5 knot, it can be ignored.

To find the stall speed at a rate of reduction of 1 knot/s, about six test stalls are carried out for each critical combination of weight, CG and configuration, varying the entry rate from 0.5 knot/s to 1.5 knot/s, so as to average scatter. Or, if in practice, one may count oneself into a stall by seconds: '*One-and-two-and-three-and-four....*' During certification tests a standard trimmed speed of 1.5 V_S was used for economy of effort and to save time. Because in the test sequence the stalls followed single-engined climbs, many of which were at airspeeds close to 1.5 V_{S1}.

The standard rate of reduction of airspeed to a test stall is intended to eliminate applied 'g', so that the results are as close to 1.0 g as possible.

Figure 8.15 shows a typical set of light aeroplane straight stall speeds, using the first method and standard cockpit instruments during a test flight for the renewal of a C of A. The test run started at 1.5 V_s (about 82 KIAS), reducing speed at a conveniently slow rate to 70 KIAS, at which the stop-watch was started. The rate was changed during a series of separate runs as shown, from 70 to the stall, at four or five intervals between about 50 s and 10 s. The radial lines are speed reductions of –0.5 KIAS/s, –1.0 KIAS/s and –2.0 KIAS/s respectively. The –1.0 KIAS/s radial intercepted the curve at 56.5 knots. The radials show that the slower the rate at which airspeed is reduced, the higher the stall speed. The faster the rate the lower the stall speed.

Indeed, in an aeroplane like a De Havilland Tiger Moth it is possible to reduce indicated airspeed so fast in the approach to the stall that the ASI can be made to indicate zero. It is a fault of many Cessna 150, 172 and 182 variants that, with two occupants and a reduction rate of –1.0 KIAS/s in the approach to the stall, it is possible to achieve stall IAS 'off the clock', i.e. with the needle below the minimum indicated airspeed of 30 knots. While one might argue that this is safe, in my opinion it should not be taken for granted. There can be large calibration errors at airspeeds less than 1.3 V_s. To be tempted to make an approach to land an aeroplane with such an error, at, say, $1.4 \times 30 = 42$ KIAS, would be foolhardy.

In any case, the Cessna aeroplanes that I have mentioned are not unique. Look back to Chapter 5 and Figs 5.10, 5.11 and 5.12. These show the way in which marked divergence between CAS and IAS appears at speeds less than about 1.3 V_s for other aircraft. The requirement is that each airspeed indicator must be calibrated to show true airspeed at sea level in the standard atmosphere, with a minimum practicable instrument calibration error when the corresponding pitot and static pressures are applied. In flight the minimum airspeed for determining the system error, including position error – but excluding calibration error of the airspeed indicator – is 1.3 V_{s1}. The error may not exceed 3% of the CAS, or 5 knots, whichever is greater. In Fig. 8.15, 1.3 V_{s1} corresponded with 74 KIAS, after instrument errors had been applied. There was no reliable way of working backwards from IAS to CAS below that airspeed, because there was no requirement for the manufacturer to provide the information.

Stall Warning

The purpose of stall warning is to give the pilot an adequate spread of airspeed between warning of the approaching stall, and the stall itself, so giving time to avoid it. Stall warning must satisfy the following criteria:

☐ *Distinctiveness*: it must be clear and unambiguous. Ideally aerodynamic buffet which starts in a light aeroplane around 1.07 V_s, say a minimum of 5 knots above the stall, is the best. A warning horn is a powerful attention-getter. A WARNING LIGHT IS UN-ACCEPTABLE, because it is easily missed when a shaft of sunlight falls across it.

Note: Whatever the warning, it must not be so distracting as to cause a hazard.

☐ *Consistency*: it must be reliable and repeatable, with flaps and gear in all of the positions used in normal wings level and turning flight. If articial stall warning (stick-shaker and/or stick-pusher) is needed for stalls in one configuration, then it must be fitted for all possible configurations.

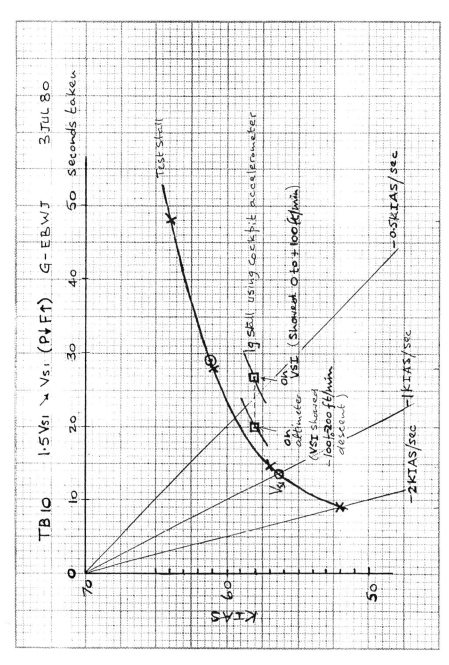

Fig. 8.15   Effect of rate of approach to a wings level straight stall in the clean configuration. The points marked with squares were attempts to stall at a constant indicated 1.0 *g* on the cockpit accelerometer, while maintaining a fixed altitude and zero rate of descent. The poor quality of the reproduction is due to this being a copy of an actual, post-test, plot.

Plate 8–5 The De Havilland Tiger Moth can be pulled into a stall so fast that the ASI indicates zero. This specimen is, in fact, a rebuilt Queen Bee, a target drone variant without anti-spin strakes, but basically a Tiger Moth in terms of its flying qualities. (Author)

Testing of stall warning is concurrent with other stall tests, noting airspeeds, quality and the amount by which it leads each type of stall.

Stall Deficiencies

When stall quality is deficient we usually find that the stall speed is too high, or there is an excessive wing drop, or there is a lack of stall warning, or that the stall translates into a worse departure, like a spin. This is rare today. There are a several fixes which, having worked on other aeroplanes, are worth trying. It is a matter of 'sucking and seeing', making each trial SMALL – like the half-steps into the unknown, cautioned earlier.

Breaker-strips and warning vanes

When aerodynamic warning is lacking it may often be provided by the addition of a small, sharp edged, breaker-strip attached to the leading edge(s) of the wings (see Fig. 8.16). Being sharp, breaker-strips trip the flow into turbulence causing buffet which, if the strip is well-placed, will be felt through the elevator by the hand of the pilot on the control column. Careful tailoring and tuning is needed, for if a breaker-strip is over-large the stall occurs at a higher airspeed, which then spoils field performance by lengthening the required distances.

A stall-warning vane is a small tab, protruding edge-forward from a slot in the leading edge of a wing. It is a flip-flop switch, free-falling to OFF, activated by movement of the

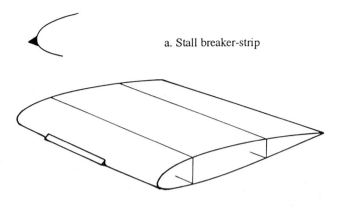

a. Stall breaker-strip

b. Adjustable electric stall-warner vane

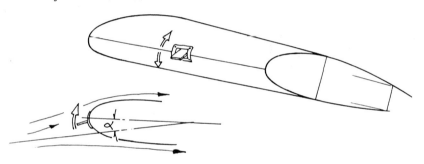

Fig. 8.16   Two common stall warning devices. A carelessly tailored breaker-strip can spoil field performance by causing higher stall speeds and longer take-off and landing distances. A vane is a form of adjustable 'flip-flop switch'. Care is needed when setting at the correct incidence. It is easily adjusted, but can be vulnerable to icing.

stagnation streamline at the leading edge of the wing. As angle of attack is increased the stagnation streamline moves downwards and, at the angle selected for warning, the vane is blown upwards. Electrical contact completes a circuit to a warning horn, stick-shaker or stick-pusher.

Example 8–4: Stall-Proneness when Turning Around a Point in a Wind

Agricultural aircraft and others used for fire-fighting, aerial and photographic survey work are particularly susceptible to stall and spin-related accidents, because of the need for low turns, and when operating over rising terrain. Clear and unambiguous warning systems are invaluable as during such operations pilots are usually reluctant to look into the cockpit to check airspeed, attention being demanded outside. The situation is not helped by protective helmets and headsets worn of necessity for sound-insulation, because these can make it near impossible to hear some of the simpler and least expensive stall warnings.

Plate 8–6   Stall breaker-strips, to provide aerodynamic warning, fitted to the leading edges(s) of a wing. (Author)

Figure 8.17 shows as an example the situations which can arise during aerial survey. Here the pilot is positioning the aeroplane in a strong wind, adjusting airspeed bank and yaw relative to the ground to give the photographer in the left-hand seat of the aeroplane the longest and steadiest shot of his target. Figure 8.17b(1) to (4) show tracks over the ground when the ratio of wind speed to true airspeed increases. This is a dangerous trap which can draw a pilot into flying too low. The slackening wind-gradient at low level causes the TAS/wind speed ratio to increase the closer one approaches the ground, and this is what the pilot seeks instinctively while concentrating on the job in hand. Figure 8.18 shows the effects of a photographer's/observer's requirements upon angle of bank, airspeed, slip and skid.

I frequently use the Cessna 172 and 182, which have many near-ideal properties for aerial photography. The cameraman is in the LH seat because the side-window is held open conveniently by the propwash of the right-hand rotating propeller. Use of flap is limited by sideslip restrictions and, as we have seen, slip and skid are a consequence of yaw. Therefore, airspeed, windspeed, height above terrain, bank, slip and skid must be carefully judged. The reason is not only that of asymmetric blade effect with yaw and sideslip (*'P'-effect*), but because many aeroplanes suffer from pressure errors which tend to produce deceptively slow indicated stall speeds. Those aircraft mentioned have notoriously weak, hard to hear, audio stall warning.

Stalls with Critical Engine Inoperative

Critical engine-inoperative handling is discussed later when we come to consider control.

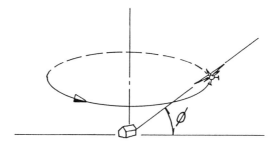

a. Path around a point target in still air

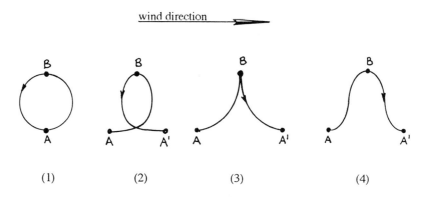

b. The effect of a wind increasing in strength (from left to right) upon the path of an aeroplane over the ground. The separate paths (which are not to scale) are in fact related by the law that the centripetal acceleration V^2 / R = a constant in each case, where V is the TAS and R the *instantaneous* radius of curvature of the path at any point. At A and A' the aeroplane is downwind, while at B it is into wind.

Fig. 8.17 Paths over the ground of an aircraft carrying out steady balanced turns at constant airspeed and angle of bank in winds of differing strengths.

For the present it is enough to say that stalls with the simulated failure of one engine are demanding, and are therefore carried out in still air. Their purpose is to demonstrate that there is no undue tendency to spin when stalled from an unaccelerated, wings-level entry with the critical engine inoperative and flap and landing-gear retracted. An undue tendency to spin exists when excessive skill and unusual control handling by the pilot is needed to prevent this happening.

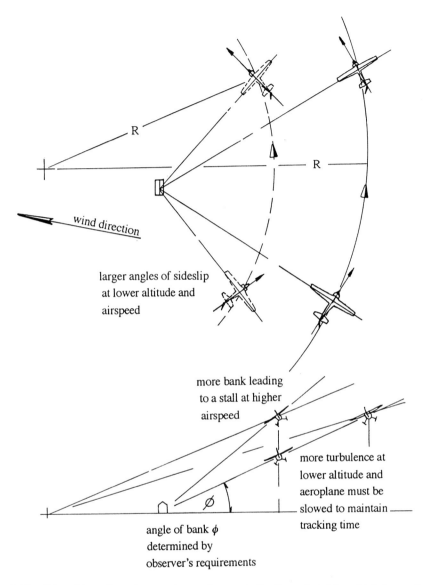

R

R

wind direction

larger angles of sideslip
at lower altitude and
airspeed

more bank leading
to a stall at higher
airspeed

more turbulence at
lower altitude and
aeroplane must be
slowed to maintain
tracking time

angle of bank ϕ
determined by
observer's requirements

Fig. 8.18 The effect of an observer's requirements for line of sight and minimum target-tracking time with a hand-held camera, upon angle of bank, airspeed, slip and skid provided by the pilot. For the best and safest results the pilot and observer must be a trained team, with the observer understanding the constraints placed upon the pilot.

In this chapter the job of the test pilot is to assess asymmetric *stall quality*, not engine-out handling. Ideally the aeroplane should have enough aileron and rudder authority at full travel to hold the wings laterally level and the heading constant up to the stall, with the live engine operating up to 75% of maximum operating power. The propeller of the failed engine is in the normally inoperative position (feathered, if designed to be so). If 75% power is excessive, then it must be reduced until a wings level steady heading is achieved. The approach is from 1.5 V_s at -1 knot/s, like other wings-level stalls.

Again, this is a test to be shown, or better still carried out, by a trained test pilot. It can be unpleasant when misjudged, like poking a stick in the ear of a one-eyed donkey – you cannot predict how it will lash out.

Stalls Inverted, or with Negative *g*

Aeroplanes can be stalled with positive or negative '*g*', erect or inverted. Unless the aerofoil section is symmetrical, then stall characteristics and speeds differ, because the camber and aspect of the aerofoil section to the air is totally different. For example, the nose-down pitch one expects in an erect stall may not occur inverted, or under negative '*g*'. A well-cambered aerofoil section, when flying at a negative angle of attack, can pitch nose-up at the stall.

Aeroplanes designed for extreme aerobatic manoeuvres (like the *Système 'D'* machine in Fig. 1.7) need comprehensive stall tests in each mode of intended operation. That aeroplane would need additional one-engine inoperative stall tests on each of its twin engines.

Stall Control by Passive Fixes ('Tuning' an Aeroplane)

It is rare to find an aeroplane which is fault-free and without shortfall in some respect. Pilots also have their problems. The commonest faults of both are:

- □ *Failure* to meet design or scheduled stall speeds.
- □ *Stall warning* excessive, or inadequate, or simply out of adjustment.
- □ *Excessive wing dropping* at the stall.
- □ *Inadequate authority* of one or more controls.
- □ *Differences in stall speed* between engine(s) idling and zero-thrust power settings.
- □ *Pilot faults:*

    (1) difficulty in determining *what defines the stall*;
    (2) difficulty in determining the actual *height loss* (altimeters often suffer lag);
    (3) difficulty in *recommending an optimum technique* for others to follow;
    (4) failing to reduce speed at − 1 knot/s;
    (5) having badly coordinated hands and feet.

We are concerned mainly with passive 'fixes' of faults, i.e. additions or modifications to airframes during development to reduce stall speeds and improve, *inter alia*, stall quality. Active boundary layer control, in the form of sucking and blowing as a means of re-energizing or removing a stagnating, separation-prone, boundary layer has no place here. Flaps too, and engine installations are left aside, being part of the basic design.

Vortices

At the core of most stall problems – and many flying quality problems – lies the vortex. One cannot live without it. Circulation around the bound vortex generates lift – and lift-dependent drag. As chopped-up and lumpy turbulence shed from surfaces and junctions, it degrades control and causes drag of a different kind. Vortices generated, or shed asymmetrically, cause pitch, roll and yaw. A vortex can surge an engine compressor, and snuff a turbine-engine like a candle. An oscillatory vortex can cause fatigue of materials, vibration and structural damage, sometimes far from its origin.

Example 8–5: Flutter caused by a Vortex

A vortex shed from a rear corner of a lead ballast weight, clamped to the nosewheel leg of a Slingsby T 67, was the source of apparent aileron flutter. The flight was to check spin characteristics at forward CG, and the ballast weight, rounded at the front with a slab trailing edge, was bolted in two halves to grip the nosewheel leg between. During spin tests the weight had rotated slightly causing one sharp rear corner to stick out further than the other. While checking handling at high speed on the run back to base, slight buffet coincided with fluttering of the ailerons, well below V_{NE}. It was discovered that a vortex, tripped by the protruding corner of the ballast weight, buffeted the bottom of the fuselage in its wake. The vibration ran outboard along the span of the long glass wing structure, shaking the ailerons. Removal of the ballast weight from the nosewheel leg cured the problem.

Example 8–6: Power in Trailing Vortices

On a larger scale, vortices left in the wakes of helicopters and heavy aeroplanes can destroy lighter aircraft through loss of control. Around 1966–67 during an RAF exercise my late navigator and I chased after an Avro Vulcan, a heavy delta V-bomber, which was making an approach to land. I did not then appreciate the danger. The powerful tip vortices and downwash scooped up our Meteor NF 14, no frail brute and heavy with warload. It rolled us upside down and dumped us, juddering and almost out of control, close to the ground. There have been many other accidents, and much Flight Safety publicity warning pilots of such dangers.

Example 8–7: Care when Flying Chase

Great care is needed when flying chase during tests of a large aeroplane. The second experimental North American XB-70 and a Lockheed F-104 were lost, in June 1966, when the fighter was rolled inboard by the tip vortex shed by the starboard wing of the bomber. It collided with the fins and, finally, the port wing.

But, to return to the smaller scale: set a thief to catch a thief. Often the cure for one troublesome vortex is another, and Fig. 8.19 shows a number of solutions to the problem. All of the fixes shed favourable, separation and stall-delaying, vortices which suppress departures in roll, yaw and pitch. However, they increase basic drag, but may decrease incremental drag from other sources. Those which, like the cuff, protrude forward of the CG, move the neutral points forward, reducing the static and manoeuvre margins and contributing to longitudinal instability.

Deliberate vortex-generation

☐ *Vortex generators* (Fig. 8.19a) are miniature aerofoil surfaces which poke up through the boundary layer into the more energetic and less disturbed relative airflow. They may be set at lifting angles of attack, so that the trailing tip vortex from each (with an intense suction at its core) draws inwards energetic air and infuses it into the decelerating boundary layer. While the VGs may turbulate and thicken the boundary layer, they effectively delay separation, because a turbulent boundary layer remains attached to and follows a surface further aft than one that is laminar. Drag is increased, but maybe not as much as it would be through laminar separation. Vortex generators also serve to maintain the effectiveness of flying-controls at large angles of deflection.

Plate 8–7a   Strong tip vortex shed by a low aspect ratio delta wing in the former RAE Bedford wind tunnel. Oil smears show flow on upper surface of the wing. The main vortex is rotating clockwise and the secondary anticlockwise as viewed. Ref.: Example 8–6. (Courtesy of the former Royal Aircraft Establishment)

Plate 8–7b   Avro type 698 Vulcan. (Courtesy of the Royal Air Force Museum)

Plate 8–8 Lockheed F-104A. One such aeroplane, in unplanned formation with the XB-70, touched and was rolled across the wing of the bomber. Both were destroyed. Ref.: Example 8–7. (Courtesy of Lockheed California via Philip Jarrett Collection)

Plate 8–9 North American XB-70. Ref.: Example 8–7. (Courtesy of USAF via Philip Jarrett Collection)

a. Vortex-generators shed miniature vortices which draw energetic air into a clinging and 'stagnating' boundary layer, helping to control it, while ridding the airframe of its adverse effects. Usually their drag-penalties are less than their aerodynamic benefits.

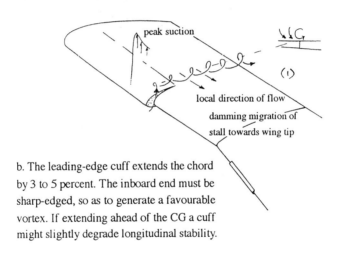

b. The leading-edge cuff extends the chord by 3 to 5 percent. The inboard end must be sharp-edged, so as to generate a favourable vortex. If extending ahead of the CG a cuff might slightly degrade longitudinal stability.

.....The cuff is cambered and has some properties of an extended leading-edge flap

Fig. 8.19 A selection of 'fixes' for generating and stabilizing favourable vortices. These are used to suppress flow separation and to control the start and development of the stall. Although mainly used on wings, they are also applied to tail surfaces. While causing increases in incremental drag, they often reduce drag from other sources.

Stick-on vortex generator kits are now available commercially, but care is needed in placing them on the airframe. When this is done they are effective in delaying separation of the boundary layer, reducing approach and stall speeds. They can greatly improve take-off weight, operational restrictions on minimum fuel weight, and one-engine-inoperative low-speed handling of multi-engined aeroplanes by reducing V_{MC}.

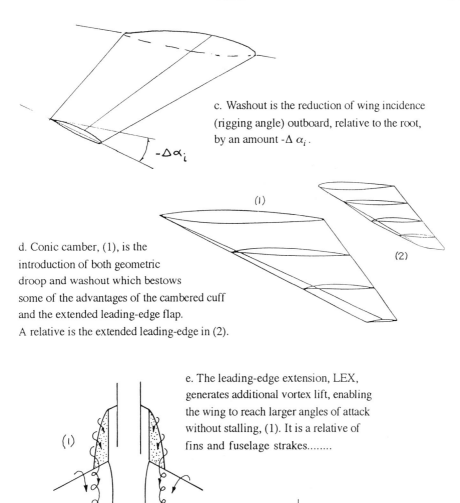

c. Washout is the reduction of wing incidence (rigging angle) outboard, relative to the root, by an amount $-\Delta\,\alpha_i$.

d. Conic camber, (1), is the introduction of both geometric droop and washout which bestows some of the advantages of the cambered cuff and the extended leading-edge flap. A relative is the extended leading-edge in (2).

e. The leading-edge extension, LEX, generates additional vortex lift, enabling the wing to reach larger angles of attack without stalling, (1). It is a relative of fins and fuselage strakes........

......as is the small, triangular, plate-strake at the roots of, e.g., the *SAH-1 / Sprint* trainer, to provide pre-stall buffet.

Fig. 8.19 – Continued.

On the other hand they cause drag, because without the energy-consuming vortex a VG can hardly be considered to work. When installed correctly speed loss is around –2 KIAS. When in the wrong place speed degradation of 10 to 12 knots has been reported to the *Aviation Consumer* (Ref. 8.8).

The angles of attack at which plate vortex-generators are set are important, depending upon whether they make the VGs co- or counter-rotating. Co-rotating VGs are used to

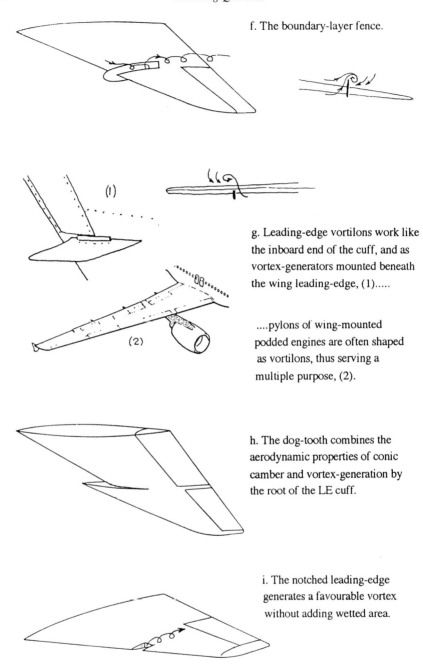

f. The boundary-layer fence.

g. Leading-edge vortilons work like the inboard end of the cuff, and as vortex-generators mounted beneath the wing leading-edge, (1).....

....pylons of wing-mounted podded engines are often shaped as vortilons, thus serving a multiple purpose, (2).

h. The dog-tooth combines the aerodynamic properties of conic camber and vortex-generation by the root of the LE cuff.

i. The notched leading-edge generates a favourable vortex without adding wetted area.

Fig. 8.19 – Continued.

produce a side-wash, down-wash or up-wash (depending upon the surface to which they are fitted) to drive the boundary layer in one direction only – i.e. inboard, towards the root so as to delay a tip-stall.

VGs to produce counter-rotating vortices are used more to turbulate the boundary layer, e.g., ahead of the ailerons, energizing the boundary layer and delaying separation.

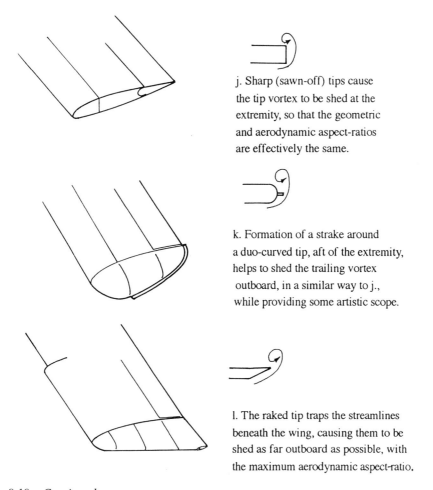

j. Sharp (sawn-off) tips cause the tip vortex to be shed at the extremity, so that the geometric and aerodynamic aspect-ratios are effectively the same.

k. Formation of a strake around a duo-curved tip, aft of the extremity, helps to shed the trailing vortex outboard, in a similar way to j., while providing some artistic scope.

l. The raked tip traps the streamlines beneath the wing, causing them to be shed as far outboard as possible, with the maximum aerodynamic aspect-ratio.

Fig. 8.19 – Continued.

Vortex generators are often attached near to the leading edges of thick flying control surfaces. When the surface is in line the VGs are hidden between the leading edge and the control surface shroud. They protrude into the airflow when the control is deflected, and are particularly beneficial on rudders for improving asymmetric handling.

☐ *Cuffs* (Fig. 8.19b) are forward (cambered) extensions of the leading edges of aerofoil surfaces. To work effectively a cuff must have a sharp edge at its root. The reason for this is that an area of suction is formed just behind the leading edge of the cuff, whereas inboard, along the adjacent leading edge of the uncuffed wing is a much higher dynamic pressure in the vicinity of the stagnation streamline. The pressure differential induces a spanwise component of flow from high to low, towards the tip, and the sharp inboard end of the cuff trips this, shedding a vortex. Sharpness is of the essence, because to form a vortex under such circumstances one needs an edge which the air cannot negotiate without separating and so forming a vortex. If a line is drawn from the leading edge of a cuff to the trailing edge of the wing, then it will be seen that the cuff provides geometrical washout – and shares aerodynamic features with an extended leading-edge flap.

(b)

(a)

(c)

Plate 8–10 (a)   Vortex generators fitted to a fin and in the junction of
the fin and tailplane to improve tail authority;
(b) and (c)   vortex generators fitted to tail of Aerostar in Example
8–8. Note differences from (a).

Warning: DO NOT FAIR THE ROOT OF THE CUFF INTO THE LEADING EDGE
OF THE WING IF YOU WISH TO AVOID AN IRRECOVERABLE SPIN.

☐ *Washout* (Fig. 8.19c) is one of the oldest devices, incorporated by twisting, or winding,
the wing along its axis in such a way as to decrease the angle of incidence of sections
outboard relative to those at the root. By changing incidence along the span of a wing the
lifting circulation and distribution of vortices shed from the trailing-edge is tailored to
approach more nearly those shed by an idealized elliptical lift distribution. Tip lift coefficients are reduced, while those at the root are increased, for a given overall working
angle of attack of the wing. The streamlines above and below the tips, which are in effect
yawed (swept inwards above the wing, outwards underneath), become less so. The trailing vortex is shed further outboard, increasing the aerodynamic span of the wing and the
induced aspect ratio, diminishing lift-dependent drag. The reduction in sweep of the
streamlines above the wing tips slows the rate at which thickening and separation of the

(a)

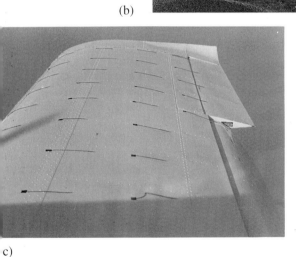

(b)

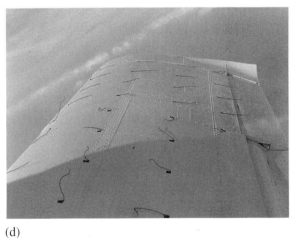

c)

(d)

Plates 8–11 Ref.: Example 8–9.
(a) Leading edge cuff on the NDN turbine-engined Firecracker. Note the sharp-edged root and undercamber.
(b) Similar cuff on a Nash Petrel, with wool tufts.
(c) Wool tufts on the unstalled wing of a Petrel (the near tuft is snagged on a paint-scuff).
(d) The same wing at the stall, showing tufts smoothed by the cuff and flow reversed inboard. The erect tuft, fifth from tip, is lifted by the vortex at the root of the cuff.
(Author)

airflow outboard can be transported inboard. Tip stalling is delayed, while making it more likely to start at the root.

☐ *Conic camber* (1) and the *extended leading edge* (2) (Fig. 8.19d) are used on high-performance swept wings with aerodynamically thin sections, enabling them to operate more efficiently off-design. Camber reduces the violence of the separation-forcing differential between the intense suction-peak, just behind the leading edge, and the sharply rising pressure downstream towards the trailing edge. Thus, the potentially fragile spanwise bound vortex of the basic wing, which would break down rapidly at large angles of attack and low speeds, is reduced in intensity, becoming more resilient to separation. Like the cuff, both have aerodynamic properties like an extended leading-edge flap.

☐ *Leading edge extensions* (LEX) (Fig. 8.19e) are highly swept surfaces ahead of the main wing. They are also referred to as strakes, gloves, fillets and apex regions, depending upon who has fitted them and to what. They have multiple benefits, in that:

(1)   at shallow angles of attack they cause little interference;
(2)   they generate vortex lift, so that for little additional weight total wing area may be reduced, or when fitted retroactively wing loading is marginally decreased;
(3)   vortices are shed chordwise and symmetrically at the kinks between the LEXs and the main wing leading-edges. This has the advantage of sustaining lift inboard, towards the wing roots, where there are often substantial losses.

Although a LEX marginally reduces the aspect ratio of the wings, this enables them to reach larger angles of attack, and so to generate more lift before any stall. For example, a properly tailored LEX with an area of 4% of the main wing has a potential for increasing the maximum angle of attack by around 20%, and C_{Lmax} by 30 to 40%.

Note: Leading edge extensions are close relatives of dorsal fins and strakes, dealt with separately below.

☐ *Boundary layer fences* (Fig. 8.19f) are used on the upper surfaces of swept wings where there are spanwise components of boundary layer drift towards the tips. There the boundary layer thickens with increasing angle of attack and becomes prone to early separation. The fence trips a chordwise vortex which rotates in the same sense as that shed by the root end of a cuff. In this respect it works in the same way and does the same job, while being tailored to a different planform.

☐ *Vortilons* (Fig. 8.19g) in their simplest form resemble boundary layer fences beneath and protruding forwards of the leading edges of swept-back wings. In a more complex from they are integrated into serving as engine-mounting pylons. At large angles of attack vortilons trip the spanwise component of flow from root to tip into chordwise vortices, which rotate in the same sense as those shed by cuffs and boundary layer fences. Care is needed in their shaping and directional camber. An engine pylon-vortilon, which rises above the leading edge of a wing to join the upper surface, can have an adverse effect if it increases local velocities near the wing leading edge. Then there will be a loss of C_L and an increase in drag (Ref. 8.7). The vortilon appears to work best when its leading edge intercepts the lower surface of the wing behind the leading edge.

☐ *Dogteeth* (Fig. 8.19h) have a similar effect to conic camber, by reducing the peak pressure and proneness to separation of the boundary layer of the wing behind and outboard of them. They are a relative of the cuff, and also fitted to the leading edges of some negatively cambered (inverted) tailplanes of high performance aeroplanes. The purpose is to increase tail authority in pitch at extreme angles of attack.

Plate 8–12   Leading edge vortilons and winglets, Rutan VariEze. (Author)

☐ *Notched leading edges* (Fig. 8.19i) are particularly useful on highly swept wings (which shed intense leading edge vortices at large angles of attack), by stabilizing the spanwise position of the vortex. In this way it is prevented from wandering up and down with yaw, primarily maintaining lateral and directional, but also longitudinal, stability and control within acceptable limits.

☐ *Wing tip shape* has a critical effect upon the distance outboard at which the trailing vortices are shed and therefore, upon aerodynamic aspect ratio and lift-dependent drag.

 (1) *Sawn-off* wing tips (Fig. 8.19j) although ugly are most effective, provided that the edges of the tips are kept sharp. The tip vortex is then shed by the sharp edge, because the air cannot sustain an infinitely rapid change of direction around it. The aerodynamic aspect ratio approaches closely the geometric, and lift-dependent drag is almost the minimum achievable for the planform at subsonic speeds.

 (2) *Modified duo-curve* wing tips (Fig. 8.19k). The aerodynamic span and aspect ratio may be improved by the addition of a sharp edge to force separation at each wing tip. Sharpening should start at the section crest and run aft to the trailing-edge.

 (3) *Raked* wing tips (Fig. 8.19l) make the trailing-edge of the wing longer than the leading, which keeps the working mass of air in contact with the wing as far aft and outboard as possible. Duo-curved or rounded wing tips fail to do this, causing the trailing vortices to be shed further inboard, nearer to $\pi/4$, or 78% of the span. The aerodynamic aspect ratio is proportional to the square of the distance between the tip vortices. The further apart they are, the lower the lift-dependent drag.

For greatest benefit the trailing corners of raked tips should be sharp. A rake angle of about 20° is near optimum for an aspect ratio of 6, increasing to 25° when $A < 6$.

Example 8–8: Vortex Generator Wrongly Installed

Vortex generators were responsible for the wildest behaviour and flight test results I ever experienced with a civil aeroplane. The solution turned out to be simple. The aeroplane was a Piper Aerostar, an elegant and potent small piston twin. The type had a reputation for liveliness at the stall but, with care and coordinated footwork, it could be shown to comply with the requirements. Operational experience appeared to be less happy because there had been a number of accidents with the type, and this particular aeroplane had been fitted, post-production, with a VG-mod. kit, which could be purchased off-the-shelf, and incorporated the following:

- a small longitudinal strake on the port side of the nose;
- a set of four stall breaker-strips along the leading edge of the port wing, between the tip and the engine nacelle;
- a single breaker-strip at the starboard wing tip;
- twenty-four vortex generators in pairs beneath each wing, immediately ahead of the ailerons;
- twelve vortex generators in pairs beneath each half of the tailplane, just ahead of the elevators;
- twelve vortex generators in pairs down the port side of the fin, just ahead of the rudder;
- a small curved 'eyebrow-like' strake on the port side of the fin, just ahead of the root of the rudder, concave surface downwards;

These were as shown in Fig. 8.20a (reproduced as sketches made at the time). With the kit as fitted this particular aeroplane had unusually erratic stall characteristics, and could not have been used from the small airfield where the operator was based. The symbols are as shown in Table 4–4, with aileron and rudder deflections, ξ and ζ, represented by personalized squiggles.

Figure 8.20b shows the straight, power-OFF (throttle closed) stalls in the three flap and gear configurations: (1) clean; (2) take-off flap, gear-UP; (3) flap and gear down for landing. In the first the scheduled stall speed was about 99 MIAS. Pre-stall buffet started at 120 and it was immediately and increasingly hard work with the ailerons (ξ) and rudder (ζ) to prevent excessive wing dropping and yaw. Full right rudder and half right aileron were needed by 100 MIAS, with a struggle to reach an indeterminate stall between 85 and 100.

As increasing amounts of flap were selected (20° and full, the latter with gear down), the aerodynamic warning buffet margin in each configuration grew smaller. The observed stall speeds, using full back stick, decreased even faster below schedule. Right aileron movements grew more erratic, right rudder calmed down, while a left nose-slice occurred in the landing configuration, against transiently full right rudder. Wing dropping was excessive.

Using 75% power and the same flap and gear configurations (Fig. 8.20c) rudder deflection was erratic, from neutral to full right, clean. Aileron demands were less initially, but became increasingle erratic up to full right when changing configuration to that for landing. Up to full back stick was used, with observed stall speeds well below those scheduled.

Without running through remaining results, blow-by-blow, suffice it to say that the cure lay in carefully checking the recommended location of each vortex generator. This revealed that some had been fitted incorrectly (Fig. 8.21). Rectifying matters involved moving the

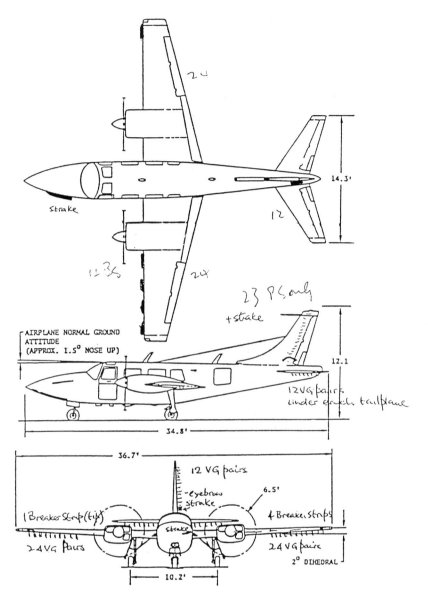

Fig. 8.20a Rough, pre-test flight, sketch of kitted vortex-generators fitted to an Aerostar, to improve stall handling. Interpretation of handwritten notes: 24× VGs in pairs, outboard under each wing ahead of the ailerons; 4×breaker-strips, port wing leading-edge; 1×breaker-strip, starboard wing leading-edge at the tip; strake on port side of nose; 12× VGs in pairs under each half tailplane, ahead of the elevators; 12× VGs in pairs down the port side of the fin, ahead of the rudder; 1× eyebrow-shaped strake (concave side downwards) at the lower port side of the fin, ahead of the rudder.

small strake on the port side of the nose upwards by about 2 in (5 cm); while the vortex generators beneath the port tailplane were moved to a new line, swept from a point about 1.0 in (2.5 cm) further aft at the root, to nil at the tip. With VGs re-located the Aerostar's stall was cured.

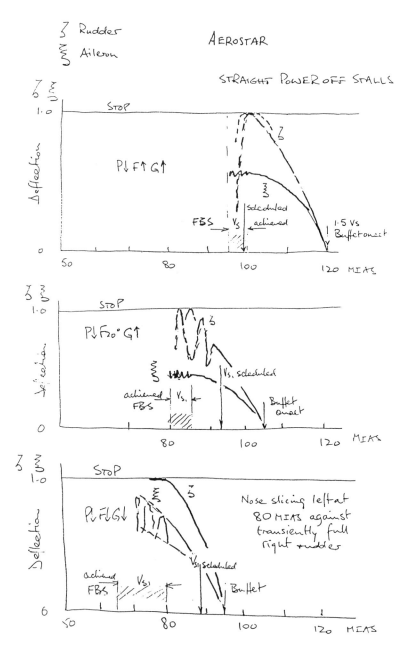

Fig. 8.20b Roughly sketched pilot impressions of straight, power-OFF (three configuration) stalls prepared for post-test debriefing.

Dorsal and ventral fins and strakes

Shapes and proportions of fins and rudders are important. There is a modern tendency to make them tall with elegant but unhealthily high aspect ratios. We know that higher aspect

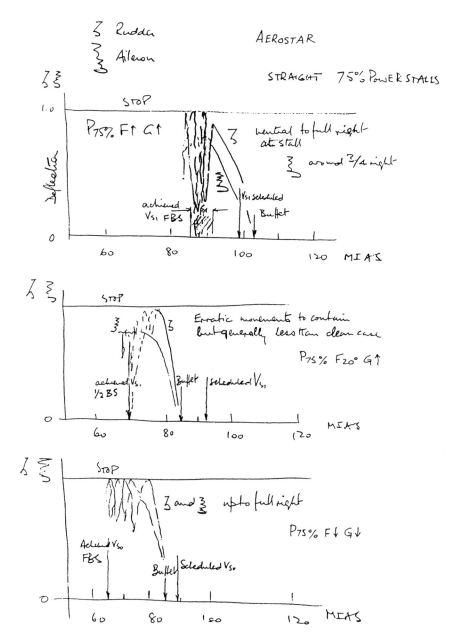

Fig. 8.20c Roughly sketched pilot impressions of straight, power-75% (same three configurations as in Fig. 8.20b) stalls, for post-test debriefing.

ratio surfaces stall at shallower angles of attack than those with low aspect ratio. In full rudder sideslips one can often feel the tremor from the direction of the tail, caused by buffet heralding a possible fin-stall around a further corner. A dorsal fillet is a quick and reasonably cheap and easy way of reducing fin and rudder aspect ratios, increasing the angle of attack before a stall occurs.

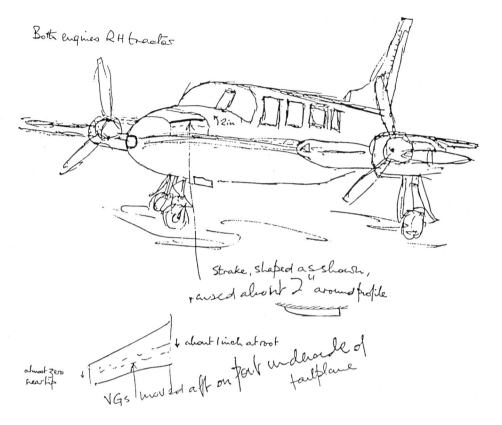

Both engines RH tractor

↑2in

Strake, shaped as shown,
raised about 2" around profile

almost zero
rear lip

↓ about 1 inch at root

VGs moved aft on port underside of tailplane

Fig. 8.21   Subsequent rough perspective sketch of the modified aircraft, to show the altered position of the nose strake and VGs under the port side of the tailplane, following the results in Fig. 8.20b and c. When the nose-strake and these VGs were re-located the problem was cured. Asymmetric fitting of VGs, breaker-strips and strakes, mainly on the port (left) side of the aeroplane was almost certainly due to propwash (which is the reason for the note that both engines were RH tractor).

   Dorsal fins and strakes ahead of the root leading edges of fins and tailplanes function in exactly the same way, and with fundamentally the same purpose, as leading edge extensions at wing roots. They generate vortex lift at large angles of yaw and pitch, favourable for the purpose intended, and delay the stall. Such devices are among the commonest of fixes today. Whatever their form, whenever they appear on an airframe they are usually 'tell-tales'. They give away the existence of an initial problem, or sometimes indicating a change of powerplant, or configuration to meet a different operational standard, resulting in airworthiness shortfall which then had to be made good.

Example 8–9: Early Fixes on a Prototype

Figure 8.22 shows proposed modifications to a first prototype aeroplane, following handling tests. They were aimed at achieving compliance with airworthiness requirements by providing:

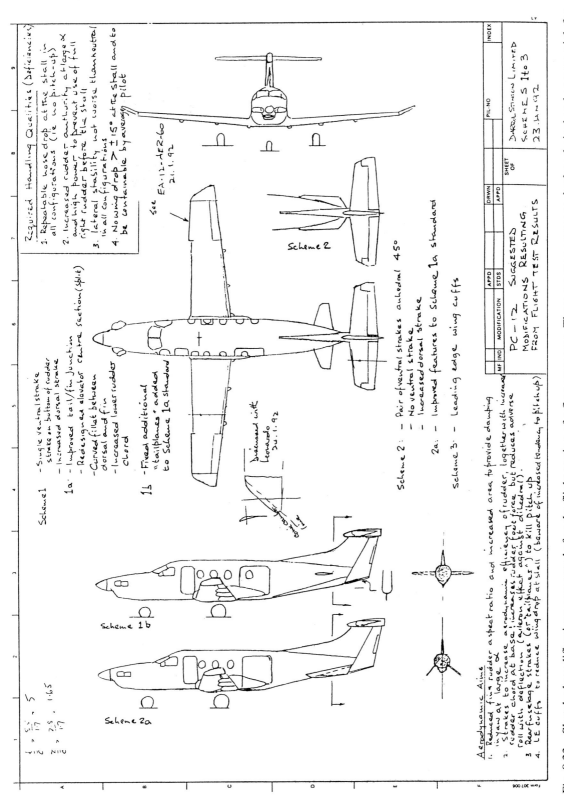

Fig. 8.22 Sketched modifications suggested after the flight testing of a first prototype. There are many reasons, both technical and commercial, for a manufacturer modifying or setting aside suggested 'fixes'.

☐ *Repeatability* of required stall characteristics in all configurations.
    Suggested fixes were to add either rear fuselage strakes, or small 'tailplanes' to cure any tendency to pitch-up.

☐ *Improved rudder authority* at large angles of attack with high power, which had involved use of full-right rudder before reaching the stall.
    Suggested fixes were to fit a dorsal extension to the fin, to improve flow over the rudder, and damping in yaw at large angles of attack, while reducing the aspect ratio of the fin and rudder as a precaution against a possible fin-stall. Rudder chord to be broadened and a sharp strake added at the base, to prevent possible separation of the airflow some distance above the base of the rudder. It was felt that, although these modifications could cause a higher foot-force, they might also help to reduce adverse roll with rudder deflection (i.e. the rudder acting as an aileron) and, therefore:

☐ *Reduction of wingdrop to < ±15°*, making it containable by an average pilot.
    Suggested fixes were to fit wing leading-edge cuffs, directly ahead of the ailerons, to reduce wing drop at the stall. These to be as small as possible because, being ahead of the CG, they could aggravate a tendency to pitch-up and work against the rear fuselage strakes or 'tailplanes'. Finally, the wing tips to be raked, with built-in conic camber, as swept tip extensions of the cuffs.

As this is written the fixes are being tested in small steps, modified in turn, and results are not yet complete.

More Complicated Fixes

The passive fixes described already tend to be relatively simple additions which follow initial flight tests. They are limited in extent and should not involve vast design effort. If they fail to improve stall quality as intended during development flight testing, then it is necessary to look to more complicated devices.

The test pilot may feel impelled to recommend fixes without realizing the expense and complication involved. If so then it helps to get some feel for the scope of what is being done elsewhere at every opportunity, before making suggestions. One example of a major fix is quoted by Frank Vann in his biography of Willy Messerschmitt, when discussing the Messerschmitt Me 210 disaster:

> "The longitudinal stability was a cause for great concern and the tendency to go into a spin did not endear it to the pilots.
>   It was reported that, after the first flight, the test pilot told Messerschmitt that the fuselage needed to be lengthened by at least a metre....'   (Ref. 7.1)

It must have taken pithy courage to put it so bluntly, for the great man raised objections.

Table 8–2 contains a list of basic references and has been extracted from Ref. 1.5 for the purpose. Both slat and slot are defined in Chapter 3. Fixed slats and slots can be installed without exceptional skill and effort. Typically they increase lift by a little more than 1/3 but also increase drag. Lift/drag is reduced by nearly 1/3. Automatic (Handley Page) slats increase lift by a little more than 2/5, and reduce lift/drag by about the same amount.

To use leading and trailing edge flaps on high-performance modern aeroplanes involves mechanical complexity and the use of power. It is now a long step from the pre-World War II Westland Lysander, which had its leading edge slats sucked open by movement of the stagnation point in the airflow, below the wing leading edge. The slats in turn lowered the

Plate 8–13   Messerschmitt Me 210 which suffered from longitudinal stability problems and a tendency to depart into a spin. (Courtesy of Deutsches Museum, Munich (Munchen))

trailing edge flaps automatically, leaving the pilot to sort out priorities between the stick, throttle and large wheel which adjusted the tailplane angle (see Example 10–3).

Note: When a fix is fitted, very often it then has to be fixed and fettled in turn!

Example 8–10: Doubts about Stall Quality – Tri-Motor Transport (Example 6–8)

In Example 6–8 (Fig. 6.16) a three-surface tri-motor project is used to point to areas of flight test concern. In practice, at every stage – and long before metal is cut – there is a burden upon company test pilots and flight test engineers responsible for a particular project to advise and encourage, and to warn in advance when they see troubling or possibly dangerous features.

TABLE 8–2

Flap and slat characteristics
(Ref. 1.5)

Description	Profile	C_{Lmax}	$\alpha^°$ at C_{Lmax}	L/D at C_{Lmax}	C_{Mac}	Reference
Basic aerofoil Clark Y		1.29	15	7.5	−0.085	NACA TN 459
0.3c Plain flap deflected 45°		1.95	12	4.0	—	NACA TR 427
0.3c Slotted flap deflected 45°		1.98	12	4.0	—	NACA TR 427
0.3c Split flap deflected 45°		2.16	14	4.3	−0.250	NACA TN 422
0.3c, hinged at 0.8c Split (Zap) flap deflected 45°		2.26	13	4.43	−0.300	NACA TN 422
0.3c, hinged at 0.9c Split (Zap) flap deflected 45°		2.32	12·5	4.45	−0.385	NACA TN 422
0.3c Fowler flap deflected 40°		2.82	13	4.55	−0.660	NACA TR 534
0.4c Fowler flap deflected 40°		3.09	14	4.1	−0.860	NACA TR 534
0.3c Nose flap deflected 30° to 40° (best for sharp nosed sections with poor C_{Lmax})		2.09	28	4.0	—	Based upon Ref 3.10
Fixed slat forming a slot		1.77	24	5.35	—	NACA TR 427
Handley Page automatic slat		1.84	28	4.1	—	NACA TN 459
Fixed slot and 0.3c plain flap deflected 45°		2.18	19	3.7	—	NACA TR 427
Fixed slot and 0.3c slotted flap deflected 45°		2.26	18	3.77	—	NACA TR 427
Handley Page slat and 0.4c Fowler flap deflected 40°		3.36	16	3.7	−0.740	NACA TN 459
0.1c Kruger flap (retracts backwards forming LE profile)		1.88	—	—	—	Estimates based on flight test results in Ref 3.11
0.1c Kruger flap 0.3c Fowler flap deflected 40°		3.41	—	—	—	

NACA References : aspect ratio 6, Reynolds number 609 000

Plate 8–14   Ref.: Examples 6–8 and 13–4: the De Havilland Comet airliner suffered ground stalling accidents, failing to take-off on two occasions ...
(Courtesy of former De Havilland Aircraft Company)

Plate 8–15   ... and tests are now carried out to ensure that this can no longer happen at V_{MU}. Here the aircraft is a Boeing 747. Ref.: Examples 6–8 and 13–4.
(Courtesy of the late Hugh Scanlan – past editor of *Shell Aviation News*)

One such dubious feature of the commuter project is the arrangement of the foreplane, mainplane and tailplane, all of which are nearly in line when the tail-bumper is on the ground – as it will be during tests of proneness to ground-stall.

The wing areas were sized and proportioned originally to cause the foreplane to stall first, by carrying a little more than a third of the weight, while the rearplane carries slightly less than two-thirds at aft CG. But all fuselages and nacelles lift, and the long body is no exception. Its aerodynamic centre is about a quarter of the length from the nose – roughly coincident with the ac of the foreplane – and it has a curved forebody bathed in a lifting propwash. Therefore it is possible that the combination of foreplane-plus-body could reverse the situation, with a risk of the mainplane stalling first.

That risk is increased by a large part of the mainplane being affected adversely by downwash and by the wake of the foreplane at the stall. The stabilizer could be similarly affected by both fore and mainplane. As propwash from the centre-engine is destabilizing, stall behaviour could be critical, leading to pitch-up and in-flight post-stall gyrations, PSGs (see Chapter 12).

Scraping the bumper may easily occur at aft CG, when a pilot is attempting to haul the aeroplane into the air from a short strip. With nearly all of the lift well forward, the aeroplane might then continue to over-rotate with the rear-fuselage still on the ground, bringing the flying surfaces into line and exacerbating the situation.

Therefore it is prudent to voice such concerns early in the project. Wind-tunnel tests are essential. The basic assumptions behind the design and the calculations based upon them should be challenged because, on grounds of sheer flexibility in operation, canard configurations have never been as universally successful as those with tails at the back, even though they might be superior when flying for range. The magnitude of possible interference with the fore and aft lift distribution could lead to cutting down the area of the foreplane, or enlarging the area of the rearplane. A modification of that kind could affect adversely at least the field performance, cruise L/D, CG position and loading-range, and longitudinal stability and control.

Within the limits of conventional practice the stall marks the slowest airspeed at which completely *controllable* flight is possible (even though advanced experimental military aircraft can now be manoeuvred to angles of attack and attitudes once thought impossible).

Controllability is the beginning of everything. An unstable aeroplane can be controlled and made to fly, as long as it is not so unstable that the pilot is unable to cope with it. An aircraft without a vestige of control can be made to fly well, like a free-flight model aeroplane, but at one design speed only – and it will be impossible to fly off-design, with no guarantee that it will land safely.

So, in the next chapter we look at controllability and manoeuvreability.

References and Bibliography

8.1 Lachman, G. V. ed., *Boundary Layer and Flow Control*. Oxford: Pergamon Press, 1961

8.2 Engineering Sciences Data Unit, Sheet 66034 (1966), *The Low-Speed Stalling Characteristics of Aerodynamically Smooth Aerofoils*. London: ESDU, 251–259 Regent Street, W1R 7AD, 1966

8.3 Abbott, I. H. and von Doenhoff, A. E., *Theory of Wing Sections*. New York: Dover Publications, 1959

8.4 Flight Commander McMinnies, WG, RN, A flight commander, *Practical Flying*. London: Temple Press, 1918

8.5 Stinton, D., *Aero-Marine Design and Flying Qualities of Floatplanes and Flying-Boats,* London: The Royal Aeronautical Society, 1987

8.6 Bertin, J. J. and Smith, M. L., *Aerodynamics for Engineers,* 2nd edn. London: Prentice-Hall, 1989

8.7 Shevell, R. S., *Fundamentals of Flight,* 2nd edn. London: Prentice-Hall, 1989

8.8 *Aviation Consumer,* Vol. XXIII, No. 20, Greenwich: Connecticut 06836-2626, 15 October 1993

CHAPTER 9

Control and Manoeuvre

'The Oxbox* was a real 1930s military – puffing, resonant, exacting and full of sharp corners.
The sun poured blindingly through the greenhouse canopy onto the matt black instruments. It
was delightful to try turns, trimming out, synchronising the throttles. Southeast at 1850 revs
and 130 mph, over the immemorial patch-work of hedgerows since bulldozed in the name of
economy, to reconnoitre Portsmouth. The *Victory* crouched like a varnished bird's nest among
modern grey monsters .... Downwind for Wallop, old fashioned Cheetah radials clopping at
each elbow and smelling strongly of petrol. Then, while still at some height, a firm holdoff into
the surface breeze where the scent of hay predominated. Steeper and steeper grew our attitude
as we waffled along, like a percheron being put at a jump too slowly, until at last a sonorous
rumble told of the perfect three-pointer ....'
* Airspeed Oxford, twin-engined trainer.
Extract from the unfinished book by the late Hugh Scanlan (past Editor of *Shell Aviation
News*) about the RAF in the 1930s

Wilbur Wright put it in a nutshell in a letter to his father in September 1900, when he
described his work as '*experiment and practise with a view to solving the problem of equilibrium ... under proper control under all conditions ....*' He was particularly concerned with
the consequences of engine failure. The modern test pilot is tasked with, among other things,
assessing the effects of flying controls to achieve equilibrium, no less. If the handling qualities provided are unsatisfactory then there are three areas upon which he must concentrate:

☐ Seeking to understand the causes behind the effects.
☐ Assessing how much the basic design, or just the configuration, might be at fault
(design changes will be time-consuming and costly).
☐ Making recommendations.

Therefore this chapter is arranged in three parts:

☐ *First*, what should one expect of the flying controls, in terms of authority and effective-
ness?
☐ *Second* follows a survey of the more conventional arrangements of flying controls.
☐ *Third* are some of the fixes one might suggest to improve control effectiveness.

What is Expected of Flying Controls?

A detailed compendium of Airworthiness or Mil Spec control requirements would be out of
place. We only need to get the flavour. For this, Ref. 9.1 is an admirable source.

Plate 9–1   Airspeed Oxfords during World War II. (Courtesy of the Royal Air Force Museum)

The *grandfather* control requirement is that an aeroplane must be safe and manoeuvrable throughout the flight envelope:

☐ From take-off to landing (power-ON and -OFF, with wing flaps extended and retracted).
☐ It must be possible to make smooth and safe transitions from one flight condition to another, including turns and sideslips, without exceeding the limit load factor (for structural strength), under any probable operating condition.

In short, an aeroplane must *never* be able to get away from the pilot. Nor should the controls be so powerful as to enable the pilot to break it through normal handling.

Longitudinal control: elevators and stabilators

It follows from the grandfather clause that there must always be adequate control authority in pitch when power, landing gear and flap changes are carried out, either separately, or in combination. This includes an ability to glide in the landing configuration with an out of trim stick force not exceeding 10 lbf (45 N). Table 4–6, tests 7 and 8, give a broad indication of configuration changes and airspeeds between $1.2\,V_{S0}$ and $1.4\,V_{S1}$. Speeds are low, so as to demonstrate adequate control authority in the take-off and landing configuration, when the tail is working in a highly disturbed wake with reduced efficiency, η_s, and low dynamic pressure, q_s (see Equation (3–50)).

It must also be possible to manoeuvre a small aeroplane to make a landing without elevator (i.e. using trimmer alone to achieve zero rate of descent on touch-down (Table 4–6, test 14)) and without exceeding the operational and structural limitations. While it is argued in some quarters that commuter aeroplanes should also have the greater capability of being able to land safely, and not merely to reach a point from which one might then flop on to or connect with the ground without breaking anything, this argument has not yet prevailed.

Minimum allowable elevator control forces in manoeuvres, to prevent overstressing the aeroplane by exceeding the positive limit load factor (n_1 in Fig. 9.1), depend upon the class of the aeroplane, and whether the pilot has a stick or a yoke. For example, the pull force on a wheel or yoke must not be less than $W_0/100$ or 20 lbf, and for a stick, $W_0/140$ or 15 lbf (whichever is greater in either case). For this the aeroplane is required to be in the clean configuration with appropriate maximum continuous power settings, as specified for either piston or turbine engines.

Note 1: *Aeroelasticity*. Although a structure may be strong it might not be rigid enough. Aeroelastic distortion at higher speeds may reduce the authority of any control by elastic distortion. The range of movement may also become limited. For example, when pulling out of a dive with a conventional tail, up-elevator produces negative lift downwards, which bends the fuselage downwards, changing the tail angle of attack, reducing negative tail-lift, or increasing further its positive lift. Either diminishes in turn the down-load produced by the elevator, while elevator deflection and stick-force are decreased.

In a similar way aileron deflection attempts to twist a wing in the opposite direction. If the wing is not stiff enough in torsion, it will reduce or even reverse the effect of the aileron. The phenomenon is called *aileron reversal*.

Note 2: *Stick force and displacement with 'g'*. The normal manoeuvring load factor, n_1, is the unique value, corresponding with gross weight, W_0, of that appearing as a general term in Figs 8.10 and 8.11 and Equations (8–1) and (8–2). A simple way of checking stick-force

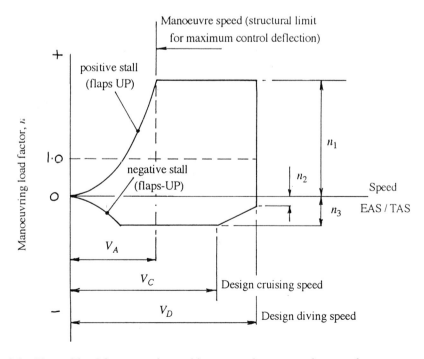

Fig. 9.1 Normal load factor vs airspeed in a normal manoeuvring envelope.

per g is to trim the aeroplane initially in straight and level flight. Then roll into a 60° banked turn which, when established and steady, *applies* 1.0 g. Measure the stick force needed to maintain the same airspeed without changing the power setting (Table 4–6, test 3 and Fig. 8.11). It may be necessary to let the aeroplane descend slightly, to maintain airspeed in the face of the increased drag. A small change in altitude can generally be ignored, as long as it does not alter significantly the indicated and true airspeed relationship. A linear 4 to 5 lbf/g is generally accepted as a comfortable force for the pilot of a highly manoeuvrable, agile aeroplane.

Example 9–1: Stick-Force to Reach Limit Load – Système 'D' (Example 1–3)

The Système 'D' triplane in Fig. 1.7 is a special case by virtue of its intended agility. If designed for $n_1 = 8.0$ g under normal airworthiness rules, then it would need a minimum of $4 \times 8 = 32$ lbf/g (142 N/g) to reach limit load. This lies within the temporary and prolonged application limit control forces for light and small aeroplanes, compare Tables 9–1 and 9–2 (Refs 9.1 and 9.3), which is far too high for such a machine of this class. Aerobatic pilots expect stick forces nearer 15 to 20 lbf at the most to reach the limit load for their aircraft (67 to 89 N/g). Compare this with the much higher limit control forces of heavier commuter category and large aeroplanes in Table 9–2 (Ref. 9.3). They argue that they are specialized pilots in highly specialized and thoroughbred aircraft – which puts their agile machines beyond the bounds of normal regulation.

Figure 9.2a and b show satisfactory and unsatisfactory stick-force/g and aileron characteristics brought about by aeroelasticity of the tail and rear fuselage, and the wing. It is a common phenomenon, but not always as extreme as shown. The acceptability of some degradation of control depends upon the degree to which final results are affected. Structural *stiffness* is essential, strength alone is not enough.

When tightening a turn, or in a pull-up, the aeroplane is also rotating in pitch about the CG. This induces an increase in angle of attack at the tail, which acts to oppose tightening. Therefore more stick deflection and control force is needed to increase the normal acceleration in a manoeuvre than in a steady turn at the same airspeed.

Note 3: *Stabilators (all-moving, 'pendulum', monobloc and flying tails).* The stabilator or all-moving tail has no separate elevators. It is useful for altering the stick-free neutral point (centre of pressure and fulcrum of aerodynamic forces between wing and tail) within broad limits. It is also used to alter the stick-force needed to change speed.

The stabilizer is usually fastened to a durable cross-tube, mounted in bearings securely bolted to the fuselage or fin structure. Some very light aeroplanes have a stabilator held on with two bolts, forming the hinges. Stabilator displacement may be by direct linkage: push rods, coupled with a trailing edge tab – the size of which and gearing is adjusted to vary the stick forces. It may also be arranged as a servo tab, driven by the pilot, which in turn drives the stabilator.

My experience leads met to *treat stabilators and their mountings with healthy respect.* On test flights I have found loose mountings – easily detected by GENTLY attempting to lift, and displace fore and aft, a stabilator by its tips. I have found slackness in the tab circuits and hinges especially – making them flutter-prone. Stabilators are, on the whole, far more susceptible and vulnerable to wear and tear than a conventional tailplane-plus-elevator.

There was a near catastrophic incident with a light twin, on its third flight after servicing and repainting. The pilot, discovering that he had no elevator control immediately after

Table 9–1

Characteristics of average pilot
(An historic table from Ref. 9.1 to guide pilots of period refurbished and replica military
aeroplane. Units are pounds force, inches and seconds.)

No.	Case		Ailerons (sideways or peripheral force)		Elevator (push or pull)		Rudder (push)
			Stick	Wheel	Stick	Wheel	
1	Greatest all-out effort of which he is physically capable for a very short time (two hands) (lb)		90	120	180	220	400
2	Maximum force it is permissible to demand of him for short while (lb)	a. 2 hands	–	80	100	110	200
		b. 1 hand	50	50	70	70	
3	Greatest force he cares to exert for a short while (lb)	a. 2 hands	–	30	–	40	60
		b. 1 hand	20	20	30	30	
4	Largest hand or foot movement to which he is accustomed for full control travel		±10″	±20″	±9″	9″	–5″
5	Smallest time, from the initiation of the control movement, in which he can reasonably be expected to apply full control. Forces as in 3.		$\frac{1}{4}$ s	$\frac{1}{2}$ s	Not applicable		$\frac{1}{2}$ s
6	Reaction time, i.e. time-lag between the occurrence of an unexpected event and the instant when the pilot starts taking corrective action on the controls.		Anything up to 3 s				

lift-off, chopped the throttles and dropped back on to the runway. An engineer had failed to fit the bolts which held the stabilator to the cross-tube, and no proper inspection had been completed. The stabilator worked on the two previous flights only because the paint secured it to the cross-tube. On the third take-off the paint let go.

Lateral control: ailerons and spoilers

AILERONS

The size and shape of an aileron is determined more by the need for authoritative low-speed handling, on take-off and landing, in turbulence and in crosswinds, than for high speed manoeuvrability. This can result in over-large ailerons for high speed, and higher stick forces when manoeuvring. Aeroplanes resist the attempt of a pilot to make them roll, although

TABLE 9–2

Limit control forces for light and small aeroplanes
(Ref. 9.3)

Values in pounds force applied to the relevant control	Pitch	Roll	Yaw
For temporary application			
Stick	60	30	–
Wheel (two hands on rim)	75	50	–
Wheel (one hand)	50	25	–
Rudder pedal	–	–	150
For prolonged application	10	5	20

the resistance may not be readily apparent. Rate of roll, p, for a number of opposing fighters in World War II are shown in Fig. 9.3 (Ref. 5.13). It was an important area of investigation. The Merlin-engined Spitfire and Hurricane with float carburetters were unable to sustain negative-g without fuel starvation and an engine cut. Their injection-engined opponents could push negative-g and dive while the Spitfire and Hurricane were forced to half-roll to follow, or escape. The rolling manoeuvre took time, so considerable attention was paid to improving their ailerons. The degree of resistance in roll depends upon aerodynamic damping which is a function of the angle of attack of the wing and, therefore, upon:

$$\text{helix angle of the path of the wing in roll} = \tan^{-1}\left(pb/2V\right) \qquad (9\text{–}1)$$

The wingspan, b, affects the local angle of attack of wing sections outboard and, hence, the wing contribution to aerodynamic damping. Long wings resist roll more than short. Many agile aeroplanes have had relatively short-spanned wings, with clipped or cropped tips for manoeuvrability (which is why, again during the war, a clipped-wing version of the Spitfire was introduced (Ref. 9.2)).

Rate of roll at low airspeeds is of paramount importance, as we see from the following. Light aeroplanes weighing less than 6000 lb (2730 kg) must be able to reverse a 30° banked turn in 5 s, with maximum take-off power in the take-off configuration, at the greater of $1.2\ V_{S1}$ or $1.1\ V_{MC}$ (one-engine-inoperative with propeller in the minimum drag position). Heavier aeroplanes must complete the turn reversal under the same conditions in not more than 10 s. In the landing configuration, with power for a 3° approach at $1.3\ V_{S1}$, the requirement for a light aeroplane is to reverse a 30° banked turn in 4 s. Heavier aircraft must not exceed 7 s.

Provided that control authority at low airspeeds is adequate there should be few problems at higher speeds. For a constant helix angle and angle of attack, Equation (9–1) tells us that rate of roll increases directly with true airspeed – as long as no aero-elastic effects arise. There are limits to the effectiveness of the ailerons caused by the lateral force a pilot is able to apply by stick or yoke.

SPOILERS

The advantage of the spoiler is that it provides roll control without the same aeroelastic side-effects as ailerons. Thus the wing structure may be made lighter. The spoiler acts by dumping lift, pushing a wing down instead of lifting it up. While this avoids adverse

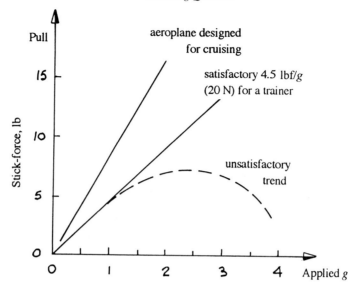

a. Pitch control by elevator.

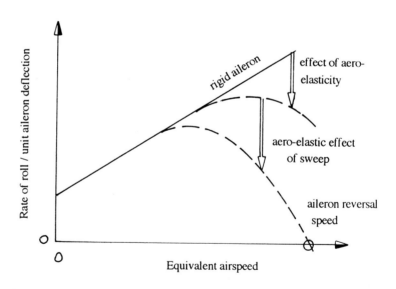

b. Roll control with aileron.

Fig. 9.2   Degraded pitch and roll performance caused by aeroelastic distortion of structure.

yaw – away from the intended direction of roll and turn – by increasing the drag of the downgoing wing, there is loss of authority as airspeed is reduced. Therefore, spoilers tend to be mixed with ailerons – the latter for use at low speed, the former taking over at high speed where aeroelastic effects become critical.

$$\text{aileron helix angle} = \tan^{-1}(p\,b/2\,V) \qquad (9\text{-}1)$$

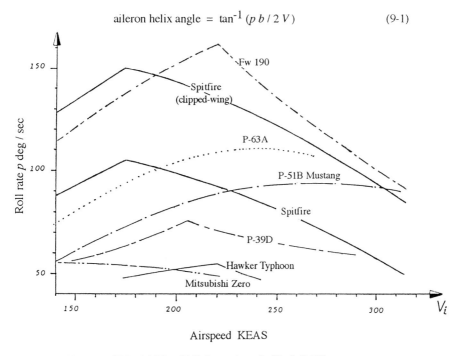

Fig. 9.3 Roll-rates of World War II fighter aircraft (Ref. 5.13).

Directional control: rudder(s)

The rudder is basically a trimming device, used to *'keep the wind blowing in the pilot's face'* by balancing turns and eliminating asymmetric consequences of propwash(es) spiralling over tail and other surfaces. In practice it is too complex in effect to consider in isolation, because it causes yaw, sideslip, skid and roll, affecting the lateral and directional qualities. Slip and skid also affect attitude and moments in pitch of aeroplanes with propellers.

Where the rudder comes into its own is when dealing with engine failure in a multi-engined aircraft, when landing in crosswinds, and as the spin-stopper. Give an aeroplane too much rudder and it is over-lively directionally and likely to drop a wing at low airspeed. Too little rudder and asymmetric forces take over at increasingly higher airspeeds, spoiling field performance.

Control Authority

The way in which an aeroplane handles: its speed of response to the pilot's hands and feet, and its apparent control 'sweetness' or harmony (see Equation (2–5)), depends upon two things. First is the quality and sensitivity of its controls. Second is their ability to overcome the inertia characteristics of the masses making up the configuration of the aircraft.

With one or two exceptions, like spoilers, most flying controls consist of surfaces which generate lift in one direction or another either as flaps, by changing aerofoil camber or, as in the case of all-moving (or all-flying) stabilators, by change of angle of attack. The fettling, by means of flight tests, of the qualities of manual controls is a continuing problem.

The aeroplanes dealt with so far split broadly into those which are *agile*, for aerobatics, fighters and trainers, and those designed for *sedate* cruise conditions. When discussing

Plate 9–2   The rudder(s) of tailfirst aeroplanes tend to lack authority. The Rutan Defiant has a rudder in a destabilizing position, beneath the fuselage and ahead of the CG. It is more effective than it might be when mounted at the mainplane tips. But with nosegear down there is an adverse interference and directional control was degraded on an example tested by the author. (Author)

control quality this is a useful distinction to keep in mind. What might be rated as good quality for one may not be quite so good for the other, even though both are designed for the same role.

Controls must enable the pilot to fulfil the function(s) for which the aeroplane was designed. The majority of manual control problems for cuising aeroplanes are those surrounding the design of the elevator(s) and rudder(s), both of which invariably affect the stability. Aileron design and modification are of equal importance when an aeroplane is designed for agility. Nor should controls flutter, because this is the path to structural failure.

Pilot effort

Manual control systems feed back forces to the hands and feet of the pilot through the control hinge-moments, affecting his likes, dislikes and opinions. Power controls are given artificial feel to simulate idealized manual force levels. Factors of importance to a pilot assessing a manual system are:

- The *greatest force* he cares to exert for any length of time.
- The *largest hand and foot movements* to which he has grown accustomed (some guidance is given in Chapter 4, Fig. 4.4a).
- His reaction time.

Table 9.1 (Ref. 9.1) shows values of control forces for the average pilot. These are results which were applicable to pilots of both sides during World War II. Not only is the table of historic interest, it is useful for pilots of refurbished and replica World War II aeroplanes. It is also relevant to the design of turboprop and other high performance trainers.

By comparison, Table 9–2 shows modern pilot control force limits for light and small civil aeroplanes, taken from Ref. 9.3. Table 9–3 (Ref. 9.4) compares design control loads and torques for commuter category and heavier transport aeroplanes. The forces in all three tables are 'ball-park' in that they do not differ greatly, yet their variation is enough to show that the larger the aeroplane the heavier are the control forces needed to prevent pilots from breaking them. Stately transport aeroplanes, though big and strong, are floppy beasts. Heavy forces provide protection, while saving structure weight needed otherwise to stiffen the structure.

Control forces affect reaction time. One should expect the reaction time on any control with a heavy force to take up to 3 s (Table 9–1). It may be that the time taken for a pilot to react to a situation exceeds the time taken to apply the necessary control. This bears upon action following an engine failure.

Note: A test pilot should 'calibrate' his or her hand and foot forces regularly. It is easy to underestimate foot forces. Stand on one leg with the knee slightly bent and the force felt is equal to one's weight: anything between, say, 70 kg (155 lb) and 90 kg (200 lb). Yet, that is no great strain for a pilot in full health, even holding it for one minute, and the leg hardly feels such a force at first.

Example 9–2: Restriction of Stick Movement – Jodel Light Aeroplane

The degree of control exerted by the pilot depends upon the working-space and ergonomic design of the cockpit, and the ease with which he can reach and move all cockpit controls through their full range. For example, I have found combined aileron and elevator

TABLE 9–3

Limit control forces and torques
(compared between commuter category aeroplanes of design
weight 19 000 lb (8600 kg) and large aeroplanes certified under
JAR-25)
(Ref. 9.4)

Control	JAR 25.397 (c) (all weights) Max.	Min.	FAR 23.397(c) Commuter Cat. (@ 19 000 lb) (1) Max.	(2) Min.
Aileron				
Stick	100 lb	40 lb	90 lb	40 lb
Wheel (3)	80D	40D	67.5D	40D
Elevator				
Stick	250 lb	100 lb	225 lb	100 lb
Wheel (symm)	300 lb	100 lb	270 lb	100 lb
Wheel (unsym)		100 lb		100 lb
Rudder	300 lb	130 lb	270 lb	130 lb

D = control wheel diameter, inches (force is produced in lbf).

movements of a stick in a side-by-side Jodel light aeroplane restricted by normal thighs. This in turn restricted the bank available to correct the effects of turbulence and crosswind on landing (which sometimes needed full-back stick, into the corners). With low wing loading, which made the aeroplane dance about, aileron control verged on the inadequate.

Control Feel and Wake Effects

What determines the control forces in Tables 9–1 to 9–3 and the way a manual control feels to the hand or foot of the pilot? The response of the aeroplane must be involved for it to be possible to say that one was light and 'squirrelly', while another felt heavy and sluggish.

Changes in the angles of attack of moveable control surfaces and the fixed surfaces to which they are attached alter their overall pressure distributions (Fig. 9.4). These, acting ahead of or behind the hinge line of a control cause hinge moments, $\pm H$, as shown in Fig. 9.5 for a symmetrical aerofoil section. If the pilot releases the control the hinge moment causes the control surface to float, or trail either along (–), or against (+), the relative wind. The sign convention is shown in Fig. 9.6. The surface could be a fin and rudder, tailplane and elevator, or wing and aileron. Being simple flaps, only the camber and proportions differ. To describe what happens we use pressure coefficient, C_p, which is a relative of the lift coefficient, C_L, such that:

   pressure coefficient = pressure difference/dynamic pressure (undisturbed)

and, in terms of the static pressure and Equations (3–12) and (3–26):

$$C_p = \Delta p / q \tag{9-2}$$

When integrated over a strip of the aerofoil of area equal to the (chord × unit width in the spanwise direction) we obtain the air load to which the strip is being subjected. This is then converted into the local lift coefficient as follows.

In straight and level flight, when weight = lift:

$$\text{local strip (lift) loading}/q = (\Delta W / \Delta S)/q = \text{local } C_L \tag{9-3}$$

which has exactly the same form as Equation (3–27).

In Fig. 9.4 the broken line is the change in pressure coefficient when the angle of attack alone is increased, the hinged portion remaining fixed. The unbroken lines are the resultant pressure coefficients per unit angle of deflection, δ, for two hinged surfaces: one at half chord (e.g., an elevator or rudder), the other around 22% (like an aileron or a simple flap). As we saw in Chaper 3, δ is the general symbol for the angle of deflection of a hinged surface. In the particular case of an elevator $\delta = \eta$, while for a tab we use $\delta = \beta$.

Hinge moments are converted into generalized coefficients by Equation (3–49). Over *small angles* (close to zero) and when unaffected by compressibility, the rate at which a hinge moment coefficient changes is almost linear with angle of attack and flap or control deflection (Fig. 9.7) Knowing that a tail working in the slower wake of surfaces and junctions ahead of it suffers a loss of dynamic pressure, as we saw in Equation (3–50), we may write:

$$\text{hinge moment coefficient, } C_H = H / q_\delta S_\delta c_\delta \tag{3-49}$$

by introducing tail efficiency and substituting:

$$q_\delta = \eta_s q \tag{3-50b}$$

When separation of flow over the control surface occurs hinge moments increase rapidly.

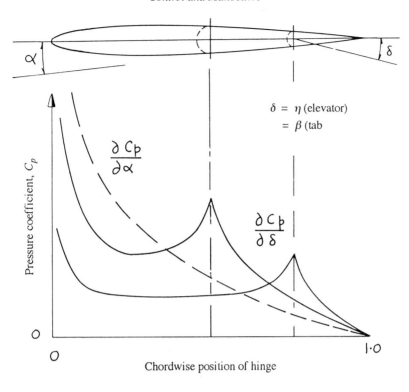

$\delta = \eta$ (elevator)

$\quad = \beta$ (tab

Pressure coefficient, C_p = pressure difference / dynamic pressure

$$= (p_1 - p) / q$$

$$= \Delta p / q \qquad (9\text{-}2)$$

(see also eqs. (3-12) and (3-26))

Fig. 9.4 Resultant pressure on an aerofoil per unit angle of deflection (Ref. 9.4). Compare with Fig. 3.19.

Often loss of aerodynamic efficiency in the wake is forgotten when calculating the size of stabilizing and control surfaces. This causes some of the difficulties which arise in handling flight tests, when control surfaces appear to be inadequate and stability is degraded. Often more tail area is needed, which is an expensive and near impossible solution late in the day. Attempts are made to increase effectiveness of a deficient control as a lifting surface by redesign, but this is rarely a great success. The most that can be expected is a mild improvement.

Example 9–3: Reduced Elevator Authority, Tail-Down – Me 262 Prototype

The Messerschmitt Me 262V-1 of 1942, which was referred to in Chapter 1, had a tailwheel undercarriage in its original form. Early flying was carried out successfully with a Jumo 210 piston-propeller engine in the nose, as the Jumo 004 turbojet engines were not then ready. When they became available the V-3 version had the turbojets fitted and it was calculated that the aeroplane should lift off around 110 mph (177 km/h). But the test pilot, Fritz

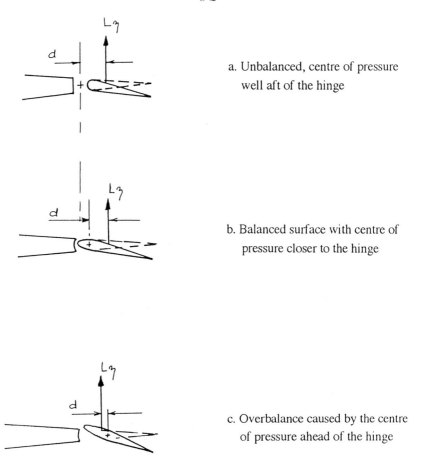

a. Unbalanced, centre of pressure
well aft of the hinge

b. Balanced surface with centre of
pressure closer to the hinge

c. Overbalance caused by the centre
of pressure ahead of the hinge

Fig. 9.5   Effect of hinge location upon control surface balance.

Wendel, found that even when taxying at 110 mph the elevators were ineffective and the tailwheel could not be lifted. The tail was in a position where it would have been adversely affected by wakes from the broad fuselage and wing junctions (see Fig. 9.8a). The tips of the tailplane were in line with the turbojets and there may also have been some wake-suction beneath the horizontal tail surfaces.

One of the test team suggested that Wendel should stab the brakes sharply at 110 mph, so raising the tail out of the wake. Wendel is reported by Mano Ziegler as saying afterwards:

> 'After I applied the brakes, the machine tilted forward. When it reached its flying attitude I felt the pressure on the elevator immediately and could take off straight away. The engines ran perfectly. It was a great joy to fly this machine. I have seldom been so inspired on a first flight ....'

The first flight lasted 12 minutes. On a second flight of 13 minutes, later in the day, Wendel discovered that during turns the airflow over the inboard portions of the wings separated early, and this was attributed to the thin wing section. The section was thickened and in-board flaps were fitted. The aeroplane then remained substantially unchanged – apart from the fitting of a nosewheel undercarriage.

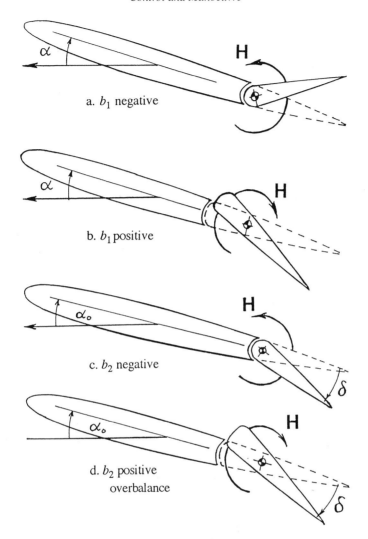

a. b_1 negative

b. b_1 positive

c. b_2 negative

d. b_2 positive
overbalance

Fig. 9.6   Floating tendencies of control surfaces (Ref. 9.4).

Control Balancing and Response Effect

Aerodynamic balancing of a control surface is the way of adjusting the area and section contours fore and aft of the hinge, so that the resultant of control lift acts on a manageable arm from the hinge and produces a hinge moment in the correct sense. In general:

$$\text{aerodynamic balance} = \text{area moment fwd of the hinge}/\text{area moment aft of} \\ \text{the hinge} \qquad (9\text{--}4)$$

when we really mean:

$$= \text{aerodynamic force-moment fwd of the} \\ \text{hinge}/\text{aerodynamic force-moment aft of the hinge}$$

The reason for saying this is that the lift generated by the control surface contour ahead of the hinge differs from the lift generated aft. In short, the lift slopes of each portion of the

$$+C_H = +H / q S_\delta c_\delta \qquad\qquad (3\text{-}49)$$

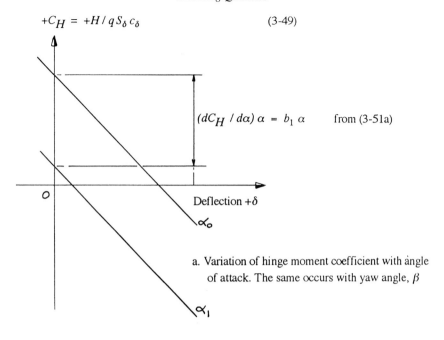

$$(dC_H / d\alpha)\, \alpha = b_1\, \alpha \qquad \text{from (3-51a)}$$

Deflection $+\delta$

a. Variation of hinge moment coefficient with angle of attack. The same occurs with yaw angle, β

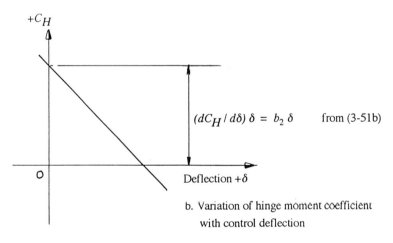

$$(dC_H / d\delta)\, \delta = b_2\, \delta \qquad \text{from (3-51b)}$$

Deflection $+\delta$

b. Variation of hinge moment coefficient with control deflection

Fig. 9.7 Slope of hinge moment coefficient with angle of attack and control surface deflection within the working range.

control surface are not the same. Care is needed to avoid overbalance, which occurs when the moment of the balance force ahead of the hinge is overpowering (see Fig. 9.5c) and helping to deflect the control instead of hindering.

To simplify what follows we construct this identity, knowing that:

$$a_1 = \Delta C_L / \Delta\alpha \quad b_1 = \Delta C_H / \Delta\alpha$$
$$a_2 = \Delta C_L / \Delta\eta \quad b_2 = \Delta C_H / \Delta\eta$$

so that:

$$a_1 / a_2 = b_1 / b_2 = \Delta\eta / \Delta\alpha \qquad\qquad (9\text{-}5)$$

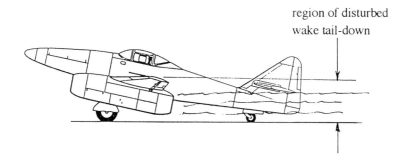

region of disturbed
wake tail-down

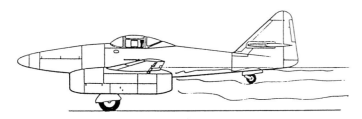

Fig. 9.8 The Messerschmitt Me 262 V-3 tail could not be raised at 110 mph during initial testing, without a jab on the wheel-brakes which jerked the tail up, out of the jet wake.

Now, from Chapter 3, lift coefficient, in terms of lift slopes and angles of attack and deflections:

$$C_L = a_1\alpha + a_2\eta + a_3\beta \qquad (3\text{--}52)$$

Consider what happens to tail lift when, say, an upgust increases the angle of attack and the elevator is free to float, compared with the change in tail lift when the pilot holds the stick firmly on meeting the gust:

$$\frac{\text{rate of change of tail lift, elevator free}}{\text{rate of change of tail lift, elevator fixed}} = \frac{a_1\,\Delta\alpha - a_2\,\Delta\eta}{a_1\,\Delta\alpha}$$

$$= 1 - (a_2\Delta\eta/a_1\Delta\alpha)$$

or, if preferred

$$= 1 - (a_2/a_1)\,(\Delta\eta/\Delta\alpha) \qquad (9\text{--}6)$$

Inserting the hinge-moment part of the identity Equation (9–5) we have obtained the bracketed term in Equation (3–55), which gave the equivalent lift slope, so let us call it the equivalent lift slope ratio:

$$\bar{a}_1/a_1 = 1 - (a_2 b_1/a_1 b_2), \quad \text{i.e. } 1 - (a_2/a_1)\,(b_1/b_2) \qquad (3\text{--}55a)$$

The equivalent lift slope ratio is of fundamental importance, as could be seen when it first appeared in Chapter 3, where it was found to affect relationships between the static and the manoeuvre margins (stick-fixed and -free) and the fundamental stability of an aeroplane. The task of the test pilot is to provide the designer, the aerodynamicists and engineers with the information needed to balance the controls for optimum handling, while achieving the degree of stability needed to satisfy the operational requirements.

Of equal importance is *response effect* (sometimes called 'b_1 *effect*'), the name given to the alteration in the control hinge moment when the aeroplane responds to movement of a control surface. It is the measure of the change in how the aeroplane feels when manoeuvring. The simplest of flap-like control surfaces, like an elevator, rudder and aileron, when deflected in the positive (lift) direction has hinge moments arising from the resistance of the air, trying to restore it to its undeflected position. The change in the restoring hinge moment with small control deflections is linear and in the negative sense. The accompanying change of lift coefficient with control deflection is linear and in the positive sense (compare Figs 9.7 and 9.9).

Pilot effort (indicated by hinge moment coefficient), caused by change in angle of attack and deflection of control and tab, is given by:

$$C_H = b_1\alpha + b_2\eta + b_3\beta \qquad (3\text{--}51)$$

Using the same argument as when deriving Equation (9–6) we shall apply it to pilot effort when moving the elevator through $\Delta\eta$, to cause a change in angle of attack, $\Delta\alpha$. Comparing it with the effort involved in deflecting the elevator through the same angle, but without a resulting change in angle of attack (as happened in Example 9–3), let us call this ratio the response factor, K:

$$K = \frac{\text{pilot effort on moving the elevator and changing the angle of attack}}{\text{pilot effort on moving the elevator without changing the angle of attack}}$$

$$= \frac{b_1\Delta\alpha + b_2\Delta\eta}{b_2\,\Delta\eta}$$

i.e.

$$K = 1 + b_1\,\Delta\alpha / b_2\Delta\eta \qquad (9\text{--}7)$$

Turning again to the identity in Equation (9–5) we may now write:

$$K = 1 + (a_2 b_1 / a_1 b_2), \quad \text{or,} \quad 1 + (a_2/a_1)(b_1/b_2) \qquad (9\text{--}8)$$

which has the same form as the equivalent lift slope ratio, except that the sign is reversed. Here pilot effort, instead of being proportional to the rate of change of elevator force with deflection, b_2, is proportional to Kb_2. As a simple control surface has a negative (elevator-up) trail when angle of attack is increased, negative b_1 lightens the stick-force, making b_1/b_2 less than unity. A positive b_1 makes the response factor greater than unity, and the stick-force feels heavier.

Usually response effects are more marked for elevators and rudders than for ailerons. The reason is that unit deflection of an elevator or rudder tends to produce changes in pitch and yaw which are about five or more times larger than the angle of roll resulting from aileron deflection.

For a given control movement at any speed, both the hinge moment and pilot effort are proportional to Kb_2. The pilot's control force increases with the scale (size) of an aeroplane and as the square of the airspeed – as long as compressibility and aeroelastic distortion of airframe and control surfaces are absent. However, the reverse occurs when building and flying small scale (replica and sport) aeroplanes. Control forces decrease for a given rate of

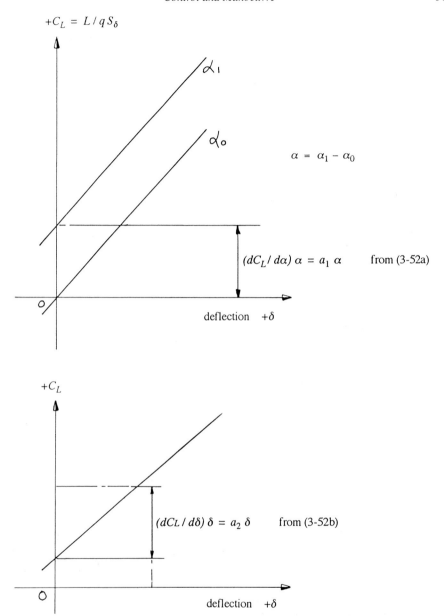

$+C_L = L / q S_\delta$

α_1

α_0

$\alpha = \alpha_1 - \alpha_0$

$(dC_L / d\alpha) \, \alpha = a_1 \, \alpha$ from (3-52a)

deflection $+\delta$

$+C_L$

$(dC_L / d\delta) \, \delta = a_2 \, \delta$ from (3-52b)

deflection $+\delta$

Fig. 9.9   Lift slope with control surface deflection within the working range.

response making them '*as quick as a squirrel*' – or a cat. Twisting around to look behind can cause a lightning-fast jerk in the flight-path.

Historic (and Later) Fixes – Horns and Set-Back Hinges

As aeroplanes grew bigger, heavier and faster through and after World War I it became a two-handed, and then a four-handed job for two pilots to control them in some turbulent flight conditions. Several decades later, in a short digest of the research during World War II which resulted in Ref. 9.1, H.H.B.M. Thomas records that out of sixteen aircraft from that period:

> '... about four fifths had Frise ailerons. Just over half had elevators with a set-back hinge balance and a third had a horn balance. On the rudders the balance was more nearly divided equally between horn and set-back hinge ....'   (Ref. 9.5)

Today such fixes are still the most conventional and useful forms of control surface balance. Their origins are lost and their forms so taken for granted that we accept them as standard. A control surface hinged a short way back from its leading edge exhibits one of the earliest types of aerodynamic balance, although one may guess that it is probably later than a primitive strip of cord along part of the trailing edge. The Frise aileron is a special example of a set-back (and lowered) hinge which provides an asymmetric horn ahead of the hinge. This is so important that it will be dealt with separately.

A simple relative of the set-back hinge is the horn at the tip of a surface (Fig. 2.3). A relatively small horn is easier to modify than it is to rebuild completely a new set of large control surfaces, with the hinges maybe only a fraction of an inch away from where they were initially.

The shape and amount of the control overhang, or horn forward of the hinge, has a powerful influence upon b_1 and b_2, as shown in Fig. 9.10a, b and c (Ref. 1.5). Modifications like those shown in Fig. 9.10d may be part of the initial configuration of a new aeroplane when it has been developed by using parts of another (see one example in the drawings of the Pilatus PC-7, Figs 1.4 and 1.5).

Non-integral surfaces

The devices shown in Fig. 9.10e and f, unlike the earlier horns, are not integral with the aileron profile. The spade is a simple plate, generating lift forward of the hinge when the ailerons are deflected. It acts to lighten the control forces and can be adjusted by altering its area.

Figure 9.10f is a novel idea by Speedtwin Developments Ltd, Trelleck, Monmouth. Claimed advantages are:

☐ Icing does not upset the mass balance of the ailerons.
☐ Changing the size of the vanes does not affect aileron mass-balancing.
☐ The size of the vanes can be adjusted to suit requirements.
☐ A streamlined plate, when mounted fore and aft in front of and aligned with the vane in its neutral position, ensures positive self-centring at small angles of aileron deflection.
☐ Similar vanes can be adapted to the rudder and elevators, e.g., to alter the elevator trail characteristics, providing stick-free stability where this is weak.

Table 9–4 summarizes the effects of modifications to balance and leading-edge forms upon lightness or heaviness of a control. If the gap between the main surface and the control is unsealed and unshrouded, then air is free to flow through it from the side with the higher static pressure where the pressure is lower. A blunted profile with a curved or bluff 'brow', which emerges into the airflow, accelerates it. The resulting suction over the crest of the horn, lying ahead of the hinge, lightens the control. A balance with a sharp leading edge, on the other hand, decelerates and may even stall the flow squeezing past, increasing the static pressure on the leeward side. Not only does this tend to decrease the overall lift of control-plus-main surface, it also has the effect of trying to push the horn in a direction *against* the flow through the gap, which has the effect of heavying it. A control of this kind with a sharp horn is highly sensitive and vulnerable to quite small tolerances occurring in production. Manufacturing costs are relatively high because tolerances must be maintained within close limits.

Putting a seal between the leading edge of a control and the wing or other surface ahead of it prevents air squeezing through the gap. This heavies the control by making b_1 more negative while also increasing the lift slopes, a_1 and a_2. So, a sealed control surface is harder for the pilot to deflect, but more efficient as a lift-generating flap.

Example 9–4: Control Horns on World War I German Aeroplanes

The effectiveness of control horns was discovered early, apparently by the Germans, and we saw in Chapter 2, Example 2–2, in which the Fokker DVII and Bristol F2B Fighter were compared, how their behaviour differed. The DVII, which had control horns on ailerons, elevators and rudder, flew with a quick, jerky response to control, while the Bristol, which had none, responded with a graceful swing. It was also reported as being relatively fast (we would say in terms of V_D/V_{S0}). Major Vere Bettington, writing from France on 13 May 1917 reported the Bristol Fighter as being able to dive vertically, and to reach a speed '*consider-ably over 230 mph*' (MIAS?).

Being flapless the maximum C_L would not have exceeded about 1.3. When coupled with a wing loading of 6.39 lb/ft^2 (31.22 kg/m^2) (see Table 2–1) in Fig. 6.15, the stall speed, V_{S0}, would have been around 48 KEAS (near enough MIAS), so that:

$$V_C/V_{S0} = 230/48 = \text{about } 4.8$$

As control forces between V_{S0} and V_C or V_D vary as $(V_C/V_{S0})^2$ and $(V_D/V_{S0})^2$ the ratios vary between 10:1 in the cruise to 50:1 for a racing aeroplane at V_D. If correctly reported, speed ratios of the biplane Bristol Fighter varied in the ratio 4.8 : 1, and control forces by 23 : 1.

The DVII was said to be not very fast, but agile at altitude, where airspeeds were slower. There, lightness of control would compensate for large deflections and otherwise relatively large hinge moments (b_2 times angle of deflection of whichever surface) needed to achieve maximum control authority in a dogfight. This could be one of the soundest reasons for the Germans leading development of balanced aileron, elevator and rudder controls for numerous types of aeroplanes of that period, and not only DVII, commented upon by Roger Vee (Ref. 2.2).

The Frise balance (Leslie George Frise, British aircraft designer (1895–1979))

The Frise balance is important. It is of asymmetric form and was devised to reduce adverse yaw in the direction of the down-going aileron (see Fig. 9.11), i.e. against the direction of roll and intended turn. The effect is unpleasant and leads to imprecision in manoeuvre. The Frise aileron has the hinge moved downwards and the aileron nose reshaped as shown in Fig. 9.12. At small angles of deflection the up-going aileron is overbalanced, which helps to deflect the down-going aileron on the other side.

The net balance of a pair of Frise ailerons is a function of the neutral rigging position of the controls. Care is needed when rigging because:

☐ Upfloat, and rigging upwards from neutral cause overbalance for small deflections.
☐ Downfloat, and rigging downwards from neutral cause heavier forces for small deflections.

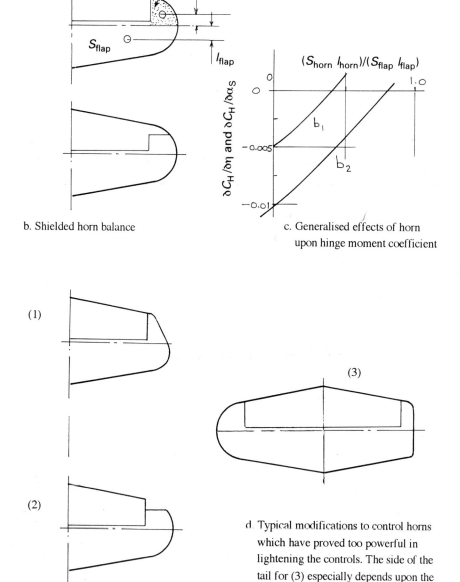

a. Unshielded horn-balance

b. Shielded horn balance

c. Generalised effects of horn upon hinge moment coefficient

d. Typical modifications to control horns which have proved too powerful in lightening the controls. The side of the tail for (3) especially depends upon the handing of the propellers and propwash

Fig. 9.10a–d   Forms of horn balance with cheap and cheerful modifications made during development test flying.

Example 9–5: Aileron Overbalance, Misrigged Upfloat – Helio Courier

During a test flight of a high-winged, single-engined monoplane (which, if I remember correctly had fabric-covered slotted ailerons) they began to snatch and overbalance,

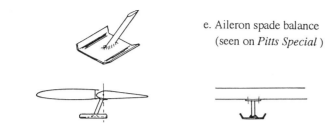

e. Aileron spade balance
(seen on *Pitts Special*)

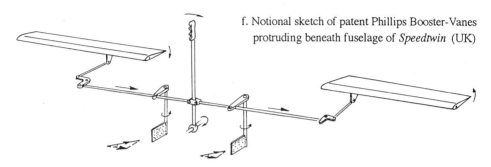

f. Notional sketch of patent Phillips Booster-Vanes
protruding beneath fuselage of *Speedtwin* (UK)

Fig. 9.10e and f　Aerodynamic balance areas which are not integral with aileron profile.

Plate 9–3　Speedtwin with patent Philips Booster – vanes beneath
fuselage. (Courtesy of Peter Philips, Speedtwin Developments)

TABLE 9–4

Basic control	Modifications		Effect
	Balanced/leading-edge shape		
Unshrouded	blunt		lightens control forces
	sharp		heavies control forces
	sealed		heavies both blunt and sharp
	(If the nose gap is less than $\frac{1}{4}$% local chord then control behaves as though fully sealed.)		
Shrouded	sealed blunt		more balance area needed than when blunt and unsealed
	sealed sharp		less balance area needed than when sharp and unsealed
	unsealed		heavies both blunt and sharp
	(If the nose gap is more than $\frac{1}{2}$% local chord then control behaves as though unsealed (Ref. 9.1).)		

50 KIAS or more before V_{NE} during a high-speed run. The flight was cut short and the ailerons examined. As required in the technical manual a length of cord was secured with fabric doped on the upper surface of both ailerons, a short distance forward of each trailing edge. Its purpose was to cause downfloat. The fabric should have been butted up against and tucked under the leading edge of the cord, at its junction with the wing. Like that it would have blocked the relative airflow, increasing locally the static pressure, pushing the ailerons downwards. Instead the fabric formed a smooth fairing over the cord which accelerated the local flow, inducing a lifting suction which caused upfloat.

When the cord was stripped off and replaced properly, the problem disappeared and the aeroplane reached V_{NE} without difficulty.

Tabs

The control tab is a primitive device with many variants. It is a separate surface, integral with the main control surface, with the task of relieving a control force, or assisting the pilot in moving a control surface. Another way of putting it is to say that as control forces vary as the square of airspeed, tabs are a way of defeating the 'speed-squared law'.

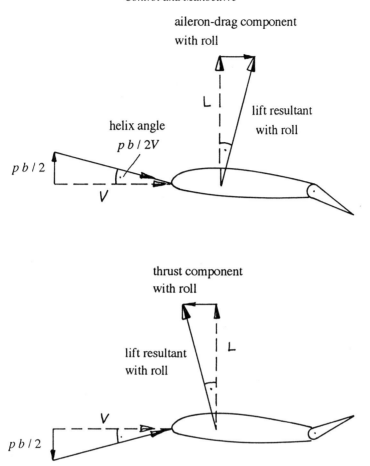

Fig. 9.11 Adverse yaw components of drag and thrust caused by roll.

The five commonest forms of tab with which we are concerned are: adjustable, balance, manually operated, spring and servo (Fig. 9.13a to e):

☐ *Trim-tab* (a) is the most basic. On the simplest aircraft it remains fixed and takes the form of a piece of plate attached at the trailing-edge of the control surface (see Fig. 9.19b). The tab is bent in the required direction between flights until the optimum setting is found.

The more advanced form of trim tab is adjusted in flight by the pilot, usually by means of a cable-operated screw-jack. The load in the jack is reacted at the control surface, through the jack mounting. To cope with high loads in a dive it is essential that the jack is efficent with an adequate mechanical advantage. The structure, mounting and cable must have low elasticity.

☐ *Geared-tab* (b) is a balance tab, which is sometimes called a *link-balanced* tab. It has the same effect as placing balance area forward of the hinge. The tab is geared to move in a given ratio to the main control surface, and is denoted:

$$m = \Delta\beta/\Delta\eta \qquad\qquad (9\text{--}9)$$

The pilot's effort is applied directly to the main control surface and not to the tab, the general effect of which is to alter the value of b_2 without affecting b_1.

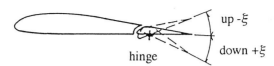

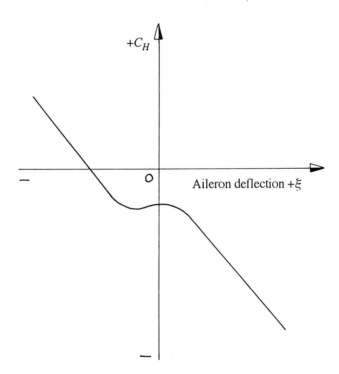

Fig. 9.12 The Frise aileron, designed to neutralize adverse yaw with aileron deflection. Excellent balance when correctly rigged.

When the geared-tab moves in the opposite direction to the main control surface it is called a *balance* or *lagging-tab*, lightening the hinge moments and control forces. When the tab moves in the same sense as the main control it heavies the hinge moments and forces. Then it is called an *anti-balance*, or *leading-tab*. The anti-balance tab is often used to increase an initially low stick-force per *g*.

☐ *Servo-tab* (c) is a small control surface attached to the trailing edge of the main control surface. The pilot's effort is applied direct to the servo-tab, not to the control surface. Tab movement provides the hinge-moment which drives the main control surface. Servo-tabs can be so effective that feel must be provided artificially for the pilot. Problems are:

(1)   loss of effectiveness and precision at low speed (approach and landing) and at the stall;
(2)   avoiding damage to the tabs caused by the main control hitting the stops when taxying and in flight;
(3)   tendency to flutter.

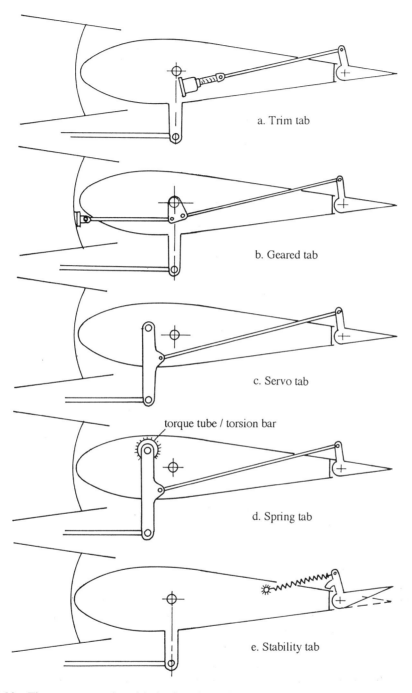

Fig. 9.13 Five common tabs with the functions, from a to e of trimming out and reducing control forces, and improving stability by alteration of hinge-moment and trail characteristics (after Vernon C. O. and Archbold E. J. N. (1946), Handley Page Ltd).

☐ *Spring-tabs* (d) were used extensively during World War II. Not only do they act as variable servo-tabs, in that the spring or a torsion-bar is incorporated in the tab circuit (so that the pilot's effort is applied to the spring/bar and thence to the control surface), they also improve repeatability of results. This enables the 'speed-squared law' to be defeated while giving a high degree of balance without overbalance, smoothing operation of the control surface over the full range of its movement. The linkage to the stick deflects the tab and the spring in proportion to the applied stick force. The spring changes the tab gearing with airspeed. The input force is less than it would be with a rigid crank and no tab.

Note: While the hinge-moment of a spring-tab about its own hinge contributes to and reduces the effort required of the pilot, such tabs have been known to flutter and should not be used with structures that are not stiff.

☐ *Stability-tab* (e) is a means of increasing the stick-free longitudinal stability of an aeroplane by arranging a spring which deflects the tab upwards. To see how it works, imagine the aircraft to be trimmed with zero control-force in steady flight, with the control surface and tab in line. A disturbance, such as a gust, increases the angle of attack and reduces the airspeed. This unloads the tab, which the spring deflects, causing the tab to act like a servo, pushing the control surface in the opposition direction.

Stability-tabs are also called *q-feel tabs* because their deflection varies as $q = \frac{1}{2}\rho V^2$. They have been applied to ailerons and to rudders – in the latter case to alter directional trim with airspeed.

Keep tabs simple

Although it is possible to combine tab functions, e.g., a geared-tab coupled with a spring-tab, by incorporating a screw-jack in the tab circuit, there can be problems. The tab angle can become too large if it gets out of adjustment for any reason. Then it can cause non-linear effects, in the form of flutter, or reversal fo the hinge-moment, if the flow over the tab suffers oscillatory separation, or stalls.

It is better to advise keeping tabs and their circuits simple by not mixing different types.

Example 9–6: Metal-Skinned Ailerons – Spitfire

H.H.B.M. Thomas describes the problems encountered with Spitfire ailerons early in World War II (Ref. 9.5). Originally they were fabric covered and subject to distortion caused by internal pressure differences, arising from gaps around the hinges. To reduce such variation they were given metal skins, with the trailing edges of the top and bottom skins brought to a sharp edge. But there remained variation in aileron heaviness caused during manufacture by small tolerances in their finishing (a consequence of extensive sub-contracting of work). Figure 9.14 shows the different hinge moment coefficients obtained from tests of six quite separate ailerons.

Aileron control forces were high and scatter of hinge moment results like those shown could make a difference of 18 lbf to 36 lbf needed to apply quarter aileron deflection at 250 MIAS. The late Harold Best Devereux told how they discovered differences caused by, in one case, the top skin overlapping slightly the bottom skin. While in another, the bottom skin might protrude further than the top. Trailing edge effects are critical and powerful, and the formation of a 'step' of this kind at the trailing edge made a significant difference to stick force.

The initial fix, recounted by Best Devereux, was to get rid of the step by fettling the trailing edges of both skins to a flat, at right angles to the chord, with a heavy-duty rasp. This made aileron forces similar between different aircraft.

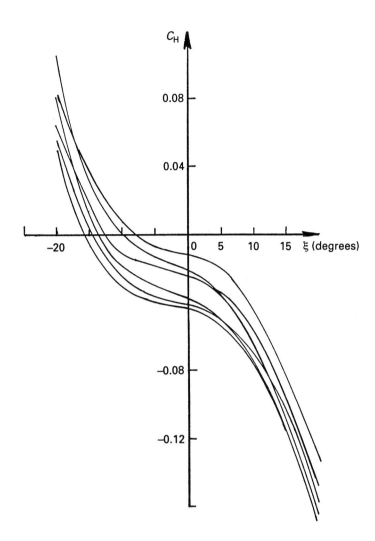

Fig. 9.14 Hinge-moment coefficients in roll of six Spitfires showing scatter and, therefore, marked difference in feel between each with ailerons covered with metal instead of fabric (extracted from Ref. 9.5, and courtesy of H. H. B. M. Thomas).

The spring tab was the ultimate fix, because it coped completely with the characteristics of the basic control, and reduced the stick forces by a third to a half (Fig. 9.15).

Note: The *trailing edges* of control surfaces designed for manual operation at subsonic speeds, have a powerful effect upon control characteristics. As they are farthest from the hinge, quite small changes of pressure and loading in their vicinity are capable of producing relatively large alterations in hinge moment. High performance aeroplanes with powered (non-manual) controls have no such feed-back to the pilot. The trailing

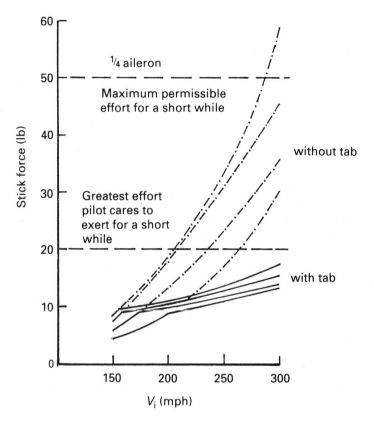

Fig. 9.15 Spitfires suffered from heavy aileron forces. These were reduced to manageable levels by fitting spring tabs, which also appeared to lessen the scatter shown in Fig. 9.14 (extracted from Ref. 9.5, and courtesy of H. H. B. M. Thomas).

Note: Broken lines show forces obtained from Table 9–1.

From Equation (3–65): equivalent airspeed, $V_i = V\sigma^{0.5}$ (3–65a)

edge of a manually-operated control surface is the place to look when difficulties are encountered.

Control Volume

The original fixes which, now proven by useage, have been adopted as standard control forms lead quite naturally to rules governing their proportions. The same has been done with the stabilizing surfaces which through accident, trial, error, research, development and calculation, now have standardized proportions. These are called control and stabilizer volumes (see Chapter 3, Equations (3–52) and (3–58) to (3–61)) which are used to guide the aerodynamicist and designer with their first approximations of workable control and stabilizer areas.

Control and stability, especially stick-free stability, are inseparable. Control takes precedence because it was the ability to control the path of an untethered lifting surface which enabled man to fly. The following features have a direct effect upon control (and stability). All can be sources of shortfall and inadequacy:

☐ *Control and trimability*: control surface area and volume, inertia distribution — rate of pitch, roll, yaw; a tendency to depart from trimmed flight; rate of departure

☐ *Stability*: fixed surface area and tail-volume — damping ratios; short and long period (phugoid) oscillations; pilot-induced oscillations (PIOs)

The control and stabilizer volumes contain all of the geometric factors which determine control and stabilizer authority. If the pilot finds that the aeroplane cannot be loaded far enough aft without it becoming light and twitchy on the controls, or that he cannot flare on touchdown, even though the CG is not at its forward limit, then it is reasonable to ask if the proportions are wrong, or if the tail and control surfaces are big enough. The answer depends upon the factors which degrade their effectiveness and efficiency (wakes and other interference, mass distributions, i.e. pitch, roll and yaw inertias, inertia coupling, type of engine installation, the balance of lift, drag, thrust weight and CG location).

In Equations (3–36) and (3–53) we encountered the term:

stabilizer (tail) volume coefficient, $\overline{V}_s = (S_s/S)\,(l_s/\overline{c})$ (3–53)

The general form of that equation is:

control and tail volume = area of working surface × distance of centroid
or 1/4 chord from the CG (9–10)

To turn such volumes into coefficients they are divided by other, relevant, aspects of the geometry of the aeroplane. For example, the greater the area of the wing, the larger must be the control and stabilizer areas. Wing span dominates yawing and rolling motions. The longer the span of the wing the larger must be the fin and rudder volumes, such that:

fin plus rudder volume coefficient, $\overline{V}_f = (S_f/S)\,(l_f/b)$ (9–11a)

while rudder volume coefficient, $\overline{V}_\zeta = (S_\zeta/S)\,(l_\zeta/b)$ (9–11b)

in which rudder area is measured aft of the hinge. As rudder area is around half the area of the fin-plus-rudder it follows that $\overline{V}_\zeta$ is roughly half the value of $\overline{V}_f$. Exactly the same thing can be done to derive elevator volume coefficient from Equation (3–53):

$$\overline{V}_\eta = (S_\eta/S)\,(l_\eta/c) \tag{9–12}$$

This too is roughly half $\overline{V}_s$.

Ailerons are treated in exactly the same way, such that if S_ξ is the aileron area aft of the hinge-line and y_ξ the moment arm of the centroid of an aileron about the CG, then introducing wing area and span:

aileron volume coefficient, $\overline{V}_\xi = (S_\xi/S)\,(y_\xi/b)$ (9–13)

Average values of control and other surface areas are given in Table 9–5. Roughly average volume coefficients are given in Table 9–6 for light and small turboprop aeroplanes: agile two-seat trainers, agricultural aeroplanes and commuters. In practice there are broad variations and it can be useful to make one or two calculations before suggesting an increase in control surface area.

TABLE 9–5

Control surface area ratios

Surface area ratio	Stabilizer plus elevator		Fin plus rudder	Aileron
$\dfrac{\text{total surface area}}{\text{wing area}}$	ancient modern canard	0.10 to 0.17 0.16 to 0.20 0.15 to 0.25	0.03 to 0.06 0.075 to 0.085	0.08 to 0.10
$\dfrac{\text{control plus tab area}}{\text{total surface area}}$		0.50 to 0.55	0.50 to 0.60	0.18 to 0.30
$\dfrac{\text{balance area ahead of hinge}}{\text{control plus tab area}}$		0.15 to 0.25	0.16 to 0.25	0.20 to 0.25
$\dfrac{\text{control tab area}}{\text{control plus tab area aft of hinge}}$		0.05 to 0.10	0.05 to 0.10	0.04 to 0.06
$\dfrac{\text{aspect ratio}}{\text{wing aspect ratio } A}$	conventional about 0.66 canard 1.20 to 1.50		0.18 to 0.30	1.00 to 1.30

Note: Expect to modify conventional tail values by $(1/\eta_s)$ for wake effects, power and configuration (see Fig. 8.3e); depending upon whether or not the aircraft is agile or sedate. The latter average $\frac{1}{3}$ larger surface area ratios.

TABLE 9–6

Average volume coefficients

Surface	Symbol	Volume coefficient
Tailplane plus elevator	$\bar{V}_s$*	about 0.48
Elevator	$\bar{V}_\eta$	0.25 to 0.30
Fin plus rudder	$\bar{V}_f$	about 0.18
Rudder	$\bar{V}_\zeta$	0.90 to 0.11
Aileron	$\bar{V}_\xi$	about 0.40

* Compare with Table 8–1.

Shortfall and Fixes

Some of the commoner faults of flying controls are due to:

- ☐ Configuration.
- ☐ Swept control surfaces.
- ☐ Control circuit friction, breakout forces and backlash.
- ☐ Residual (out of trim) hand and foot forces.
- ☐ Insufficient control authority.
- ☐ Control locking.
- ☐ Flutter and 'buzz'.

Configuration

Control surface configuration, like the configuration of a wing: hinge location and disposition of area, chord distribution (taper), choice of section and maintenance of contour (impossible with fabric ailerons at high speed), trailing-edge angle, compound or simple sweep, aspect ratio, twist or wind from root to tip, and fairing, affect control efficiency and authority.

Ailerons, for example, are best located in the region of maximum lift generated by the wing. This is around two-thirds of the semi-span outboard from the wing root. Interference between wing, fuselage and engine nacelles reduces lift in their vicinity. The lift distribution finally tapers off towards the tip. We saw in the last chapter that tips to avoid are those which encourage the trailing tip vortices to be shed inboard of the tip itself.

The wing generates lift by forcing downwards the mass of air it encounters in unit time. The stronger the trailing vortices shed from near the tips the greater the outboard sweep of the streamlines in that region. It follows that a rounded wing tip, shedding tip vortices about one-fifth of the span inboard from its extremity, should have the ailerons further inboard, which is less efficient, because the moment arm of each about the CG is less, and with it roll power and rate of roll, p (unless aileron area is increased). Therefore, and as noted earlier, the wing should be longest in span at the trailing edge. This not only traps the outboard flow of the streamlines as far aft as possible, lengthening the aerodynamic span, it enables smaller, lighter and more efficient ailerons to be fitted.

When take-off and landing is required with heavy loads (historically, short decks of aircraft carriers, and out of short fields) flaps are needed which are large in area and as long in span as it is possible to make them. This reduces the length of trailing edge available for mounting the ailerons and threatens their effectiveness.

The amount of chord occupied by a control suface varies. Rudders and elevators are around 50%, or a little less. Ailerons average 20 to 25% of the wing chord for an agile aircraft. A transport category aeroplane with long span flaps may combine spoilers with ailerons which are short in span but occupy 33 to 35% wing chord.

Example 9–7: Drooped Leading-Edge Modification – Cessna 172M

As we saw in the last chapter short take-off and landing (STOL) modifications to improve field performance must be taken with a pinch of salt. An increase in lift coefficient and slower stall speeds do not necessarily follow. According to the *Aviation Consumer* this was found by Cessna when the wing leading edges of the 1973 172M variant were recontoured by increasing the radius, to incorporate 'camber lift' droop. The object was to enable the aeroplane to operate at larger angles of attack and reduced airspeed. Aileron effectiveness in the stall and spin recovery suffered. Spin entry was also made more complicated (Ref. 9.6).

Swept control surfaces

Swept-back surfaces may be fashionable, but control surfaces with swept-forward hinges are more efficient than those with swept-back. As angle of attack is increased there is a tendency for lift to peak outboard, especially when wing and tail surfaces are backward-swept, which induces a spanwise component of boundary layer flow, from root towards the tip. A control surface loses efficiency when a component flow begins to run along its length (see Fig. 9.16). A forward-swept hinge line places the surface more nearly normal to the local airflow, which helps to maintain control authority to larger angles of attack. Ailerons on a forward-swept wing continue working deep into a stall.

For precisely the same reasons a swept-back fin and rudder is less efficient than one which has some forward sweep. This is self-evident from Fig. 9.16.

Control circuit friction, breakout forces and backlash

Breakout force is one which has to be overcome before the control surface can be moved. Backlash is lost motion between the pilot's hand or foot and the control surface. Three-axis microlights tend to have large amounts of backlash. One result is that a control may begin to flutter if it is improperly balanced and the backlash is considerable. Cables suffer lost motion through their elasticity. Couple backlash with friction and one has an unpleasant recipe for imprecision and possible disaster.

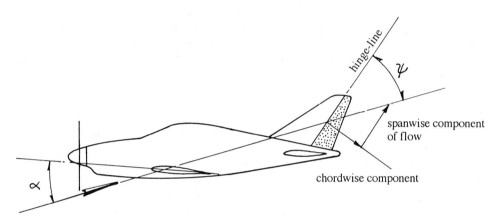

Back-swept rudder hinge-lines do not help:

- ▫ The rudder begins to cause forces and moments in pitch, like an elevator.

- ▫ Rudder effectiveness is reduced, roughly as the sine of the angle ψ, with a component of airflow increasingly established along its length.

- ▫ The drag coefficient of the aeroplane increases with angle of attack, α, and the wake thickens, decreasing the dynamic pressure recovery at the tail (*see* fig. 8.3e), exacerbating the adverse effect of sweep-back.

- ▫ The probability of the rudder becoming fully immersed in the wake of the tailplane and elevator in a developed spin is much increased (*see* especially fig. 12.5b).

Note: Forward-sweep of the hinge-line improves rudder effectiveness as the stall is approached (forward-swept hinge-lines of other control surfaces have the same effect). The reason is that with increasing angle of attack the rudder increases in span normal to the relative airflow.

Fig. 9.16 Fashion can be counter-productive when it satisfies no aerodynamic or structural need.

Control circuit friction is a force hard to eliminate without careful design. It is shown in Fig 9.17, making it impossible to trim that aeroplane within a band of ±10 knots. Too much friction makes it hard to move a control accurately and smoothly, without jerking. If strong enough to neutralize control surface float stick-free, it gives a false impression of stick-free stability, by substituting an element of stick-fixed stability. Alternatively, if the stick is displaced to change airspeed without retrimming and then released, friction may in the extreme prevent the control floating from the position to which it has been displaced. The false impression given is that the aeroplane does not tend to return as it should to the original trimmed speed, and it must therefore be unstable.

Agile aeroplanes are manoeuvred for much of their airborne time and rarely remain in a trimmed condition for long. The effect of control circuit friction is less adverse during manoeuvres in an agile machine than when setting up and attempting to trim an accurate cruise condition in one designed for that purpose alone.

A breakout force, unlike friction, only resists the initial deflection of a control. When excessive, it can be just as annoying, because it causes jerkiness in manoeuvre. Even so, a breakout force can be useful. Figure 9.18 shows the same circuit as in Fig. 9.17, but with a breakout force added to increase the precision with which a pilot may then achieve an accurate trimmed speed.

Residual (out of trim) hand and foot forces

The earliest control fix appears to have been the addition of a strip of cord about 3/16 of an inch (5 mm) in diameter (like a length of old-fashioned clothes line) doped on to the trailing edge of an elevator, rudder or aileron, to help push the surface up or down. The authority of the cord is adjusted by adding more, or cutting it shorter. It has the disadvantage that it is a single speed only device, although its presence helps to reduce to manageable proportions the magnitude of residual hinge moments at speeds either side of the trim point.

Such a length of cord is sketched in Fig 9.19a. As we have seen it works as an aerodynamic dam which pushes the control surface down, or up, depending upon whether the cord

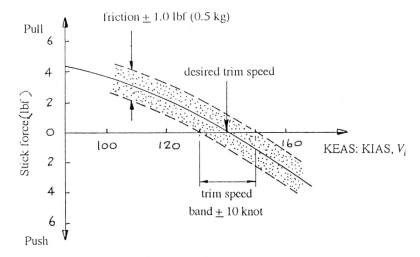

Fig. 9.17   Effect of circuit friction upon trim speed.

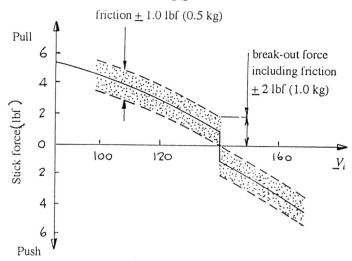

Fig. 9.18   Effect of adding ±1.0 lbf breakout force to the control circuit in Fig. 9.17.

is above or beneath the trailing edge. The cord must be well tucked under at the leading-edge (see Example 9–5).

Note 1: Modern metal and glass skins must use other than cord doped on to their surfaces, and so resort is made to bits and pieces of bent plate, and strips or wedges of wood or GRP, like those shown in Fig. 9.19b and c. However, the vulnerable corners of a wedge of wood or GRP are susceptible to damage. This may change the aerodynamic characteristics as dramatically as a damaged trailing edge.

Note 2: As a warning, a bevelled trailing edge is shown separately in Fig 9.20. It is a powerful anti-flutter device, but it is tricky, vulnerable to hangar-rash and, to avoid a non-linear b_2. should never be used without sealing the nose of the control surface. If damaged, a bevelled trailing edge can cause flutter.

Example 9–8: Aileron Flutter Caused by Damaged Trailing-Edge – AA-5

A case of aileron flutter was reported on one particular (now Gulfstream Aerospace) AA-5, which had metal ailerons with bevelled (wedge-shaped) trailing edges. When pulling *g* in the clean configuration, and also with one-third flap, flutter started around 68 KIAS (about 1.3 V_s). Backlash was discovered in the aileron control circuit. They had been shimmed, but shims were not called for by the manufacturer. A high frequency fluttering oscillation of the port aileron could also be seen and felt through the control yoke at moderate IAS in a 45° to 60° banked turn (around 0.7 *g* applied, see Fig 8.11). The test was terminated and on landing a small, localized, 'hangar-rash' dent was found in the upper bevel at the trailing edge. It was only narrow in width, a few millimetres deep, and looked like a hit by the edge of a tool–but not hard enough to have damaged the paint. When the aileron was changed the flutter disappeared.

The modifications shown in Fig 9.19 can be applied to either top or bottom or both surfaces. When applied to both, they heavy the surface in both directions, and have the same effect as a thickened (slab) trailing edge, which also improves the lifting effectiveness of a control

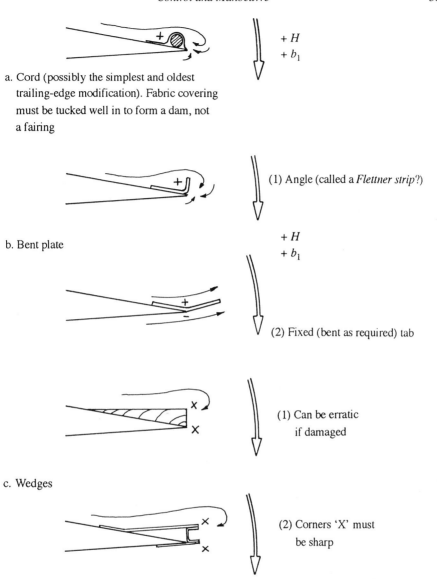

a. Cord (possibly the simplest and oldest trailing-edge modification). Fabric covering must be tucked well in to form a dam, not a fairing

+ *H*
+ *b*₁

(1) Angle (called a *Flettner strip*?)

b. Bent plate

+ *H*
+ *b*₁

(2) Fixed (bent as required) tab

(1) Can be erratic if damaged

c. Wedges

(2) Corners 'X' must be sharp

Fig. 9.19 Three trailing-edge fixes for making a control surface float or trail in one direction. Care is needed because such fixes, especially b and c, which are vulnerable to 'hangar-rash' and other superficial damage, can badly mar stick-free stability.

by increasing a_2 (lift slope with deflection). An explanation which is reasonably adequate is that a region of dead air adheres to and forms a near static wake behind the slab, past which less retarded air slides. This has the effect of increasing aerodynamically the apparent chord of the control surface in a similar way to drawing tangents to upper and lower surfaces until they meet, beyond the trailing edge, and then filling the space between with control skin and structure. The downstream extremity of the region of dead air forms a new 'aerodynamic trailing edge'.

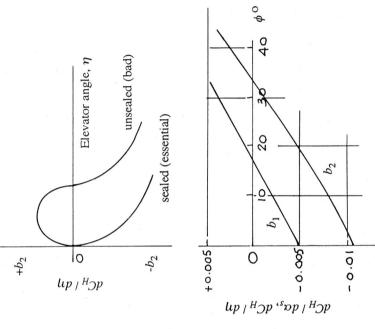

Fig. 9.20 Bevelled trailing-edge, a sensitive and tricky device which must be sealed. A slightly dented bevel has provoked aileron flutter with increasing angle of attack. It becomes more effective with increasing ϕ. Good quality control during manufacture is hard to achieve.

Note: Reference 9.1 provides advice that, as a general rule, the included angle of the trailing-edge of a control should never exceed 16°. Ideally it should be less. Trailing-edge angles have powerful effects upon hinge moment characteristics. A large angle makes a control sensitive to the position of the transition point from laminar to turbulent flow in the wing, fin or stabilizer boundary layer, making hinge moments non-linear.

Therefore, if the pilot discovers non-linearity in a control force, a good place to start looking is the included angle (and for any other feature or discontinuity) at its trailing-edge.

Insufficient control authority (leaks, gap-sealing and vortex-generators)

A leak of air from one side of a surface to another distorts the boundary layer into which it emerges, by forming an 'equivalent surface' of near static air, like that described in the wake of a slab trailing-edge. Such leaks increase parasite drag and cause losses in performance. Control surfaces suffer leaks from the higher pressure (concave) side to the lower pressure (convex) side, especially around hinges. These reduce local lift coefficients of controls and flaps (and attendant hinge moments), as well as contributing to the total parasite drag by causing vortices.

The rule which applies to wings, tails, canards and flying control surfaces is: *do everything reasonable and with minimum penalty upon performance to prevent a leakage of pressure differential between the opposite sides of any surface.*

Figure 9.21a(1) and (2) show satisfactory and unsatisfactory aileron characteristics, such as might be caused by an absence of or inadequate sealing. Three fixes worth suggesting (which depend upon the detailed form of the control system) are shown in Fig. 9.21b (1), (2) and (3).

A quite different way of making an airflow behave and maintain the highest possible pressure differential is to mount vortex-generators on the leading-edge of the control surface, arranging for them to be hidden within the control gap until it is deflected. The greater the deflection, the further the VGs protrude. In this way they delay separation on the convex (suction) side of the control surface, so maintaining lift.

Vortex-generators are most easily installed where a gap is ample, as on the leading-edge of the rudder, behind a thick fin.

Control locking

Although the problem of control locking is really a matter of configuration, shape and disposition of areas, it is a phenomenon warranting separate discussion. It comes about most commonly because of fin stall at a large angle of yaw, and a b_1 so powerful that the pilot cannot overcome it. A similar thing can also happen with an elevator when spinning, or when working in strong downwash from powerful high-lift flaps. An overpowerful b_1 can prevent the elevator being moved to effect the control needed to achieve recovery. Aileron locking, while possible for similar reasons, is far less common. But never assume beforehand that it will not be found at some stage during prototype testing. Anticipate the unexpected.

Example 9–9: Fin Fixes – Handley Page Halifax and Boeing B 17

The Halifax was one of three heavy bombers used by Bomber Command of the RAF during World War II (Fig. 9.22). It was four-engined, had twin fins and rudders, and nose mid and tail gun turrets. Many of the accidents it suffered were attributable to rudder overbalance.

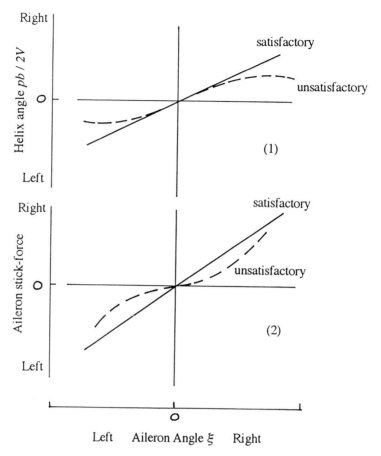

Fig. 9.21a Satisfactory and unsatisfactory aileron characteristics (the latter often caused by poor sealing).

The fix significantly altered the size and shape of the fin surfaces, which made the aeroplane instantly recognizable. Although as a flying replica the Halifax might never be revived, there were many interacting related effects discovered during tests and operations which caused catastrophe for young and inexperienced pilots and their crews and these are still worth knowing.

Reference 9.7 records that during propeller feathering tests in 1940 the test pilot reported that overbalance of the rudder limited the speed at which the Halifax could be kept straight with two propellers feathered on the same side. If deflection of the rudder trim tab was reduced to avoid overbalance, the foot force became excessive. The initial attempt to fix the problem was to reduce rudder deflection by 3°, by adjustment of the rudder stops. Although this alleviated the problem it was only partially successful. The stops had design faults which, through high forces and consequent wear, enabled the rudders to reach larger angles than intended instead of them being restricted. In one accident during sideslip tests the top half of a rudder broke away as a result of the force with which overbalance caused the rudder to hit the stop.

Several modifications were tried, including leading-edge slats on the fins. The set-back hinge was cut back, trailing-edge cords were fitted, experiments were carried out with three different sizes of horn balance, to no avail.

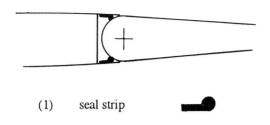

(1)   seal strip

(2) fabric strip doped on with
    with *small* gaps, to clear hinges

(3)   seal strip for
      Frise aileron

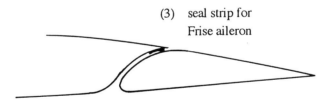

Fig. 9.21b   Three simple fixes to reduce equalization of pressure between upper and lower sides of a control surface.

Bulbous noses were fitted to the leading edge of the balance portion of each rudder. These were arranged to protrude into the airflow at large angles of attack and should have helped the pilot if he managed first to apply a significant amount of corrective rudder. But, clockwise (RH) rotation of the propellers caused the aeroplane to yaw and roll left when the critical port engines were cut. This was exacerbated by the rudder overbalancing and locking over to port, which appears to have nullified the effect of the bulbous modification. The symptoms were rudder oscillation at 170 mph with both port engines cut, followed by over-balance at 160 mph. Control could not be regained until the starboard engines were throttled back and speed increased to 180 mph.

The mixture of leading-edge and balance tab modifications, coupled with 10° bank towards the live engines almost solved the overbalance problem at speeds down

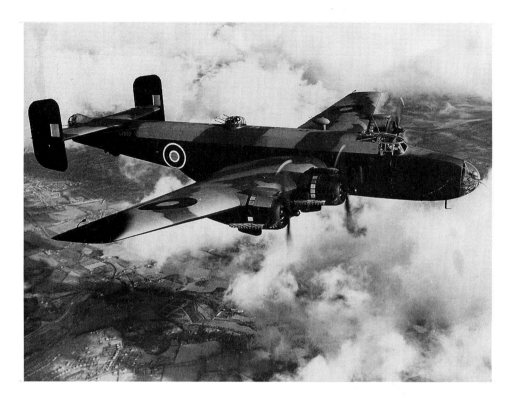

Plate 9–4 Ref: Example 9–9: Handley Page Halifax III heavy bomber of World War II suffered rudder locking with yaw and needed re-design of the vertical tail surfaces to those shown. This shows the later streamlined nose which improved the aerodynamics. (Courtesy of Philip Jarrett Collection)

to 130 mph. But foot loads remained excessive and the rudders were sluggish at low speed.

Further modifications carried out included the fitting of large asbestos tunnel shrouds over the engine exhausts. While shielding the flames from the eyes of prying night-fighters, these caused excessive turbulence over the inboard nacelles, vibration of the wing trailing-edge and a marked loss of directional stability. Take-off performance became critical. Attempts were made to reduce drag, among other things, by sealing engine nacelle and body leaks, and cleaning up the airframe and paint finish. The front and upper turrets were removed and the nose fitted with a streamlined Tollerton Aircraft Services fairing, the efficiency of the radiator cooling flaps was improved. The inboard engine nacelles were dropped to a slightly lower position on the wing.

The most significant modification from our point of view was the reshaping of the fins and their increase in area. Without its fin(s) a conventional aeroplane is directionally unstable. A fin-stall has the effect of removing working fin area. Yaw, when it starts, increases uncontrollably. The hinge moment curves of the Halifax were initially non-linear, leading to a hinge-moment(coefficient)-for-trim curve as shown in Fig. 9.22a and b (from

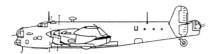

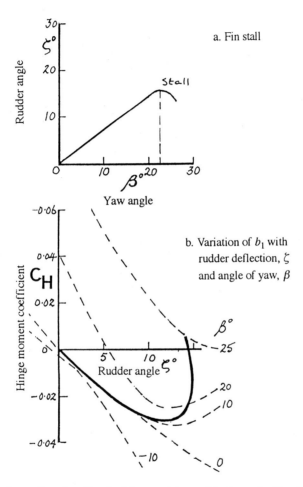

a. Fin stall

b. Variation of b_1 with rudder deflection, ζ and angle of yaw, β

Fig. 9.22 The fin-stall and rudder-locking problem with the early Handley Page Halifax of World War II (Ref. 9.5 and courtesy of H.H.B.M. Thomas). (Side elevation courtesy of Ref. 9.8.)

Ref. 9.5). The curve should be compared with Fig. 9.7a, while substituting yaw angle, β, for angle of attack, α. The fin-stall appeared to take place between 20° and 25° yaw.

The reshaped fins had a lower aspect ratio and, as we saw in Chapter 8, the effect of leading-edge extensions and strakes is to decrease aspect ratio so increasing the angle at which the main surface stalls.

The Halifax in Fig. 9.23 (top) was not alone. Figure 9.23 (bottom) shows the dorsal strake added to the Boeing B 17, the Flying Fortress, also from World War II, to prevent fin-stalling, which is always aggravated by a tendency to rudder locking. As the fin-stall

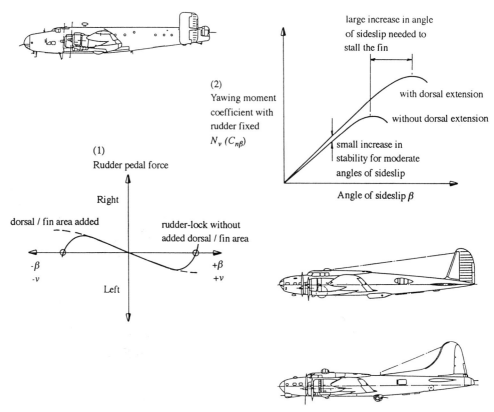

Fig. 9.23 Modified fin and rudder surfaces to prevent fin-stall and rudder-locking. Top: Handley Page Halifax with final 'billiard table' fins and rudders. Bottom: the Boeing B 17 Flying Fortress 1 and 2, which had a large dorsal added for similar reasons, Graph (2) is from Ref. 9.1, and courtesy of H.H.B.M. Thomas; side views are from Refs 9.8 and 9.9, courtesy of C. H. Barnes and Owen Thetford.

sideslip (yaw) angle is approached the sign of the rudder hinge moment changes, making the rudder deflect in the direction which increases the sideslip. The pilot usually diagnoses this as overbalance. If he is pushing left rudder to sideslip to starboard, the first hint of a fin-stall is lightening of the load in his left leg. This is followed by the rudder pedal running away from the foot, to full port deflection. If he cannot push it back again with the opposite foot, the rudder is effectively locked over. The broken lines show the effect of a dorsal or, in the case of the Halifax, the 'billiard table-type' of fin with additional area which solved the problems of overbalance and fin-stalling. A dorsal can increase the fin-stalling angle from less than 20° to 30° or more.

Note: The aspect ratio of a tailplane and elevator behind the wing should be no more than about two-thirds of the wing aspect ratio, to encourage the tail to stall later than the wing. A fin and rudder needs an aspect ratio nearer a half of this, say, a third that of the wing.

Never be afraid to suggest the addition of a dorsal early in a project, in anticipation of trouble later. This is always likely if the rear fuselage is beautifully curved in section, with 'slippery cross-sections' which provide little or no cross-force and damping in a sideslip.

Plate 9–5   The enormous dorsal strake of the Flying Fortress (here a Boeing B 17G in 1943) helped to reduce rudder locking and risk of fin stall with yaw. Ref.: Example 9–9. (Courtesy of Philip Jarrett Collection)

Flutter and 'Buzz' (see control surface profiles in Fig. 9.25)

In the early days when speeds were comparatively slow it was reasonable to cover control surfaces with tough, lightweight fabric. This could be sewn into place (sixteen and also thirty-two stitches to the inch (2.54 cm)) and then doped. With increasing airspeeds fabric-covered surfaces became distorted by increasing pressure differentials. It bulged where it was sucked or pushed out, and became hollow when the internal pressure was less than that outside. This changed trailing-edge angles and the aerodynamic qualities of the control surfaces, and we have already seen that Spitfire ailerons were metal-covered in due course to provide the necessary strength and stiffness and control quality in manoeuvre at higher airspeeds.

Flutter is a high-frequency oscillation of the aerofoil surfaces caused by a struggle between the aerodynamic forces and stiffness of the structure. It is dynamic and arises when a wing, a tailplane or fin are relatively free in bending and torsion.

Inertia and trail characteristics of the control surfaces can modify the overall aerodynamics of the system, so that an initial disturbance is increased, or the tendency to overshoot the original condition is increased. There are three kinds of flutter:

☐ *Torsional-flexural flutter* of the main surfaces in Fig. 9.24a. The motions often have separate period times of torsional and flexural oscillation. They can be excited by flutter of the control surfaces.

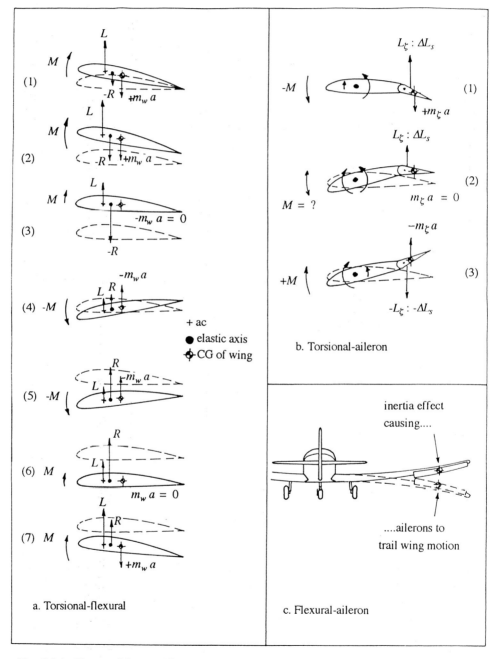

Fig. 9.24   Types of flutter: (fin + rudder) and (tailplane + elevator) modes are similar. In the diagrams, *a* is the applied acceleration.

☐ *Torsional aileron flutter*, shown for a first half-cycle in Fig. 9.24b. It is akin to aileron reversal and can be delayed, or prevented, by mass balancing to bring the CG of the aileron on to, or slightly ahead of, the hinge – or by making the aileron control irreversible. Of the two methods, mass-balancing is usually the lightest.

☐ *Flexural aileron flutter* (see Fig. 9.24c) is similar to torsional aileron flutter, but it is caused by the inertia of the aileron excited by and lagging behind the cyclic flexing and twisting of the wing.

Note: The location of the mass-balance weights is important. The nearer the wing tip they are the smaller and less penalizing they tend to be. In some installations the mass-balance is distributed along the leading edges of the ailerons, within the profile, so that concentration of the mass at any one point does not excite separate torsional flutter of the aileron.

The same comments apply to fins and rudders, tailplanes and elevators. Springiness of the fuselage in bending and torsion can excite flutter in both sets of surfaces.

Control surface profiles, especially those which are fabric-covered or otherwise distort easily, affect flutter characteristics by altering the b_1/b_2 ratio. A comfortable value of which ratio appears to be around -0.70 to -0.75 (Ref. 9.1).

A small and high powered agile aeroplane, with low rudder-fixed stability caused by destabilizing effects of a propeller well ahead of the CG, could suffer from an undamped directional oscillation if given a rudder horn, to make b_1 positive, and a geared tab to reduce the numerical value of $-b_2$. Such an oscillation is called *snaking* and it can become uncontrollable. Being more or less proportional to speed squared, the only way to stop snaking and breaking the aeroplane is to slow down. The cure is to make b_1 zero or negative, and to mass-balance the rudder statically.

The quickest way to reduce b_1 is to attach a length of cord, sometimes referred to as Flettner strips, on each side of the trailing-edge of the rudder (see Fig. 9.19a and b(1)).

Note: Thickening the trailing-edge of a control surface, adding cord or Flettner strips, increases the lift slope, and with it control effectiveness and authority. But make sure that fabric, dope, or other means of adhesion are compatible with the skin to which it is attached.

Example 9–10: Historic Aileron Flutter – Supermarine S4

An historic case of aileron flutter resulted in the loss of the Supermarine S4, the British 1924 Schneider Trophy entrant, a clean monoplane seaplane (Fig. 9.26a). The aeroplane was so streamlined and elegant that a reporter is alleged to have commented:

> 'One cannot help feeling a certain amount of surprise that a British designer has had sufficient imagination to produce such a machine.' (Ref. 9.10)

Although the cantilever wing had been tested to destruction at the RAE, Farnborough, the pilot, Henri Biard, reported slight vibration from the region of the wing tips when setting a new speed record for seaplanes. Subsequently, and before it could compete in the race, the S4 crashed. Biard, who survived, said that flutter was responsible. Nothing appears to have been proven, but significantly all of the monoplanes, British and foreign, designed for the Schneider Trophy thereafter incorporated external wire bracing. By 1931 the Supermarine S6B, outright winner (by non-appearance of other competitors) went on to set a new world speed record, initially it suffered rudder-flutter and finally had rudder and aileron mass

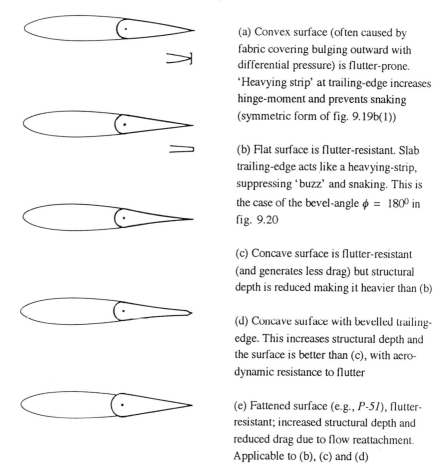

(a) Convex surface (often caused by fabric covering bulging outward with differential pressure) is flutter-prone. 'Heavying strip' at trailing-edge increases hinge-moment and prevents snaking (symmetric form of fig. 9.19b(1))

(b) Flat surface is flutter-resistant. Slab trailing-edge acts like a heavying-strip, suppressing 'buzz' and snaking. This is the case of the bevel-angle $\phi = 180^0$ in fig. 9.20

(c) Concave surface is flutter-resistant (and generates less drag) but structural depth is reduced making it heavier than (b)

(d) Concave surface with bevelled trailing-edge. This increases structural depth and the surface is better than (c), with aero-dynamic resistance to flutter

(e) Fattened surface (e.g., *P-51*), flutter-resistant; increased structural depth and reduced drag due to flow reattachment. Applicable to (b), (c) and (d)

Fig. 9.25   Control surface profiles which affect flutter characteristics.

balances in addition to wire bracing between wings, floats and fuselage of the earlier S5 and S6A.

Figure 9.26b shows the Supermarine S5, designed for the 1927 contest. Although smaller than the S4, it was heavier and more powerful and became the first finale British winner. The extent of the bracing shows how seriously Biard's report was taken by its designer, R. J. Mitchell.

Note: All Schneider Trophy seaplanes were light aircraft by modern definition.

Control-Feel Trickery

Mechanical devices

It is at this point that control and stability merge. Corrective devices inserted in elevator circuits, the *downspring* and *bobweight*, are used primarily to improve longitudinal stability. Like the primitive elastic bungee before them, they are also used to alter stick-force per *g* characteristics, making a small and light aeroplane feel like one that is much heavier. Being part of the control system they are most easily dealt with at this point.

Vanes and *spades* are related aerodynamically and, like control horns, are used to lighten aileron loads.

CENTRING AND DOWNSPRINGS

The springs and their effects are shown in Fig. 9.27a and b and they work in the same ways as primitive bungee elastic. Displacement of the stick is opposed to an extent by the spring, which increases the pull or push force needed to change speed. Thus, it gives an impression of speed-stability which, as we see in the next chapter, is most naturally interpreted by the pilot as an indication of positive longitudinal stability.

BOBWEIGHTS

The bobweight and its effect are shown in Fig. 9.28. Applied *g* alters the moment of the bobweight, which feeds back to the pilot's hand as an apparent change in hinge-moment and, thus, stick-force.

A control that is not mass-balanced, like those of slower flying light aircraft and gliders, has the CG of the control behind the hinge. This makes the elevator droop downwards when the aircraft is at rest on the ground. By pulling the elevator downwards the CG of the control is acting in exactly the same way as a bobweight, by adding an increment, H_{mech} (due to the mechanical arrangement of the system). In flight the increment is added to the hinge-moment, H_η, which is caused aerodynamically. Thus, the total hinge moment may be written as the sum of the increments due to the basic configuration, angle of attack, moment imposed by the CG, deflection of the control surface, and deflection of the tab:

$$H = H_0 + H_\alpha + (H_{mech} + H_\eta) + H_\beta \qquad (9\text{--}14)$$

When manoeuvring the mass of the control – the bobweight – is acted upon by applied *g*, pulling the elevator downwards. The impression, again, is of increased speed-stability and stick-force per *g*. The disadvantage, of course, is that its mass is also acted upon proportionally by negative *g*. So, while bobweights work well in smooth conditions, they can be miserable in turbulence, magnifying changes of stick-force and leading to possible pilot-induced oscillations.

Aerodynamic Damping and Relative Aircraft Density

A phenomenon not mentioned so far is the effect of the relative density of an aircraft upon dynamic response to control at altitude, and therefore upon the dynamic stability. The *relative aircraft density parameter* is non-dimensional, a pure number, and is expressed in two forms. The first is the *longitudinal* relative aircraft density, μ_1, the second is *lateral*, denoted μ_2. It is the second with which we are concerned in this chapter.

The relative aircraft density parameter describes the 'density' of the aircraft relative to the density of the surrounding air. As the air density falls the relative density of the aircraft increases. If the aeroplane is flying at a given EAS at altitude, the forces generated by the control surfaces are the same as they would be at the same EAS lower down. Thus, in the absence of extraneous compressibility effects, for a given control input and rotational inertia the aeroplane responds at a rate which does not change with altitude. But, the TAS is faster at high altitude than it is nearer to the ground, so that the *change in angles of attack* of the wing, stabilizer and other surfaces in contact with the air *are less at high altitude* than they would be at low (for much the same reason as is shown in Fig. 10.4c, in the case of a faster component of propwash). Consequently, *aerodynamic damping forces decrease with altitude*, making response to control, or a disturbance, or an upset, much faster.

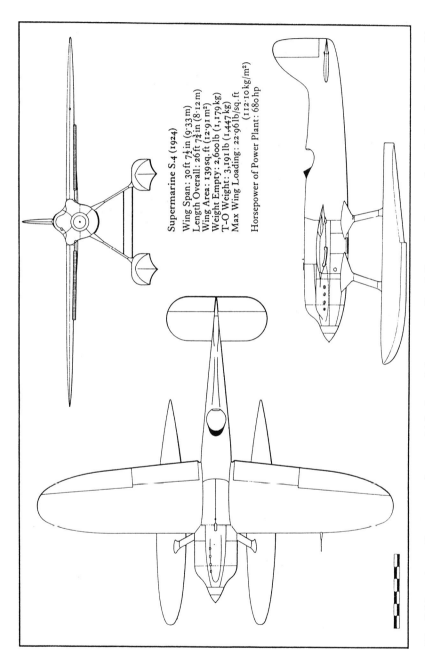

Supermarine S.4 (1924)

Wing Span: 30 ft 7½ in (9·33 m)
Length Overall: 26 ft 7¼ in (8·12 m)
Wing Area: 139 sq. ft (12·91 m²)
Weight Empty: 2,600 lb (1,179 kg)
T-O Weight: 3,191 lb (1,447 kg)
Max Wing Loading: 22·96 lb/sq. ft
(112·10 kg/m²)
Horsepower of Power Plant: 680 hp

Fig. 9.26a A cantilever experiment which altered the course of anti-flutter racing seaplane design for the Schneider Trophy (Ref. 9.10 and courtesy of Robert Hale and Company).

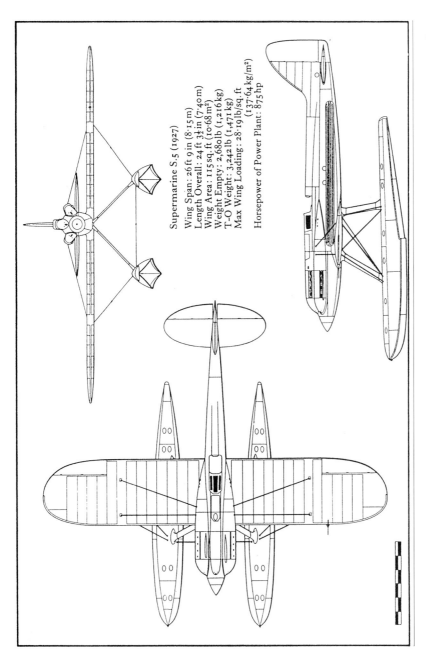

Supermarine S.5 (1927)

Wing Span: 26 ft 9 in (8·15 m)
Length Overall: 24 ft 3½ in (7·40 m)
Wing Area: 115 sq. ft (10·68 m²)
Weight Empty: 2,680 lb (1,216 kg)
T-O Weight: 3,242 lb (1,471 kg)
Max Wing Loading: 28·19 lb/sq. ft
(137·64 kg/m²)
Horsepower of Power Plant: 875 hp

Fig. 9.26b The externally-braced Supermarine S5 showing the lesson had been learnt. This aeroplane was the first finale British winner (Ref. 9.10 and courtesy of Robert Hale and Company).

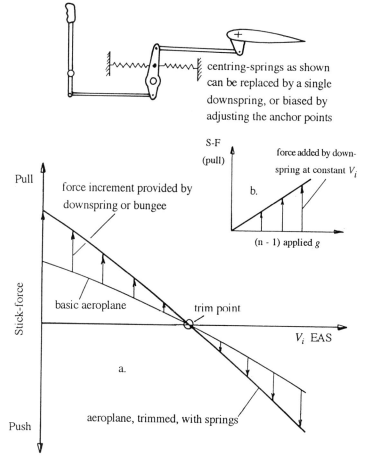

Fig. 9.27 Mechanical trickery in the form of a downspring or bungee (elastic) to increase stick-force per *g* and apparent stick-free stability. When the spring pulls the elevator downwards it increases the static-margin stick-free. The arrangement shown also increases the stick-force to change speed, producing an apparent increase in speed-stability and improving trimability (see Figs 9.17 and 9.18). The strength of the spring (its rate) must not be overdone.

The longitudinal relative aircraft density parameter, μ_1, is directly proportional to the mass of the aircraft, and inversely proportional to the ambient density of the air, the moment arm of the horizontal stabilizer about the CG, and the wing area:

$$\mu_1 = W/[g\rho \, (l_s \, S)] \tag{9–15a}$$

For lateral motions the relative aircraft density in roll uses the semi-span, $b/2$, for a length:

$$\mu_2 = W/[g\rho \, (b/2)S] \tag{9–15b}$$

As altitude is gained, aerodynamic damping decreases in the ratio $(1/\mu_2)$. If the pilot 'winds-up' the aeroplane about its roll axis, reduced aerodynamic damping at altitude makes it harder to stop the motion with the same accuracy as one might low down. Thus aircraft become livelier in response to control as altitude is gained, even though the IAS remains constant.

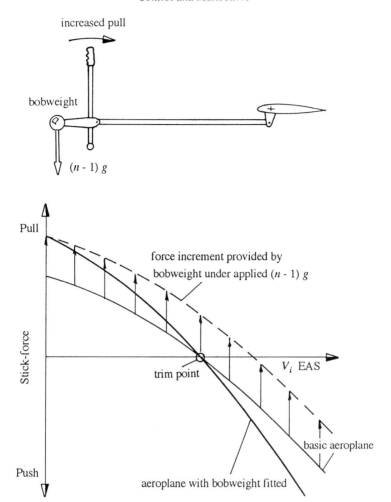

Fig. 9.28 Another mechanical trick is the bobweight which provides a dynamic download on the elevator which increases rapidly with rate of pull (dynamically applied g). In smooth conditions it gives one the impression of increased longitudinal stability. But in turbulence, suffering $\pm(n-1)g$, it can be an unmitigated nuisance. Mounting is also important. Beware of a mechanical arrangement in which the bobweight passes dead-centre (especially top-dead-centre) as the stick is displaced, causing a reversal of stick-force (it has happened).

Example 9–11: Aileron, Swinging Weight – Junkers Ju 88

The World War II Junkers Ju 88 (Fig. 9.29) was versatile and saw service in many forms. It reflected what remains sound technical policy of starting with a good airframe and then adapting it for different roles. The aeroplane featured slotted ailerons which were coupled with flaps, but high aileron forces. These were reduced by an unusual system of swinging weights in the aileron circuit.

Figure 9.29 (Ref. 9.5) shows that the aircraft suffered initially from exceptionally heavy aileron forces in roll. They decreased somewhat with altitude due, possibly, to improved response with increasing relative density, μ_2.

Plate 9–6   Today a Junkers Ju 88 like this would be worth a king's ransom, could one be found. Ref.: Example 9–11. (Courtesy of Philip Jarrett Collection)

Superimposed from Table 9–1 by the same author, are lines showing the limiting control forces for contemporary aircraft. Peripheral forces applied at the rim of a control wheel or yoke have broken lines. They show that even five degrees of aileron needed as much force as the pilot cared to exert with both hands on the wheel. A single handed 20 lbf would only have achieved something like three degrees aileron deflection.

The arrangement of the weights would appear to have eliminated dynamic effects due to turbulence as long as the pilot held the ailerons fixed. When the ailerons were applied centrifugal forces in roll acted on both weights. The weight which had moved farthest ahead of its fulcrum produced a more powerfully favourable moment than the one on the opposite wing, helping the pilot to deflect the ailerons in the required direction.

Aerodynamics versus Inertia (the Square-Cube Law)

Over-control of masses in motion can be a problem with small and very light aeroplanes. Such aircraft (scale replicas, small homebuilts, microlights and ultralights) are dominated by their aerodynamics. Inertia plays the lesser part in their handling qualities. The 'square-cube law' asserts itself, in that aerodynamic forces are dependent upon area, i.e. *length*

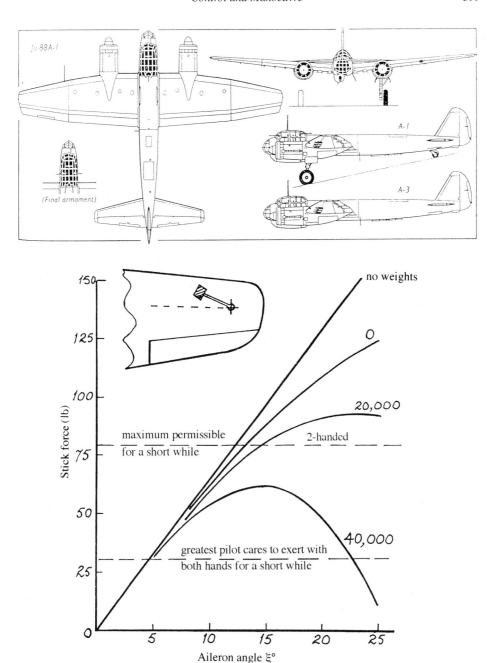

Fig. 9.29 Swinging weights in the aileron system of the World War II Junkers Ju 88 showing the magnitude of the pilot effort required in roll, even with the dynamic assistance of centrifugal force acting on the weights. The reduction in force with altitude is due to μ_2, the relative density effect of reduced damping (see Equation (9–15b)) (Ref. 9.5 by courtesy of H. H. B. M. Thomas; and Ref. 2.21 by courtesy of William Green).

squared, while mass depends upon density of the material and the volume which it occupies, in short, upon *length cubed.*

If we wish to assess response to control – the relative authority of the controls over the inertia characteristics, in terms of the *'aerodynamics/inertias'* – by measuring rates of pitch, roll or yaw, we find that:

$$\text{aerodynamic force/inertia varies as } (\text{length}^2/\text{length}^3) = 1/\text{length} \tag{9–16}$$

So, wishing to compare two aircraft made from substantially the same materials and drawings, but to different scales, Equation (9–16) is better stated as:

$$(\text{aerodynamics/inertias}) \text{ vary as } 1/\text{scale} \tag{9–17}$$

Thus, as aircraft are built smaller, their aerodynamics increasingly dominate their inertias and they become quick-reacting and over-lively. Very small homebuilts can become treacherously fast in response to disturbance in less than ideal flying conditions, as shown in Fig. 9.30 (Ref. 1.5).

Control surfaces of miniature aeroplanes have such small chords that, from the point of view of the pilot, they are more 'pressure-controls' than 'movement-controls', and it is easy to cause pilot-induced oscillations. The BD-5, VariEze, and the German Gyroflug SC01 Speed Canard (Fig. 9.31) have wrist-operated side-sticks. These help to scale down and so reduce otherwise grossly powerful pilot control inputs from a normal centre-stick.

Another problem for pilots of miniature aeroplanes is that of Reynolds number, which decreases directly with reduction in scale. Figure 8.6 showed that at full-scale R_x, the stall could be gentle, while using a surface of much smaller scale could make the stall 'peaky' and sharp-edged. Drag also increased, spoiling lift/drag in smaller scale. The result is that more power/weight is also needed for comparable performance, and as we shall see, power is destabilizing, making tricky handling qualities still more unpleasant.

Example 9–12: BD-5B Simulator

Many years ago I was given the opportunity to test the BD-5B at the factory, using the long runway at Newton, in Kansas. The aeroplane was diminutive, like the nose of a sailplane with wings and tail, and a belt-driven pusher propeller behind. To accustom pilots to the smart handling qualities, Bede had a BD-5 carcass mounted as a flight simulator at the end of an articulated arm, in front of a pick-up truck.

The truck driver responded to the opening of the throttle in the cockpit, by putting his foot down and accelerating along the runway. Apart from the lack of engine noise, one flew the aeroplane from take-off to touch-down, adjusting the flight path up, down and side-to-side, with aileron, elevator and rudder. It was possible to achieve three to five take-offs and landings in the length of the runway. The experience was invaluable, as the quick reactions of the simulator were similar to those of the real aeroplane.

Figure 9.32a shows variation in rolling, pitching and yawing moments of inertia, A, B and C, for a number of single and twin-engined aeroplanes up to about 22 000 lb (10 000 kg). They illustrate the way in which inertia grows faster than weight. They are indicators of the flywheel characteristics of aircraft, and are proportional to radius of gyration about each axis, squared.

If the mass of the aeroplane is:

$$m = W/g$$

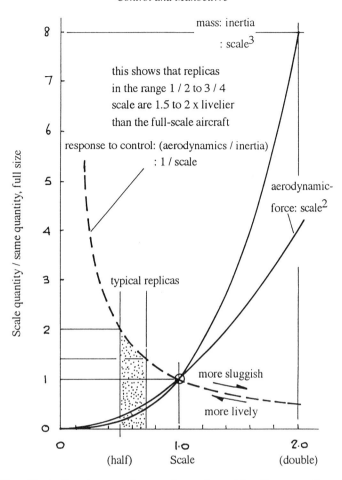

Fig. 9.30 The effect of scale upon the aerodynamic and inertia properties of aeroplanes. Those which are small in scale, or have very low wing and power loadings, are likely to be found on the left (livelier) side of scale = 1.0 (full size) (Ref. 1.5).

then:

$$A = I_{xx} = m\,k_{xx}^2 \quad \text{in roll} \tag{9–18a}$$

$$B = I_{yy} = m\,k_{yy}^2 \quad \text{in pitch} \tag{9–18b}$$

$$C = I_{zz} = m\,k_{zz}^2 \quad \text{in yaw} \tag{9–18c}$$
$$\leq A + B \tag{9–18d}$$

In the equations k is the radius of gyration of an imaginary 'equivalent flywheel', rotating about each axis, as shown in Fig. 12.2a. The larger the moment of inertia the greater the effort on the part of the controls to wind-up, and then to wind-down an aeroplane, and stop it rotating.

An aeroplane does not have to be designed for cruising to be more sedate than another, as the following quotation shows while describing the effects of the differences in inertia distribution between two fighting scouts: the long-engined SE5 and the more compact Sopwith Camel in World War I:

Plate 9–7   The clean Gyroflug SC01B-160/UK Speed Canard has a side-stick and flying controls which are more 'pressure-controls' than 'movement-controls'. Canard aeroplanes with manual controls are more sedate than agile and best-suited to cruising. (Courtesy of IFT Cranfield)

'In the SE 5 the long stationary engine, the tanks, and the pilot's cockpit behind them meant that the load was spread over more than half the length of the body. It was this that made the machine impossible to turn sharply and unwieldy in aerial combat. The Camel, whose engine was a flat rotary, and whose tanks and pilot were all packed close together, had its weight concentrated, and was thus far lighter and handier in the air. (The best trick flying during the war was put up on Camels by pilots such as Armstrong or Banks.) The machine was not particularly fast on the level, but it climbed well and could best any other scout in a fight. Like all Sopwith productions, it was a bit on the light side; but for actual flying, next to the Triplane it took first place with me.'   (Ref. 1.8)

Microlight aircraft

There are three kinds of microlight: weight-shift, three-axis and paraglider – *parapente* (French). Three-axis are almost conventional aeroplanes in configuration. Weight-shift are different. The parapente is a form of navigable parachute. Engines are invariably two-stroke, which can be temperamental, liking either to run flat-out, or to idle, but not both. They are high-revving and noisy, unless properly (and more expensively) silenced. Yet, there have been improvements in noise suppression by microlight builders which are well in advance of those undertaken by designers and constructors of conventional aircraft.

Weight-shift microlights

One of the most revolutionary developments in aviation has been the weight-shift microlight

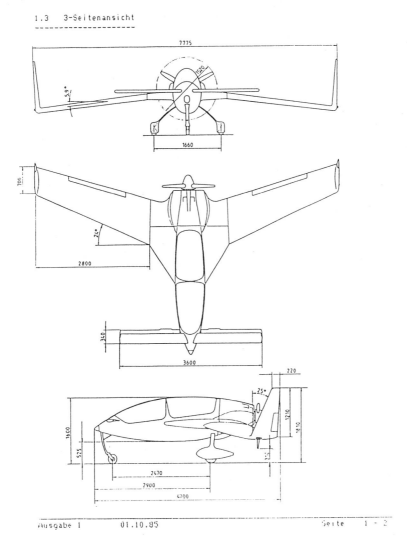

Fig. 9.31 Gyroflug SC01B Speed Canard: light and lively, with a side-stick (drawing reproduced by courtesy of Gyroflug mbh).

aeroplane, developed from the original delta-shaped membrane 'parawing' concept of Francis and Gertrude Rogallo in the late 1940s. In due course the hang-gliding movement developed as a bid for personal freedom. The weight-shift microlight arrived when tricycle-wheeled units were hung beneath wings, with small, high-revving chainsaw and other engines behind, driving pusher propellers. The trike is suspended from a single articulated attachment point on the central keel. Both trike and wing are free to tilt longitudinally and laterally. Control of the tilt is by means of a triangular A-frame, made from tube, with a cross-bar

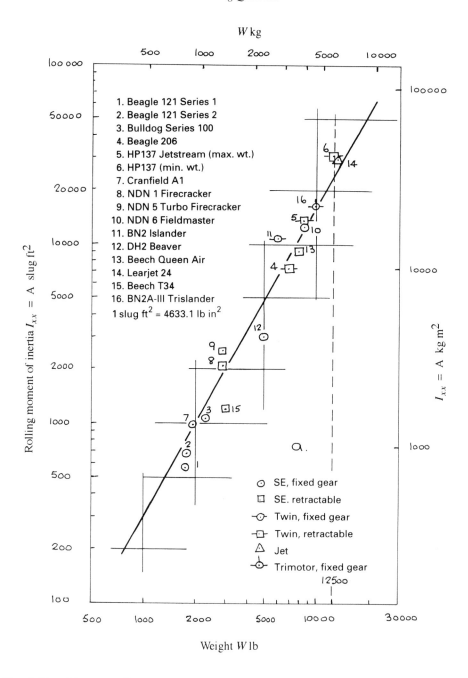

Fig. 9.32a Moment of inertia in roll, $I_{xx} = A$ (Equation (9–18a)), of mixed single and twin-engined aeroplanes including commuters, of MTOW below 10 000 kg (22 000 lb) (Ref. 1.5).

at the bottom, which is rigidly braced to the wing. Today the delta shape has become less acute, trike units are now more streamlined, and the aircraft are often described as *flexwings*.

How control is actually achieved is arguable and apparently the reverse of a three-axis system. With the weight shift machine, pushing forward on the control-bar of the A-frame

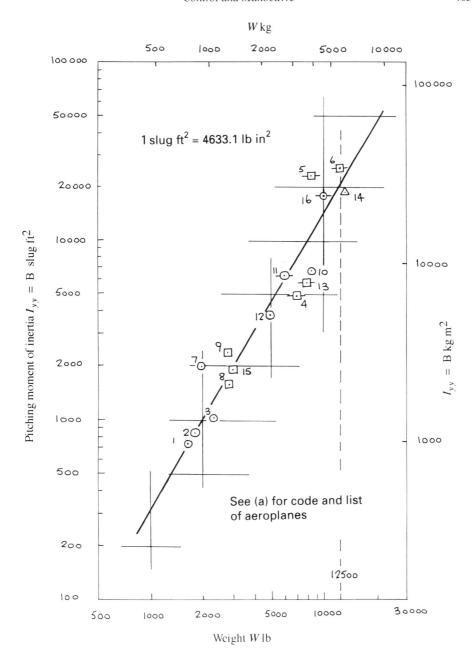

Fig. 9.32b Moment of inertia in pitch, I_{yy} = B (Equation (9–18b)) of the same group of aeroplanes as in Fig. 9.32a (Ref. 1.5).

increases angle of attack and reduces airspeed. Pulling back has the reverse effect. Push the bar to the left to turn to starboard, and to the right to turn port. One view is that the pilot, pulling and pushing himself relative to the wing, shifts the CG, which tilts the wing nose-up to reduce speed, nose-down to increase it, and side to side, to turn. The other view is that he uses the A-frame to tilt the wing, pointing the lifting cross-force resultant in the direction that he wants to go.

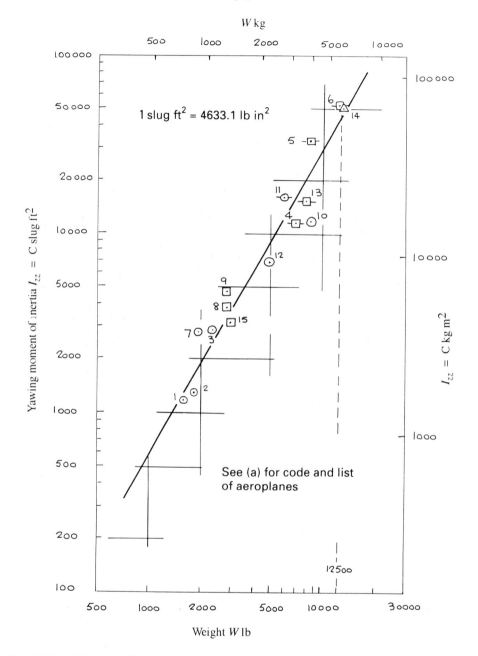

Fig. 9.32c Moment of inertia in yaw, $I_{zz} = C \leq A + B$ (Equations (9–18c) and (9–18d)) of the same group of aeroplanes in Fig. 9.32a (Ref. 1.5).

It is probably a mixture of both, but the degree of the one as against the other depends upon the relative masses of the trike-plus-fuel-plus-crew and that of the wing.

The modern wing is rigged with a form of conical camber which washes out towards the tips, and a reflex to the trailing-edge of the keel member. In this way the necessary vee is formed between the higher incidence portion of the wing ahead of the CG, and the parts –

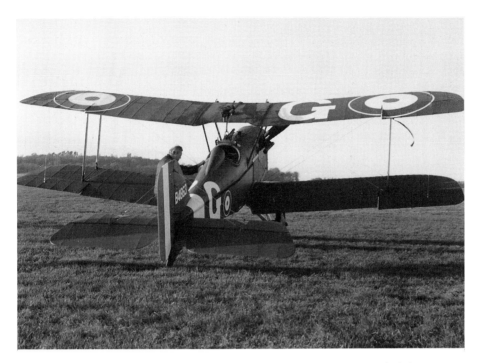

Plate 9–8 Replica SE5A with a long, stationary water-cooled, in-line vee engine and therefore more sedate inertia and radius of gyration in pitch and yaw than ....
Note: The engine is believed to be an original Wolseley Viper. (Author)

including the tips – which lie behind (see Fig. 3.17, particularly b and c, and Fig. 10.6, for O'Gorman's explanation of why this works).

Weight-shift microlights, when flown properly, are safe. When trimmed they can be flown hands-off. The trike can be moved fore and aft on the keel. If too far forward it tends to dive and may 'tuck', i.e. pitch nose-down and tumble. The pilot must fly hands-on for pitching and rolling moments to be transmitted to the wing.

S-TURN ACCIDENTS

In recent years there have been a number of broadly similar accidents with current flex-wing microlights (and hang-gliders) arising from what might best be described as steep S-turns or 'Dutch-rolling' at high speed. At best the motion may take the form of mild snaking with roll. At worst the result can be yawing and rolling to the point of a divergent departure into a wing-over. The Dutch-roll, discussed in Chapter 11, is a poor lateral-directional handling quality resulting from strong roll with sideslip and weak directional stability.

Weight-shift flex-wings are tailless and rely upon sweep-back and washout for favourable roll with sideslip, directional and longitudinal stability. They yaw and roll naturally into the relative mind. Each oscillation is accompanied by an oscillatory change in direction of the relative wind, first from one side, then from the other. At low speeds and large angles of attack dihedral-effect is enhanced by sweep (see Fig. 11.4a). As speed is increased the

Plate 9–9   ... this original Clerget rotary engine replica of the Sopwith
Camel which was off in another direction each time one blinked ....
Ref.: Example 10–4. (Author)

angle of attack flattens and whatever 'weathercock stability' there was is diminished as the
keel surface area aft rises into the main wake of the wing and A-frame. Roll with sideslip
increasingly dominates the motion.

Lack of lateral control of the roll with sideslip is at the root of the problem. I have measured a maximum lateral force on the cross-bar of the A-frame without my grip slipping, as
a modest 20 lbf (9 daN) – which nevertheless seems heavy because of inadequate human
ergonomic design in the lateral mode. The period time of swing of the flex-wing, rotating
about an instantaneous centre some way above it, is close to the rate at which the average
pilot can apply such lateral loads to the bar. Therefore, just as a miniature battery is able to
run a watch or clock for a long time, by giving a tiny pulse to the mechanism at the critical
moment, so too may the pilot find that a swing develops with the microlight acting like a
pendulum. Indeed he may use much smaller lateral forces than those needed to start the
motion. But, from this point onwards sideslip intrudes and the rolling characteristics of the
wing take over, leaving the pilot increasingly out of phase and unable to prevent a rolling
departure and the trike tumbling into the wing.

This is raw pilot-opinion – for what it is worth in the absence of anything better as this is
written. The remedy is to prevent sideslip developing and becoming oscillatory.

Three-axis microlights

The three-axis microlight has aileron, elevator and rudder controls. Though marginally
heavier than the flexwing, its more efficiently shaped wings make it more skittish in turbulence. Climb performance may also become negative, as I discovered in curl-down from

Plate 9–10 ... or the Sopwith Triplane. This replica has a fixed radial engine of much later vintage and the certification pilot would regard it as much more of a look-alike with a not exactly to scale nose. Originally this replica had the CG much too far aft. Ref.: Example 10–5. (Courtesy of John Lewis)

trees some distance away upwind. Because structures are elastic and often floppy, controls require full-scale deflection at times, and responses can be sluggish.

Roll control by ailerons may be imprecise, because of susceptibility to aileron reversal by twisting the wing in the opposite direction. This makes the use of spoilers attractive, because of their smaller torsional effects. Unfortunately spoilers are less effective than ailerons at low airspeeds. Also as we have seen, they fail to enhance lift as the flap-like aileron does. They work simply, by dumping lift, but they have the advantage of countering adverse yaw without resort to precision devices, like the Frise aileron.

Parapente (paraglider wing) microlights

The *parapente* (French) is the name given to a high aspect ratio paraglider wing: a parachute, with a double skin, stitched at the trailing-edge, which has open leading-edges forming intake ports. The skins are held in shape by aerofoil-section-shaped gussets. These run fore and aft, and form cells which are linked laterally internally, through holes, held in shape by fine netting. Ram air, entering at the leading edges, equalizes pressure in the cells and so shapes the aerofoil sections.

Plate 9–11   Typical flexwing microlight, the Hornet Dual Trainer
Raven controlled by weight-shift. (Courtesy of Hornet Microlights)

The paraglider is probably the simplest and potentially the cheapest of all flying ma-chines. The pilot hangs beneath it and controls his flight by pulling on one set or other of the cords which connect to the trailing-edge of each wing-tip. Hauling on the right-hand cords pulls downwards on the starboard trailing-edge, increasing the drag of that half of the wing, and the aircraft turns to starboard. Pulling both sets of cords together lowers the trailing-edges of both wings, increasing their camber and lift, which enables the pilot to flare and touch down on his feet at low forward speed. The paraglider can be soared in rising air over great distances.

The microlight parapente (Example 13–3 describes one version) has a weight-shift trike unit, with a two-stroke pusher engine, suspended beneath the wing. The trike may be single seat or a two seater, depending upon the wing area. Take off and climb are with power, using a foot-operated throttle. Steering is by cords tied to handles, gripped in each hand. Climb, level and descent are adjusted by throttle-setting (like Mignet's Flying-Flea). If the engine stops, the wing descends as a steerable parachute.

Testing Microlights

In the UK microlights are tested to BCAR Section S – *Small Light Aeroplanes* (Ref. 9.12). Required flying qualities are listed in Table 9–7. The British Microlight Aircraft Associa-tion, BMAA, uses a flight test schedule for investigating the flying qualities of microlights against the requirements of BCAR Section S, and a second for Issue/Revalidation of a Permit to Fly for Series Production Aeroplanes. Consideration has been given to S-turn accidents. The requirement for control and manoeuvrability now includes the ability to stop

Plate 9–12 (a) PowerChute parawing (parapente) microlight. (Courtesy of PowerChute)
(b) PowerChute Raider with wing deployed. Ref.: Example 13–3. (Author)

a banked turn at 60° and reverse it to 60° bank in the opposite direction.

If any pilot contemplates weight-shift flight testing he should:

☐ Contact the CAA and BMAA for advice (the CAA will recommend the BMAA anyway).

☐ Remember what was said above and *never assume that a weight-shift (or other) microlight is safe* because it is simple. They have as bad a bite as their larger sisters.

☐ Get yourself BRIEFED AND CHECKED OUT before you are bitten.

TABLE 9–7

BCAR Section S – Flying qualities

Weight and CG: determination of limits; empty weight and corresponding CG.

Performance: stalling speeds; take-off distance to 15 m in zero wind at not $<1.3\ V_{S1}$ or $V_{S1} + 10$ knot; climb from ground to 1000 ft; rate of descent; landing distance on short grass in zero wind at not $<1.3\ V_{S0}$ or $V_{S0} + 10$ knot at scheduled power setting.

Controllability and manoeuvrability: must be safe in all phases of flight up to 60° angle of bank, with ability to stop and reverse turn to 60° in the opposite direction, between $1.6\ V_{S1}$ and V_A, achieving smooth transitions with acceptable pilot effort; longitudinal control at not $<1.3\ V_{S1}$ to V_{DF}; lateral and directional control, from 30° to 30° bank in <5 s at $1.3\ V_{S1}$ and V_{NE}; pitch control force not <7 daN to proof load; trimable from $1.3\ V_{S1}$ to $2.0\ V_{S1}$ at all powers and extreme CGs.

Stability: static longitudinal in the climb, cruise and approach (approach power and throttle-CLOSED); lateral and directional (with crossed controls and sideslip corresponding with deflection of lateral control); dynamic stability, damped short-period at all powers, with stick-fixed and -free; combined lateral and directional oscillations between V_{S1} and V_{DF} must be damped with primary controls fixed and free.

Stall quality: at fwd and aft CG limits and extreme weights (gross and light), with speed reduction of 1 knot/s to determine V_{S1} and V_{S0}, wings level; turning stalls, 30° bank; behaviour in recovery.

Spinning: Authority must be consulted when spinning is intended.

Ground handling: directional stability and control; control in crosswinds on take-off and landing.

Miscellaneous: freedom from excessive vibration to V_{DF}; no buffeting severe enough to interfere with control or fatigue crew, or result in structural damage during normal flight.

Warning: It is dangerous to attempt to flight test a flexwing, from the erect stall to the stall inverted. A suitable dynamic test rig is needed. In the UK the Civil Aviation Authority can advise, and also the British Microlight Aircraft Association.

Example 9–13: Testing a Three-Axis Microlight with Spoilers

Some years ago it was necessary to test a dainty parasol three-axis microlight, with a slightly swept wing, spoilers for roll control, and a two-stroke engine. The spoilers were held IN by bungee elastic, and were operated by cords running in the open air to each wing, from an extension of the bottom of the stick. Moving the stick to the right or left pulled on the right or left cord and operated the spoiler in the top surface of the appropriate wing. Centralizing the stick, or moving it the other way, the bungee pulled the spoiler back into its housing.

At the end of the run to V_{NE} at 3000 ft (about 1000 m) two things happened: the engine stopped and a gust struck, tipping the aircraft on to its starboard wing-tip. An attempt to pick up the wing with lateral control failed, for it appeared that equal drag on both cords had deployed both spoilers. Fortunately, left rudder and the dihedral effect of sweep-back, coupled with pendulous stability, tilted the aeroplane back on to an even-keel. The engine could not be restarted and the flight ended along (not across) a furrowed field, head-on to an irate farmer with tractor, hauling cabbages. Bearing down upon the microlight he bellowed: '*Move the bloody thing!*', but accepted the explanation, in good time, with grace.

In the next chapter we discuss longitudinal stability, needed originally to enable the pilot to release the controls and attend to something else – such as writing a note and dropping it to people on the ground (an early operational requirement) – without risk of the aeroplane running amok.

References and Bibliography

9.1 Morgan, M. B. and Thomas, H. H. B. M., *Control Surface Design in Theory and in Practice*. London: the Royal Aeronautical Society, August 1945

9.2 Andrews, C. F. and Morgan, E. B., *Supermarine Aircraft Since 1914*. London: Putnam, 1981

9.3 *JAR-23, Post-Consultation*. London: the Civil Aviation Authority, 25 March 1993

9.4 *JAR-23, Paper 26/6, Structural Requirements for Commuter Category Aeroplanes*. London: the Civil Aviation Authority, January 1992

9.5 Thomas, H. H. B. M., Aircraft controls, *The Historic Aircraft Association, Third Symposium at the Shuttleworth Collection*, Old Warden Aerodrome, Bedfordshire, 1982

9.6 Used aircraft guide: Cessna 172/Skyhawk, *Aviation Consumer*, 15 May 1993, Vol. XXIII, No. 10. Greenwich, Connecticut 06836-2626

9.7 Merrick, K. A., *The Handley Page Halifax*. Bourne End, Bucks, Aston Publications, 1990

9.8 Barnes, C. H., *Handley Page Aircraft Since 1907*. London: Putnam, 1976

9.9 Thetford, O., *Aircraft of the Royal Air Force Since 1918*. London: Putnam, 1962

9.10 *ARC Report for the Year 1928–29*. London: Aeronautical Research Council

9.11 Mondey, D., *The Schneider Trophy*. London: Robert Hale, 1975

9.12 CAP 482, BCAR Section S, *Small Light Aeroplanes*, London: Civil Aviation Authority

CHAPTER 10

Longitudinal Stability

'Stability and control and its interference with the human being is a people-thing rather than a machine-thing ... it's the way the human pilot interprets the aeroplane's effect upon him .... Squirrelly aeroplanes are better than monstrously overstable ones.'
Hugh Scanlan, past editor of *Shell Aviation News*

'I have however learnt from those who flew this machine that the fatigue of three or four hours balancing was very great indeed, and the Wrights at present fit a tail.'
Lt Col. Mervyn O'Gorman, Superintendant, Royal Aircraft Factory, Farnborough (1909–16)   (Ref. 10.1)

In Chapter 3 the bones of stability were bared. It is a difficult subject at the best of times, not just because of the mathematics. Today stability is a requirement, a principle set in stone. From the point of view of airworthiness and operational regulation, aeroplanes must be stable. It is irrational to suggest otherwise. Indeed, to do so is to encourage the raising of eyebrows and the questioning of sanity. Other kinds of aeroplanes – those not involved in public transport operations and flight for hire and reward by the innocent fare-paying passenger – must also be stable. Granted, the requirements may be slightly more relaxed in their observance, but the grandfather clauses are substantially the same in FARs Part 23, 25 and JARs-23, and -25:

'The aeroplane must be longitudinally, directionally and laterally stable .... In addition the aeroplane must show suitable stability and control "feel" (static stability) in any condition normally encountered in service, if flight tests show it is necessary for safe operation.'
(Refs: FAR 23.171 and JAR 23.173)

In this chapter we concentrate upon longitudinal static and dynamic stability. Tests for static stability (basic free-flight model aeroplane stability) are relatively simple and by no means exhaustive, as long as no marginal conditions are discovered. There must be sufficient change in control force, when displaced from the trimmed condition, to produce suitable feel for safe operation. Concentration upon limited aspects of stability by regulatory authorities has brought criticism. The following example makes the point.

Example 10–1: Implied Criticism of an Airworthiness Requirement

We saw in the last chapter (Example 9–11, Ju 88) how response to control, or to a disturbance of some kind, depended upon the *relative aircraft density*. The longitudinal relative aircraft density, μ_1, was given in Equation (9–15a). When this is large then pitching oscillations of the mass of the aircraft (W/g) are less well damped by the air, density ρ, than when the aeroplane is small. It is upon μ_1 that the longitudinal dynamic response and dynamic stability depend. When disturbed in pitch an aeroplane with a high value of μ_1 needs more powerful aerodynamic damping to slow it down again than one which has a smaller mass.

The following quotation is from the US *Aviation Consumer*:

> '… the Piper Cheyenne, despite its wildly divergent instability, did in fact meet the letter of Part 23 stability regulations, which require positive static stability (i.e. a tendency to return to trim speed), but not positive dynamic stability (i.e. a damped phugoid).'
> Editor (Ref. 10.2)

It points to a deficiency in the airworthiness requirements as well as in the aeroplane.

Part 23 (and JAR-23) calls for: '23.181(*a*) Any **short period** oscillation … *must be heavily damped …*'. The requirements then go on to specify the conditions. A phugoid is a long-period oscillation, not short. It would seem reasonable to urge such a requirement to be reinterpreted and restated as: '**Any short or long period** *oscillation … must be heavily damped …*' (we deal with short and long period (phugoid) oscillations later in this chapter).

Having said that, the FAA *Engineering Flight Test Guide for Small Aeroplanes*, Ref. 10.3, by mentioning both would appear to imply that both need to be checked by the test pilot, while making a clear distinction between them. Mil Specs, on the other hand, particularize minimum levels of stability and control qualities. These were reflected in the Cooper Harper Handling Qualities Rating Scale in Table 5–5. They break down broadly into:

☐ *Level 1:* Flying qualities which are clearly adequate for the flight phase of the mission.
☐ *Level 2:* Flying qualities which are adequate to accomplish the flight phase of the mission, while accepting some increase in pilot work load, or degradation in mission effectiveness, or both.
☐ *Level 3:* Flying qualities such that the aeroplane can be controlled safely, but the pilot workload may be so high, or the mission effectiveness so degraded, or both, that the non-terminal phases which demand utmost precision must be abandoned:

   (1) If portions of a mission cannot be completed, then it must be possible to terminate them safely.
   (2) Phases which do not involve precision tracking, but which involve launch and recovery of the aircraft, and some functions like in-flight refuelling as a tanker, can be completed.

Ironically, in spite of published airworthiness requirements one often found private owners of homebuilt aeroplanes sometimes questioning rigid compliance with, e.g., Part 23 criteria. There is a preference for less – or even neutral – longitudinal static stability in certain flight phases, and this has been a common fault among a number of homebuilts flown at Oshkosh. Weakest longitudinal stability is found at aft CG, where there is less trim drag from the stabilizer, and this bestows improved lift/drag upon which high performance depends. Sailplane pilots too use aft CGs as a matter of course, ballasting their aircraft to achieve the flattest and farthest glide.

Longitudinal Stability Considerations

Static stability is most easily understood by the pilot as speed-stability, encountered in the last chapter, when discussing springs and bobweights in elevator control circuits. This means that when trimmed, a push on the stick is needed to increase speed and a pull to decrease it. If a disturbance causes the nose to drop without the elevator angle being altered, the speed increases and, if the aeroplane is longitudinally stable, the nose will rise and the increase in airspeed will then decay. A loss of airspeed causes the nose to fall and speed to increase.

We saw in Chapter 3, Fig. 3.15, that the resultant of the aerodynamic forces on an aircraft acts at the neutral point between wing and tail, foreplane and rearplane. If the CG is arranged to lie at the neutral point, NP, then static longitudinal stability is neutral – in theory anyway, as long as the airframe is rigid and propwash effects remain constant with airspeed. The magnitude of the longitudinal static stability is determined by the length of the CG (static) margin, measured between the centre of gravity and the neutral point. For a wholly rigid aircraft that margin is the *CG margin*. But for a real – flexible and floppy – aeroplane it is called the *static margin*.

For static stability the CG must lie ahead of the NP, and this is true for any planform configuration: tail first, tail last, and tailless. The CG range available is dependent upon the size and location of the rearplane surface, and upon the local pressure recovery, q_s.

Dynamic longitudinal stability characteristics also depend upon planform configuration and location of the CG. Their importance lies in the possibility of oscillatory divergence in angle of attack and airspeed when disturbed, like that mentioned above in the case of the Cheyenne.

Measurement of Static Stability by Static Margin

In Chapter 3, Equations (3–37a) and (3–37b), we saw that stick deflection and control-force needed to change speed provide measures of stick-movement and stick-force stability, as long as there are no springs, bobweights or other tricks incorporated in the control system. When an aircraft structure is effectively rigid we talk of CG margins, stick-fixed and -free H_n and H'_n respectively. These are the distances between the CG and the neutral points NP and NP', stick-fixed and -free. When the structure is floppy the CG margins stick-fixed and -free become K_n and K'_n:

$$\begin{aligned}\textit{stick-travel } \textit{to change speed} &= \text{stick-fixed static margin, } K_n \\ &= \textit{stick-fixed stability} \\ &= \psi_1 \text{ (CG margin), } H_n \quad \text{from: (3–39a)}\end{aligned} \tag{3–37a}$$

$$\begin{aligned}\textit{stick-force } \textit{to change speed} &= \text{stick-free static margin, } K'_n \\ &= \textit{stick-free stability} \\ &= \chi_1 \text{ (CG margin), } H'_n \quad \text{from: (3–39b)}\end{aligned} \tag{3–37b}$$

The tendency for the stick to move and push or pull against a pilot's grip with an increase or decrease in airspeed is an indication of positive longitudinal static stability, because the pilot must then push against the stick to hold a higher speed and pull back on it to hold one that is lower. Moreover, as we saw in the last chapter (Fig. 9.2) stick deflection and force are linear when unaffected by aeroelasticity. In practice, instead of being straight lines they are curved, because deflection of the structure alters increasingly control angles and angles of attack of otherwise fixed surfaces. The greater the curvature, the more elastic is the structure.

Figure 10.1 (Ref. 1.5) shows the origin of the changes in pressure distribution (suctions only for simplicity) which are the origin of stable and unstable moments in pitch. When the centre of incremental pressure moves rearwards relative to the CG, only then is static longitudinal stability present (see insets (1), (2) and (3)).

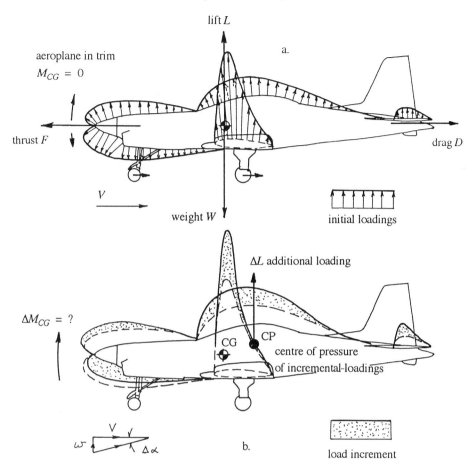

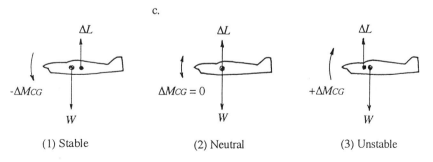

Fig. 10.1 The origin of stable and unstable moments when disturbed.

Figure 10.2a indicates the way in which a stable curve of pitching moment coefficient, $-C_{MCG}$, changes with angle of attack, lift coefficient and airspeed, as compared with neutral and unstable slopes in the inset (1). Figure 10.2b shows the wing, body and tail contributions to the curve like that in the preceding drawing, for one particular aircraft. The

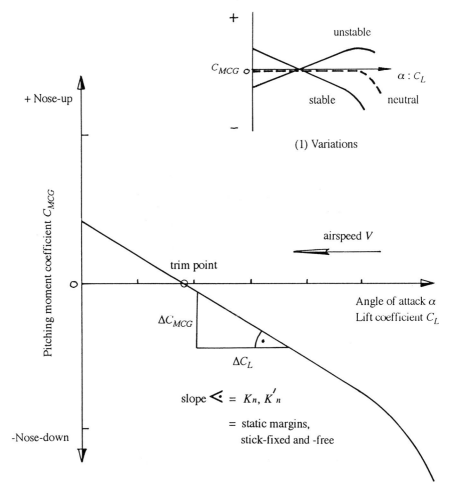

Fig. 10.2a   Static longitudinal stability is indicated by the slope of the $\Delta C_{MCG}/\Delta C_{L}$ curve (see Equation (3–39) and for C_{MCG}, Equation (3–29)).

negative, downward, slope of a stable curve confirms that a stable pitching moment is nose-up with increasing speed (decreasing angle of attack) and nose-down with decreasing speed and increasing angle of attack. The steep hook downwards at the right-hand end of the curve represents an ideal state of affairs, in that the nose pitches downwards at the stall to regain flying speed, as we saw in Chapter 8. Were the hook to turn upwards, showing pitch-up and instability, this would be unacceptable.

Static longitudinal stability is less important for agile aeroplanes than it is for those intended to cruise over long distances for many hours at a stretch. Indeed, there was research carried out after World War I when a Farnborough test pilot, then Squadron Leader (later Air Marshal Sir), Roderick Hill concluded (in the case of a Sopwith Camel with the wings moved backwards by 7 in (17.78 cm) to improve longitudinal stability) that:

'Stability destroys the liveliness of the control, and to a certain extent the ease of landing, and it is against this change of feel that the main criticism is directed.'   (Ref. 10.4)

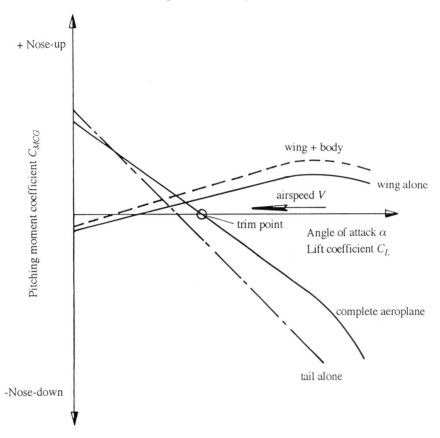

Fig. 10.2b   The way in which parts of the aeroplane contribute to static longitudinal stability. Each increment of $\pm C_{MCG}$ when added to the rest makes up the whole.

Figure 10.2 shows the way in which parts of the aeroplane contribute to the static longitudinal stability. Figure 10.3, which is after Gates (Ref. 10.5), goes further by showing the effects upon the static margin of body and nacelles, power (propwash) and control surface trail characteristics stick-free. Thus, the latter figure links control and stability, which is the point at which we left matters in the last chapter.

Figure 10.3 reveals the contribution of the tail to static longitudinal stability. This is proportional to its effectiveness and therefore, to tail volume and volume coefficient (Equation (3–52), working values of which were given in Table 8–1) and, of course, upon tail efficiency, η_s (Equation (3–50)).

Agile aeroplanes should have small force variations with change of speed, because speed changes are likely to be rapid and constant retrimming must be avoided. A sedate cruising aeroplane needs quite large variations in stick-force with airspeed to alert the pilot, especially when flying on instruments without an autopilot.

Why Power is Usually Destabilizing

The same Fig. 10.3a(2) shows that the effect of power is usually destabilizing. Such effects were mentioned early in Chapter 8, and Fig. 10.4, taken from Ref. 1.5, shows

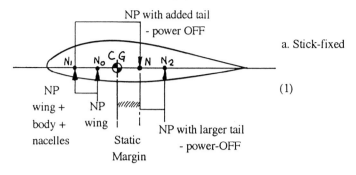

a. Stick-fixed

(1)

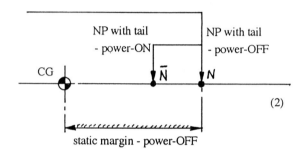

(2)

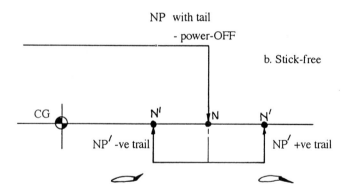

b. Stick-free

Fig. 10.3   Effects upon Static Margin (SM), hence, longitudinal stability, of tail size, power (propwash) and control surface configuration (stick-free) (after Gates (Ref. 10.5)).

them diagramatically. All six apply to propeller units. All except the third and sixth are also present to an extent with the rotating cores and blades of turbojet and fan engines. Remember too that jet engine intakes generate lift so, while talking about propeller units, the distinctions between the effects of propellers and jets are not simply black and white.

Countering 'P'-effects

Adverse power effects are awkward to calculate in advance. When their magnitudes are revealed it is usually too late to introduce major redesign or modification. The simplest of all forms of relief for the pilot of a small aircraft is to bias a control surface aerodynamically

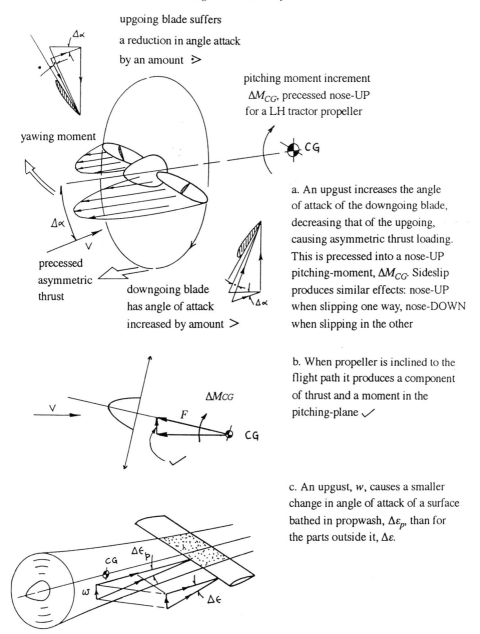

upgoing blade suffers a reduction in angle attack by an amount >

pitching moment increment ΔM_{CG}, precessed nose-UP for a LH tractor propeller

yawing moment

$\Delta\alpha$

precessed asymmetric thrust

downgoing blade has angle of attack increased by amount >

CG

a. An upgust increases the angle of attack of the downgoing blade, decreasing that of the upgoing, causing asymmetric thrust loading. This is precessed into a nose-UP pitching-moment, ΔM_{CG}. Sideslip produces similar effects: nose-UP when slipping one way, nose-DOWN when slipping in the other

b. When propeller is inclined to the flight path it produces a component of thrust and a moment in the pitching-plane ✓

ΔM_{CG}

F

CG

c. An upgust, *w*, causes a smaller change in angle of attack of a surface bathed in propwash, $\Delta\varepsilon_p$, than for the parts outside it, $\Delta\varepsilon$.

Fig. 10.4 Some of the destabilizing effects of propellers. The tractor propeller as shown is more adverse in this respect than the pusher.

by sticking a length of cord on one side or the other of a rudder, aileron or elevator, as shown in Fig. 9.19a.

A better solution which is less vulnerable to alteration, can usually be provided by *slightly* tilting the thrust line of the engine in the required direction, as shown in Fig. 10.5b(1) and (2). If the tilt is excessive then the propeller axis will no longer be aligned with the airflow, causing out of balance air loads, vibration and loss of propulsive efficiency.

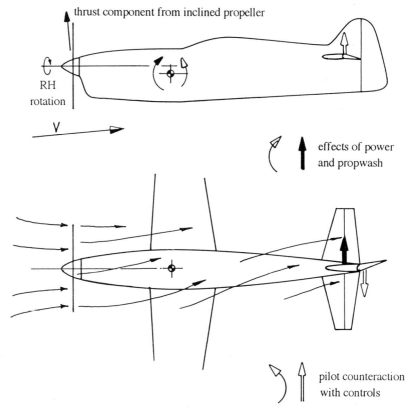

Fig. 10.5a Out of trim effects of power and propwash. Unnecessary displacement of flying controls is to lose motion, some authority and to increase trim-drag.

A small amount of downthrust provides an increment of nose-down pitching moment with power, which counters the reduction in longitudinal stability shown in Fig. 10.3a(2). In a similar way sidethrust is used to reduce the amount of rudder deflection lost to the pilot in Fig. 10.5b(2), improving too the directional stability.

If downthrust and sidethrust are provided during subsequent development, check to ensure that there is no significant loss of clearance between engine and cowling. There is a requirement for minimum engine clearances and if these are lost then redesign of the cowling – and possible adjustment to the run of engine exhausts – may be necessary. No exhaust gases, especially from turboprop engines, should come into direct contact with any unshielded part of an aeroplane.

Importance of the Trim Point: CG–CP Relationship

The slope of the $\Delta C_{\text{MCG}}/\Delta C_{\text{L}}$ curve determines whether or not an aircraft is stable or unstable. Stability viewed in such plainly mathematical terms will not satisfy pilots used to pushes and pulls on the control column. It is academic to talk of decreasing push force, or increasing pull force when one is trapped into holding an untrimmable push or a pull and is rapidly growing tired of doing so.

To talk common sense in a real world an aeroplane must have a trim point which, as we can see from the diagrams, only exists when the stability curve cuts the axis where $C_{\text{MCG}} = 0$.

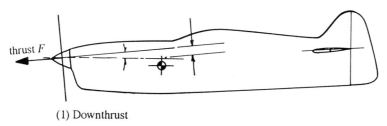

(1) Downthrust

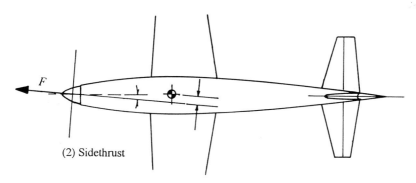

(2) Sidethrust

Fig. 10.5b Provision of (1) downthrust (or upthrust) and (2) sidethrust to recover lost motion, control authority, and cut trim-drag. Thrust offsets should balance aircraft at the design point which is usually the cruising speed, weight and CG position. Amounts shown are diagrammatic and replace the counteracting moments from the flying controls in Fig. 10.5a.

This means that there must be no residual push or pull needed to fly straight and level, between the stall and the maximum permissible speeds, in any configuration. Further, this condition should occur at the design cruising, or operational speed.

To be trimmable we need what Cayley, 'the father of aeronautics', discovered in 1809 (see Fig. 3.16), and what the humble builder of free-flying model aircraft knows, that there must be a vee in the incidence between that part of a lifting surface ahead of the CG and the remainder which following behind is set at a shallower angle of attack. While the connection between the cone and an aircraft might not be readily apparent, break for a moment and make a paper cone as described, hold it vertically by the rim, and then release it, edge downwards. The lower edge of the cone will *rise* and level itself until, after a few damped oscillations, it descends point first to the floor. The descent is stabilized, even though the angle of attack is large.

It is easy to show by calculation, as O'Gorman has done in his historic sketch in Fig. 10.6a (Ref. 10.1), that when such a vee exists then an upgust, or increased angle of attack, moves the centre of pressure of the system aft of the CG, causing a corrective nose-down pitching moment. Conversely, a downgust, or decrease in angle of attack moves the CP ahead of the CG, which brings the nose up. What O'Gorman described as the CP of the system is the trim-point when steady. Several configurations in Fig. 3.17 show how the vee fore and aft of the CG appears in practice. Figure 10.6(b) is his useful sketch of the CG position on several contemporary 'S'-type (*Duck*, now canard) aircraft of the period 1913 (Ref. 10.19).

Plate 10–1   Sir George Cayley's man-carrying glider c1852 in towed flight, showing the vee between the planes. Built by John Sproule and flown by Derek Piggott. (Courtesy of John Sproule)

It has been argued that use of a vee fore and aft is control, not stability. Maybe – but without the vee no trim point would exist, and without it a pilot could not demonstrate stability in practice to the satisfaction of anybody, least of all to himself.

Effect of CG upon Static Longitudinal Stability

Figures 10.1 and 10.3 show quite clearly that if an aircraft has a trim point (at which it can be flown 'hands-off'), then any increase in speed *without altering the trimmer setting* involves a push on the stick or yoke, to move the elevator down, preventing the nose from rising. A loss of speed involves pulling back on the stick to prevent the nose dropping. Although both the force and elevator (stick) displacements are indicative of static longitudinal stability, force changes are of greater importance to the pilot than stick movements – which are fiendishly hard to judge by eye with any accuracy. Some device is needed, either to provide axes the pilot can see, to which the hand can be related or, less awkwardly as a piece of cockpit equipment like a desyn.

Stick force and stick movement, as indicators of longitudinal static stability, show what are often called *stick force stability* and *stick displacement stability*.

The further ahead of the neutral point the CG is located, the larger the static margin and the more stable the aeroplane. This is shown in Fig. 10.7 for the same aeroplane as that used for Fig. 9.17. Note that the ordinate shows PULL upwards, i.e. to maintain angle of attack. They could have been drawn the other way up to correspond with Figs 10.1 and 10.2, but it is easier to appreciate this way. As the CG is moved aft by changing the loading with ballast, payload, or both, the stability becomes neutral when it coincides with the neutral point of the combination (thus CG = NP = ac). Moving the CG further aft makes the aircraft unstable.

When an aeroplane is short of longitudinal stability it is sometimes possible, during development, to move the neutral point aft by means of sweeping the whole wing, or cranking part of it. When an otherwise clean and elegant (more often than not subsonic) aircraft has an unusual sweep or kink in the wing planform, this is the most likely cause.

The limit on forward CG is authority of control in pitch. An aeroplane that is too nose-heavy, quite apart from losing agility, is hard to take off and land at low enough airspeeds for the required WAT field lengths needed for safe operations. With a CG too far forward it is easy to 'wheelbarrow' – followed by a ground-loop and broken nosegear leg at least. Numerous low-winged light aeroplanes with nosegear units have been mishandled in this way, by the pilot *'adding 5 to 10 knots (or mph) for the wife and kids'* and then putting the aeroplane on to the ground, because he is running out of airfield. Therefore always anticipate such a risk with low-winged aeroplanes fitted with nosegear, especially if they have a tail-down sit when at rest.

Figure 10.8 illustrates stick position stability against increasing airspeed (with elevator UP in the same sense as PULL in Fig. 10.7). Both diagrams suggest ways of measuring stability by means of a meat-hook spring balance, or with a device which shows elevator position, such that:

$$\text{static longitudinal stability} = \Delta\eta_{\text{trim}}/\Delta C_{\text{L}} \tag{10–1}$$

which applies to any aircraft, rigid or flexible, at high or low speed.

Example 10–2: Measuring Stability by Elevator Trimmer Position Against IAS

When an aeroplane is fitted with a calibrated elevator trimmer this can be used as an analogue for $\Delta\eta_{\text{trim}}$ in the above equation. Nose-up elevator and trim is needed at low speeds. With increasing speed more nose-down elevator and trim is needed to fly level. Provided that the ASI is calibrated accurately, reasonable curves can be obtained. At higher speeds aeroelastic distortion of tab and elevator, and cable stretch must be expected to cause non-linearity in results.

Figure 10.9a shows the way in which this test was carried out using indicated elevator trim position against IAS, for the De Havilland Chipmunk, during my final *Preview*, at the Empire Test Pilots' School. The elevator trim scale was marked NOSE UP and NOSE DOWN, and graduated with elevator trim position (equivalent to elevator angle, β). It must be remembered that for the elevator to be trimmed UP, the trim tab had to be DOWN, and vice versa. So, to visualize what is going on, above zero the elevator is UP (in the negative sense), but the trim tab is DOWN (which is positive by the conventions shown in Fig. 3.19).

The curves showed that the Chipmunk was stable, but decreasingly so as airspeed increased, and as the CG moved aft.

A rough approximation to the shape of the M_{CG} versus C_{L} curve may be constructed, as sketched in Fig. 10.9b, knowing that C_{L} varies as $(1/\text{IAS}^2)$. Here $f(C_{\text{R}})$ (which is a function of the lift coefficient, C_{L}) was expressed in terms of the stall speed in the relevant configuration so that:

$$f(C_{\text{R}}) = 1/(V_{\text{i}}/V_{\text{Si}})^2 \tag{10–2a}$$

or simply:

$$= 1/(V_{\text{i}}/100)^2 \tag{10–2b}$$

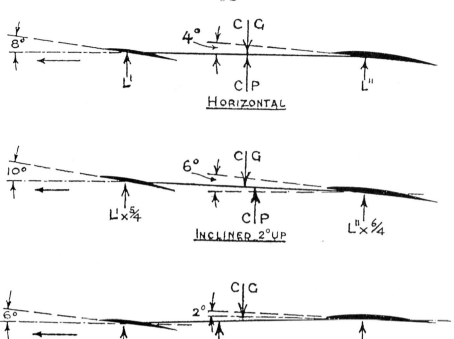

FORE & AFT STABILITY

Historic explanatory insight by Mervyn O'Gorman (Superintendant of the Royal Aircraft Factory, Farnborough in 1911) of the role of the vee between the planes (decalage) in achieving longitudinal (fore and aft) stability (ref.10.1):

Here the foreplane is half the area of the rearplane. Lift is proportional to area x angle of attack at small angles and constant speed. In the upper case the fore-plane lifts twice as much per unit area as the rearplane and the centre of pressure lies midway between. *The CG is arranged to coincide with the CP.*

In the centre (2° UP case – assume an upgust has produced this effect) the foreplane is at 10° and the rearplane at 6°. The foreplane has now increased its lift in the ratio 5/4, and the rear-plane to 6/4, moving the CP aft of the CG, tilting the nose down in a stabilising sense.

In the lower (2° DOWN case – a downgust) the foreplane lift becomes 3/4 and the rearplane 2/4 respectively, moving the CP forwards of the CG, raising the nose.

Fig. 10.6a Compare with Fig. 10.1c, and that in (my) italics with Figs 10.11 and 10.12 – all are relevant to the problem of balancing historic and accurate full-scale replica aeroplanes.

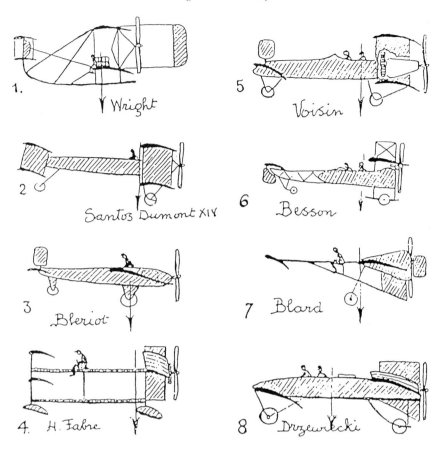

Fig. 10.6b   O'Gorman's drawings of contemporary aircraft he classed as the Duck, or 'S' type, with the arrow marking the position of the CG (and therefore the CP) (Ref. 10.19). Rutan has also used a forward-set but ventral 'rhino' rudder on his tail-first, push-pull twin, four-seat Defiant, which first flew in 1978.

which makes the calculation easier to handle without changing the shape of the curve. Trimmer positions were plotted the other way up, i.e. trimmer nose down being in the positive sense of elevator nose down, as a function of $(1/IAS^2)$, using whichever of the above equations is preferred (for the final report I used Equation (10–2b)).

Other basic tests are listed in Table 4–6, item 10, which gives three significant low speed configurations. Item 6 is also useful, checking trim against IAS from V_x, or 1.4 V_{S1} to V_{MO}, while using a constant power setting. In each case one looks for a nose-down push and stick movement with increasing airspeed, and vice versa. The tests in item 10 of Table 4-6 involve trimming the aeroplane at each speed in turn, in the required configuration, and then increasing and decreasing airspeed using the elevator, by the amounts shown, *viz.*, $\pm 0.1\ V_S$.

Dynamic Longitudinal Stability: Oscillatory Motions

Dynamic oscillatory motions can be troublesome in any mode. There are two in pitch, one short period, the other much longer. The short period oscillation has a period of around

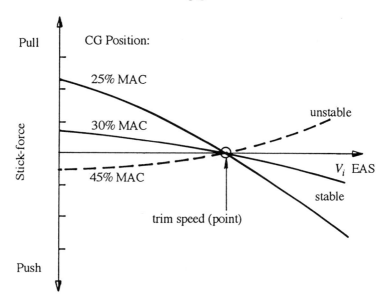

Fig. 10.7 Effect of CG position on stick-force to change speed (i.e. stick-free static margin and stick-force stability, see Equation (3–37b)). Compare with Figs 9.27 and 9.28.

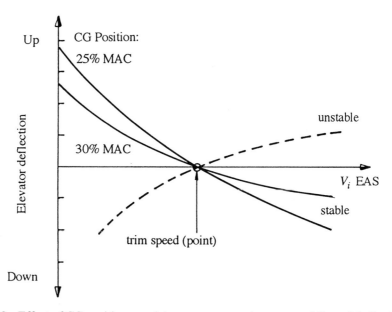

Fig. 10.8 Effect of CG position on stick-movement to change speed (i.e. stick-fixed static margin and stick-movement stability, see Equation (3–37a)).

one to two seconds, or a little more. The longer, a phugoid oscillation, has a period nearer 30 s.

The damping ratio of an oscillation is the length of time for one cycle divided by the time taken for the total number of cycles to decay:

damping ratio = t/Tn (10–2)

A heavily-damped oscillation has a damping ratio of 0.3 or more, and there should be no residual oscillations of which the pilot is aware.

In-flight motions (longitudinal, lateral and directional) depend upon the particular mixture of static and dynamic characteristics, for no two aeroplanes are completely alike. Motions of an aeroplane result from the size and radii of gyration of its total mass when excited by a disturbance, responded to by the static stability (which can be likened to a system of springs and the calming of aerodynamic damping possessed by its configuration, as shown in Fig. 10.10a):

☐ *Mass and radius of gyration*. No two aeroplanes of the same type ever have precisely the same weight and CG, even when brand new. As they grow older, suffer wear and tear, maintenance and repair, they diverge one from the other – even though they still look alike.

Note 1: This is significant when we have to deal with rebuilds and replicas which are intended to be flown during air shows to simulate the use and handling qualities of their historic originals.

Note 2: Every aircraft has two shapes, one inside, the other outside. The latter depends upon the combination of moments of inertia, A, B and C (see Fig. 9.31a, b and c – and look too at Fig. 9.29). A test pilot never assumes that because on the outside an aeroplane looks like what it purports to be, then what is inside must be alright. On the inside it could be a wolf in sheep's clothing when wound up, and with its dynamics challenging its aerodynamics, as when spinning.

☐ *Static stability*. The stronger the stable response to a disturbance, the stronger the rating of the imaginary spring in the system in Fig. 10.10a.
☐ *Damping*. This arises from increased angles of attack of various portions of the airframe, and by increased airspeed. Relative aircraft density modifies the picture (as noted in the last chapter when discussing increased liveliness of response to control at altitude).

Thus, an aeroplane may be represented by a model which has the characteristics changed by altering the resistance of the dampers and rates of the springs. Figures 10.10b and 10.10c show, in b(1) and c(1), static stability by a return to the undistorted condition. However b(1) takes longer and develops a dynamic mode because aerodynamic damping is weaker.

There are three conclusions:

☐ A body may be *statically stable* and also *dynamically unstable*.
☐ If a body is *statically unstable* it cannot be *dynamically stable*.
☐ *Oscillations*, both stable and unstable, are indications of the *degree of dynamic stability* possessed.

Short period oscillations

A short period longitudinal oscillation involves both angle of attack and normal acceleration. It has a period of a few seconds and occurs at more or less constant airspeed. Stick-fixed the oscillations are usually well damped. Stick-free there may be oscillatory hunting of the elevator, which can be dangerous, as shown by the first two items in Table 10–1 which lists the factors which affect damping.

Oscillations are excited by a sharp deflection of the elevator under test conditions, and/or by rough air, and occur as the aircraft adjusts itself to the altered angle of attack.

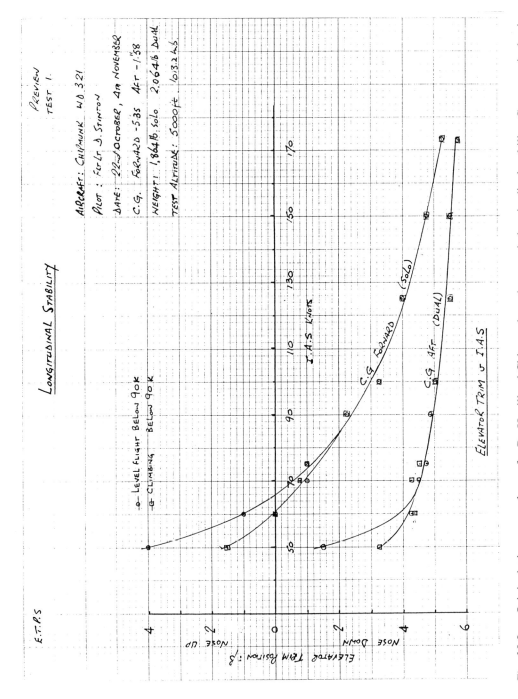

Fig. 10.9a Original trim curves drawn for De Havilland Chipmunk to show static longitudinal stability. Curvature is due in part to wake effects with changing attitude, and to aeroelasticity in the tail structures and control runs.

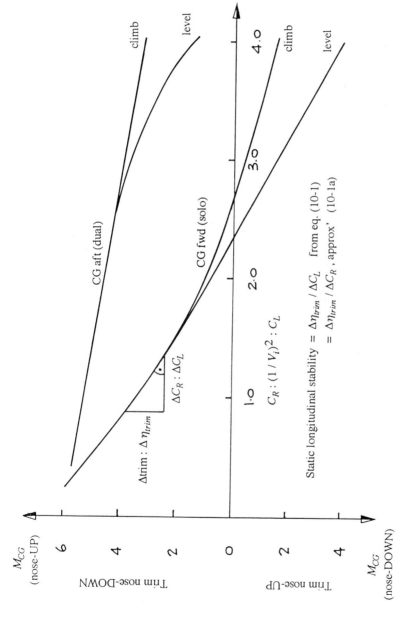

Fig. 10.9b Crude original conversion of the trim curve in Fig. 10.9a to show shape of M_{CG} against C_R (proportional to the lift coefficient, C_L). This derives from nose-DOWN elevator trim opposing M_{CG} nose-UP. These curves confirm that the De Havilland Chipmunk is stable; that full power for the climb is destabilizing; and that the aircraft is less stable with CG aft (dual configuration).

Plate 10–2 Early De Havilland Chipmunk with original narrow-chord rudder and no anti-spin strakes. Ref.: Examples 10–2 and 12–4. (Courtesy of former De Havilland Aircraft via Nigel Price)

The danger, when attempting to check a short period oscillation by opposing control movements, is of getting out of phase. Then it may be transformed into a pilot-induced oscillation, a PIO. Usually the only course open to effectively check a short period oscillation which shows no sign of decay is to grab and immobilize the control column.

Sometimes a long period phugoid oscillation is excited at the same time as the short period. Great care must be taken to ensure that the input by the pilot is neither excessive nor abrupt, and that the hand hovers close to the stick so as to grab it if the response of the aircraft becomes too excited.

Long period (phugoid) oscillations

In his book *Aerodonetics*, F. W. Lanchester considered at length the general principles of stability and control, and of the flight path of small amplitude, which he called a *phugoid* (meaning flight-like) (Ref. 10.6). From such considerations he evolved a theory by means of which he was able to suggest how an aircraft might best be proportioned. Although his work came to a dead end, he was able to use his phugoid charts to point to danger zones around the cusps, as in the swoop-stall-swoop-stall-swoop flight path of a pigeon in mating flight (easily simulated with a tail-heavy free-flight model chuck-glider). Today, the often (but not always) innocuous phugoid excites little interest, because it tends to be gentle and slow and controlled without difficulty. It may be hardly noticeable in normal flying conditions,

Plate 10–3   Later De Havilland Chipmunk with broad-chord rudder
and anti-spin strakes. Ref.: Examples 10–2 and 12–4. (Author)

and has to be induced by trimming the aeroplane to fly accurately hands-off, and then displacing the stick to hold a new speed for a short time, about 10 knots above or below the trimmed speed. Then either:

☐ bring the stick back to the trimmed position and hold it there to demonstrate the *stick-fixed phugoid* (which rarely exists), or
☐ release the stick when in the displaced position to produce a *stick-free phugoid*.

Note: Remember the Cheyenne in Example 10–1. On releasing the stick, the pilot must keep his hand close to it, so as to ensure that nothing unexpected occurs that cannot be stopped in time.

The factors which affect the damping of a phugoid oscillation are given in Table 10–2.

Checks for Dynamic Stability

Dynamic stability is checked at points at which static longitudinal stability is assessed. There is no need to do so at every point, although if those basic points already mentioned in Table 4–6, items 6 and 10 only are evaluated, then checks of dynamic stability should be carried out at the same time.

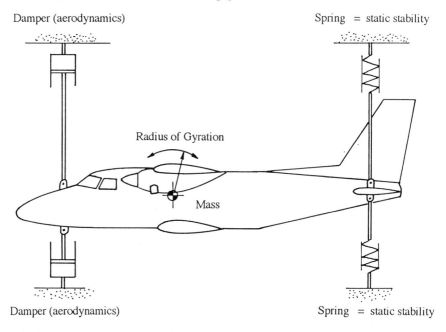

Damper (aerodynamics) Spring = static stability

Radius of Gyration

Mass

Damper (aerodynamics) Spring = static stability

Fig. 10.10a Simulation of an aircraft as a combined mass + spring + damper system which, together, determine the oscillatory characteristics when disturbed.

As a rule it is best to carry out both tests in sequence: change the speed using a small stick movement – note whether push or pull is in the correct sense and the magnitude of the forces – then CAREFULLY release the control column to see how the aeroplane responds. Great care is needed, especially at high speed, no matter how small the control jerk might be.

Note: In item 10 of Table 4–6 the incremental changes in airspeed in the climb and on the approach *are less than 10%* of the trimmed values. So BE WARY – never go at it like a bull in a china shop!

Tail-First (Canard) and Three-Surface Configurations

With the exception of tailless and flying wing configurations, all aeroplanes have tandem surfaces. A wing at the front and a stabilizer following behind is a close relative of an aeroplane like the Gyroflug SC01B in Fig. 9.30. The only difference lies in the relative areas, lift-slopes and CG positions. All share the physical property described by O'Gorman in Fig. 10.6. The greatest advantage of the tail-first arrangement is that (within limits) stalls are innocuous, lift being lost over only a small portion of the total lifting area. The greatest danger is from an overloaded foreplane. Then a hammerhead stall may follow, with the aeroplane pitching inverted nose-down. Loss of directional stability caused by foreplane wake-interference aft may also cause the aircraft to attempt to change ends.

The primary attraction of the canard, as noted earlier, is that both fore and aft surfaces provide lift in the same sense. When properly trimmed they are capable of higher lift/drag ratios, with superior range to conventional aircraft carrying around tail at the back, which is often providing an anti-lift download for trim.

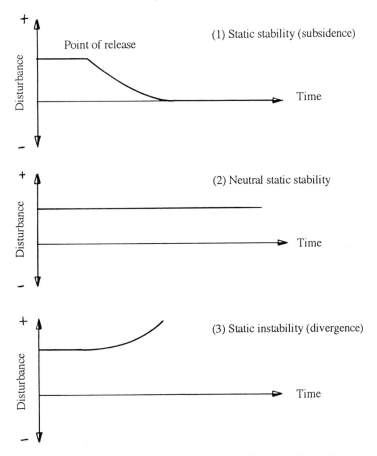

Fig. 10.10b Three alternative aspects of static longitudinal qualities. Compare (1) with Fig. 10.10c(1), both of which show a greater return to the original, but (1) has more powerful damping than Fig. 10.10c(1).

If the CG lies between the planes, each is induced to lift only a proportion of the total weight of the aircraft. A pure tandem, formed by cutting a wing in half and then arranging the areas equally front and rear can, in theory, look like an attractive solution. By providing each half wing with a set of differentiated control surfaces a tail might be done away with. Unfortunately it does not work. Free-flying models cannot decide which of the planes is dominant. A large foreplane produces powerful, nose-up, destabilizing moments about the CG. The equally large rearplane produces powerful, nose-down, stabilizing moments.

The result is a phugoidal flight path like that investigated by Lanchester, shown in Fig. 10.10c(2). The oscillations have their longest period when the CG is midway between the planes. The CG also has the largest range of movement fore and aft without upsetting the static stability unduly. Move the CG forward and the period shortens and damps out as the foreplane becomes the wing. The stall speed of the foreplane is increased, and it has to be made larger to reduce the wing loading and the speed at which it ceases to be of any use. The available CG is diminished, and the now excessive rearplane must be reduced in area to save structure weight and cut drag.

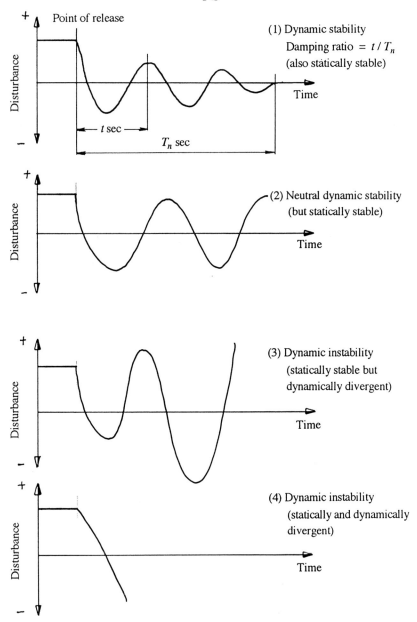

Fig. 10.10c   Four aspects of dynamic stability and instability.

A similar thing happens if the CG is moved aft, to make the rearplane the wing. The period of the oscillation shortens and damps out. The available CG range is diminished as the rearplane grows into a wing, and the foreplane is reduced in size for economy of effort. As the CG is moved forward, making the foreplane into a wing, the vertical tail surfaces, which provide weathercock stability, can be made economically smaller than when the rearplane becomes the wing of a canard aeroplane.

With a tail behind the wing it is easier to sweeten the harmony of the controls, making A : E : R = 1 : 2 : 4, whereas a canard tends to be the other way around with ailerons

TABLE 10–1

Damping of short period oscillations

Factor	Effect of increasing the damping factor	Notes
Elevator aerodynamic balance, b_2	Increasing balance area (reducing b_2) reduces damping	If b_2 is small and the control closely balanced, poor damping, or even unstable oscillations, may occur when mass-balancing is also insufficient. The effect is worsened if b_1 is positive and can then occur even with mass-balance, or over mass-balance.
Elevator (trail hinge-moment coefficient, b_2	Reduces damping	
Mass-balance of elevator	Increases damping	
Elevator moment of inertia	Reduces damping	
Aircraft moment of inertia	Increases damping	
Airspeed	Reduces damping	If poor damping is encountered at low airspeed, oscillations may become violent and possibly dangerous at higher speeds.
Friction in the elevator circuit	Reduces damping	

heaviest and harmony tending towards A : E : R = 4 : 2 : 1. Further, tail-first aeroplanes that I have flown have not had satisfying aileron authority, and rudders have been over-weak. When combined with heavyness compared with over-light rudders and longitudinal sensitivity in turbulence, my impression is of them being back to front in more ways than one.

Three-surfaces

At first sight the configuration of fore, main and tailplane behind looks uneconomical - and it might prove to be no more than a fashion in time. The aeroplane in Fig. 6.16 is a *'libellula'* (i.e. of the dragonfly family – a form of tandem-biplane, but with extreme back-stagger) fitted with a tailplane for trim and control in pitch. The adjustable tailplane, which has an elevator, is powerful and has the theoretical advantage of freeing both the fore and mainplanes to get on with the business of providing longitudinal stability and lift efficiently. In this way a wider than normal CG-range might be used, while avoiding one major problem – that of running out of elevator on the foreplane as flap is lowered on the rear mainplane.

Three-surfaces promise reductions in trim-drag in the cruise, provided that the CG position is optimized by pumping fuel between tanks in the foreplanes and mainplanes. Clearly, this is an area to be explored in flight tests, to determine the critical range of CG movement and its effect upon flying qualities, particularly in the approach and landing with one engine inoperative, combined with failure of the fuel-transfer system.

TABLE 10–2

Damping of long period (phugoid) oscillations

Factor	Effect on damping of increasing the factor	Notes
CG position	No general rule	CG position affects the type of motion (see below) but damping of the stable phugoid does not necessarily increase with forward movement of the CG.
Profile drag	Increases damping	The most important factor after CG position.
Tailplane size	Increases damping	The damping coefficient in pitch is proportional to the tailplane area and the square of the length of the fuselage from CG to ac of the tailplane.
Fuselage length	Increases damping	
Inertia weight or spring in elevator circuit	Generally reduces damping	Particularly with small tail or at high altitude (relative aircraft density effect, μ_1).
Mass-balance of elevator	Increases damping	Even with mass-balance, or over mass-balance.
Friction	Reduces damping	Particularly with small static-margins when the restoring aerodynamic forces are small.
Increased airspeed	Increases damping	By increased aerodynamic forces.
Propwash	Usually decreases damping	(See Fig. 10.4c).
Altitude or wing loading	Reduces damping	Effect small with mass-balance and little friction but important with weights or springs in circuit.

Early and Historic Aeroplanes

Rebuilt and replica early and historic aeroplanes, particularly those World War I up to the end of the big piston-engined era, after World War II, pose some of the most interesting problems for pilots. It is a feature of the historic aircraft movement that builders and owners find information wherever it may be discovered, accurate or unreliable, true or false, for such is the substance of legend. One has to test the evidence to find out if it hangs together and presents an authentic, coherent picture. It also involves the certification pilot and engineer in considerable research into technical history and eyewitness accounts of flying qualities, to ensure authenticity. We saw in Chapter 2 how even the most scant aside by a pilot may be significant.

For history to be useful one must have a respect for it. The physical problems people encountered then in aeronautics are the same as those waiting to trap the unwary among us now. It pays to try to get inside the mind of our pioneer ancestors and look at things through their eyes.

For example, during the period between the Wrights and World War I people were struggling with performance, control and stability every bit as single-mindedly as we, except that they were without the tools we have been fortunate enough to have been handed to us by others. There were vast and rapid changes afoot. The first Superintendant of the Army Aircraft Factory, Lieutenant Colonel Mervyn J. P. O'Gorman (one of whose quotations appears above in this chapter) was a far-sighted man whose initiative introduced formal scientific method into aeronautical research and development. In November 1911 he prepared *R&M No.59* which classified configurations as follows:

SE: Santos Experimental, after Santos Dumont, the originator of the 'canard' or tail-first type of aeroplane. Only one of this class was built, and the classification was later used for tractor single-seaters of the Scout Experimental formula.

FE: Farman Experimental of pusher biplane formula.

BE: Bleriot Experimental of two-seat tractor biplane general purpose formula.

RE: Reconnaissance Experimental of two-seat tractor biplane reconnaissance formula.

TE: Tatin Experimental monoplane with pusher propeller behind the tail.

BS: Bleriot Scout, a combination of the BE and SE formulae in single-seat scout form.

These indicate how scientific attention first turned towards the problem of the longitudinal stability of tail-first versus tail-last configurations, for that is what lay behind such a system of classification.

There is felicity in the descriptions of problems of aircraft of that period. O'Gorman advised against attaching too much importance to classification such as he had advocated, urging instead that attention should be paid to:

'… the method adopted in each case to secure fore and aft stability, which is throughout by some attempt to get a Vee between the surfaces back and front.'   (Ref. 10.1)

He went on to show that the front plane, regardless of size, must be the more heavily loaded of the two (to produce an upright vee, echoing Cayley's experiments with a weighted cone):

'Any more lightly loaded auxiliary plane, such as is placed in front of the carrier planes or wings, diminishes the longitudinal stability… by giving rise to an inverted Vee. This is instanced by the front elevator planes of the Farman – the effect of which has to be countered by increasing the size of the tailplanes.'   (Ref. 10.1)

In the quotation at the head of Chapter 2 the HAA definition of historic aircraft stressed their display and use '*in representative working order*'. This is especially important when an original aircraft exists, or one which is predominantly original, for then one must rely upon guesswork as little as possible. Remember too that 'original' aircraft today have been so repaired and rebuilt that nearly all are replicas to a degree. The test pilot must have a good idea of just how much of a flying-machine is no longer original and truly representative. If not, then ASK – or dig out what you can. If you are to write a report, this vital information must be included.

One hundred percent replica and 'look-alike' aircraft may be less important in this respect – unless the pilot invites trouble by attempting to carry out a display which mimics one of which the original was capable, but which the replica is not. One thinks, for example, of attempting to use a wooden full-scale look-alike aeroplane in precisely the same way, using the same speeds and manoeuvres as the all-metal production original.

Example 10–3: Rebuild/Replica – Westland Lysander

The Westland Lysander was a high-winged Army cooperation aeroplane with an original Bristol Mercury III radial engine. It is now remembered as the machine used to fly agents into and out of occupied France during World War II. The bulk of the one that I flew was what the model-maker would call 'scratch-built' from released materials, in that parts had come from Westlands, or had been hand-made elsewhere, using drawings prepared for aeromodellers, together with sundry bits and pieces of an original Lysander. The aeroplane looked right, smelt and sounded right (but was to suffer acute engine trouble which resulted in around nine forced and emergency landings for a number of pilots, six of which were mine).

There was available a reprint of *Air Publication 1582C, Pilot's Notes for the Lysander III and IIIA*, with the same engine. The notes gave reliable guidance on airspeeds, procedures and handling, which was invaluable when it came to clearing the aeroplane as possessing representative Lysander flying qualities.

The inboard and outboard leading edge slats opened automatically. Those outboard were independent of the inboard – the latter connecting with the trailing edge flaps, movements of which were governed automatically by airspeed. The lower the speed the more the flaps were deflected and the inboard slats opened.

On take-off the Lysander was airborne after a short run at 45 KIAS in a three-point attitude. Accelerating to 145 KIAS without retrimming by means of the tail actuating wheel involved high stick forces. The tail actuating wheel, mounted vertically on the left–hand side of the seat, altered the angle of the tailplane. The range of movement indicated adequate longitudinal static stability.

The aeroplane could be trimmed to approach comfortably with power at 70 KIAS, or glide (throttle closed) at 74 KIAS. 'Creeper' approaches could be made with power.

The trickiest feature at low speed was the combination of tail actuating wheel and throttle (also operated by the left hand while the right hand held the stick). The effect of such a powerful nose-up trimmer was exacerbated by a nose-up trim change with power, which in combination could produce a dangerously nose-high attitude. Pilot's Notes warned against this:

> 'MISLANDING
>
> Open the throttle enough to maintain flying speed while the tail actuating wheel is wound forward.
> It may then be opened fully.
> On no account must the throttle be opened fully with the tail actuating wheel wound back.'

This feature was the cause of a considerable number of accidents with Lysanders during World War II. The aeroplane had a mind of its own but, when mastered, it engendered confidence in its unusual low speed qualities.

When dealing with early writings and reports one finds that there is a constant interchanging of the concepts of control and stability. In the case of the Lysander the pilot adjusted the vee between the wing and tail surfaces by means of the tailplane actuator wheel. Is that a

Plate 10–4a   A modern full scale flying replica of what looks, smells and sounds like a World War II Westland Lysander. But does it fly like one, and how can you know, half a century later? Ref.: Example 14–5. (Author)

form of control to restore equilibrium after a change of speed – or, in the case of aircraft with a fixed vee, is it a physical way of providing a *real* trim point and so a stable configuration? I like to think that Gates, who saw stability and control as two sides of one coin, might have replied '*Both, depending upon which way one looks at it*'.

The area of greatest difficulty lies in the planning of flight tests for historic aeroplanes about which little is still known. As in Chapter 5 one normally starts with weight and balance: but what is the test weight, where ought the CG to be – and in which direction will it move as fuel is consumed? With early aeroplanes the shape of the aerofoil sections of high or low, front and rear planes is of vital importance, because camber and trailing edge angle tend to determine where the centre of pressure of each lies. Historically the CP fixed the location of the CG (which often involved fore and aft adjustment of the planes, followed by adjustment of the tail setting angle – with down- and side-thrust added). If that did not do the trick then maybe an elevator trimmer was fitted, in the form of a spring, or a bungee elastic cord to relieve control forces for the pilot.

It is always surprising to discover how much our forefathers knew. Cayley, for example, started experimenting in 1804, one year before the Battle of Trafalgar, with gliders, in the form of free-flying kites. There is evidence of him having flown a man and a boy. He also built a man-carrier in 1843, and had his coachman airborne, who thereafter gave notice. A replica has flown in recent years (Ref. 10.8). Cayley was a polymath whose mind ranged widely, from problems of a local squire to Member of Parliament, from making honey to ballistics. Of his achievements most relevant to us were these:

Plate 10–4b The Edgley Optica, with crew compartment forward of the ducted propulsor, generated lift from the bulged forebody with power at low airspeed. A component of downthrust was provided by the eye-lid lip (unpainted) at the duct outlet ...
Ref.: Examples 5–2 and 5–3. (Author)

☐ He discussed movement of the centre of pressure of a surface.
☐ He considered stability and control by means of dihedral and moveable tail surfaces (the italics are mine):

> 'The rudder (*i.e. elevator*) gives most stability to the course of the flight, when slightly elevated, so as to receive a small degree of pressure downwards; but when truly balanced by the weight of the prow, it flies farthest, when the rudder is in the same plane as the sail (*wing*) ....
> The side steerage by the small separate upright rudder is always perfect when other adjustments are complete.' (Ref. 3.5)

☐ He suggested superimposed biplane and triplane wings to achieve maximum lift for minimum weight.
☐ He found that curved (cambered) surfaces produced more lift than flat, and knew that low pressure (vacuity) on the upper surfaces provided the main lift.
☐ He knew that aerodynamic force was a function of the density of air and the square of velocity.
☐ He knew about streamlining and solids for least resistance, and proposed a shape which had almost the same profile as a symmetrical NASA aerofoil section.

Sproule, a biographer of Cayley, posing the question about how much Sir George Cayley really knew, answered rhetorically: '... he (Cayley) *was so far ahead in both theory and practice that no-one really knew what he was talking about*.' (Ref. 10.8).

Plate 10–4c ... To use the full design CG range, which varied with cockpit load (1 to 3 occupants ahead of the wing) ballast had to be exchanged between nose and tail. The door is open on the ballast stowage in the port fin.
Ref.: Examples 5–2 and 5–3. (Author)

So it is the same with many other pioneers – Cayley is but one example. The Frenchman, Alphonse Penaud (1850–80) was another. He knew about the efficacy of dihedral and having a negatively angled (fixed) tailplane to provide the vee between the fore and rearplane surfaces. What was significant is that from Penaud onwards in Europe aviators searched for inherent stability. In the USA the Wrights considered it and then settled, initially and successfully, for aircraft that were all control and no stability to speak of.

We know now that the danger of an excess of inherent stability is ineffectiveness of control. This was realized too late for many in the opening stages of World War I. The emergence of the agile armed scout and two-seater easily dominated the slow, stately and stable observation and reconnaissance aeroplanes, like the BE2C. To achieve agility with a manually-controlled aircraft the CG must lie closer to the NP than in the case of a sedate cruising aeroplane.

How Were CG and CP Arranged?

It took much effort to discover where designers of the period sought to place the CG on their early biplanes, triplanes and monoplanes. This information was needed to find out how past designers thought and worked out their problems, to cross-check modern rebuilds and replicas. Two clues were significant, one from Barnwell, designer of the Bristol F2B (Example 2–2), in the book written in 1915, which was based upon a paper he read to the Glasgow University Engineering Society in the winter of 1914 – eleven years after the Wright Brothers had flown successfully at Kitty Hawk. Barnwell's clue is in Fig. 10.11

Plate 10–5   An early production version of the Royal Aircraft Factory BE2C of 1914. Inherently stable and sedate, it was no match for agile scouts, having been designed for reconnaissance and being flown hands-off for lengthy periods. The aircraft is fitted with an RAF pitot head and engine. (Courtesy of John Bagley, formerly of RAE and Science Museum)

(Ref. 2.9). The second clue came indirectly from NACA Report No.1 of 1915 (Ref. 10.9), showing a wind tunnel model of the Curtiss JN2 (Jenny), via the EAA and reproduced in Fig.10.12. Barnwell advised (and the asterisks are mine):

> 'We must now draw out a side elevation of the body of the machine with seats, tanks, motor and tail skid, keeping all the weights as close together as possible ....' (*see Fig. 10.11**) 'We shall employ a "non-lifting" Tail plane, that is to say, a form symmetrical about its central horizontal plane and with this plane parallel to the axis of the propeller.
>   This form is perhaps the safest to employ, as it will give no difference in lift or depression, whether in the propeller slip stream (when the motor is running) or not (when the motor is stopped). We shall set the chord of the aerofoils at 3° to the propeller axis.
>   We now require to place our Aerofoils and Landing Gear, less Tail Skid, of course, on the body in such a manner that the total reaction on the Aerofoils, at 3° for i (*incidence**), passes through the CG of the whole machine (of this more anon), and that the centre of the wheel axle of the Landing Gear is about 12″ (*inches**) ahead of it.'   (Ref. 2.9)

He went to say that, guided by (wind tunnel*) model figures for the aerofoil section, mark on each wing of the biplane the position of the CP at an incidence angle of 3°. Join the CPs of the top and bottom planes. Then mark a point on the line constructed a distance 4/7 of the gap between the planes, above the bottom plane. This last point is that of the CP of the

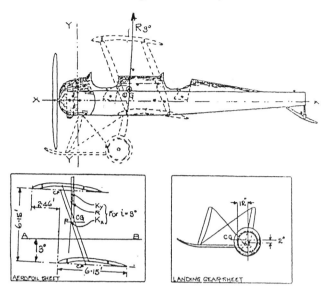

TABLE FOR HORIZONTAL & VERTICAL C.G

ITEM	W	l	h	$W \times l\,(+)$	$W \times l\,(-)$	$W \times h\,(+)$	$W \times h\,(-)$
Propeller	28	+2·0	·–	56	–	–	–
Motor	250	+·7	–	175	–	–	–
Cowling	32	+·4	+·4	13	–	13	–
Motor Mounting	36	–·2	–	–	7	–	–
Oil & Tank	86	–·6	+1·3	–	52	112	–
Passenger	175	–2·5	+1·0	–	390	175	–
Passenger's Seat	10	–2·8	+·4	–	28	4	–
Petrol & Tank	294	–5·2	+1·4	–	1530	412	–
Body	90	–6·7	–	–	602	–	–
Instruments	30	–7·1	+1·5	–	213	45	–
Controls	30	+7·5	–	–	225	–	–
Pilot	175	–8·7	+1·0	–	1520	175	–
Pilot's Seat	10	–9·1	+·4	–	91	4	–
Tail	86	–19·0	+1·0	–	1630	86	–
Tail Skid	7	–19·7	–1·0	–	138	–	7
Aerofoils complete	430	–4·9	+2·8	–	2107	1208	–
Landing gear	129	–2·5	–3·9	–	322	–	503
TOTAL (loaded)	1898	4·53	+·91	244	8855	2234	510

W = wgt of Item in lbs

l = Normal distce of CG of Item from line Y–Y .   + ahead, – behind .

h =   "   "   "   " " "   "   "   X–X   + above, – below

Fig. 10.11 An historic drawing from 1916 by F. S. Barnwell, designer of the Bristol F2B Fighter (see Fig. 2.2) showing arrangement of CG and CP (*R* at 3° incidence) as a preliminary to calculating tail size (Ref. 2.9).

biplane cellule. A normal drawn through the point is the '*line of Lift reaction*', while a line constructed through the same point parallel with what we would now call the horizontal datum of the aircraft is '*the line of Dynamic Resistance*' (drag*) of the biplane at an incidence, *i*, of 3° (in this book we use i_w for wing incidence and i_s for that of the tail stabilizer (see Fig. 3.14)).

The side view of the Curtiss 'Jenny' in Fig. 10.12 showing fundamentally the same procedure as that of Barnwell, confirms that the resultant crossforce between about 2° and 4° passes through the centre of gravity. The reason for angles of incidence (here a synonym for

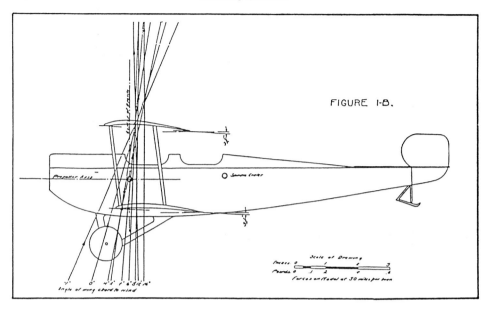

FIGURE I-B.

Fig. 10.12   An historic drawing from 1915 of a model of the Curtiss JN2 'Spinning Jenny' from NACA Report No. 1. This confirms the procedure described by Barnwell in Fig. 10.11 (Ref. 2.9). It is doubly historic because it is from the first report of the newly formed National Advisory Committee for Aeronautics, which is now NASA, the National Aeronautics and Space Administration.

angle of attack) being around 3° is that this is the angle for best lift / drag – of the wings, give or take a degree or two. Moments were then taken about the CG to calculate the area of the tail surfaces. A non-lifting section was used for the tail to provide no out of balance force when properly rigged, assuming it to work in undisturbed air. Curtiss appears to have mounted the tailplane of the Jenny on the top longeron, parallel with the propeller axis, as Barnwell also advised.

Barnwell went on to say that if it was possible to get accurate model figures for the air reactions on the body of the machine, then these should be used to find the necessary area of tailplane to counter the instability of the body. If such figures were not available, and as the destabilizing effect of a narrow (tandem-seat) body was comparatively small, then one should *'merely add a small amount to the calculated tail surface necessary for the aerofoils alone – say 1/10th'*.

Note: Calculations made by designers of the period used coefficients like k_x and k_y for resistance (drag) coefficient and lift coefficient respectively. These equate to their modern forms:

$$k_x = C_D/2 \tag{10-3a}$$

$$k_y = C_L/2 \tag{10-3b}$$

The evidence led to a search through numerous period aerofoil sections to enable predictions of likely centres of pressure to be made for aerofoil sections 'eyeballed' for modern replicas of historic aircraft. It was found that CP appeared to depend upon two geometric features: camber and the trailing-edge angle. Figure 10.13 shows a number of sections

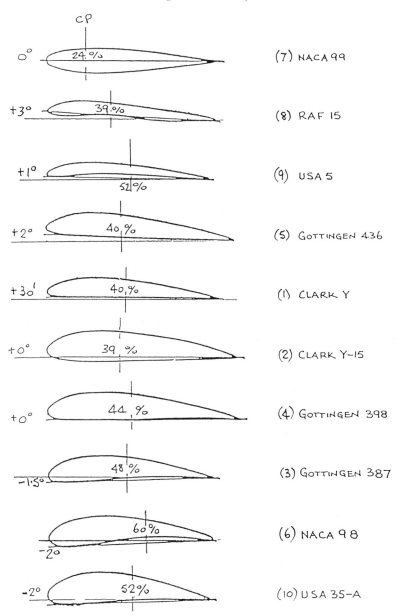

Fig. 10.13 A rough sketch made during a search for information on mixed British, German and American wing sections which reflected technology by the end of World War I. This was needed as a guide to calculation of CGs for test flying rebuilt and (often eyeballed) replica aeroplanes for the issue of UK Permits to Test and to Fly. The centre of pressure, CP, which moves aft with increasing camber, corresponds with the reported angle of incidence for the best lift/drag ratio of a section (numbering is relevant to Fig. 10.14).

investigated, and Fig. 10.14 the results. Both sketches are copied direct from my technical notes, to show that when confronted with test flying an unknown aeroplane, interesting truths can leap out of the simplest and crudest of sketches. The aerofoil section details are

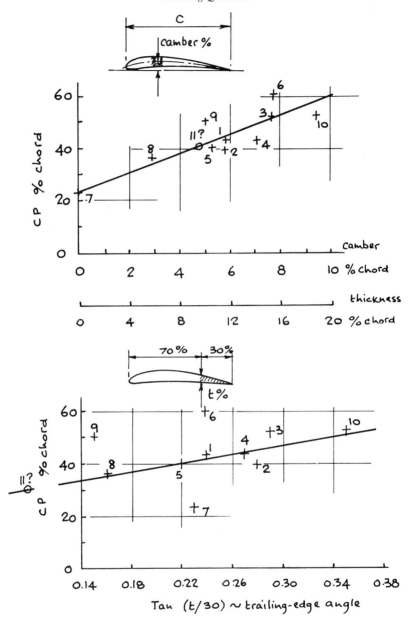

Fig. 10.14 Another rough sketch which summarized the information accumulated for the
wing sections in Fig. 10.13. The upper graph plots CP position in terms of percentage camber. The lower graph, CP position in terms of trailing-edge angle. The deeper and more
cambered the section the further aft moves the CP, which accounts for the apparent short-
ness of the nose (wings well forward) of many World War I aeroplanes – the Fokker DVII of
Fig. 2.3a being a particular case in point. Numbering is as in Fig. 10.13, except for No. 11
which is an estimate of the wing section of the Albatros DVa in the Smithsonian Museum in
Washington.

from Ref. 10.10. The formula in the top left-hand corner of the latter gives the approximate
location of the CP in terms of fraction of the chord aft of the leading edge, x/c, pitching
moment coefficient about the leading edge and lift coefficient:

$$CP = x/c = \text{approximately } C_{mLE}/C_L \qquad (10\text{--}4)$$

The arrangement of the symmetrical tailplane in line with the propeller thrust line provides the vee between the wings (set at incidence $i_W = +3°$ to the tailplane which has an incidence $i_S = 0°$). So, the CG and CP still coincide, exactly as shown by O'Gorman in Fig. 10.6, drawn around 1911. The CP, in the absence of a contribution from the stabilizer, is the neutral point (NP) of the whole system.

Modern aeroplanes have the CG arranged to lie in a range either side of the aerodynamic centre of the wing-plus-body-plus-nacelles, which is a few percent of the standard mean chord forward of the quarter chord of the wing alone:

$$\text{CG coefficient} = \text{CG/wing chord} \qquad (10\text{--}5)$$

$$\text{modern aeroplanes} = 0.16 \text{ to } 0.28 \text{ SMC} \qquad (10\text{--}5a)$$

The results in Fig. 10.14 show that typical World War I sections had the CP, and therefore the CG of the aeroplane, around 38% to 45% of the chord of the equivalent plane (which Barnwell said was at 4/7 of the gap of a biplane above the bottom plane). In fact the range of CGs were:

$$\text{early aeroplane CG coefficients} = 0.25 \text{ to } 0.50 \text{ SMC (equivalent plane)} \qquad (10\text{--}5b)$$

This made them more twitchy than modern machines. Indeed, let Barnwell have the last word:

> '... we want to ensure that our machine has a *slight* margin of stability and that ample controlling power is afforded to the pilot to enable him to quickly alter at will its attitude in any direction.' (Ref. 2.9)

So, the test pilot tasked with flying a rebuilt or replica of a World War I aeroplane must expect it to demand a hands-and-feet-on effort, with characteristics unlike modern production light aeroplanes.

We know that the centre of pressure on an aerofoil is a hard thing to chase around (which is why we now resort to use of the more or less fixed aerodynamic centre, ac). On the wing(s) alone the CP moves forward with decreasing airspeed and increasing angle of attack, to a limit close to the quarter chord at the stall. With increasing speed the CP moves back to disappear somewhere behind the trailing edge. Add a tail and the reverse happens, because we are now considering the CP of the *whole* aeroplane and, as the drawing by O'Gorman showed, the CP of the whole moves in a stabilizing sense: aft with increasing angle of attack, and forward as it is reduced. A tail also increases stability when manoeuvring (see Fig. 10.15).

Today we simplify matters by using the aerodynamic centre of the aeroplane, about which the nose-down pitching moment coefficient remains constant with angle of attack (see Fig. 10.16). The tail (or foreplane) surfaces are then stuck on at the angle needed to reduce the moment to zero at the operational design point, defined by airspeed. Wings near the front and foreplane surfaces all work in a similar way. O'Gorman has already made the distinction for us, quoted earlier. A canard foreplane is only a small wing in front with a large tail (the wing) trailing along behind. A tail at the back has a large foreplane (the wing) up ahead. The difference between them in real terms is that a small foreplane-plus-large rearplane has the overall CP further aft than one with the wing in front. Inevitably this results in larger and heavier and draggier fin and rudder surfaces aft than one finds with a conventional aeroplane.

Look at the proportions of our next example, and also at Table 9-5, which at the top gives total surface area/wing area.

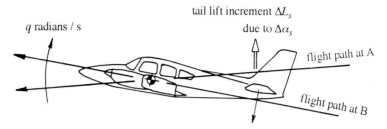

a. Rotation of an aeroplane in pitch between points A and B which are a small time interval apart on a flight path curving around some instantaneous centre of rotation....

...(1) downward pitch rate causes upward airflow component, w , at tail which alters tail angle of attack by $+\Delta\alpha_s$, and in turn increases tail lift by $+\Delta L_s$ ......

....so that $+\Delta L_s$ as drawn moves the overall CP of the aircraft aft, producing an apparent increase in the static margin and the longitudinal stability....

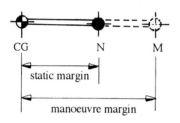

....(2) the increased static margin is re-named the *manoeuvre margin* (which can also be either stick-fixed or stick-free)....

b. Change in angle of attack at the tail, $\Delta\alpha_s$ is decreased by increasing TAS as altitude is gained. The tail lift increment is reduced in direct proportion - and this is the origin of livelier response at altitude (*see* relative density effect in chapter 9 and eq. (9-15)). While a fore-plane ahead of the CG suffers a change in angle of attack in the opposite sense, lively effect upon response is the same

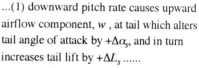

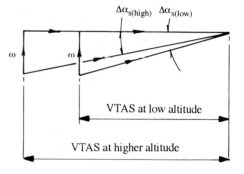

Fig. 10.15 Forces and motions when manoeuvring. Increased tail lift, which moves the neutral point N further aft in Fig. 10.3a and b, changes it into a *manoeuvre point*, M. The static margins, stick-fixed and -free, K_n and K'_n in Equations (3–58) and (3–59) become the *manoeuvre margins*, H_m and H'_m, stick-fixed and -free (see Equations (3–60) and (3–61)).

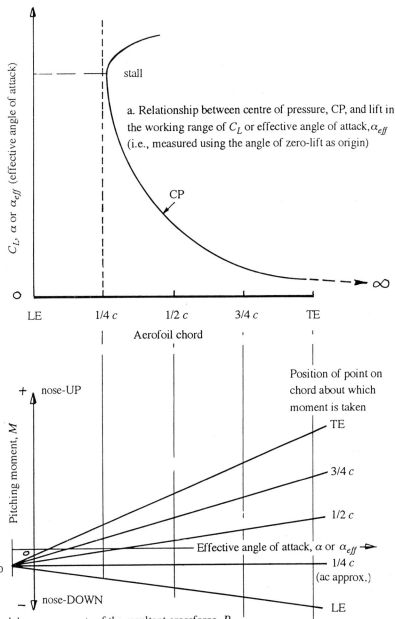

a. Relationship between centre of pressure, CP, and lift in the working range of C_L or effective angle of attack, α_{eff} (i.e., measured using the angle of zero-lift as origin)

b. Lift and drag components of the resultant crossforce, R, acting through the CP, produce a pitching moment which varies in magnitude and sign, depending upon the point about which it is taken to act.

Fig. 10.16   In the case of a modern conventional aeroplane the most convenient location for the CG is at the aerodynamic centre, ac, of the whole aircraft (which is usually about 2% or 3% of the mean chord ahead of the aerodynamic centre of the wing – which lies at or close to its quarter-chord as shown here). Thus, there is a significant difference between modern CG locations and those of ancient aircraft seen earlier.

Example 10–4: Replica – Sopwith F1 Camel

The Camel was rated by Victor Yeates as:

> 'a wonderful machine in a scrap ....'   (Ref. 1.7)

and by Major Oliver Stewart as:

> 'a treacherous aeroplane – one sided, feverish, vicious ... although it killed many people, it evoked fanatic loyalty from those who flew it ....'   (Ref. 10.11)

The replica I flew with an original Clerget rotary engine was a wholly remarkable little aeroplane resembling in detail the drawing in Fig. 10.17. The ratio of (tailplane-plus-elevator)/(wing areas) was only about 9%, marginally less than the 10% given in Table 9–5. The ailerons occupied more than average area, while the fin and rudder were also diminutive – roughly 3% of the wing area tail-down and in the wake of the fuselage decking and engine. In terms of geometry alone, when coupled with long wings, features of this kind should alert any test pilot to the likelihood of qualities of a different, possibly unconventional, nature.

Power was controlled by separate air and fuel levers beside the seat on the left-hand side. Because of my long back I was hunched on top of the gun breeches, and unable to see the ASI and tachometer without screwing my head sideways through 90°. The aeroplane had a fixed tailskid and taxying needed bursts of power to increase the wash over the rudder.

Maximum RPM were slow, around 1100 to 1200 – no more than half those of a modern Lycoming or Continental. The Camel wafted into the air with three skips and a hop, accompanied by a quiet buzz from the large propeller, exhausts, and the friendly smell of hot castor oil. The aeroplane had no stability to speak of about any axis. It was all control. Blink, and it was away in another direction, with instant response to the smallest movement of any control to bring it back again. It could be pointed and sideslipped with ease in any required direction – what was needed of a gun-platform by a scout pilot.

At high speed it was longitudinally and directionally skittish, too much so by modern standards. Turning to the right the Camel whipped round like a dog chasing its tail, appearing to accomplish this in a diameter little more than its own length. Gyroscopic precession caused the nose to drop, and a conscious touch of elevator was needed to level the turn. Turning left, all was balanced and sedate. The stall was docile; with spin recovery in less than a quarter turn, left and right.

After the first take-off, at 800 ft the engine failed. The warm odour of castor oil was gone. A wire to a spark plug had broken, snagged its free end in the soft aluminium cowling and, as the engine-plus-propeller rotated, like a tight cheese-wire the lead cut the others in turn. The aeroplane floated gracefully on its long wings. The ensuing forced landing had to be made in a crosswind from the left. The Camel landed itself, three-point, emphasizing the advantage of control over stability when it comes to the moment of truth.

The next part belongs in the next chapter, but is relevant to what has been said already about wakes. With the small fin and rudder affected adversely by the clutter of wires, struts, engine and fuselage, plus the protruding bulk of the pilot ahead of them, they were ineffective tail-down. The aeroplane swung left into wind against full right rudder and stopped within a few feet of a stone wall. The rudder bar had no pedals, no stirrups, and was coated with glossy varnish. My right foot slid off the end of the bar and became jammed beneath and between it and the fuselage structure. It then took several hard blows with my left heel on the right-hand end of the bar to free my right foot. Only later, after recommending the

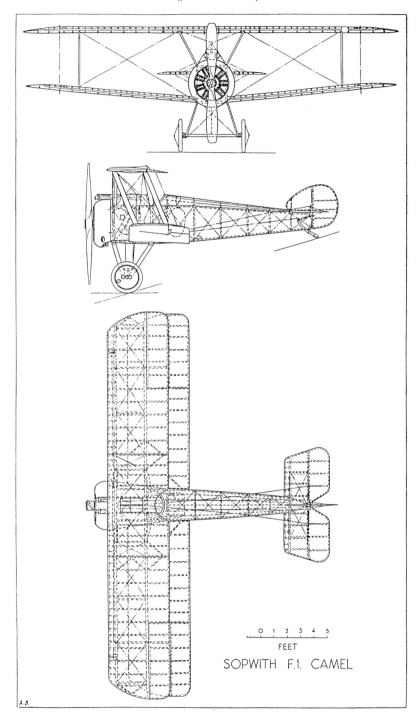

Fig. 10.17 Sopwith Camel, courtesy of Arthur Bowbeer and J. M. Bruce (Historic Military Aircraft No. 10, *Flight*, 22 April 1955). It shows the proportions of this agile, rotary-engined, unstable little scout which had a tailplane and elevator about 9%, fin and rudder only 3% and ailerons around 15% of the total wing area. Compare with Table 9–5 to see that, with the exception of the ailerons, such values are the bottom limit of what is now recognized as current practice.

addition of stirrups, someone discovered notes from World War I which advised the pilot to mount the Camel and '*place the feet firmly in the stirrups*'. The problem had reared its head and been solved already.

As mentioned earlier – and the comments made by the (then) Squadron Leader Hill, on the adverse effects the increase in static longitudinal stability had on landing a Camel, as a result of moving the wings aft by 7 in (about 18 cm), reducing the:

$$\text{CG coefficient from } 0.385 \text{ to } 0.27 \tag{10–6}$$

– the incident confirmed how right he was about the ease of landing a Camel without static longitudinal stability.

Lack of stability in scouts was less important than it was for larger reconnaissance and bomber aeroplanes, because sorties were comparatively short. But there were other reasons, such as small inertia in pitch, implicit in the configuration of the Camel, which contributed to its agility. This was dealt with in Chapter 9, when discussing the interplay of aerodynamics and inertias, and the comments of Cecil Lewis on the merits of the SE5 and the Camel, quoted from Ref. 1.8.

Another observation of Hill's is relevant. By 1922 Farnborough test pilots were exploring the 50% higher number of accidents to pilots learning aerobatics in unstable aeroplanes. Frequently the aeroplane could not be righted when inverted:

'It should be noted that the standard fighting scout of the Royal Air Force is longitudinally unstable ....* The chief difference lay in the absence of self-righting properties in the longitudinally stable type .... Though if anything more longitudinally stable than the modified Camel, the SE5A was the more controllable aeroplane in inverted flight because of its effective tail.'   (Ref. 10.12)
* See Equation (10–6).

For comparison the SE5A is added in Fig. 10.18. Note the different proportions: wings of lower aspect ratio (more raked tips on the original aeroplane, giving it a longer span, were cut back as a result of their structural failure – which modification accounts for the differing rake of the wing tips and the tips of the tailplanes and elevators). There is marked dihedral on upper and lower planes. The tailplane and elevator combination is proportionally larger than that of the Camel. The nose is much longer, and the cockpit farther back, because of the geared in-line engine and fuel tank (see Cecil Lewis, Ref. 1.8), when discussing *Aerodynamics vs Inertia* in Chapter 9). The fin and rudder combination is also larger, and the aeroplane had a ventral fin surface.

In many ways the Camel and SE5A were the equivalent in World War I of the British Hurricane and Spitfire of World War II.

The Equivalent Monoplane

Another source of difficulty arises when a replica of an historic aeroplane – which is otherwise 'right' in every detail but has the 'wrong' engine – fails to have the correct relationship between the CG and CP. The rotary moment of inertia about each axis can make the flying qualities wildly unrepresentative.

To sort out the CG–CP we have to start by finding the centre of pressure of the *equivalent monoplane* wing of a biplane or triplane. This involves estimating the lift of each wing and calculating its moment about a convenient datum, usually the leading edge of the bottom plane. The sum of the moments is then divided by the sum of the lift of all of the planes. It

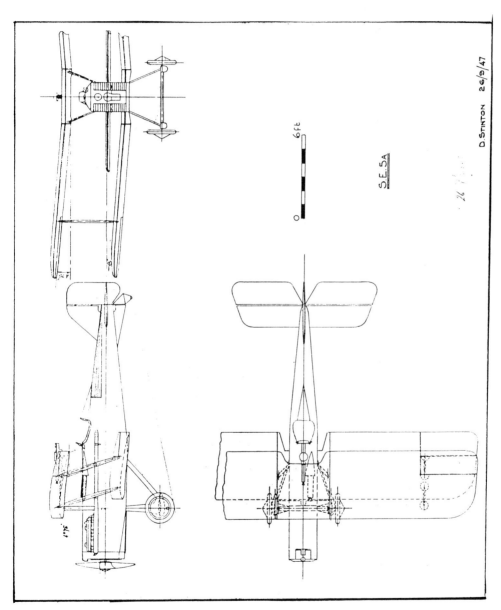

Fig. 10.18 SE5A designed at the Royal Aircraft Factory, Farnborough, 1916, its proportions should be compared with the preceding drawing of the Sopwith Camel. Tail surfaces are relatively much larger in area. The engine is a geared in-line Vee, making the nose longer, increasing the pitching moment of inertia (B).

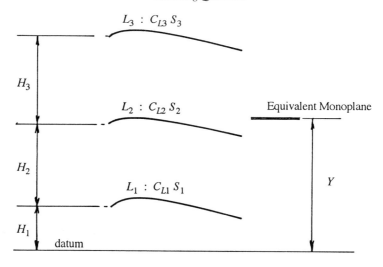

$L_3 : C_{L3} S_3$

H_3

$L_2 : C_{L2} S_2$ Equivalent Monoplane

H_2

Y

$L_1 : C_{L1} S_1$

H_1

datum

Taking moments about the datum, to find the equivalent monoplane of a biplane or a triplane (for a biplane $L_3 = 0$):

$$L_1 H_1 + L_2 (H_1 + H_2) + L_3 (H_1 + H_2 + H_3) = Y (L_1 + L_2 + L_3) \qquad (10\text{-}7)$$

On dividing throughout by L_1:

$$H_1 + (L_2 / L_1) (H_1 + H_2) + (L_3 / L_1) (H_1 + H_2 + H_3) = Y [1 + (L_2 / L_1) + (L_3 / L_1)] \quad (10\text{-}8)$$

When biplanes and triplanes are orthogonal (planes of equal size), $S_1 = S_2 = S_3$, and

$$(C_{L1} / C_{L2}) = (L_2 / L_1) = p, \text{ while } (L_3 / L_1) = (L_3 / L_2)(L_2 / L_1) = p^2 = (C_{L3} / C_{L1})$$

such that: with forward stagger $p =$ about 1.35

and with back-stagger or none $p =$ about 1.00

The reason why the lift of the upper plane is higher than the one immediately below is that the upper interferes adversely with the lifting airflow over the plane beneath it.

Fig. 10.19   General treatment of triplanes and biplanes to find the equivalent monoplane wing, which is needed for correct location of the CP and CG.

helps to have an idea of the general method, so this is shown in Fig. 10.19 and Equations (10–7) and (10–8), enabling either wing arrangement to be worked out as follows. It also helps to know that with one plane above another and with both of equal area, the one above has about one third more lift than the plane beneath. If they are not of equal area, then their area-ratio must be taken into account as an additional factor:

☐ For an *orthogonal triplane* with forward stagger, assuming:

gaps, $H_1 = 0$, $H_2 = H_3$, $(H_2 + H_3) = G$, and relative lift, $p = 1.3$, so that Equation (10–8) becomes:

$$[(1.3/2) +1.3^2]/[1 + 1.3 + 1.3^2] = Y/G = \text{about } 0.59, \text{ or } 7/12 \qquad (10\text{-}9)$$

☐ For an *orthogonal biplane* with forward stagger, making the same assumptions as in the case of the triplane, but adding that $H_3 = 0$ (so that $H_2 = G$), then:

$$[1.3/(1 + 1.3)] = Y/G = \text{about } 0.57, \text{ or } 4/7 \qquad (10\text{--}10)$$

which is the same as the value recommended by Barnwell.

Example 10–5: Replica – Sopwith Triplane

The triplane, like the biplane, was conceived originally by Cayley. The elegant little Sopwith Triplane of 1916, described by an RFC pilot as '*an intoxicated staircase*' (Ref. 10.15), was derived from the earlier and equally dainty Sopwith Pup. For a time Triplanes operated by the Royal Naval Air Service were the top scoring fighters on the Western Front. They were reported to be able to overtake an opponent underneath before rearing nose-up, hanging on the large slow-turning propeller and opening fire from below. Today, in the early 1990s, this would possibly be recognized as a relative of the modern 'Cobra manoeuvre', devised by pilots of agile Russian fighters.

In this case a problem arose with a near accurate Triplane replica fitted with the standardized larger tail, shown in Fig. 10.20, but with the rotary engine replaced by a lighter fixed Siemens radial of a much later date. Figure 14.3 shows that the World War I rotaries were heavier per horsepower than later air-cooled engines.

Tail heaviness was the main concern on the test flight. An already-reported tendency to ground-loop pointed to a CG much too far aft. In flight the aeroplane had to be held down all of the time. Releasing the stick at altitude resulted in a loop.

We knew where the CG was but not where it should have been. It involved finding the equivalent plane, described in Fig. 10.18, and estimating the likely CP, using the information in Figs 10.13 and 10.14. Figure 10.21 shows part of the investigation involved, using the most representative and accurate drawings to hand – in this case, taken from Ref. 10.13.

Beware of 'experts'. When attempting to locate the most likely CP–CG of the original aeroplane advice of one such expert was that '*All Sopwith Aeroplanes were tail heavy, and this* (replica) *is no exception*'. It was only through extensive reading that a source was found which stated that the original Triplane could be trimmed to fly hands-off, unlike contemporary types from the same Sopwith stable.

From Fig. 10.13 the aerofoil section used for the replica appeared to resemble USA 5, so that at an incidence of 2° one might reasonably expect the CP to lie near the 50% chord. A cross-check, using the point of contact of the mainwheels with the ground, and then striking lines at 10° and 13° aft of the perpendicular (which is typical of the compromise needed between ease of lifting the tail on take-off and not tipping nose-over on landing), confirmed the CG estimate to be reasonable. To achieve it nose ballast was needed.

At a test weight of 1554 lb (706 kg), with ballast of 53 lb (24 kg) attached to the engine mounting, plus a 22 lb (10 kg) machine gun, the CG coefficient was at 51.8% of the equivalent monoplane chord – and 42.6% of that of the middle plane. The original gross weight of a Triplane is quoted as 1415 lb (643 kg), so that the replica was nearly 10% heavier. Even so, it stalled at a slow 42 MIAS, and could be trimmed 'hands-off' at 90 MIAS with elevator trim $\frac{1}{2}$ nose-down (tail lift upwards). But, when pitched nose-up at that speed by a gust, there was no recovery, hands-off, until the speed had dropped to 60 MIAS, at 2000 RPM.

Note: An original Triplane would have had around half these RPM and much slower propwash. The power of the Siemens radial would therefore have been far more destabilizing

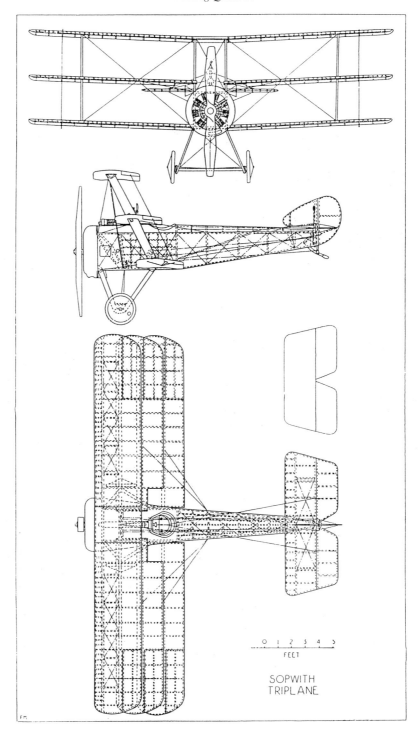

Fig. 10.20 Rotary-engined Sopwith Triplane of 1916, courtesy of F. Munger and J. M. Bruce (Historic Military Aircraft, No. 16, *Flight*, 19 April 1957). An aeroplane with the alleged ability to 'hang on its prop'.

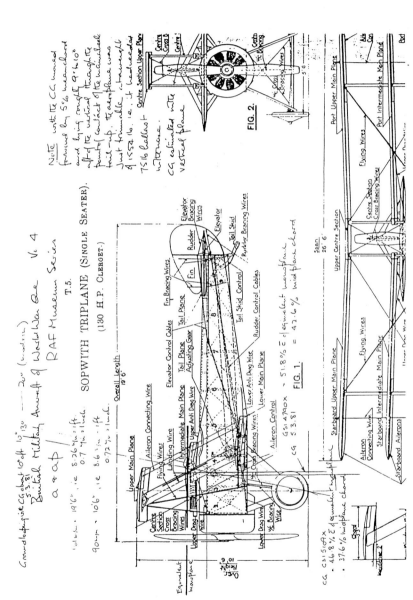

Fig. 10.21 Sopwith Triplane drawing from a source which is a close contemporary of the original aeroplane (Ref. 10.13) used for the purpose of flight testing a full-scale replica to determine an acceptable CG position for a Permit to Fly (see Fig. 10.21 Notes/Commentary for interpretation of handwritten annotations). The replica had an unrepresentative radial engine instead of a rotary, was uncharacteristically tail-heavy (unlike the original Triplane) and was reported to ground-loop with ease.

Working from top to bottom and left to right of drawing of 'Sopwith Triplane' copied from
a contemporary source:

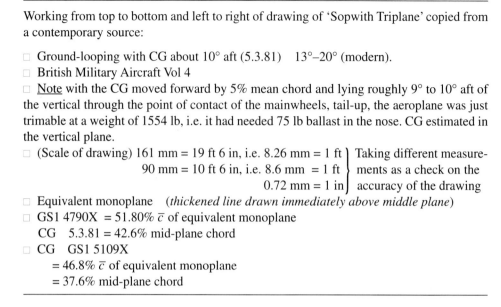

☐ Ground-looping with CG about 10° aft (5.3.81)   13°–20° (modern).
☐ British Military Aircraft Vol 4
☐ <u>Note</u> with the CG moved forward by 5% mean chord and lying roughly 9° to 10° aft of
the vertical through the point of contact of the mainwheels, tail-up, the aeroplane was just
trimable at a weight of 1554 lb, i.e. it had needed 75 lb ballast in the nose. CG estimated in
the vertical plane.
☐ (Scale of drawing) 161 mm = 19 ft 6 in, i.e. 8.26 mm = 1 ft ⎫ Taking different measure-
                  90 mm = 10 ft 6 in, i.e. 8.6 mm = 1 ft ⎬ ments as a check on the
                          0.72 mm = 1 in ⎭ accuracy of the drawing
☐ Equivalent monoplane   *(thickened line drawn immediately above middle plane)*
☐ GS1 4790X = 51.80% $\bar{c}$ of equivalent monoplane
   CG   5.3.81 = 42.6% mid-plane chord
☐ CG   GS1 5109X
      = 46.8% $\bar{c}$ of equivalent monoplane
      = 37.6% mid-plane chord

Fig. 10.21 Notes/commentary

for the replica, with reduced aerodynamic damping. It was pragmatic to settle for what we
could get and for what we would have liked.

The replica became longitudinally unstable, but flyable, at 105 MIAS and 2000 RPM.
All told, it was considered to be acceptable for the purpose of demonstration, as intended.

During the investigation an error was discovered in the weight schedule that had been
provided. This makes the point that *it is a responsibility of the test pilot to check and discover these things, and to see that they are noted and rectified.*

S. B. Gates and aeroplanes from 1915 to 1945 (the Classic and 'Warbird' era)

S. B. Gates worked at the Royal Aircraft Establishment from 1915, when it was still the
Royal Aircraft Factory, until around 1972, long after I had been posted onwards. During
the 1970s he was given the title 'The Dean of Stability', because of the outstanding contributions that he made to the subject. His work has been referred to more than once
in this book, for the good reason that what he has to say is in the language of the test
pilot as well as the engineer. His brevity and clarity is attributed to the unfortunate
stammer from which he suffered. It was such an effort to enunciate his words that those he
spoke were short rather than long, and selected with precision. Gates was a mathematician, a 'wrangler' and joined what in time became the Aerodynamics Department. He
recalled in a biographical memoir that it was (later Sir) William S. Farren (1892–1970)
who first interested him in Flight Dynamics. In quite a different direction, it was W.S.
Farren to whom I owe having been taught to fly, as the result of a scheme he introduced for
those on the design staff, when Technical Director of the Blackburn Aircraft Company.

Gates argued that trim and control force characteristics are the clearest measure of static
stability, in that the magnitude of the forces depend upon the CG and the relevant Neutral

Point. From this he arrived at the idea of *stick-force per g* (see Fig. 3.20), which is the force felt by a pilot per unit increment of applied normal acceleration during a pitching manoeuvre (Ref. 10.14). Whether or not this idea came in a blinding flash we shall never know, but it has that mark of authenticity.

Starting with the Neutral Point as a fixed entity he went on to develop its application by stages, each depending upon contributions from different parts of the aeroplane (see Fig. 10.3).

Note: Gates uses N_0, N_1, N and $\bar{N}$ to define each particular NP. We have added an extra one, N_2, when a larger tail is fitted. It is also useful at this point to introduce his concept of the Manoeuvre Point, MP, which, for simplicity, has not been shown in Fig. 10.3. The Manoeuvre Point, which is explained below, is also displaced aft, like N_2, for the reason given in Fig. 10.15a(2), in which the neutral and manoeuvre points are shown as N and M.

Steps taken using Gates' neutral point:

☐ The *neutral point*, N_0, of the wings alone, their aerodynamic centre, is close to the quarter chord point.
☐ The *nacelles and body* shift the neutral point forward to N_1.
☐ The *tail* shifts the neutral point back from N_1 to N, aft of the CG.
☐ When the aeroplane manoeuvres (e.g., as when pulling out of a dive, as in Fig. 10.15), the tail stabilizer is moving downwards about the CG. This increases its angle of attack and the lift it generates, so moving the neutral point, N, to a new position further aft, called the *Manoeuvre Point*, MP. The aft movement of NP to MP is stabilizing. Like the neutral points, stick-fixed and free, there are manoeuvre points, stick-fixed and -free, H_m and H'_m, given in Equations (3–60) and (3–61).
☐ A *larger tail surface* has the powerful effect of moving N_1 further aft of the CG, to N_2.
☐ *Slipstream (propwash)*, when appreciable, is usually adverse, involving a forward shift of the neutral point, to $\bar{N}$. Thus the neutral point, N (without propwash), represents the case when gliding.
☐ At *high speed* the neutral points engine ON ($\bar{N}$) and OFF (N) tend to coincide. As speed decreases $\bar{N}$ moves forward of N (but this movement was not established as being progressive as the stall is approached). The distance ($\bar{N} - N$) is always greater in the climb than at top speed. The distance is probably largest just before the stall.
☐ The *unfavourable effect of the propwash* increases with propeller disc-loading at a given airspeed, and decreases with height of the thrust line above the CG.
☐ *Aeroelastic distortion* of the airframe and tail control and stabilizing surfaces moves neutral points N and $\bar{N}$ forward, reducing stability.

Gates concepts were evolved for low speed 'rigid' aeroplanes. As performance improved with the coming of the jet engine he sought to extend his methods to include the conditions of high-speed flight. His neutral point is approximately the same as the manoeuvre point during a short period oscillation, when the aeroplane is pitching in a similar fashion to that shown in Fig. 10.15b.

A recurring question is that of longitudinal manoeuvrability which, in theory, can be adjusted in two ways. The first is by giving an aircraft comparatively large static and manoeuvre margins, with a closely-balanced elevator to keep stick forces low. The second is to maintain small stability margins while relaxing elevator hinge moment balance (Ref. 9.5).

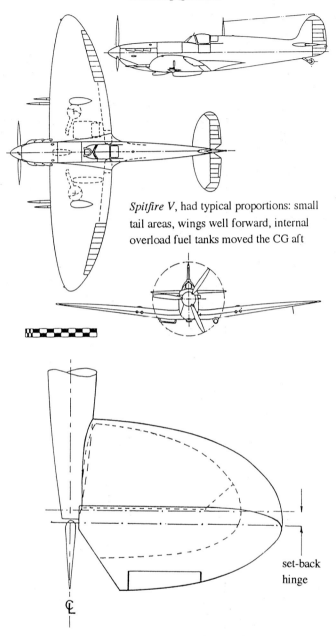

Spitfire V, had typical proportions: small tail areas, wings well forward, internal overload fuel tanks moved the CG aft

set-back hinge

Fig. 10.22   To increase the range of the Spitfire by adding internal fuel (aft of the CG) had an adverse effect upon longitudinal stability. One proposed cure was to have been the fitting of an experimental large tail, to explore the theory that the longitudinal stability could be achieved by complementary routes of increased static and manoeuvre margins, while reducing elevator hinge moments by close balancing (Refs 9.5 and 9.9). The modification was shelved.

Example 10–6: Large-Tail Spitfire

Early marks of Spitfire suffered from a CG that was too far aft, which made it difficult to extend the range and the increasing demands on fuel consumption made by more powerful and thirsty engines. The only obvious spare internal volume for fuel was in the rear fuselage, so far behind the CG that it made matters worse.

RAE Farnborough was given the task of designing an elevator with a wide range of hinge-moment characteristics, but which fitted the existing attachment points on the tail (Ref. 9.5). This is shown in Fig. 10.22, compared with a drawing of the Spitfire V (from Ref. 10.16). The larger elevator (Fig. 10.22) had a balance tab and, initially, a horn balance was intended to give a more extreme range of experimental hinge-moment characteristics. Location of the tail attachment points limited the amount of set-back hinge balance that could be employed. Because of the structural depth needed for the modified hinge it also broke RAEs design rule: '*that trailing edge angles should be kept below 16°*', to avoid steep slopes to the trailing edge which cause uneven and premature separation, accompanied with buffet.

What the experimental tail showed was that increased area to extend the static and manoeuvre margins could be accommodated without proportional increases in stick-force per g, by closely balanced elevators, provided that $-b_2$ was not less than 0.05 and the manoeuvre margin not greater than 0.09 MAC (Ref. 9.5).

Flight testing of the modified tail was delayed and finally abandoned. Later marks of longer nosed more powerful Spitfires, with Griffon engines, did not present quite the same CG and loading problems. Later marks, like the F 22, had larger tail surfaces, but kept the original form of horn balance.

Gates' Advice on Piloting Requirements for Longitudinal Stability

The following simple rules stated by Gates are still relevant today for every test pilot, designer and home-builder of an aeroplane:

☐ The *stick-free stability should be at least neutral* at the aft centre of gravity in the climb, but at the same time:
☐ At the *forward centre of gravity the stick force* to hold a speed 30% above the top speed when trimmed for top speed should not be prohibitive.
☐ At the *forward centre of gravity there must be enough elevator* to get the tail down when landing with flaps down.

This advice is fundamental to all aeroplanes, whether the control system is manual, or electronic. A pilot is happiest when the aeroplane abides by these simple rules.

In the next chapter we consider lateral and directional flying qualities.

References and Bibliography

10.1 O'Gorman, M., *Problems Relating to Aircraft*. London: Incorporated Institution of Automobile Engineers, 1911

10.2 Letters, *Aviation Consumer*, Vol. XXIII, Nos 11, 12. Greenwich: Connecticut 06836-2626, June 1993

10.3 8110.7 *Engineering Flight Test Guide for Small Aeroplanes*. Washington: Department of Transportation, Federal Aviation Administration, 20 June 1972

10.4 Hill, Squadron Leader R., *Experiments on a Modified Camel Aeroplane*. Farnborough: RAE Report No BA 340, 25 January 1920

10.5 Gates, S. B., *Longitudinal Stability and Control*. London: Aircraft Engineering, September 1940

10.6 Lanchester, F. W., *Aerodonetics*. London: Constable and Co, 1908

10.8 Sproule, J. S., *Checking up on Sir George*. London: Shell Aviation News 405, 1972

10.9 Parks, D., *From the Archives NACA – 1915*. Oshkosh: Sport Aviation, Experimental Aircraft Association, October 1987

10.10 Warner, E. P. and Johnson, S. P., *Aviation Handbook*. New York and London: McGraw-Hill, 1931

10.11 Pudney, J., *The Camel Fighter*. London: Hamish Hamilton, 1964

10.12 Hill, Squadron Leader R.M., *The Manoeuvres of Inverted Flight*. R & M No. 836 (Ae 86), London: HMSO, September 1922

10.13 RAF Museum Series, *British Military Aircraft of World War One, Vol 4*. London: Arms and Armour Press, 1976

10.14 Gates, S. B., *An Analysis of Static Longitudinal Stability in Relation to Trim and Control Force*. R & M No. 2132, London: HMSO, April 1939

10.15 Morris, A., *Bloody April*. London: Jarrolds Publishers (London), 1967

10.16 Anderson, S. B., 'A look at handling qualities of canard configurations', Paper 85-1803, *AIAA 12th Atmospheric Flight Mechanics Conference*, 19–21 August 1985, American Institute of Aeronautics and Astronautics

10.17 Butler, G. F., An analytical study of the induced drag of canard-wing-tail aircraft configurations with various levels of static stability, *Aeronautical Journal of the Royal Aeronautical Society*, October 1983

10.18 Howe, Professor D., Aircraft that fly backwards? The application of forward-swept wings, *Aeronautical Journal of the Royal Aeronautical Society*, October 1983

10.19 O'Gorman, M., Stability devices, *Aeronautical Journal of the* (then) *Aeronautical Society*, now the RAeS

CHAPTER 11

Rolling Moment with Sideslip, and Directional Stability

'… Ailerons ride a bit too high. The fin needs slight adjustment. I note these items down on my data board, and push the stick over to one side. The wing drops rather slowly. The ailerons on the *Spirit of St Louis* aren't as fast as those on the standard Ryan. But we expected that. Hall made them short to avoid overstraining the wing under full-load conditions, and he gained a little efficiency by not carrying them all the way out to the tip. The response is good for a long-range air-plane ….'

Charles A. Lindberg, *The Spirit of St Louis* (New York: Charles Scribner's Sons, 1953)

In the conventional course of events this chapter would have been entitled, simply, '*Lateral and Directional Stability*' except that, as said elsewhere, *there is no such thing as lateral stability*. To be stable longitudinally or directionally, an aircraft must have a tendency to return to a previously-trimmed state following a disturbance, to take up a flight path pointing into an altered wind direction at the same angle as before. What happens in the lateral case is that (ideally) the aeroplane is caused by dihedral to roll *away* from the component of relative wind due to sideslip, not to adjust and point into it. Our interest is in *dihedral effect* – the rolling moment due to sideslip – and those factors which affect its magnitude and direction.

Sideslips

Sideslipping is flight in which the aeroplane is yawed, causing a component of airflow to come from one side. It is an easy manoeuvre to perform, but it takes skill to execute accurately, involving coordination of eye, hand and foot.

There are two kinds of sideslip used in flight testing: straight and steady; and full-rudder. All sideslips, particularly those with full rudder, must be carried out with care. Until the flying qualities have been determined, a test pilot does not apply rudder quickly or harshly. Control movements are measured with deliberation. A rapid boot-full of rudder converts a static case into one which is dynamic, with an aircraft response that might be hard to check before it becomes excessive. Later, however, when the pilot is more sure of the nature of an aeroplane, it will be necessary to ensure that rapid entry into and recoveries from sideslips do not result in loss of control.

The angle of sideslip is that between the relative wind and the plane of symmetry of the aircraft. The angle has no accurate correlation with what is shown on a turn and slip or turn and bank indicator. Even so, with practice it is possible to produce repeatable and accurate results, cross-referring between original compass heading or heading shown on a direction indicator, and the change of angle caused by application of control.

Sideslipping is a neat and useful manoeuvre, carried out at relatively low speeds – not more than V_A, the structurally limiting design speed beyond which full control must never be applied. It is used operationally for getting rid of excessive height, and when carrying out crosswind approaches and landings. It is especially useful in forced landings, when there can be far more grief touching down short and hitting the near hedge at 50 KTAS, than over-running into the far hedge at 5 knots. To glide approach and arrive with height in hand, which can be slipped off, makes that kind of difference after the loss of one's single or final engine.

In flight sideslips are most useful for checking and measuring the interactions between dihedral effect and weathercock stability, neither of which can be entirely separated. Tests in Table 4–6, items 11 and 12 include seven basic configurations, sideslipping to left and right, at 1.2 V_S for each. Before contemplating sideslips the pilot should know the design manoeuvring speed, V_A, and, of course, the proven (demonstrated) stall speed and weight in the selected configuration.

A sideslip test starts with picking a datum to slip towards: e.g., a distant feature on the ground, or a particular edge of cloud visible through the front windscreen which, being in the same air mass, has no movement relative to the aeroplane. Set the test configuration and trim at the required airspeed. Point the nose a few degrees to the side of the datum OPPO-SITE to the one in which it is intended to slip.

Note: The aeroplane is trimmed in the desired test configuration before pointing the nose to one side, because this gives the opportunity to detect tendencies when the rudder is used on its own in a slight flat turn.

When settled, the test is started by applying bank in the intended direction of slip, as shown in Fig. 11.11. Angles of sideslip used for tests must be appropriate to the type of aeroplane, but in no case should the constant heading sideslip angle be less than that obtainable with 10° of bank. If, however, it turns out to be less, then the sideslip angle for the test should correspond with the maximum bank angle obtainable with full rudder deflection or a 150 lb (66.7 daN) pedal force (Table 9–2).

As the aircraft starts to slip sideways it begins to lose height, because the lift resultant is canted sideways and its vertical component, which opposes weight, is reduced. Ideally the aircraft also starts to weathercock into the sidewind – as long as the fin surface is adequate – by rotating about the CG. The nose goes down, in addition to which drag is increased, and with it the rate of descent. To restore the required rate of descent, ease back on the control column to increase the angle of attack, just enough to make the lift component again equal the weight, while maintaining direction of slip with rudder and angle of bank with aileron. Because the machine is slipping sideways this is the reason for pointing the nose offside initially. If one does not, then the aeroplane slips in a different direction, and may also turn as it does so, through lack of a datum against which to maintain a straight and steady flight path. If that happens, then results and subsequent deductions will be inaccurate.

Ideally, the static lateral rolling moment in a sideslip should tend to reduce the angle of bank with landing gear and flap in any position, using symmetrical power up to 75% MCP, at speeds above 1.2 V_{S1} in the take-off configuration, and at speeds above 1.3 V_{S1} in other configurations. Any tendency for the angle of bank and sideslip to increase is a non-compliance.

Turns Using One Control

Lateral and directional flying qualities cannot be separated with ease. Nor is the approach to their analysis quite the same as that of the last chapter, in which we dealt with longitudinal static and dynamic stability. There could be historic reasons for this being so. The description of sideslipping is evidence of the complexity. Consider what happens when attempting to turn using either the rudder or the ailerons separately.

Start with rudder alone, by pushing the rudder pedal to the left with the wings level. If there is positive dihedral effect the aeroplane will start to roll to the left:

☐ As the aeroplane yaws the starboard wing, travelling faster than the port, generates extra lift and this rolls the aircraft to the left.
☐ The aeroplane is now skidding to the right with the nose possibly turning farther to the left, with the aircraft attempting to negotiate an uncomfortable flatt-ish turn to port.
☐ Positive dihedral effect causes the starboard (upwind) wing to rise, which rolls the aircraft to the left, and the port wing, which is also retreating and travelling more slowly than the starboard wing, descends. This increases the angle of attack of the port wing which increases its drag. If the speed is slow enough the port wing might stall, followed by rolling and yawing departures to the left …

So, moving the rudder alone causes both yaw and roll, and this could lead to much else.

Now consider the situation with rudder fixed while attempting to turn using aileron alone. Movement of the stick to the left depresses the starboard aileron and raises the port. The aeroplane rolls to the left, but almost certainly suffers some adverse yaw to the right at least, for reasons given in Fig. 9.11 (which led to the design of the Frise aileron). Yaw to the right causes sideslip to the left, such that the relative airflow is from the port side. If the area of the vertical tail surfaces is adequate, the nose will gradually yaw to left with the aeroplane sideslipping around a turn to the left, with its nose yawed to the right, away from the centre of the turn. It is an uncomfortable maneouvre.

If the area of the vertical surfaces is barely adequate, the angle of sideslip will be increased by the drag of the starboard aileron, with the nose yawing farther to the right. Eventually the port wing might generate enough roll with sideslip to balance the aeroplane in an uncomfortable sideslipping turn to port. If the keel surface aft of the CG is inadequate, directional stability will be lost and the aeroplane might well roll against aileron. One can only speculate.

All of this is too full of words. Shorthand is needed which enables us to analyse the various contributions.

Derivatives (Shorthand Descriptions of Lateral and Directional Responses)

Figure 3.11 and Table 3–1 showed the general system of axes and notations, whereas Fig. 11.1 has extracted the particular lateral and directional elements of Fig. 3.11, as used in the UK, to create the derivatives listed in Table 11–1. These enable us to break down the origins of forces and moments in terms of the control inputs and responses causing them. They are named after the letters used in their construction, for example: L_ξ (el-xi), N_ζ (en-zeta), L_p (el-pi), L_v (el-vi), and so on. A corresponding set applies to the longitudinal case, but they are not used in the UK to anything like the extent of the lateral and directional cases.

In Table 11–1 those listed along the top might be labelled CAUSE, while those down the side are EFFECT. For example, moving the stick to the left with rudder fixed introduced aileron deflection, ξ, yawing moment, N, and sideslip velocity, v. When discussing

Elements shown in +ve sense
(see fig. 3.11 and table 3-1)

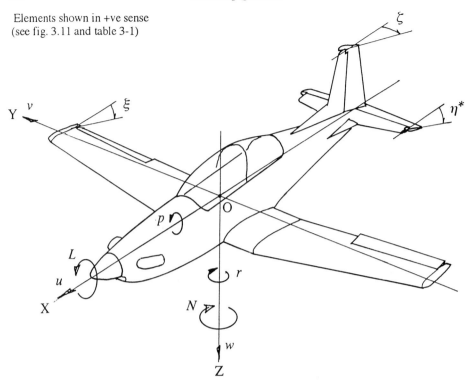

*the effect of elevator is not unknown in the directional case, although its is unlikely to
affect other than a propeller-driven T-tail configuration at low airspeed and high power
(*see* example 12-9). The sign of any moment depends upon handing of the propeller.

Fig. 11.1   Elements involved in generating the lateral derivatives in Table 11–1.

handling qualities in general and for the purpose of design, we resort to non-dimensional
coefficients – like the lift and drag coefficients and others with which we are already familiar. Examples are:

☐ *Rate of change of rolling moment with aileron deflection*:

$$\delta L / \delta \xi = L_\xi \text{ for the moment derivative} \tag{11–1}$$
$$= l_\xi \text{ for its non-dimensional coefficient} \tag{11–1a}$$

☐ *Rate of change of yawing moment per unit aileron deflection*:

$$\delta N / \delta \xi = N_\xi \text{ for the moment derivative} \tag{11–2}$$
$$= n_\xi \text{ for its non-dimensional coefficient} \tag{11–2a}$$

☐ *Rate of change of yawing moment per unit sideslip velocity*:

$$\delta N / \delta v = N_v \text{ for the moment derivative} \tag{11–3}$$
$$= n_v \text{ for its non-dimensional coefficient} \tag{11–3a}$$

in which the non-dimensional coefficients are obtained by dividing the basic force and
moment derivatives by the factors in Table 11–2, e.g.:

Table 11–1

Lateral derivatives

Derivatives caused by moment or force	Control displacement		Angular velocities		Sideslip velocity v	Weight* W
	Aileron ξ (xi)	Rudder ζ (zeta)	Roll rate p	Yaw Rate r		
Rolling moment, L	$[L_\xi]$	L_ζ	L_p	L_r	L_v (dihedral effect)	0
Yawing moment, N	N_ξ	$[N_\zeta]$	N_p	N_r	N_v (weathercock effect)	0
Side force, Y	Y_ξ (usually negligible)	Y_ζ	Y_p (negligible)	Y_r (negligible)	Y_v	$W \sin \phi^*$

* Weight and its effect, due to angle of bank, ϕ, has been added for completeness, even though $W \sin \phi$ is not a derivative.
[ ] Primary effects of aileron and rudder controls.
Note: A derivative N_η, reducing the authority of the fin and rudder of a T-tailed propeller-driven aeroplane is rare, but not unknown (see Example 12–9).

TABLE 11–2

The factor by means of which the derivatives in Table 11–1 must be divided to make them non-dimensional

	Control displacement	Angular velocity	Sideslip velocity
Side force	$\rho V^2 S$	–	$\rho V S$
Moment	$\rho V^2 S \, (b/2)$	$\rho V S \, (b/2)^2$	$\rho V S \, (b/2)$

b = wing span ρ = ambient density S = wing area
V = TAS in compatible units, e.g., ft/s or m/s
Note: $(b/2)$ in some texts is written as s.

$$n_\xi = N_\xi / [\rho \, V^2 \, S \, (b/2)] \tag{11–2b}$$

in which $(b/2)$ is the semi-span of the wing. Thus the total yawing moment on a particular aeroplane, caused by the pilot doing nothing beyond moving the control column to the left is the sum of N_ξ and N_v. But, as a general statement affecting all aeroplanes one would say that it is the sum of n_ξ and n_v.

In test flying there is considerable cross-referencing between UK and US sources. Table 11–3 lists a number of corresponding British and American symbols which crop up during transatlantic discussions of stability and control qualities in roll and yaw.

Table 11–3

Transatlantic correspondence between terms used when
discussing control deflections and their non-dimensional
roll, yaw and sideforce coefficients

Item	UK	USA
1. Control deflections:		
Aileron	ξ	δ_a
(Elevator	η	$\delta_e)*$
Rudder	ζ	δ_r
2. Non-dimensional rolling moment coefficients		
Aileron	l_ξ	$C_{l\delta a}$
Rudder	l_ζ	$C_{l\delta r}$
Roll rate	l_p	C_{lp}
Yaw rate	l_r	C_{lr}
Sideslip velocity	l_v	$C_{l\beta}$
3. Non-dimensional yawing moment coefficients		
Aileron	n_ξ	$C_{n\delta x}$
Rudder	n_ζ	$C_{n\delta z}$
Roll rate	n_p	C_{np}
Yaw rate	n_r	C_{nr}
Sideslip velocity	n_v	$C_{n\beta}$
4. Non-dimensional side force coefficients		
Aileron	y_ξ	$C_{y\delta a}$
Rudder	y_ζ	$C_{y\delta r}$
Roll rate	y_p	C_{yp}
Yaw rate	y_r	C_{yr}
Sideslip velocity	y_v	$C_{y\beta}$

* Added for consistency with text (see Example 12–9).

Sign conventions

Examination of the sign convention used for dealing with axes, control deflections, forces
and moments in Figs 3.11 and 11.1 reveals a 'Right-Hand Rule', which is that of the wood-
screw, advancing when turned clockwise. Start by looking in the direction X from the origin, O (which is often the CG) the linear motion along it and clockwise rotation are both
positive. Now, put out the right arm to the side and we have axis O–Y, at right angles to O–
X. Linear motion along the Y-axis and clockwise rotation are again positive. Return again
to looking along O–X and roll clockwise through 90° with the right arm out. It now points
downwards along the O–Z axis. As before, motions along and clockwise rotation about
O–Z are positive.

Control deflections are positive when rotation is clockwise about the hinge, relative to
each relevant axis – but note that POSITIVE control deflections produce opposite, NEGA-
TIVE, reacting moments about each axis.

Derivatives, and their coefficients, are valuable tools – so much so that they now form
part of the common vocabulary of test pilots, scientists and engineers. Their origin lay with
Professor G. H. Bryan FRS and W. Ellis Williams, both of the Mathematics Department
of the University College of North Wales, in Bangor. Bryan was interested in flight and,
in the mid 1890s, wrote about the sailing flight of birds, in 1901 on the history of aerial
locomotion, and by 1904, together with Ellis Williams, presented before the Royal Society

(Ref. 3.4), on *The Longitudinal Stability of Aerial Gliders*. Williams, who was a lecturer in physics as well as a mathematician, was writing primarily about the stability of flying machines by 1904 (he called it balancing), and had built an aeroplane by 1910. Historically, Ref. 3.4 is possibly the first evidence of derivatives being used in the analysis of stability.

Roll and Yaw Qualities Described by Derivatives

Perhaps the first thing to appreciate is that, in spite of the mathematically coded squiggles, the lateral and directional modes are not fundamentally different from the longitudinal. This point is made in Fig. 11.2, which shows that in each plane the forces which act may be resolved so as to find a neutral point. NP is the longitudinal neutral point, already familiar in longitudinal static stability. Two more hypothetical points are devised: NP_D is the corresponding CP, the resultant of all forces contributing directional (weathercock) moments in the X–Y plane; while NP_L in the Y–Z plane is the vertical location of NP_D and, hence, the CP of the lateral forces which contribute rolling moments. Neither are used in practice, simply because the stability game starts from a different viewpoint and has been developed otherwise. But, I find it useful to keep the sketches in Fig. 11.2 in mind when analysing a test result, and for teaching. It is usually easier to visualize a problem or to put across an idea by drawing it, than by juggling with equations.

Lateral and directional causes and effects

Each lateral and directional derivative denotes a particular lateral or directional input, a cause, with unique effects. These includes the magnitude of the force fed back to the pilot, the nature and degree of response of the aircraft, and any cross-coupling in the event of a departure (e.g. yaw–roll coupling of high performance aeroplanes with short wings (moment of inertia, $I_{xx} = A$) long bodies ($I_{yy} = B$) and B/A > 1.0).
  Taking each in turn:

$-L_\xi$: rolling moment due to aileron

With the starboard aileron down, $+\xi$, the lift of the starboard wing is increased, this causes a negative rolling moment to port, denoted by $-L_\xi$ and its coefficient $-l_\zeta$. Consequently:

(1) *Roll rate with time and airspeed.* The rolling moment is a maximum when the aileron is first deflected, but aeroplanes resist any attempt of a pilot to roll. The degree of resistance depends upon inertia in roll, A, and upon aerodynamic damping, which is a function of the helix angle in roll (see Equation (9–1)). Long span wings resist roll more than short and, as noted much earlier, agile aircraft like fighters for low-level operations and aerobatic aircraft have short or clipped wings.

(2) *Effects of dihedral and weathercock stability.* Figure 11.3a shows how roll rate builds up and steadies. The broken lines illustrate how it may be affected by weak or strong dihedral and correspondingly strong or weak directional stability. This is a useful example of the utility of derivatives, two of which replace many words with the simple ratio of their coeffcieints, l_v/n_v. The slope of the curves at any point in Fig. 11.3a is the measure of acceleration in roll at that instant.

(3) *Aeroelasticity.* We saw in Chapter 9 how control surfaces may lose their effectiveness. Figure 9.21 compared satisfactory and unsatisfactory aileron characteristics due possibly, it was suggested, to poor sealing of gaps. Similar curves are the result of aeroelastic distortion of the control surface itself. Torsion applied by the aileron to the wing, especially when swept back, may cause the wing itself to twist in opposition. Either can produce results like those seen in Fig. 9.2b, which may lead to aileron reversal.

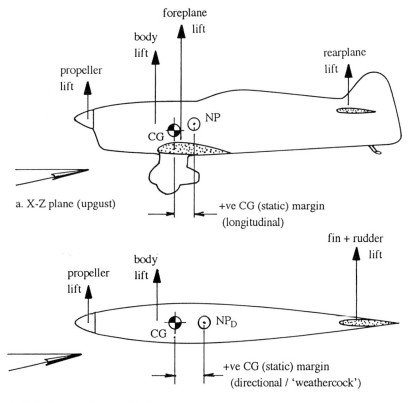

a. X-Z plane (upgust)

b. X-Y plane (sidegust / sideslip)

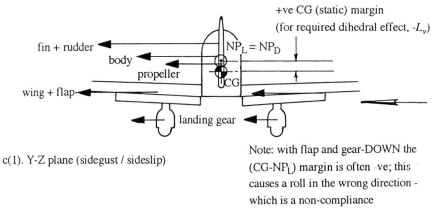

c(1). Y-Z plane (sidegust / sideslip)

Note: with flap and gear-DOWN the (CG-NP$_L$) margin is often -ve; this causes a roll in the wrong direction - which is a non-compliance

NP, NP$_D$ and NP$_L$ are the neutral points, the CPs of each set of resultant cross-forces; a +ve CG (static) margin means positive stability and rolling moment with sideslip.

Fig. 11.2   When seeking a cure for a problem it can often help, in exactly the same way as in the longitudinal case, to visualize the changes in the CG-neutral point relationships with alterations of configuration in the lateral and directional planes.

(4)   *Forward swept wings.* These tend to twist the wing in the opposite direction to those which are backward swept, increasing the incidence of the tips. It needs careful design to arrange the torsional and flexural axes of a wing in such a way as to resist changes

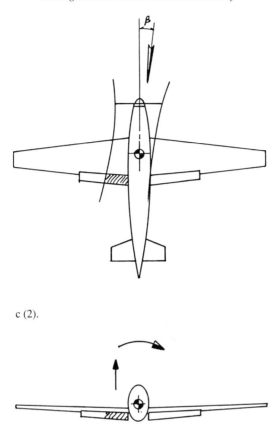

c (2).

Fig. 11.2 – Continued   All figures are purely diagrammatic. Figure 11.2c(2) shows how the simple c(1) explanation may well be modified by a yawed propwash over the flaps, in the take-off and landing configurations (especially landing with full flap selected). All depends upon the number of engines, the handing of the propeller(s) and the direction of the side-gust or sideslip.

of incidence under load. Today we use man-made fibres increasingly. The correct laying-up of woven and stranded materials is an important part of the manufacturing process, to provide the required torsional and flexural characteristics of an elastic structure.

(5) *Manual controls.* The control forces involved in applying manual aileron at high airspeeds can be considerable. Roll rate increases until the pilot is unable to hold the control deflection. Assuming the control force he applies then remains constant it is increasingly insufficient and aileron deflection and roll rate reduce asymptotically, as shown in Fig. 11.3b and c.

(6) *Power-assisted (manual) controls.* These usually (but not always) apply adequate force and hold on to the required aileron deflection in the face of increasing airspeed, until the structurally limiting speed, V_A, is reached – or until the control-jack stalls.

$+L_\zeta$ *:rolling moment due to rudder (see Fig. 11.4a)*

Modern fin and rudder surfaces are often tall, with higher aspect ratios and steeper lift slopes than those used in the past. End-plate (T-tail) surfaces further increase lift slopes and raise the centre of pressure of the vertical surfaces higher above the CG than it would be

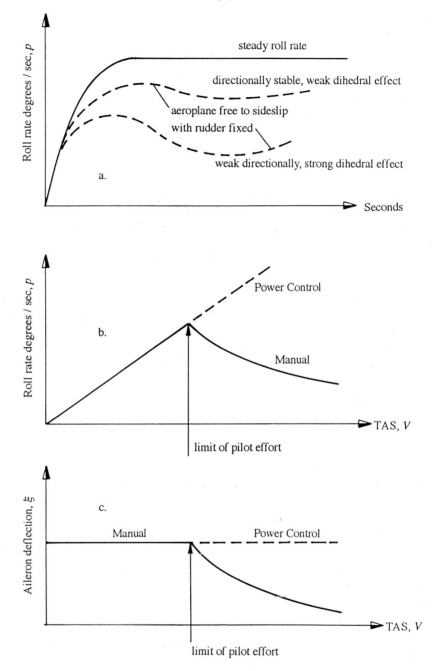

Fig. 11.3   Roll quality with time and true airspeed (see also Ref. 11.1).

without end-plates. Thus, acting like an upright aileron, rudders become more prone to causing appreciable rolling moments in the opposite direction. In terms of derivatives, rudder to port, $+\zeta$, applies a yawing moment to port about the CG, $-N_\zeta$ (and $-n_\zeta$), while simultaneously applying a rolling moment to starboard, $+L_\zeta$ (and $+l_\zeta$).

Note: The configuration of the Système 'D' triplane in Fig. 1.7 has the rudder surface more or less equally disposed above and below the CG to reduce L_ζ to a minimum. In knife-flight, the rudder becomes the elevator, and the elevator the rudder.

$-L_p$: rolling moment due to rate of roll (see Fig. 11.4b)

This is the roll-damping factor, which contributes the resistance of the wings and tail surfaces to rolling (the down-going surfaces being at greater angles of attack than those which are up-going). It is negative as expected, because it opposes the rolling motion, and causes the aircraft to settle rapidly to a steady rate of roll.

$+L_r$: rolling moment due to rate of yaw (see Fig. 11.4c)

This moment is due almost entirely to the wing. The outer wing advances faster through the air than the inner, retreating, wing and so generates more lift. For example, applying rudder to starboard, $-\zeta$, causes positive yaw to starboard, accelerating the port wing, which rolls the aircraft in the positive direction, to starboard.

The rolling moment is every bit as much an opponent as an ally in its possible consequences:

(1) *Inadvertent spin entry.* Failure of the pilot to fly a coordinated turn without slip or skid at a low airspeed (i.e. such that $L_r = 0$) is the commonest cause of all stall/spin related accidents.

It is possible to hold an aeroplane on the verge of a stall and apply full aileron in either direction without causing a wing to drop *as long as there is no trace of yaw*. If the feet are allowed to relax enough to allow the smallest amount of yaw to appear, then the retreating wing will drop ....

Note: There is no substitute for fast and well-coordinated feet!

Yet:

(2) *Flick (snap) roll.* This manoeuvre involves horizontal autorotation and is initiated with smart, almost simultaneous application of full up elevator and full rudder – the rudder leading slightly. Leading with the rudder induces L_ζ and asymmetric lift as in (1). In effect a flick is a horizontal spin, entered from a much higher speed. Aileron is rarely needed.

(3) *Crosswind handling.* On take-off and landing a gusting crosswind can be a problem with dihedral effect rolling the aeroplane in the direction of drift. Kicking on rudder to oppose the gust checks downwind drift, while roll due to yaw-rate opposes dihedral effect.

$-L_v$: rolling moment due to sideslip ('dihedral effect', see Figs 11.2. 11.4 and 11.5)

Historically and before dihedral became established as the most effective means of averting the danger of the spiral dive, other devices were tried. For example, vertical 'curtains' filled the gap between the interplane struts supporting the upper and lower planes of Voisin and Santos Dumont biplanes. Their purpose appears to have been threefold. First was to provide a side-force which opposed slip. Second, if acting above the CG, they would tend to roll an aeroplane upright again, away from the direction of slip. Third, by providing some area ahead of the CG, they might reduce the tendency to weathercock into the sideslip – which starts a spiral dive. They disappeared rapidly and cannot have been a great success.

Lanchester, Besson, Bleriot, Santos Dumont, Voisin, the Wrights and the then Army Aircraft Factory at Farnborough, with its SE 1 aeroplane (constructed by Frederick Michael

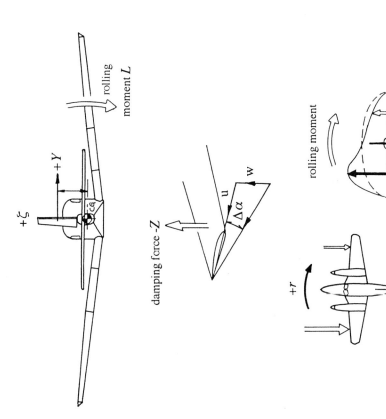

a. Roll caused by
rudder deflection: L_ζ

b. Damping effect
in roll: $-L_p$

c. Rolling moment
due to rate of yaw: L_r
(powerful in long-winged
aircraft, causing pilot to
hold-off bank)

rolling
moment L

$+Y$

$+\zeta$

damping force $-Z$

u

$\Delta\alpha$

w

rolling moment

$+r$

Fig. 11.4   Roll derivatives caused by rudder, change of angle of attack, and yaw rate ....

d. Dihedral, γ, increases angle of attack of upwind wing (shown ░░░) so causing roll with sideslip: $-L_v$

e. Sideslip increases lift and drag of the upwind wing (when backswept), enhancing $-L_v$ (and $+N_v$). Foreward sweep has reverse effect

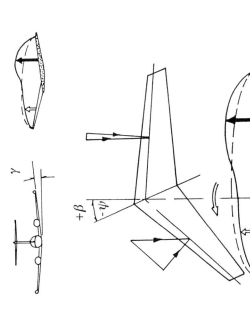

Fig. 11.4 – Continued   Roll derivatives arising from wing dihedral and sweep.

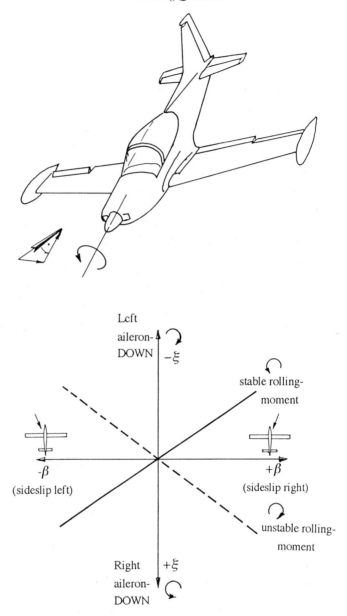

Fig. 11.5   Aileron to hold a steady sideslip as an indicator of dihedral effect.

Green, Chief Engineer and Geoffrey de Havilland as 'designer-pilot' out of a canibalized Bleriot XII and other bits and pieces) experimented with forward mounted fins and rudders. Lanchester's elegant forward fin on glider and rubber-powered models, is what the model sailplane builder today would call a 'rhino' fin, because of its resemblance to a rhino horn.

The purpose of an upright keel surface mounted ahead of the CG was to provide $-N_v$ when a wing dropped. This yawed the nose away from the direction of sideslip, leaving the pilot free to concentrate on bringing his machine back into equilibrium using the lateral controls.

Plate 11–1   Slingsby T-67 experimental 'slinglet', a delta-shaped lift-ing surface intended to depress the upwind aileron in a sideslip, aug-menting dihedral-effect by means of control deflection. (Author)

(1)   A *positive dihedral angle*, γ, gives the advancing wing a larger geometric angle of attack than the retreating wing, causing it to generate more lift, which rolls the aero-plane away from the direction of slip. Negative dihedral, or *anhedral*, has the opposite effect.

(2)   *Yawed wakes and propwash* from nacelles, body surfaces and propellers affect the lift of portions of wing, foreplane and rearplane surfaces lying within them.

   Body and nacelle wakes being more sluggish than the undisturbed flow decrease the lift of wing (and tail) surfaces downwind of them. Yaw causes wakes to trail asym-metrically over downwind surfaces, producing rolling moments which may be favour-able or unfavourable – usually the latter.

   Propwash being more dynamic than air outside it increases lift on wing and other surfaces (see Fig. 11.2c(2)). Yaw may cause a rolling moment in the opposite direction (e.g., nose to port, roll to starboard) known as *anhedral effect*. Thus, dihedral effect is weakest in the balked landing configuration, with flaps DOWN and with increased power producing the strongest propwash.

   The rolling moment coefficient, $-l_v$, is not a simple derivative, and an actual value of $-L_v$ is not easy to calculate under such circumstances, being dependent upon the location of the wings on the fuselage (high, mid or low), and upon the engine nacelles on the wings (above, in-line, or below), as sketched in Fig. 11.6.

(3)   *Sweepback* (see Fig. 11.4e) enhances lift of the upwind wing, making $-l_v$ overpowerful, so that it often has to be countered with negative dihedral, which is why many high-performance aeroplanes – especially the mid- and high-winged – have marked anhedral.

Plate 11–2   Slingsby T-67 as ordered for the United States Air Force.
Made from GRP it is a much modified, licence-built variant of the
French Fournier RF6B.
(Courtesy of Slingsby Aviation Limited)

(4)   *Raked (swept-back) wing-tips* act in a similar (weaker) way to sweep-back (Fig. 11.4e)
and usefully increase the effective span and aspect ratio of a wing, further augmenting
$-L_v$, by making the aerodynamic span and the geometric span almost the same. This
pushes the CP of each wing outboard, increasing the rolling moment.

We saw in Chapter 8 that a wing does its work most efficiently when the length of
the trailing edge is equal to the span of the wing.

Having said this, backwards-raked wing tips while improving $-L_v$, might worsen any
tendency to Dutch-roll, by emphasizing a deficiency in N_v. Therefore, it may be prudent to expect a need for a dorsal or ventral fin to augment the weathercock stability.

(5)   A *T-tail configuration* improves the effectiveness of a fin and rudder by increasing
its aerodynamic aspect ratio enhancing sideforce and rolling moment due to sideslip.

Aeroplanes like the Lockheed F-104, and an increasing number of T-tailed,
high winged transports are now designed with marked anhedral to reduce the adverse effect of excessive L_v/N_v (l_v/n_v). The discovery during flight testing that there is
insufficient fin area for the dihedral provided has often led to modification by the
addition of a large dorsal strake at the base of the fin – and sometimes ventral strakes
as well.

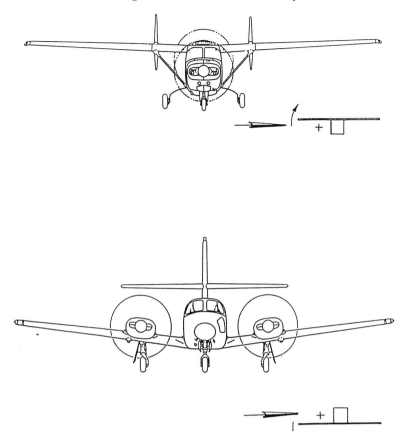

Fig. 11.6 Rolling moment due to sideslip is enhanced or reduced by wing + body and nacelle arrangement. High-winged aeroplanes tend to need less dihedral for this reason. Typical values are 5° to 7° for low-, and half these for high-winged aeroplanes.

In some cases tailplanes are provided with anhedral.

(6) *Insufficient dihedral* results in an excess of weathercock stability and proneness to spiral instability.

(7) In a straight *sideslip* the rolling moment produced by the remainder of the aeroplane is balanced by the equal and opposite rolling moment from the ailerons.

+N_ξ: yawing moment due to aileron (and taileron)

(1) As we saw in Fig. 9.11 this is a secondary, but sometimes powerful, effect of an aileron control. Normal deflection of the aileron, $+\xi$, increases aileron drag with roll and introduces a component of yaw, $+N_\xi$, in the direction of the up-going wing. The Frise-aileron is designed to counter it.

(2) Aeroplanes with highly swept, or with short thin wings which rely upon differential operation of slab tailerons – a combination of elevator and aileron – may experience a similar effect, $-N_\xi$, or $-N_\eta$, as shown in Fig. 11.7.

(3) In a straight sideslip the rolling moment due to the ailerons is balanced by the rolling moment produced by the remainder of the aeroplane.

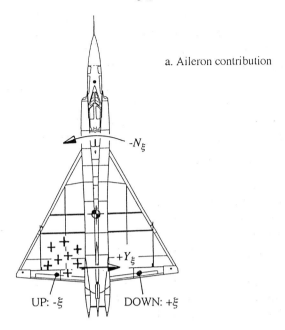

a. Aileron contribution

$-N_\xi$

$+Y_\xi$

UP: $-\xi$ DOWN: $+\xi$

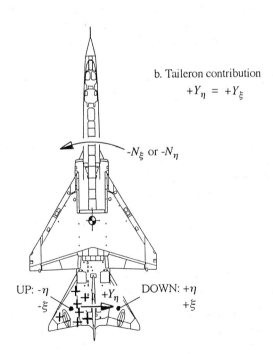

b. Taileron contribution

$$+Y_\eta \; = \; +Y_\xi$$

$-N_\xi$ or $-N_\eta$

UP: $-\eta$ DOWN: $+\eta$

$-\xi$ $+Y_\eta$ $+\xi$

Fig. 11.7   Yawing moment with aileron or taileron deflection.

$-N_\zeta$: yawing moment due to rudder

(1)   Positive rudder deflection produces a negative yawing moment. As we have already
seen, it may also be accompanied by a positive rolling moment, $+L_\zeta$.

(2)   The yawing moment is produced by the sideforce, $+Y_\xi$ or $+Y_\eta$, on the vertical tail and fuselage surfaces due to aileron or taileron deflection (see Fig. 11.7).

(3)   In a straight sideslip the yawing moment due to rudder is equal and opposite to the yawing moment produced by the remainder of the aeroplane.

Note: N_η, *yawing moment caused by elevator*, though rare is not unknown (see Example 12.9). In that particular case we found that a T-tailed light aeroplane, with the elevator mounted directly above the rudder, suffered reduced rudder authority with power, when the yoke was pulled aft, as on take-off or in the approach to 75%-power stalls.

$-N_p$: yawing moment due to rate of roll (see Fig. 9.11)

(1)   The lift/drag axes of each wing (and tail surfaces too) are rotated relative to the aircraft axes, the lift of the down-going wing being rotated forwards, and the up-going rearwards, thus contributing a negative yawing couple $-N_p$.

(2)   The induced and profile drag of the down-going wing is greater than that of the upgoing, generating a small positive yawing increment, $+\Delta N_p$, which slightly reduces the magnitude of $-N_p$.

(3)   The net result is that positive roll to starboard causes negative yaw to port. Its value increases numerically with aspect ratio, angle of attack, roll rate – and is decreased by rate of growth in profile and induced drag.

$-N_r$: yawing moment due to rate of yaw

This is the yaw-damping factor, the magnitude of which is the indicator of resistance of the aircraft to yaw. In this respect it is akin to the roll-damping term, $-L_p$. Yaw-damping is provided mainly by the fin surfaces, although the rear body sections make an important contribution.

(1)   When the aircraft is rotating in yaw the angle of attack of the fin is increased and it produces a lift force, the moment of which opposes the rotation.

(2)   Sideslipping has exactly the same effect upon $N_{r(tail)}$ and $N_{v(tail)}$ and the derivatives:

$$n_{r(tail)} = n_{v(tail)} \tag{11–4}$$

(3)   'Slippery' rear fuselage surfaces, with sections which fail to produce much damping in yaw cause trouble with spin-recovery. Indeed, with the wrong arrangement of fin and rudder such sections can be positively dangerous.

   The sharper cornered the fuselage section, the better able it is to generate turbulence in a cross-flow.

(4)   There is a small contribution, $-\Delta N_r$, caused by the drag of the faster-moving wing.

$+N_v$: yawing moment due to sideslip-weathercock stability (Fig. 11.8)

This is the tendency of the aircraft to pivot about the Z-axis, to meet the relative wind. Contributions arise from the body, wing and tail (see Fig. 11.9).

(1)   The body contribution is destabilizing and depends upon the amount of forebody ahead of the CG and afterbody behind it. It is affected by body cross-sections, especially those aft of the CG, and varies widely between different types of aircraft.

(2)   The wing contribution is slight, unless well swept. Sweepback is stabilizing, producing a component $+\Delta N_v$, while forward sweep is destabilizing.

(3)   The vertical fin surfaces are dominant. Their moment about the CG, being positive, acts to reduce sideslip. But,

(4)   As we saw much earlier, the actual tail lift depends upon the extent to which the tail or stabilizer efficiency factor, η_s (Equation (3–50)), is degraded in the wake.

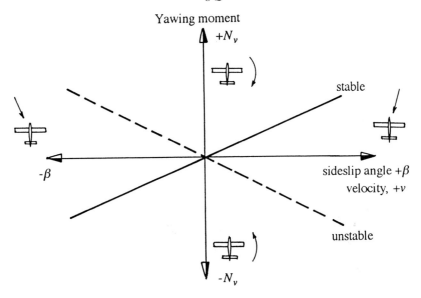

Fig. 11.8   Static directional stability.

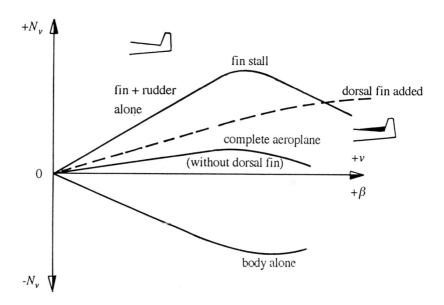

Fig. 11.9   Contributions of parts to the static directional stability of an aeroplane. (Note: Keel surface area and the shape of cross-sections of the body aft of the CG can be of critical importance to the contribution from the body alone.)

(5)   In a straight sideslip the yawing moment due to sideslip and that due to rudder are in equilibrium.

$+Y_\xi$: *sideforce due to aileron (and $+Y_\eta$, due to taileron) (see Fig. 11.7)*

Although with 'normal' configurations sideforce due to aileron is negligible, this is not the case when the wing is a highly swept delta and the aileron is in fairly close proximity to the fin. Taileron controls may also introduce a significant sideforce component, $+Y_\eta$.

$+Y_\zeta$: *sideforce due to rudder*

Positive (port) rudder deflection generates a positive sideforce to starboard. It is worth noting:

$$+y_\zeta = \text{approximately } -n_\zeta \qquad (11\text{--}5)$$

$+Y_v$: *sideforce due to sideslip (see Fig. 11.10)*

This is the side-resistance of the aeroplane to sideways motion. It depends primarily upon the vertical surfaces, body shape, sections and finish, as noted in discussion of $-N_r$. Look back to Fig. 11.2b. The location of the directional neutral point, NP_D, at which the resultant of all of the sideforces acting on the aeroplane acts, depends primarily upon the efficiency of the tail surfaces (keel surface area of the rear fuselage, and effectiveness of the body sections in local crossflows) working in the wake shed by the surfaces forward of them. A wake which deepens and becomes increasingly turbulent and sluggish with angle of attack causes the resultant cross-force, Y_v, and with it NP_D, to migrate from the normal position aft of the CG to somewhere ahead of it in an extreme manoeuvre. This is a risk with a configuration of low aspect ratio, like that used in Example 1-1, although – as commented upon below Fig. 11.11 – others may be similarly affected to a degree.

Note: The situation shown in Fig. 11.10 is complicated by a number of factors, not least of which is that slicing of the nose will be accompanied by roll. The nose-high attitude raises the sideforce and the position of the neutral point NP_D in the X–Z plane, at NP_L, higher above the CG. This enhances any tendency to roll to port, while the starboard wing, which

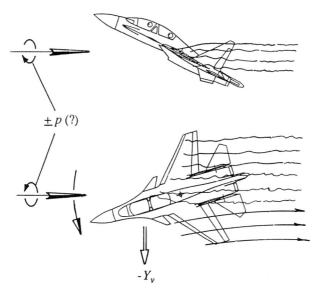

Fig. 11.10   An asymmetric stall combined with sideslip may lead to a nose-slice (departure in yaw), cross-coupled with roll about the wind-axis, depending upon aspect ratio, wing-sweep and other aspects of the configuration.

here is more stalled than the port, will tend to roll the aeroplane to starboard. Which rolling component wins is a matter for conjecture.

Weight Components – effect upon sideways acceleration and sideslip (see Fig. 11.11)

As moments are taken about the CG, wings level, the weight makes no contribution to either rolling or yawing moments. However, when banked the sideforce due to weight produces the sideways acceleration and sideslip, which give rise to so much else.

Lateral Modes: Spiral and Oscillatory Instability (Dutch-Roll)

A conventional aeroplane possess three lateral modes:

- A *rolling mode*, which depends upon the nature of the response of the aircraft to lateral control.
- A *spiral mode*.
- An *oscillatory mode*, which, for similar physical reasons is a relative of the short period and phugoid modes encountered in our discussion of longitudinal instability.

Of these the spiral and oscillatory modes are of greatest importance to the pilot.

The spiral mode

Although the uncontrolled spiral dive was an early cause of flying accidents, spiral instability is not a problem unless the time to double amplitude is less than about 20 seconds. It is caused by an excess of weathercock stability over the rolling moment due to sideslip: too

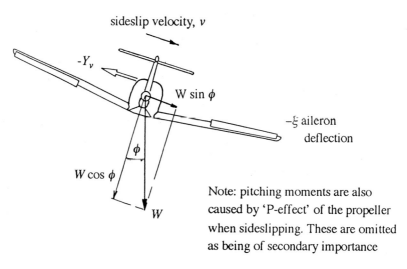

Note: pitching moments are also caused by 'P-effect' of the propeller when sideslipping. These are omitted as being of secondary importance

Fig. 11.11   The diagram shows the contribution to sideslip due to $W \sin \phi$ acting in the plane of the wings. The outcome depends upon where the sideforce, $-Y$ (and directional neutral point), lies relative to the CG. If behind, then the aircraft weathercocks into wind. If ahead, as in the preceding figure, then the nose slices away downwind. Other configurations may also be vulnerable. Example 12–9 is a case in which up-elevator on a T-tail light aeroplane reduced fin and rudder authority significantly, though not completely.

much fin and rudder and insufficient dihedral. If the aeroplane is disturbed laterally the rolling moment, $-L_v$, being insufficient fails to roll the wings level. Sideslip velocity causes the fin to contribute $+N_v$, which weathercocks the nose into the wind. Yaw increases the speed of the outer wing which, generating more lift than the inner, rolls the aeroplane more steeply into the direction of yaw. The nose goes down further, roll increases and a spiral dive ensues.

This is one of the most dangerous of motions, because it is marked by rapidly increasing departures in airspeed and angle of bank. To roll out and regain level flight takes time to deflect the controls and the aeroplane to respond. The longer it takes to recover the greater the risk of exceeding the design manoeuvring speed, V_A, and of structural failure.

Oscillatory instability

An excess of $-L_v$ causes oscillatory instability – a combined roll–yaw motion, during which the aeroplane appears to be wagging its tail, with a period between about 1 and 15 seconds. It is variously described as 'wallowing', 'Dutch-rolling', or 'snaking'. The term Dutch-roll is reserved for the motion which is dominated by rolling. Snaking describes a more marked directional oscillation. In what follows, N_r (the yawing moment due to rate of yaw) being fin-dependent is usually considered to vary with and may be substituted for N_v (see Equation (11–4)).

I recall slight oscillatory instability with the Boeing 707 at altitude, when travelling in 'steerage' near the tail. *En route* at the cruise Mach Number with the tail deeper in the wake, N_v was probably less than at lower altitude and a higher IAS. The geometric aspect of the back-swept wings would be increased for the same reason and with it apparent dihedral, so that $-L_v$ increasingly dominated the ratio of actual $-L_v/N_v$. Although the oscillatory instability was barely felt through the seat of the pants (the most sensitive part of any trained and conditioned pilot), liquid in a cup held in the hand would start a gentle rotary swilling motion, showing that the aeroplane was beginning to wallow as $-L_v/N_v$ grew in magnitude.

In fact the motion was the faint tell-tale left of the earlier divergent Dutch-roll problems encountered with the Boeing B-47 bomber. The civil Dash-80 airliner which followed it in formula if not in detail, and the Boeing 707 itself had similarly swept wings, podded engines and a single fin and rudder. One or two 707s suffered fatal loss of control when a divergent Dutch-roll had the chance to develop. The cure was to fit a yaw-damper – 'Little Herbie' to the B-47 and all subsequent jets (Ref. 11.2). Those 707s certificated for operation on the UK Register also had a ventral fin. The latter, while giving away the existence of a problem to the discerning eye, increased directional weathercock stability, $+n_v$, while the rolling moment with sideslip either remained the same or was very slightly decreased. In this way the coupled motion was shifted sufficiently to the right in Fig. 11.12, away from the oscillatory stability boundary, towards the spiral mode. The combination of a mechanical yaw damper and static ventral fin surface is a classic 'belt-and-braces' solution to a problem.

Spiral and Oscillatory Interactions

A test pilot must explore oscillatory and spiral characteristics as part of the search for dangerous features. The boundaries of oscillatory and spiral instability, in terms of their coefficients, are shown in Fig. 11.12 (Ref. 10.15). Too small an n_v (insufficient fin) combined with too much $-l_v$ (an excess of dihedral) reduces oscillatory stability. The nearer one is to the oscillatory boundary the worse damping becomes, indeed, damping decreases faster with a reduction in n_v than with numerical increase in $-l_v$. The period of the oscillation

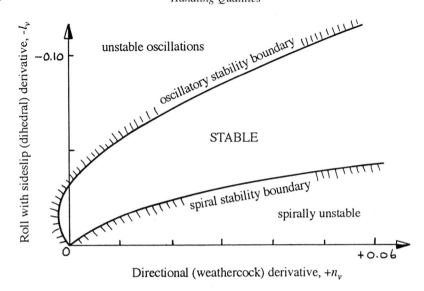

Fig. 11.12   Interaction between dihedral effect, $-l_v$, and weathercock stability, $+n_v$, directly affects oscillatory and spiral stability (Ref. 10.14).

increases with decreasing weathercock stability, breaking down into divergence for small negative values of n_v.

High wing loading, high altitude (relative aircraft density in roll, μ_2, Equation (9–15b)), large rolling inertia A, (see Equation (9–18a)), low roll-damping, L_p (and l_p), have the effect of rotating the oscillatory stability boundary clockwise (and so downwards) in Fig. 11.12.

At low airspeeds there is an opposite effect with increasing angle of attack, as the tail dips deeper into the wake, reducing weathercock stability. The spiral stability boundary is rotated anticlockwise (upwards) and may cross the oscillatory boundary when dihedral effect is weak, making it impossible to avoid spiral instability of low airspeeds.

Such motions may occur controls fixed or free. Often freeing the ailerons has little effect. But freeing the rudder may reduce directional damping to the point at which unpleasant short-period snaking occurs. It can become dangerous if corrective actions by the pilot are out of phase with the motion (driven by the control-free derivatives, b_1 and b_2 (Figs 3.19 and 9.6), which influence oscillatory and spiral stability, as shown in Fig. 11.13). So powerful are the rudder-free characteristics that they may dominate the whole picture.

It is a good rule to require oscillations to damp to half their original amplitude in not more than one cycle, following a small disturbance in sideslip and release of the controls.

Common lateral and directional faults and fixes

Table 4–6, items 11 and 12, listed basic control, lateral and directional stability tests. Five different types of single-engined light aeroplanes flown in recent years were low-winged. All suffered shortfall in the amount of rolling moment generated in straight steady sideslips, in the landing configuration especially. The reason, in my view, was to be found in Fig. 11.2c and the NP lying below the CG in the Y–Z plane. The negative 'CG-margin' caused the lower (upwind) wing to go down further when the aileron control was released. The requirement was that the corrective rolling moment should be not less than neutral,

Plate 11–3 The Boeing B707 showing the now firmly established formula of highly swept back high aspect ratio wing with plenty of dihedral effect and podded engines. These are placed in positions along the span where they 'fly' efficiently with least interference, while functioning as mass-balances against flutter. Even so, a weakness of this configuration has been a tendency to 'Dutch-roll' which, although now understood and cured, caused past grief (Ref. 11.2). (Courtesy of the late Hugh Scanlan, former editor of *Shell Aviation News*, 398, 1971)

and for the wing to go down meant that the rolling moment was positive, resulting in a non-compliance.

The ratio of $-l_v/n_v$ for each aeroplane was weak. All tended to have fairly tall (high aspect ratio) fins and rudders. The strong qualitative impression given by one was that an additional increment of rolling moment, L_ζ, caused by the amount of rudder to produce and maintain straight, steady sideslip was having the same effect as an aileron, pushing the lower wing further down.

The following shortfall can be obtained from turns on one control, or more often than not, in straight, steady sideslips.

☐ *Poor oscillatory stability.* Rudimentary oscillatory stability tests in Table 4–6, carried out at a speed of 1.2 V_S in the required configuration, will reveal evidence of inadequate damping, of Dutch-roll or snaking. The time and number of cycles to half amplitude are recorded. The results should be compared with any published requirements.
Common causes of shortfall:

(1) *n_v (or n_r) too small, or $-l_v$ too large.* The values of n_v and $-l_v$ may be checked qualitatively against Table 11–4, which gives an indication of the magnitude of the $-l_v/n_v$ ratio, controls fixed. Controls free the table still applies, but rudder and aileron forces must be considered instead of their deflections.
Fixes are increased fin area, or reduced dihedral.

(2) *Rudder $-b_2$ too small, or b_1 too large* (leading to snaking, which has a small component in roll and a short period).
Fixes are either to add cord or Flettner-strips to the trailing edge of the rudder (Fig. 9.19), or to reduce the authority of the horn balance (Fig. 9.10).

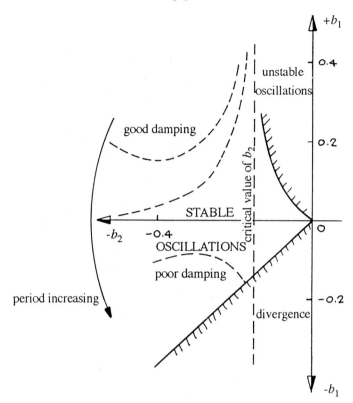

Fig. 11.13   Rudder-free trail characteristics affect both oscillatory and spiral stability (Ref. 10.14).

(3)   *Excessive rudder friction.* Tests are largely qualitative.
Fixes can involve a redesign of control hinges and cable runs – or maybe lubrication. Be careful with homebuilt aircraft control circuits especially. Check materials used. Some bearings, if other than metal, may swell when greased or oiled.

Spiral instability. Unless severe, this is not generally regarded as a serious fault. However, it can lead to trouble when flying in 'goldfish bowl' conditions, with no horizon and inadequate scanning of the flight instruments by a pilot who is out of practice, or overstressed in some way. Tests, which are more qualitative than quantitative, are carried out using one control at a time:

(1)   on aileron only with rudder held fixed;
(2)   on aileron only with rudder free;
(3)   on rudder only with aileron held fixed.

Record the magnitude and direction of the control deflection (control-fixed stability) or control-force (control-free stability) in a steady turn. For turns on aileron alone the aileron will be with the turn for spiral stability, central for neutral, and against the turn for spiral instability. Similarly for the rudder, during rudder-only turns.
Common causes of shortfall:

(1) n_v *too large and* $-l_v$ *too small.* Check by using Table 11–4.

Table 11–4

Control-fixed indicators of strength or weakness of roll and yaw derivatives, $-l_v$ and n_v

Derivative	Straight sideslips	Turns on rudder only	Turns on aileron only	Flat turns
n_v small	Large sideslip Large bank Small rudder deflection Large aileron deflection	Small rate of turn Large rudder deflection (with the turn)	Small rate of turn Large aileron deflection (with the turn)	Large rate of turn Large sideslip Small rudder deflection Large aileron deflection
n_v large		the reverse of small n_v		
$-l_v$ large	Large sideslip Small bank angle Small rudder deflection Large aileron deflection	Large rate of turn Small sideslip Small rudder deflection (with the turn)	Small rate of turn Large aileron deflection (with the turn)	Small rate of turn Small sideslip angle Small rudder deflection Large aileron deflection
$-l_v$ small	the reverse of large $-l_v$			

Fixes are to increase dihedral (but look at Figs 11.12 and 11.13 and Example 11–1 which follows later before making the suggestion). There can be difficulties. Most aeroplanes are designed with fixed dihedral and, commonly, one can do little more than turn up the wing-tips. Sometimes it is possible to increase crank at a transport joint, but this is a more expensive solution than a modification to a wing-tip fairing.

☐ *Tendency to rudder overbalance*. Tests in steady straight sideslips are intended to reveal a tendency of the rudder to overbalance, as shown in the case of the Handley Page Halifax in Fig. 9.22 and also the Boeing B17 in Fig. 9.23. From such tests it is possible to plot rudder force and also rudder position against angle of sideslip.

Note: If proper instrumentation is lacking it is often possible to run a length of cord up the windscreen. Anchor the cord securely with masking tape at its base. Its movement can be wild, and it does not work well behind a tractor propeller, but it can be interesting and is always instructive.

Common causes of shortfall:

(1) n_v *(rudder-free) is too small*. Check against Table 11–4.
    Fixes are either to increase the area of the fin surface(s); or to increase rudder b_1 by enlarging the horn balance.
(2) *Rudder* $-b_2$ *too small, or even suffering positive (overbalance)* with large deflections. Check rudder geometry – hingeing and balance.
    Fixes are either to reduce rudder balance by adding cord at the trailing edge (ensure that any basic symmetry in trim is not destroyed by so doing); or, more expensively, fit an anti-balance tab. Such a tab moves in the opposite sense to that drawn in Fig. 9.13d.

☐ *Fin-stalling*. It is easy to confuse fin-stalling at large angles of sideslip with rudder overbalance. The fin-stall (which is to be avoided at all costs) causes a loss of weathercock stability. There is then the danger of the aeroplane attempting to change ends, which is usually accompanied by overbalance of the rudder and loss of control which, with luck, is only temporary.

Fin-stalling is marked by a reversal of slope of the rudder deflection/angle of sideslip curve. Rudder overbalance (without fin-stalling) appears as a reduction in the slope of the rudder-force/angle of sideslip curve.

As a safety measure it is reasonable to suggest that balance of the rudder should be such that the maximum foot-load in Table 9–2 is reached at, say, an angle which is two-thirds of that at which the fin is likely, or is calculated, to stall.

Note: Therefore urge the design of fin and rudder surfaces which have lower aspect ratios than high (see Table 9–5).

Common causes of shortfall:

(1) n_v *too small*. Check against Table 11–4.
    Fixes are to add a dorsal fin (which helps to reduce fin and rudder aspect ratio), ventral fins or otherwise increase fin area. Alternatively increase the size of the rudder horn-balance to make b_1 positive.
(2) *Rudder over-effective or over-light*. Rudder effectiveness must be assessed in straight. sideslips, and in turns on one control. Be careful, though, because large control deflections or forces are not a certain indication of ineffectiveness. Ineffectiveness of controls is discussed below.

Fixes are either to reduce the rudder size, or the rudder balance.

☐ *Rolling into a sideslip (negative dihedral effect)*. Check symptons against Table 11–4. In turns using rudder alone if $+l_v$ is discovered this can seriously affect crosswind handling on take-off and landing, as it affects adversely the ability to kick-off drift and lift the downwind wing by use of rudder alone.

In tests plot aileron deflection against steady sideslip angle.

Common causes of shortfall:

(1)   Insufficient dihedral.
      Fixes involve increasing the dihedral of the wing, or turning up the wing-tips, or modifying the aileron of the downgoing wing to cause it to float downwards $(+b_1)$ and so increase its lift.

      Note: Cord or a Flettner strip fitted to the trailing edge of an aileron to cause downfloat is largely unsuccessful when circuit friction is present. Friction is hard to eliminate from the control circuits of long, floppy wings which bend easily. For this reason I have found trailing-edge cord and strips to be almost useless on motor gliders and light aeroplanes with high aspect ratio wings – unless such fixes are deep enough to penetrate beyond the thickening boundary layer aft. Unfortunately they are then deep, draggy and unsightly – and might be accompanied by unpredictable and undesirable non-linear effects.

(2)   Overpowerful rolling moment from the rudder.
      Fix by increasing dihedral-effect.

☐ *Weathercocking in turbulence*. This is caused by n_v which is too powerful, resulting in jerky oscillations in yaw about the CG. Do not confuse this with snaking, which involves a more distinctive displacement of the CG of the aircraft along a sinuous path. Assessment is qualitative, in turbulent conditions.

Common causes of shortfall:

(1)   Overpowerful fin and rudder surfaces. Check symptoms against Table 11–4.
      Fixes are to reduce fin and rudder area, if possible; or to reduce the magnitude of rudder b_1 by a reduction in the authority of the rudder horn balance.

☐ *Ineffective controls*. By this we mean deflections and forces which are too large for the required response of the aircraft. These are checked in straight, steady sideslips. As we see in Table 11–4, a small n_v involves large *aileron* deflections. A large $-l_v$ involves large *rudder* deflections. The same is true of control forces with control-free derivatives. Large control deflections and/or forces throughout the various tests are the reliable indications of control ineffectiveness or heaviness.

Common causes of shortfall:

(1)   Inadequate, or faulty control surface geometry. Fitting a more powerful engine than the configuration of the aeroplane was originally designed for is a common source of control inadequacy.
      Fixes involve (costly) redesign.

(2)   If $-b_2$ is excessive then the pilot will have difficulty in deflecting the control far enough.
      Fix by fitting a balance or spring tab.

(3)   Lift slope of the control, a_2, too small (often a consequence of the control working in a highly disturbed or stagnating wake).

  Fix either by increasing the size of the control, or sealing the gap between it and the main surface (see Fig. 9.21b).

☐ *Excessive change of trim with airspeed.* This is caused by inadequate trimmer control surfaces. Item 13 in Table 4–6 lists checks of minimum trim speeds and stick forces with full NOSE-UP trim. To test under other conditions it is generally sufficient to ensure that the aeroplane can be trimmed in the landing configuration, between $1.2\ V_S$ and $1.6\ V_S$ with throttle closed and at maximum continuous power (the average is $1.4\ V_S$, which will be found to be not very far away from the best climb speed). And, between $1.4\ V_S$ and $0.9\ V_H$, V_C or V_{M0}/M_{M0}, whichever is the lowest, in the *en-route* (clean) configuration, with throttle closed and also at MCP.
Common causes of shortfall:

(1) Loss of dynamic pressure in sluggish wakes, due to faulty aircraft geometry.
  Fixes are hard to recommend, beyond looking in detail at ways of improving fairing upwind of a deficient trim-tab.
(2) Loss of trimmer effectivess, for similar reasons to those causing ineffective controls, discussed above, in this case, faulty trimmer control geometry.
  Fixes are to increase the area of the trim-tab, or by sealing to increase the lift slope, a_3.

☐ *Excessive change of trim with power.* Large changes of lateral or directional trim with power lead to difficulties when manoeuvring, especially during balked landings.
Common causes of shortfall:

(1) Adverse wake effects, faulty configuration.
  Fixes are again hard to recommend, they can involve careful fairing of sources of turbulence and other wake-disturbance; and/or repositioning trimmer surface(s).
(2) Ineffective trimmer(s) for similar reasons to those affecting control surfaces.
  Fixes are to increase tab area, or by sealing to increase $-b_3$.

Example 11–1: Wing-Tip and Related Modifications – Aerospatiale Epsilon?

Figure 11.14 shows wing-tip and related modifications carried out on the French Aerospatiale TB 10 Epsilon, during development test flying. These include: altered fairings of improved form in the head-on view; and a lowered tailplane with dihedral added. Comparing the original, higher-set, position of the tailplane with Fig. 8.3d, shows that it lay on a line inclined at 14° or 15° above the longitudinal datum – which means it could have been adversely affected by the wing-body wake during the approach to the stall, in a region of pitch-up without warning.

  Other changes were a complete revision of the vertical and horizontal tail-surface arrangement. The fin and rudder were moved forwards, and a ventral fin was added. Rounded and dihedralled wing tips were also added.

  It is fair to guess that the drawings suggest the Epsilon might have run into snags during stall, or spin or perhaps in straight, steady sideslips, or some combination of them which led to significant modifications to the tail surfaces. The redesigned tail may then have so improved weathercock stability and N_v, that the interaction between $-L_v$ and N_v (and $-l_v/n_v$) was degraded, causing spiral instability. If correct, then to rectify this the wing-tips were then inclined upwards to provide the required amount of dihedral effect.

  Learning to read drawings to such a purpose was the point of Chapter 1.

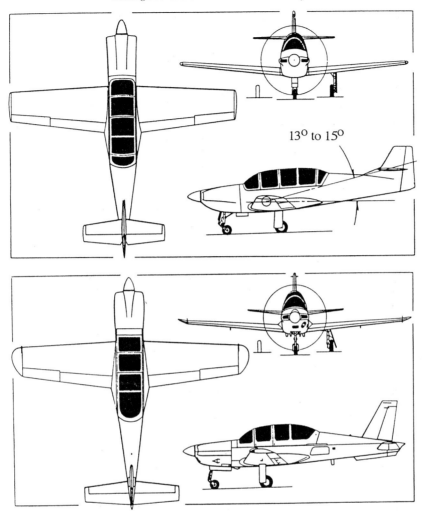

Fig. 11.14 Piston-engined Aerospatiale TB 30 Epsilon is shown in its original form (top) at the time of initial flight testing. The angular range of 13° to 15° relates angles of attack to Fig. 8.3d, to show that there might have been longitudinal problems due to the stabilizer being immersed in a critically adverse wake. Major design changes to the tail and wing-tips revealed later (bottom) that there were – either before rectification or afterwards – undesirable lateral and directional problems too (drawings courtesy of William Green).

Example 11–2: Modifications to an Experimental Prototype (Example 8–9)

Modifications are invariably made to experimental aeroplanes following first tests, as we saw in Example 8–9. The same prototype suffered shortfall in dihedral effect, leading to the following suggestion:

☐ *Winglets*. These, in Fig. 11.15, had two purposes. The first was simply to provide increased $-L_v$. The second was an attempt to win some benefit from a modification by making it work in other ways. The winglet appeared to have the potential to do this by increasing dihedral while also reducing lift-dependent drag.

Plate 11–4   Aerospatiale Epsilon after tail and wing-tip modifications.
Ref.: Example 11–1. (Author)

Winglets, which are a refined form of aerodynamic endplate, are attributed to Dr
Richard T. Whitcomb of the NASA, Langley Aeronautical Systems Division. They work
by shedding vortices from their tips. Being cambered in the same sense as the wing, they
oppose, diffuse and weaken the strength of the main trailing wing-tip vortices by 'un-
winding' them. They are claimed to reduce the drag of the bare wing by 14%, but can be
tailored for maximum efficiency at one speed only. Whitcomb's general rule is that
'*winglets give twice the increase in span efficiency as wingspan extension for the same
increase in root bending moment*'.

Tests continue as this is written.

Example 11–3: Lateral and directional deficiencies – Tri-motor transport
(Example 6–8)

The tri-motor commuter in Example 6–8, Fig. 6.16, has dihedral on the fore and mainplanes
and the project test crew wonder if it might affect adversely the lateral and directional
qualities of the aeroplane, for a number of reasons. When yawed, dihedral increases geometrically the angle of attack – and therefore drag of each plane. This, acting along the
relative wind, will produce a directionally destabilizing (anti-weathercock) moment from
the foreplane, and a stabilizing turning moment about the CG from the rearplane. Which is
likely to be the stronger? Much depends upon the airspeed and attitude of the aircraft. One
must consider, of course, the wake of the foreplane interfering with the effectiveness of as
much as half of the mainplane, and this grows worse with increasing angle of attack as the
pilot reduces speed, aiming for 1.2 V_s.

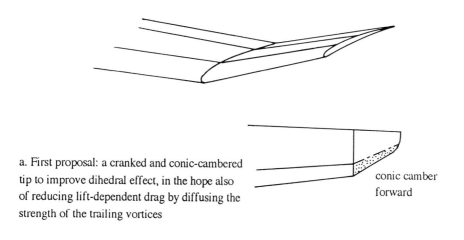

a. First proposal: a cranked and conic-cambered
tip to improve dihedral effect, in the hope also
of reducing lift-dependent drag by diffusing the
strength of the trailing vortices

conic camber
forward

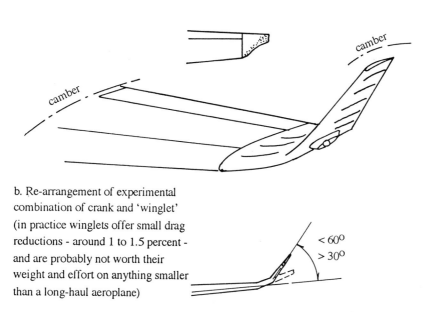

b. Re-arrangement of experimental
combination of crank and 'winglet'
(in practice winglets offer small drag
reductions - around 1 to 1.5 percent -
and are probably not worth their
weight and effort on anything smaller
than a long-haul aeroplane)

$< 60^{\circ}$
$> 30^{\circ}$

Fig. 11.15 Wing-tip modifications to experimental Pilatus PC-12 following flight testing
of some of the modifications suggested in Fig. 8.22 (both courtesy of Pilatus Aircraft Ltd).

Were dihedral on the foreplane to be reduced to zero, say, then mainplane dihedral,
acting alone, would produce its required rolling moment with yaw – but it will also generate
a nose-down pitching moment, because it is aft of the CG. Whereas if the foreplane has all
of the dihedral, then in addition to rolling the aeroplane with yaw, it will cause nose-up
pitch.

If directional and longitudinal effects of this kind occur, where might the solution
lie? It all depends upon how powerful are the pitching and yawing moments. The simplest solution could be to employ differential amounts of dihedral between fore and

Plate 11–5a and b   Winglets, extended dorsal and different ventral fin surfaces, and strakes, of the Pilatus PC-XII are typical of modifications explored during development test flying. Ref.: Example 11–2. (Courtesy of P. Siegenthaler and Pilatus Aircraft Limited)

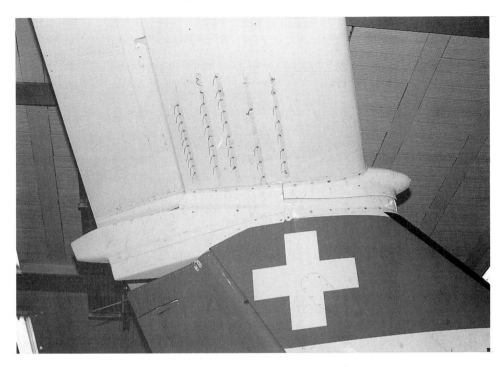

Plate 11–5c Tufting beneath inboard underside of PC-XII tailplane,
to photograph flow conditions in flight. Ref.: Example 11–2. (Author)

mainplane – anticipating a need for less on the foreplane, in inverse proportion to the area moments of each about the CG. The crew might also consider the way in which the high mainplane and its junctions will almost certainly degrade fin and rudder authority more than the foreplane. It will cause the tail surfaces to suffer more than the example shown in Fig. 8.13, because they are all so close-coupled.

Note: Interference must always be expected to be worse when a wing is high in relation to the fin and rudder surfaces. As a general rule expect an aeroplane with a high wing to need around $\frac{1}{5}$ to $\frac{1}{4}$ more vertical surface area aft than one of the same size and weight, with a wing that is set low(er). As a practical first shot, to see how bad interference might be between foreplane, mainplane and tailplane, one might use Fig. 8.3d to check each following plane in turn.

There were significant examples of the need for increased fin and rudder area in the 1940s and 1950s, particularly in the UK, when a new family of multi-engined transport aeroplanes and airliners were either designed, or modified out of existing bombers, for post-war civil use. They had high wings, bringing fuselages closer to the ground to facilitate side-loading of passengers and freight, deep wing sections, and moderately sized twin fins and rudders. The heavy bomber of the period had a low or mid wing to ease loading of the bomb-bay underneath. As far as I can remember many of the new airliners needed, or grew, a third fin because directional qualities were degraded by wakes of the high wings and junctions which interfered with their tails.

Finally, because *'P'-effect* of the centre-engine acting on a long moment arm ahead of the CG will be powerfully destabilizing with yaw, the tri-motor could need considerably more fin and rudder area than it has been given to cope with failure of the critical engine on take-off.

The subjects of the last three chapters next come together in spinning, in which control quality coupled with the longitudinal, lateral and directional properties of an aeroplane, determine how it handles when departing in pitch, roll and yaw, and during subsequent recovery.

References and Bibliography

11.1 Brown, P. W., *Airplane Handling Qualities for the Designer, Builder and Test Pilot*. Oshkosh: EAA Sun'N Fun Forum, 1984
11.2 Irving, C., *Wide-Body, the Making of the 747*. London: Hodder and Stoughton, 1993

CHAPTER 12

Spinning (and Mishandling)

'The best spin of all is that off a steep turn with engine on. An aeroplane stays in the air because of lift it obtains from the wing area, this lift counteracting the pull of gravity. Now incline your wings to the vertical and you've practically no wing area parallel to the ground. Therefore none of your 'lift' opposes gravity and .... "We endeavour to keep the nose up by applying top-rudder. This is our yawing couple and – she spins." And as you say it she suddenly woofs over the top without any warning and goes slap into a really tight and beautiful spin with engine on ....'

Pilot's Summer, an RAF Diary (c 1935?)   Frank D. Treadrey (1939)

Mishandling

Numerous accidents are caused by mishandled manoeuvres. Examples are not always as obvious as, for example, flying too low and slow, close to the ground – disregarding the old adage '*Speed not height keeps Johnny right*'. Almost any emergency can lead to mishandling of engine or flying controls, or of the knobs and switches controlling a system. While there is danger in not knowing or remembering correctly the necessary emergency procedures, there is also that of doing something in a rush, of jumping to the wrong conclusion, of getting things in the wrong order, of being out of practice, of suffering oxygen deficiency or being otherwise incapacitated, and then finding oneself out of control.

The CAA rightly insists that aeroplanes cleared for aerobatics are also cleared for spinning. Flight testing involves mishandling during aerobatics, during which flying controls are 'abused' (misused – which does not mean that they are used carelessly, in ways which lead to overstressing of control surfaces or airframe). Because mishandling of controls leads so often to stall and spin accidents it is a major reason for the certification test pilot looking for dangerous features.

For most pilots spinning remains the least enjoyable aspect of flying yet, in spite of discomfort, complexity and unpredictability, it is a profoundly interesting area to explore. A developed spin in an aeroplane is akin to a horse bolting. Although the stall followed by a spin causes the greatest number of accidents, I do not believe that the right route for any regulatory authority is to discourage stall and spin training. There is no way that I would have my children turned out as finished by a riding instructor who had not taught them handling and bringing to a halt a horse at full gallop. Nor do I much favour design for stall-spin avoidance. Not only is performance too limited but, by the same token, there is little skill and insufficient polish to be gained from riding a horse which cannot gallop because it is blinkered and hobbled. Having said that, I understand schools wishing to attract students and not frighten them away.

Never Tackle Spinning when Out of Practice

One cannot separate stalling and spinning. A wing drop at the stall will, unless corrective action is taken, lead almost inevitably to the unique combination of departures in yaw, roll and pitch which we call a spin. Never tackle spinning cold. Refresh yourself with an in-date instructor or test pilot. Above all, if you are a product of the *stall-avoidance-spin-avoidance school* then seek spin training before you try pushing yourself beyond your present limits.

The root of the spin is autorotation: self-sustaining rotation of the aeroplane about some axis which is driven by the aerodynamic and dynamic characteristics of the aeroplane itself (Fig. 12.1a). The drive comes predominantly from the wing of a conventional subsonic aircraft. Historically, clearance for spinning involved testing military aircraft to determine the best means of recovery, today there is a tendency to find out how susceptible an aeroplane is to spinning inadvertently, and how easily this might be avoided. Models are tested in spinning tunnels, and full-scale aircraft at large angles of attack and yaw, before any spin programme is started.

Training aeroplanes, both civil and military, and aeroplanes certificated for aerobatics must satisfy time-honoured: *Spin Entry* and *Standard Recovery Procedures*. All such procedures contain profoundly interesting points, each one of which could have a long tail of woe behind it.

Features Affecting Spin Behaviour

All the following features are capable of affecting an aeroplane in the ways in which it behaves in the entry, the spin and in the recovery:

☐ *Configuration* (one, two or three-surface planform arrangement).
☐ *Wing arrangement* (monoplane, biplane or triplane).
☐ *Tail type and arrangement.*
☐ *Tail size* (shape and authority of the surfaces).
☐ *Dihedral or anhedral*, of both wing and tail. Autorotation of the McDonnell Douglas F-4, Phantom revealed that at sideslip angles of 10°, the vortex on the leading side of the anhedralled stabilizer interfered with the flow past the fin, forming an asymmetric, pro-spin vortex pattern (see Fig. 12.1b, Ref. 12.2).
☐ *Angles of incidence and decalage*: the way in which the wing surfaces are rigged; e.g., in a biplane which wing has the larger angle of incidence; and the magnitude of the longitudinal dihedral between wing and stabilizer surfaces.
☐ *Engine type and arrangement*: piston, turboprop or jet; single or multi; wing- or fuselage-mounted, nose or tail.
☐ *Propeller rotation ('P'-effects)*.
☐ *Rotational moments of inertia*: A, B, magnitude of (B/A) and C (see Figs 9.32a to 9.32c).
☐ *Relative aircraft density and its effect upon aerodynamic damping* was dealt with in Chapter 9 (Equation (9–15a) and (9–15b)). The apparent increase in liveliness in response to control at higher altitude can be tricky when it comes to spinning and spin recovery. One variant of a training aeroplane, fitted with a more powerful engine (which introduces more destabilizing ('P'-effects) refused to respond to a reversed-recovery

procedure above about 10 000 ft, but did so at lower altitudes where aerodynamic damping had increased.

Combinations of these features make spinning complicated. It is impossible to make any hard and fast rules about spins and recoveries before test flying starts – although good advice can be given using representative models in a spinning tunnel, by organizations like the French ONERA (Office National d'Études et de Recherches Aerospatiales) in Lille, and NASA (USA).

Attempting to make a general rule about spinning is like making a rule about a shark: there is no rule. The reason is that every aircraft has an internal shape (intertias) and an external shape (aerodynamics); and almost certainly more than one spin mode. Whether or not more than one can be found during a controlled test programme is another matter. Unfortunately, if another mode exists then one day someone will find it – and we shall come to this in due course.

Spin Phases and Characteristics

A spin has four phases:

☐ *Entry*. Although *entry* is self-evident it is not always simple. With the stick hard back, throttle closed, at an airspeed, say, 5 or 6 knots above the stall, and usually with ailerons neutral, rudder is applied smartly in the direction of the intended spin (this is not done at the moment of the stall, because the controls may then not be effective enough to trigger the spin which follows readily, at a slightly higher speed).

Rudder causes the yaw which is needed to provoke a spin. The outer wing speeds up, the inner wing slows down and roll with yaw rate, L_r, causes the aeroplane to roll in the direction of yaw, followed by the nose dropping. The angle of attack of the inner wing increases and a stall follows (the drag of the inner wing being higher than that of the outer, augments the yawing moment from the rudder and, if all the conditions are favourable, autorotation (the spin) starts).

There is not always such sweet compliance with a boot-full of rudder in the required direction, with throttle closed and ailerons neutral. Some light aircraft, like the Cessna 150 and 152, may need a touch of power (say, 1400 to 1600 RPM) to provoke spin entry (following which the throttle should then be closed). Other types may enter a spin with less hesitation if the aileron of the inner wing is depressed (outspin-stick). This is described as crossing the controls: rudder in the required direction of spin, stick full back and in the rear corner opposite to the intended spin (coupled with throttle-CLOSED).

Yet another aeroplane may refuse to enter a spin, even at aft CG, regardless of the most unusual and adverse combinations of flying and engine controls a pilot can conceive.

Considering that there is an infinity of control combinations possible between the full-scale deflections either way, of aileron, elevator, rudder, power-lever or throttle, and the propeller pitch control, a pilot cannot be expected to reproduce all of them. That is the reason why it is impossible to guarantee that a spin will be provoked. However, when you find an aeroplane which *appears to be spin resistant*, be very very careful. IF SUCH AN AIRCRAFT DOES DECIDE TO SPIN IT MIGHT BE IMPOSSIBLE TO RECOVER. It has been trying to tell you that it lacks enough control authority to enter the spin, which means that it will probably not have enough for a recovery – so, take heed and do not press on regardless. It knows itself and has no more wish to end up all of a heap than you.

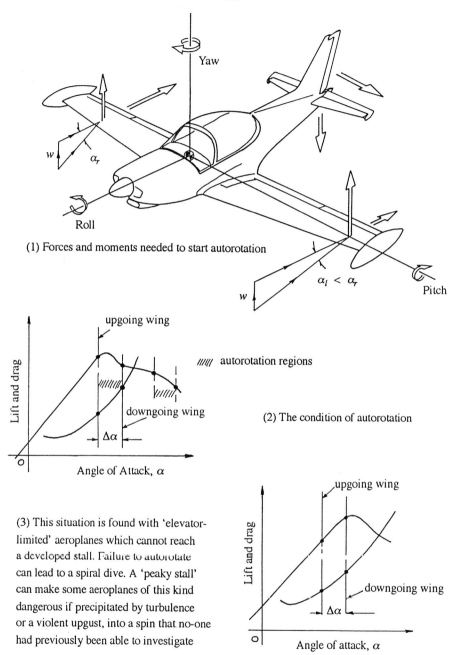

Fig. 12.1a   By autorotation the wing drives the spin (Ref. 1.5).

☐ *Incipient spin.* This is the motion between spin entry and a developed spin, in which the outcome of the struggle between growing and changing pro-spin and anti-spin moments has not been resolved.

Initially an incipient spin is unsteady, with the rate of rotation increasing, while following a ballistic flight path. The incipient spin lasts anything from two to six turns

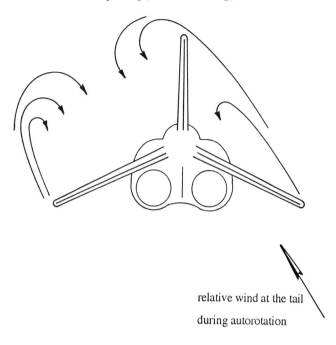

relative wind at the tail

during autorotation

Angle of attack, $\alpha = 90^{\circ}$

Angle of sideslip, $\beta = 10^{\circ}$

Fig. 12.1b In a spin the tail surfaces of the McDonnell Douglas F-4 Phantom are said to shed asymmetric pro-spin vortices (Ref. 12.2).

before steadying. Sometimes the spin may continue to be unsteady and oscillatory in nature, due to post-stall gyrational properties (PSG).

☐ *Spin.* The self-sustaining (autorotational) spiral motion of an aeroplane about a vertical axis, not necessarily through the CG (except when flat), during which the mean angle of attack of the wings is beyond the stall, with the inboard wing being more stalled than the other. A spin follows departures in roll, yaw and pitch from the condition of balanced flight. The developed spin is achieved when there is general (sometimes oscillatory) equilibrium between the predominantly pro-spin moment due to the wings and the generally anti-spin moments due to other parts of the aircraft.

Note: In a stabilized spin a light aeroplane loses 250 ft to 300 ft and takes about 2.5 s to 3 s per turn. A rate of one turn per second is fast and can be disorientating. An accelerometer registers only 1.0 g, or slightly more, showing that normal stresses are not high.

☐ *Recovery.* This is the most difficult phase of all. The aim is to stop autorotation. Just as spins vary between different types of aircraft, so do optimum recovery procedures differ. They may also vary between examples of the same type of aircraft, depending upon the spin mode. One particular aeroplane may have more than one spin mode: a primary, a secondary and maybe even a tertiary, and any one of these may differ from others of the same type.

Note 1: There is one military trainer which appears to have a perfectly benign primary mode, a high-rotational secondary mode, and an irrecoverably flat tertiary mode. Fortunately the latter is rare and was reported by one crew to be so sedate that they were able to stand on each side of the centre-section, wearing their parachutes and then, '*without the wind ruffling the hair*', they simply stepped off the trailing-edge.

Often unexpected and apparently irrational actions by the pilot are needed to regain control enough to effect a recovery, even in the primary mode. These may be quite different from what is called *Standard Recovery Procedure* or *Drill*.

Note 2: Structural damage, if it occurs, will be in the recovery if the airspeed is high, or control deflections are excessive and/or the aircraft is in an unusual attitude relative to the airflow.

Standard Recovery Drill, or Procedure

(1)   Throttle CLOSED.
(2)   Ailerons NEUTRAL.
(3)   CHECK that you are in a spin, not a spiral, and also the DIRECTION of rotation.
(4)   Stick BACK (i.e. conventional elevator-UP).
(5)   Rudder FULL against the indicated direction of turn.
(6)   **PAUSE** (say, long enough to count *one hundred–two hundred–three hundred*) allowing the rudder to bite and take effect. THEN:
(7)   Move the stick progressively FORWARD (elevator NOSE-DOWN) until rotation stops.
(8)   EASE OUT of the ensuing dive.

Note: Although a spin and a spiral are often confused, because the turn indicator points in the direction of both spin and spiral rotation, they are very different manoeuvres.

A *spin*:

(i)   involves large angles of attack;
(ii)   is a consequence of yaw at the stall so that there is skid away from the direction of rotation, as shown by the slip ball or pointer being opposite to the indicated direction of turn;
(iii) airspeed is generally low and, although there may be a tendency for the ASI to oscillate or hunt, it will do so about a mean position instead of increasing (if the airspeed does begin to increase it is a sign that the aeroplane is trying to come out. This will be confirmed if the rate of rotation also increases).

A *spiral*:

(i)   involves much shallower angles of attack than a spin;
(ii)   the slip ball or pointer tends to indicate no skid and almost zero slip;
(iii) airspeed increases, often rapidly (and therein lies the danger of structural failure to follow unless early corrective is taken).

The purpose of the spinning requirements is to ensure that the aeroplane will not become uncontrollable when flown through the applicable manoeuvre (Ref. 10.3). The purpose of spin testing is to ensure that an aeroplane will recover from a spin in the required number of turns, by normal use of the controls, without exceeding the design limitations.

All normal category single-engined aeroplanes – even those not designed for spinning and aerobatics – must be able to recover from a one turn spin, or a three-second spin, whichever takes longer, in not more than one additional turn, after initiation of the first control action for recovery. The conditions are:

- For flaps retracted and extended configurations, without exceeding the airspeed and positive limit manoeuvring load factor for each.
- It must be impossible to obtain irrecoverable spins with any use of the flight or engine power controls at the entry into or during the spin.
- If flaps are extended they may be retracted during the recovery, but *not* before autorotation has ceased.

Aerobatic category aeroplanes are, in addition, expected to be able to recover from any point in a spin up to and including six turns (or any greater number for which certification is sought) in not more than one-and-a-half additional turns, after initiation of the first control action for recovery. Turns for recovery are extended, not only because after six (or more) turns the kinetic energy of rotation is higher and takes more killing, but because regulation must be tougher.

Reversed Recovery

Example 12–1: The Sly (Undiagnosed) Spin

If caught unexpectedly by a spin, when concentrating on some other manoeuvre at low speed, it can cause total confusion if the pilot fails to realize what has happened. Then he will often take the wrong action.

Some years ago, while carrying out an investigation into why a Piper PA 38 Tomahawk had suffered a fatal spinning accident, the conclusion after extensive re-testing with the manufacturer's test pilot in the USA, was that the Tomahawk was an honest aeroplane, without complications.

Back in the UK a pilot, showing me in goldfish-bowl conditions how to stall a Tomahawk allowed the right wing to drop slightly and the aeroplane began to yaw gently to the right without him realizing it. Spin entry was slow and sly. By the second turn, with full up-elevator and low airspeed no proper stall had appeared. Still not realizing what was happening he pushed the stick forwards to unstall the aeroplane and regain speed, whereupon it pitched nose-down and started a fast rotation to the right.

This was just one more piece of evidence of the ease with which a spin can creep up unawares and the aeroplane bite, without realizing one has entered an incipient spin. Then the wrong action can follow – in this particular case, what is called *Reversed Recovery* action, i.e. applying elevator-DOWN before rudder, instead of after the pause which follows rudder, in the Standard Recovery Drill outlined above.

When such things happen, as they do to everyone at some time or another, testing is needed to discover first of all whether a spin will happen, and then what procedure is best for recovery. Ideally any conventional aeroplane should recover using the reliable Standard Recovery Drill, even though that might not be the *best* for the type, i.e. the one which achieves the *quickest recovery* with the *least height loss*. The reason is that the Standard Recovery Drill is taught in early training, and we tend to revert to early training habits when over-stressed, as we saw in Chapter 4.

Spinning Test Programmes

Table 12–1 is based in the main upon the useful Ref. 10.3 and shows the range of

(a)

(b)

(c)

(d)

Plate 12–1   When judging a horse look at its teeth – and an aeroplane
that might spin, at its tail and rear fuselage sections. Here, at random,
are some tails for you to judge for yourself. You could be right, you
might be wrong.

(e)

(f)

(g)

(h)

Plate 12–1 – Continued.

(i)

(j)

Plate 12–1 – Continued.

airworthiness spinning test configurations required by the FAA for a piston-engined aero-plane with flaps and retractable gear. The table shows clearly that the spin programme is first carried out at the safer forward CG. Later the spins are repeated at aft CG.

Reference 10.3 states that only those tests in the programme which are thought critical need be performed. This can be a difficult decision to make, not only when dealing with a civil prototype, but also with an aeroplane which has been significantly modified. How-ever, a number of the features listed in the table might not apply. For example, fixed land-ing-gear, no cowl-flaps. The task can be daunting. A spin clearance of a prototype civil aeroplane at one CG position and with all of the features listed in the table could involve the following programme:

☐ *Throttle-CLOSED, entries from straight flight*
  Pro-spin controls (rudder-LEFT, then rudder-RIGHT): $3 + 3 = 6$

Table 12–1

Piston-propeller spin-test configurations
(Based upon Ref. 10.3)

Flight condition Spins from wing-level attitude	Spin number	Flaps-Up	Flaps appch. (as approp.)	Flaps landing	Gear-Up	Gear-Down	Cowl flaps closed	Cowl flaps as required	Power Off	Power On	Foward CG
Pro-spin controls (STRAIGHT)	1	X			X		X		X		X
	2		X			X	X		X		X
Entry to left (1 through 6)	3			X		X	X		X		X
	4	X			X			X		X	X
Repeat to right	5		X		X			X		X	X
	6			X	X			X		X	X

Repeat pro-spin control entries from LEFT and then RIGHT TURNING flight.

Flight condition	Spin number	Flaps-Up	Flaps appch.	Flaps landing	Gear-Up	Gear-Down	Cowl flaps closed	Cowl flaps as required	Power Off	Power On	Foward CG
Tests for uncontrollable spins	7	X			X		X		X		X
Entry from LEFT turn, aileron AGAINST (7 through 12)	8		X		X	X	X		X		X
	9			X	X	X	X		X		X
	10	X			X			X		X	X
Repeat from RIGHT turn aileron AGAINST	11		X			X		X		X	X
	12			X		X		X		X	X

Flight condition	Spin number	Flaps-Up	Flaps appch.	Flaps landing	Gear-Up	Gear-Down	Cowl flaps closed	Cowl flaps as required	Power Off	Power On	Foward CG
Entry from LEFT turn, aileron WITH (13 through 18)	13	X			X		X		X		X
	14	X			X	X	X		X		X
Repeat from RIGHT turn, aileron WITH	15			X	X	X	X		X		X
	16	X			X			X		X	X
	17		X			X		X		X	X
	18			X		X		X		X	X

Repeat all of the above spins at AFT CG.
Total spins in package, excluding further repeats = 72.

  ☐ *Throttle-CLOSED, entries from turning flight*
      Pro-spin controls (rudder-LEFT, then rudder-RIGHT):   3 + 3 = 6
  ☐ *Power-ON* spins, entries from straight flight, repeat above:*   3 + 3 = 6
  ☐ *Power-ON* spins, entries from turning flight, repeat above:*   3 + 3 = 6
  * See **Note 3** below.

The total is a possible 24 spin entries and recoveries using pro-spin controls, before starting to look at what happens when the controls are misused (abused).

In the lower part of Table 12–1 are shown the flying control abuses, and the configurations for the same piston-engined aeroplane total:

☐ *Throttle-CLOSED, entries from straight flight*
  Ailerons out of spin (rudder-LEFT, stick-RIGHT, then vice versa):   3 + 3 = 6
☐ *Throttle-CLOSED, entries from turning flight*
  Ailerons into spin (rudder-LEFT, stick-LEFT, then vice versa):   3 + 3 = 6
☐ *Power-ON*, entries from straight flight*
  Ailerons out of spin (repeat rudder-LEFT, then rudder-RIGHT):   3 + 3 = 6
☐ *Power-ON*, entries from turning flight*
  Ailerons into spin (repeat rudder-LEFT, then rudder-RIGHT):   3 + 3 = 6
  * See **Note 3** below.

This gives another 24 configurations, and a grand total of 48, at one CG position.

When these are repeated at FWD and AFT CG, it could mean 96 different configurations, which are not the same as attempted spins. Failure to achieve entry to a spin does not mean that the pilot then gives up. The burden is upon him to keep on trying until he is quite certain that the aeroplane either will not spin, or that he has found the repeatable entry procedure.

For reasons given below a test pilot expects to try everything three times at least, as a check, and to gather numbers he did not get the first or second time around. When an aeroplane is reluctant to spin, so that different throttle settings must be tried, or a result is dubious, a programme can easily extend to 500 spin entries. If significant problems necessitate major modifications, in the form of strakes, or the addition of dorsal and/or ventral surfaces, then 500 can easily be multiplied four and five-fold, as happened in the case of one particular trainer which is now said to have carried out some three thousand spins during its development.

Questions Lawyers Ask – After an Accident

The following list is excessive for anyone planning a test-card, but these are some of the technical questions one finds lawyers asking after an accident. Test pilots never know when such questions might come, so it is as well to bear in mind the extent of what one might be faced with. Therefore, write up what happened in notes of greater length as soon as possible after the spinning sortie, so that details are not forgotten or jumbled in the memory.

Usually spins must be repeated several times in each configuration because of the difficulty of noting things like: *altitude of initiation* of control movement to enter the spin – *airspeed* at that time – what *control movements* were made (*how many* in *magnitude* and *direction*) – what were the *control forces* involved – *what happened* next (did it *roll inverted*, how far *did the nose go down* (if it did)) – what were the *control forces* in the developed spin – how *erratic* or *steady* was the *behaviour of the ASI* – what was the *range of ASI readings* – what was the *altitude lost per turn* – how many *seconds per turn* – how much *wing tilt* was there *into or out of the spin* when steady – *how many degrees nose down* were estimated – was there *any tendency to flatten the turn*, and *after how many turns* – how many *turns before initiating recovery* action – *at what altitude* was that action taken – *which control* was used to initiate recovery – *how much movement* – *what forces* were involved – was there a *pause* before moving any other control (if so for *how* long) – *which other control* was then moved – *how far* – *what force* was needed – how many *turns to recover* – *height lost* in recovery – any *tendency to enter a spin the other way* – *total height lost* – did the *propeller stop* – were there any symptoms of *fuel starvation* – *how many attempts* were made in that configuration – *was power tried* in the spin and if so *what were the effects*? About forty items are here, and the most anyone can cope with per spin are two or three when out of practice, and maybe five when very sharp.

Recording Results

A practical list to be recorded on a tape and/or a kneepad would include:

- configuration;
- altitude at entry;
- direction of spin;
- attitude(s) during entry;
- airspeed at end of first turn;
- turns to stabilize rotation;
- control positions;
- airspeed when spin stabilized;
- time in seconds per turn;
- slip, skid and wing-tilt (in which direction);
- height lost per turn;
- number of (stabilized) turns completed;
- height at which recovery initiated;
- number of turns taken to recover;
- altitude at which rotation ceased;
- speed; and
- altitude when straight and level;
- unusual behaviour during entry, developed spin and recovery (e.g., did it ever settle down, or did roll, pitch and yaw rates change?).

And even here there are nearly twenty. To attempt to complete such a list manually, in one spin, is impossible. To cover these every spin needs three or four repeats. Therefore, a spin test-card is one needing careful layout and a special shorthand. A suggested shorthand was shown in Table 4–5. Table 12–2 combines one possible format with an example of a two-turn spin.

It may be that during the spin the pilot discovers something unusual: the propeller stops, or the aeroplane reveals another spin mode. So, there must be plenty of space on the card to write, and don't forget that, in flight, the hand will be far from steady if the spin is erratic or rapid.

Note 1: If an aeroplane is claimed to be characteristically incapable of spinning, the same tests at heavier weight with expanded CG range, control angles and the rest, must be carried out for the relevant configurations selected. Aim to record the same items.

Note 2: Because of the change in B/A, the ratio of moment of inertia in pitch to that in roll, alters with CG position, I have found spins at forward CG (when these can be provoked) to be oscillatory and far more uncomfortable than any at AFT CG. Do not be put off if the aeroplane is reluctant to enter a spin at FWD CG. Try small changes in power setting, and in aileron position. It may be that, although reluctant to spin with full aileron in one or the other direction, the aircraft will do so with partial aileron. Although rare, it happens.

When working out test CGs (see Chapter 5), look carefully at items such as a heavy occupant in the front cockpit, a light one in the rear and vice versa. Look too at wrong, or the most adverse, order of selection of fuel tanks, and anything else likely to make a significant difference to rotary moments of inertia.

TABLE 12–2

Suggested test card with example of a two-turn spin
(The arrangement leaves room for numbers and shorthand remarks, while the card may be used for a single spin as shown.)

W/X: 3/8 , 2000ft 1012·3 30 nm $^{270}/_{20}$				CARD No.: 1–4
ATC: Fairfield 120·35 (idling 720)				PROJECT: 59
A/C: G–BD	T/O: 10·15 LAND: 11·23	DATE: 5 oct 93		T/O WT: 1328 kg
CREW Jones / Alderton			FUEL START: entry: 93 litres	
CONFIG: P↓ F↑ G↑			FUEL END: recovery 93 litres	
SPIN No: 4			CG: 27%	

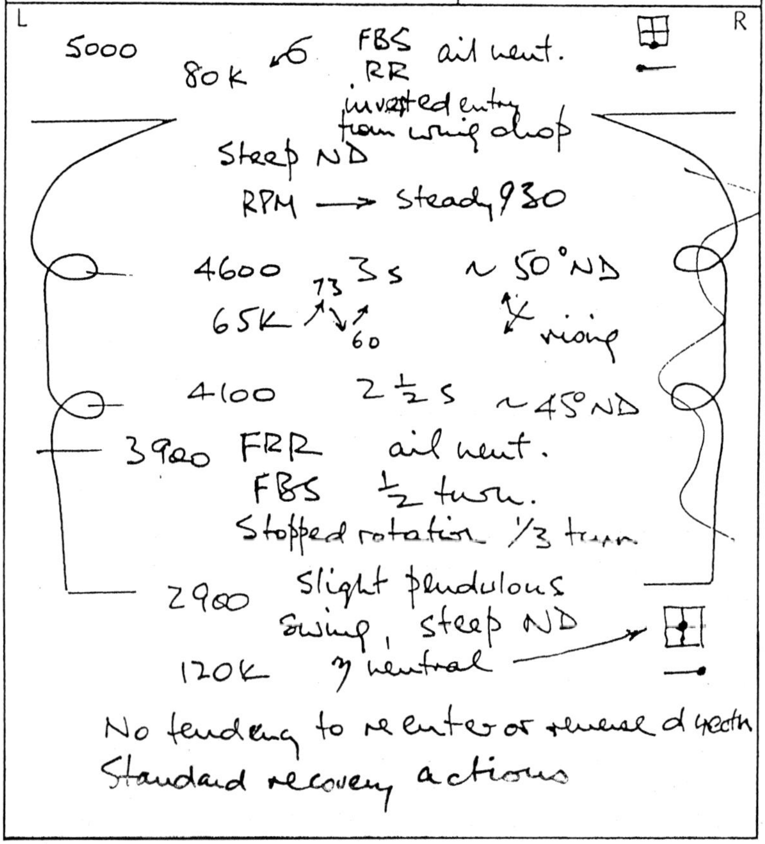

L   R

5000   80 K   FBS ail neut.
                     RR

inverted entry
from wing drop
Steep ND
RPM ⟶ Steady 930

4600   3 s   ∼ 50° ND
65 K   73 60   rising

4100   2½ s   ∼ 45° ND
3920 FRR   ail neut.
     FBS ½ turn.
Stopped rotation ⅓ turn.

2900 Slight pendulous
swing, steep ND
120 K   ↗ neutral

No tendency to re enter or reverse d irecth
Standard recovery actions

Note 3: The power settings for the spin tests are with throttle closed (power-OFF) and 75% maximum continuous power or thrust (or full throttle where altitude effects are predominant). Before take-off note the direction of rotation of the propeller and engine idling RPM.

T<small>ABLE</small> 12–2 – *Continued*

(see Table 4–5)

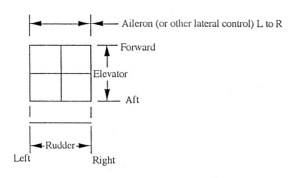

With this system the complete control positions for, e.g., an upright left spin with ailerons against can be described with fewer written notes:

and include a recovery investigation of elevator moved to neutral, while holding aileron and rudder, after four turns with it in its starting position:

Recovery control action can also be included to to cover the possibility of the investigation of elevator to neutral failing to bring about recovery. In which case stick full forward, ailerons with and rudder against to be applied after four more turns.

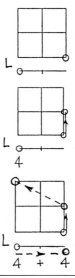

If on the slow side the propeller might stop in a spin. If idling faster than normal, this could affect the spin recovery and give a misleading impression of the effectiveness of the flying controls in the recovery.

Example 12–2: Spin Programme – Système 'D' (Example 1–3)

The triplane in Fig. 1.7 was intended to spin about any axis, regardless of angle of bank. We saw in Example 8–3 that it had stall speeds varying between 75 KEAS, power-OFF (i.e. throttle-CLOSED) erect and inverted; and 85 KEAS, power-OFF in knife-flight. This would affect the speed for spin entry, and it would also affect behaviour in a spin and in the recovery, because of the change in the ratio of B/A. This has its normal meaning in erect and inverted flight, but with 90° of bank in knife-flight it becomes the significantly larger:

$$C/A \le [1 - (B/A)] \qquad\qquad (12\text{–}1) \text{ from } (9\text{–}18\text{d})$$

In spite of doubts that the aeroplane will spin at all, a point to watch is that the heavy landing gear, out to one side or the other (depending upon which way one is banked) will make a difference to the inertia axis. This could introduce possible yaw-roll coupling effects, and these could be exacerbated with asymmetric '*P*'-*effects*.

Therefore, to explore the characteristics properly the test pilot would point out that as the aeroplane is experimental one might expect an unusually extensive spin programme, with several hundred attempted spin entries, through 360° bank – even if each does not develop. Further exploration is needed with changes of power. As the aeroplane is twin-engined to balance torque effects, or to produce asymmetric effects at will, how extensive a spinning programme must be planned with both engines together, followed by one throttled back, and then the other?

The aeroplane is a nightmare – almost certainly the dead-end of an impracticable idea.

Clearly, a spin programme is a daunting task, as the pilot must repeat many of the entries and recoveries where necessary. Modifications to improve spin behaviour also involve repetition. Re-engining an aeroplane to introduce a change in power – which may also change the direction of rotation of the propeller involves a fresh test-package of spins. One British test pilot for the manufacturer of a light training aeroplane for military use has lost count of the total number of spins, which must now be in excess of 5000. These do not include the quick 'suck-it-and-see' (and probably unrecorded) attempted entries, taking a few seconds only, to discover some small change in control combination which could be more promising that the last.

Beware of Sluggish Spin Entry and Recovery

Equal faults are those factors which prevent a pilot from entering an intentional spin in an aircraft cleared for spinning. Such faults are as important as those which delay or prevent recovery from a spin. The point was made earlier, that if an aircraft appears to be spin-resistant it might be telling the pilot that it will be irrecoverable if it does spin.

We know that an aeroplane consists of two dominant configurations, one external consisting of the aerodynamics, the other internal, represented by the agglomeration of masses and their inertias. There is a constant struggle between the two configurations whenever any kind of acceleration or deceleration occurs. In straight and level unaccelerated flight there is equilibrium, by definition. It is the outcome of the struggle which determines the differences between those aeroplanes which are sedate and others which are agile. Sedate aeroplanes should never be spun deliberately.

Aeroplanes 'Characteristically Incapable of Spinning'

Designers try to make spin-proof aeroplanes and it is claimed that they exist. But of those I have tested, not one proved to be characteristically incapable. This needs explanation. I once thought that we had found one of a type which might turn out to be so, because neither I nor four or five other test pilots could make that particular example spin within the range of *normal* loadings and control rigging angles. Unhappily another example of the type, although reluctant, did finally spin but only after much determined provocation by a club pilot who felt impelled to demonstrate that it would – in spite of a published warning that the type had not been cleared for full spinning. Never assume that because one aircraft of a type refuses to spin then others will be the same. Never ignore advice given in the AFM, POH, or Pilot's notes.

To demonstrate that an aeroplane – or a type – is characteristically incapable of spinning one must carry out full spin test with: the weight increased beyond that for which certification is sought; the CG ballasted beyond the aft certification limit; up-elevator travel expanded beyond the normal limit; and the rudder angles also expanded beyond their normal limits in each direction. The expanded limits are specified as minima in the relevant airworthiness requirements. Implicit too is the requirement that it must be impossible to obtain uncontrollable spins during the tests with any combination of the flying and engine power controls.

Mechanics of Autorotation

Autorotation is driven mainly by the aerodynamics of the wings, stalling asymmetrically and suffering unequal forces. The longer and heavier the wing, the more dominating is the autorotational moment over those opposing rotation.

The conditions required for autorotation are as shown in Fig. 12.1a(1), (2) and (3), and Fig. 12.1b. The downgoing starboard wing is more stalled than the port, such that the relevant angle of attack $\alpha_r > \alpha_l$.

Figure 12.1a(3) is the situation when the elevator lacks enough authority to provoke a stall (as when designed for stall-spin avoidance). If the up-elevator stop is not positive enough, as might be the case with certain kit-built aircraft (it happens), or if its mounting bracket is able to bend (it also happens), then one might apply more up-elevator than intended, reaching a condition like that shown in Fig. 12.1b. Result: an unintentional and untried spin with an aeroplane originally thought to be spin-proof.

Contributions from rotary inertias

Figure 12.2 shows how an aeroplane may be represented inertially by three interlocking hoops, like bicycle or flywheels, each of which has a radius of gyration about its respective axis. The moments of inertia in roll, pitch and yaw, A, B and C were given in the parts of Equation (9–18a) to (9–18d). Pitch and yaw inertias vary with CG location and configuration (tail-first or tail-last and three-surface). The moment of inertia in yaw, C, is the largest, being roughly equal to the sum of the other two. The yaw and roll rates, r and p, are added vectorially to produce the spin rate, Ω, as shown in Fig. 12.2a(1).

The way in which rotary inertia influences motion through precession is seen most easily when we think of something most owners of a bicycle have done when young. Make a chalk mark on a bicycle wheel and then, holding it upright, spin the wheel away from you, so that the chalk mark moves up and away from the eye. Now try turning the wheel to the right. This introduces a component of motion to the chalked portion of the tyre which, added to its peripheral motion, would cause it to curve to the right when viewed from above. A curve means a centripetal acceleration of the chalk mark towards an instantaneous centre on the right. This produces an equal and opposite centrifugal reaction, causing a torque which tilts the wheel to the left, in a plane 90° onwards from the plane of the intended turn. The effect is powerfully augmented by every particle of the mass of the wheel, which at any instant contributes to the torque, by providing a couple, to the left in the upper half of the wheel, to the right in the lower half.

Precessed moments are called *imposed inertia moments*.

Imposed inertia moments: Coriolis effects

Imposed inertia moments are not the same as the rotary moments of inertia. Rotary moments of inertia are indicators of the magnitude of rotary momentum locked into each

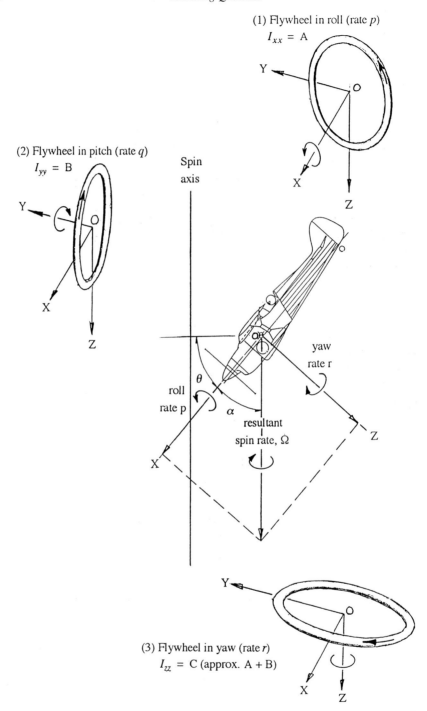

Fig. 12.2a The autorotational motion of an aeroplane is the resultant of three coupled flywheels acting in roll, pitch and, most powerfully, yaw (see Equations (9–18a) to (9–18c)).

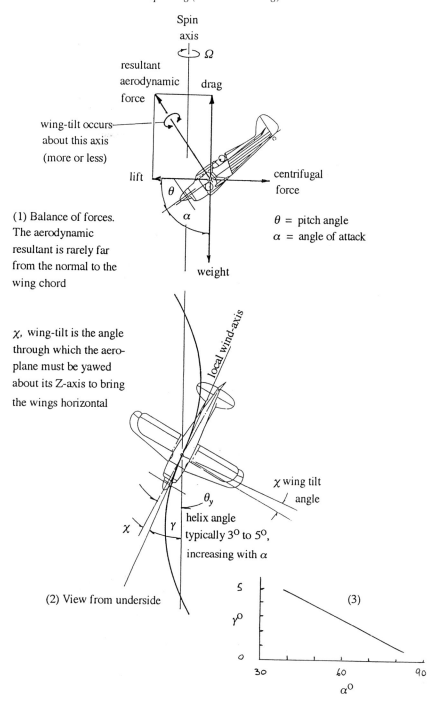

Spin axis

Ω

resultant aerodynamic force

drag

wing-tilt occurs about this axis (more or less)

lift

θ

α

centrifugal force

(1) Balance of forces. The aerodynamic resultant is rarely far from the normal to the wing chord

θ = pitch angle
α = angle of attack

weight

χ, wing-tilt is the angle through which the aero-plane must be yawed about its Z-axis to bring the wings horizontal

local wind-axis

χ wing tilt angle

θ_y

helix angle typically 3° to 5°, increasing with α

χ

γ

(2) View from underside

5

$\gamma°$

0

(3)

30   60   90

$\alpha°$

Fig. 12.2b   Balance of forces in the spin, showing also helix angle and wing-tilt.

respective mode by the product of the moment of inertia and rate of rotation about the relevant axis: Ap, Bq and Cr, respectively.

The imposed inertia moments, on the other hand, are the products of rotary momentum and angular velocity, which together cause gyroscopic precession. They arise from inertia and the attempt of masses to remain fixed in space. Relative to the pilot in a spinning aeroplane they appear as distorted motions – *Coriolis effects*.

Note: The commonest Coriolis effect, observed by the aviator and meteorologist, is the rotation of wind direction to the right with increasing velocity in the Northern Hemisphere, and to the left in the Southern.

Now look at the spinning aeroplane, rolling at rate p, with rolling moment of inertia A, in Fig. 12.2a(1). The rolling momentum is $(A \times p)$. If the aeroplane is rolling to the right, sideslipping right, with weathercock stability turning it to the right about its Z-axis, an angular velocity of rate r is imposed. The effect is to precess the motion through 90° in the direction of roll which (like the right-hand propeller) brings into play a nose-down pitching component of motion, $-\Delta q$.

But the aeroplane is yawing with a powerful rotary momentum, $(C \times r)$. As there is a rolling torque to the right, this is precessed through 90° to the right, which acts in a sense to tilt the nose upwards at a component rate $+\Delta q$ (which is stronger than the nose-down $-\Delta q$ already described). The sum of their components, $+q$, tends to increase the angle of attack, increasing the rate of autorotation and flattening the spin.

Thus, the imposed inertia moment:

$$\text{in pitch, } M_i = (Cr)p - (Ap)r = (C - A)pr \qquad (12–2)$$

Using a similar argument to that of the bicycle when in the rolling and yawing planes, we have:

$$\text{in roll, } L_i = (Bq)r - (Cr)q = (B - C)qr \qquad (12–3)$$
$$\text{in yaw, } N_i = (Ap)q - (Bq)p = (A - B)pq \qquad (12–4)$$

The ratio B/A

Whether or not the imposed inertia moments will be pro-spin (tending to keep the aeroplane in the spin) or anti-spin (tending to help it to recover) depends upon the relative magnitudes of A, B and C in Equations (12–2) to (12–4).

Because C is much greater than either A or B then $(C - A)$ is positive. In a spin to the right both p and r are positive, all three making the pitching inertia moment, M_i, positive – which is in the nose-up sense and pro-spin. As far as the rolling inertia moment is concerned, $(B - C)$ is negative, making L_i negative and anti-spin.

Yaw (slip and skid) are of fundamental importance in determining whether or not a spin will follow. The inertia yawing moment which governs so much depends upon the relative magnitudes of A and B. If B is greater than A then $(A - B)$ is negative and anti-spin. But if B is less than A, then $(A - B)$ is positive and N_i is pro-spin.

Dividing $(A - B)$ by A, this can be restated:

☐ *When $(1 - (B/A))$ is positive* (i.e. A > B) the aircraft tends towards being 'wing-dominated' and *spin-prone*. This is the case when a wing is either long and heavy, or contains large masses of fuel, engines, landing gear, or stores. So, if an aeroplane has a value of pitching inertia/rolling inertia, $(B/A) < 1.0$, then pilots will need well-coordinated feet.

☐ *When $(1 - (B/A))$ is negative* (i.e. B > A) the aeroplane tends towards being 'fuselage (or pitch)-dominated', with *spin-resistant inertia characteristics*. A value of $(B/A) = 1.3$

is thought to mark the beginning of true spin-resistance, and this is typical of a number of current training aeroplanes, as well as of jet fighters of the 1950s, which had smaller engines than the more powerful and heavy modern fighter.

Aerodynamics versus Inertias

We saw earlier that a spin is defined as the result of a struggle between the aerodynamics and the inertias. Whether the outcome results in a smooth balance between the two and a steady spin, or rotation which is erratic and oscillatory, or motions which are completely wild, depends upon the relative strengths of the two sets of moments.

In general a very light and slow aeroplane is dominated by its aerodynamics. It has plenty of wing area, has a long wing stretched out across the wind to reduce life-dependent drag, low wing and power-loadings, and it usually slips into a steady spin within one turn without fuss. Provided the configuration is such that favourable *aerodynamic* out-spin yawing moments are generated with ease, then it is unusual to encounter undue difficulty in recovery from a primary mode. If a secondary mode is lurking around a corner then it can be expected to be somewhat less predictable in occurrence, nature, and in the actions needed for recovery.

Post-stall gyration (PSG)

As aircraft grow heavier, faster and more powerful, wing and other surface areas must be kept as small as possible for economic operation. Component and structure weights rise together with increased fuel weight for powerful and thirsty engines. Configurations are increasingly dominated by fuselages stretched along the wind, combined with wings short in span. Wing and power (or thrust) loadings increase. Aircraft density, moments of inertia in pitch and yaw grow, and (B/A) becomes much greater than 1.0. While this may well be spin-resistant, if departures do occur they are likely to be far more erratic and oscillatory post-stall. Inertias dominate the scene, and the role of the aerodynamic control surfaces is to induce favourable inertia moments which in turn cause the outspin yaw needed for recovery to be generated *dynamically*. Further, the presence of secondary and even tertiary spinning modes is more the rule than the exception.

Note: The Système 'D' triplane, although intended for aerobatics, suffers not only from a large B/A, but from one which varies significantly when banked into another spinning plane. The fuselage structure, coupled engines-plus-gearbox-plus-rotors, fuel and pilot, are all contained within the relatively large and long body. This is mated with light and small wings, covered with lightweight fabric, which is transparent over large areas. Therefore, not only will the spin characteristics vary, there could easily be more than one highly erratic or oscillatory mode in each spinning plane.

Figure 12.2b shows the balance of forces in the spin. The resultant aerodynamic force is roughly normal to the plane of the wings. When the pitch angle is large ($65° > \theta > 45°$), then rate of rotation, Ω, is moderate, radius of rotation is large, velocity in the descent is moderate (as could be the IAS) and so is the angle of attack, $\alpha = (90° - \theta)$. When the pitch angle is small ($10° < \theta < 30°$) the rate of rotation is fast and the radius of rotation small. Rate of descent is relatively slow, and the angle of attack, $\alpha = 60°$ to $80°$.

The helix angle of the spin path, γ, is always small, around $3°$ to $5°$ at a pitch angle of $45°$. It varies as shown in Fig. 12.2b(3) and decreases with increasing angle of attack.

Significance of wing tilt angle, χ

The wing tilt angle, χ, is the angle through which the aeroplane is rotated about its normal Z-axis. Although it is not the same as bank angle, ϕ, bank and wing tilt are inseparably linked. Wing tilt affects the pitch-rate, q, the sign, direction and magnitude of sideslip and, hence, spin-proneness. It is not measured relative to the horizontal, but to the normal to the relevant wind-axis. In Fig. 12.2b(2) the aeroplane is shown as yawed with nose to the left (as seen by the pilot) through the wing-tilt angle, $-\chi$, relative to the local wind-axis. Figure 12.3 shows more clearly the relationship between the relative wind, the axes based upon it, and the aircraft axes. This time the nose is to the right of the relative wind axis and the wing tilt angle seen by the pilot is port wing down.

In Figs 12.2a and 12.2b(1) imagine that the aeroplane is pitched nose-down so that $\theta = -90°$ (i.e. $\alpha = 0$), as in a downward roll. Rotation about the spin axis is all roll ($p = \Omega$). Any wing tilt will then raise the nose, causing sideslip and introducing a pitch rate component. Simultaneously the roll rate will be diminished until, in the extreme case of wing tilt angle $\chi = 90°$ the rotation will be all pitch and no roll. Finally, return to the first condition, only this time decrease the pitch angle to zero, as in a flat spin, so that $\alpha = 90°$. If there is no bank and no wing tilt the motion about the spin axis is all yaw, with no roll. If wing tilt and bank are present and increase towards $90°$, then a growing pitch component is introduced and the yaw rate is correspondingly diminished.

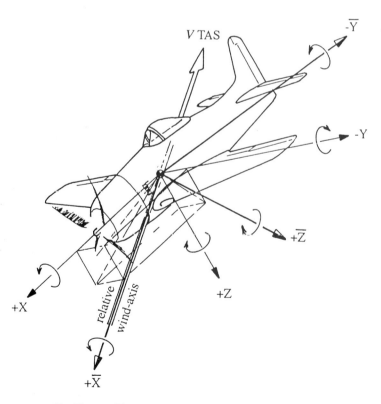

Fig. 12.3   Axes $+\overline{X}, -\overline{Y}$ and $+\overline{Z}$ are based upon the wind-axis, i.e. the direction and velocity of the relative airflow at any moment.

Figure 12.4a shows pro- and anti-spin inertial couples when B/A < 1 and > 1. Here their representation as interlocking wheels, adopted earlier, is replaced by dumbbells, but the effect is the same. When a wheel or dumbell is rotated about an axis which is not normal to the natural radius of gyration, the dynamics of the system – obeying the law of conservation of momentum – attempt to drive it into a plane of rotation in which the radius of gyration lies in dynamic equilibrium, i.e. where the principal radius of gyration is a maximum. This is usually prevented, of course, by the aerodynamics, shown in Fig. 12.4b.

Conservation of Momentum and ...

In Fig. 12.4a(1) and (2) are shown two very different mass distributions and couples. An increase in rate of rotation of each aeroplane increases the centrifugal force on every element of mass. This causes the dominant masses (the wings in the case of (1), the fuselage in (2)) to yaw each aircraft about its Z-axis, so as to orbit the CG in a plane normal to the axis of rotation. In this way angular momentum is conserved. The long-winged aircraft, with B/A < 1, yaws in-spin; the long-bodied, with B/A > 1, yaws out-spin.

Opposing the inertias are the aerodynamics, shown in Fig. 12.4b(1) to (3). To win the game of spin-recovery one must use the flying controls to alter the equilibrium condition and induce the rotary inertias to collaborate in producing 'out-spin yaw'.

The effects of flying controls can be mixed, depending upon the ratio of B/A.

Note: In what follows it is better to use the terms 'in-spin' and 'out-spin stick' instead of 'in-spin' and 'out-spin' aileron because it is less ambiguous. Although 'in-spin' and 'out-spin' aileron may describe aileron *direction*, it can also be interpreted as *effect* and so mean arguably different things to different pilots.

Aileron

Ailerons act in the normal sense, i.e. stick to the left produces a rolling moment to the left. This changes the wing tilt. When B > A and the fuselage inertia is dominant, applying in-spin stick (starboard aileron DOWN in a spin to the left), raises the starboard wing above the horizon. Yaw and roll are coupled and some of the rotary momentum is converted into roll, the enhanced rate of rotation, acting on the fuselage masses, produces out-spin yaw which favours recovery. Not only that, but down-aileron on the wing farthest from the spin-axis causes drag which slows spin rate and generates a component of out-spin yaw.

When A > B and the wing is dominant, in-spin stick couples roll with yaw. The rate increases and the masses in the wings cause yaw in-spin as they rotate about the Z-axis, in an attempt to take up a common plane of rotation. Simultaneously the spin tends to flatten. Out-spin stick has the opposite effect, decoupling roll from yaw. The radii of gyration of each wing-mass increase, reducing the spin rate and angle of attack, with the nose going down, assisting recovery.

The effects of changes in the B/A ratio with aileron deflection into the spin (in-spin stick) are shown in Fig. 12.6a.

Elevator

Forward movement of the stick, elevator-DOWN, opposes the tendency of the fore and aft dumb-bell masses to seek a common plane of rotation, by displacing them vertically in Fig. 12.4a(2). This, by reducing their radii of gyration about the spin axis, causes an increase in

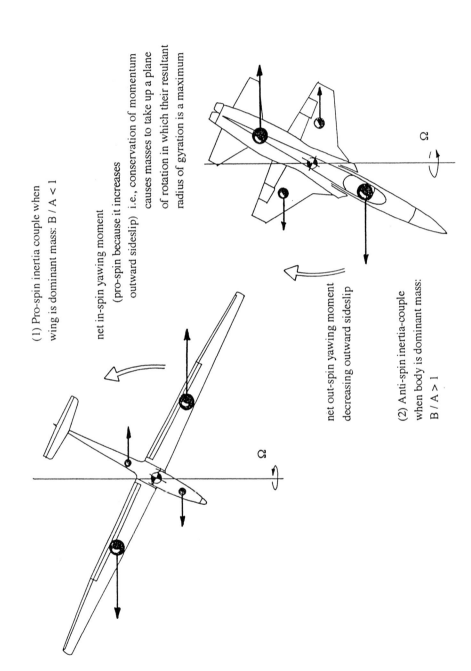

(1) Pro-spin inertia couple when
wing is dominant mass: B / A < 1

net in-spin yawing moment
(pro-spin because it increases
outward sideslip) i.e., conservation of momentum
causes masses to take up a plane
of rotation in which their resultant
radius of gyration is a maximum

net out-spin yawing moment
decreasing outward sideslip

(2) Anti-spin inertia-couple
when body is dominant mass:
B / A > 1

Fig. 12.4a   The effects of inertia when (1) B / A < 1, and (2) B / A > 1. Here the direction of spin rotation is to the left.

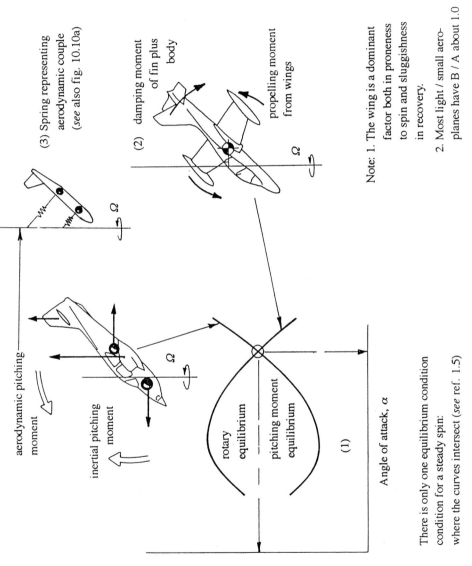

(3) Spring representing aerodynamic couple (*see* also fig. 10.10a)

(2) damping moment of fin plus body

propelling moment from wings

aerodynamic pitching moment

inertial pitching moment

rotary equilibrium

pitching moment equilibrium

(1)

Angle of attack, α

Rate of rotation, Ω

There is only one equilibrium condition condition for a steady spin: where the curves intersect (*see* ref. 1.5)

Note: 1. The wing is a dominant factor both in proneness to spin and sluggishness in recovery.

2. Most light / small aero-planes have B / A about 1.0

Fig. 12.4b   Interaction between inertias and aerodynamics.

the rate of rotation like a dancer or a skater speeding up in a pirouette by drawing in the arms. What happens next depends upon the ratio of B/A. When B > A the effect is pro-spin – the nose tends to rise and the spin to flatten. When A > B elevator-DOWN is anti-spin and assists recovery, as shown in Fig. 12.6b.

Note 1: There is a danger that full nose-down elevator will bunt an agile aerobatic aeroplane into an inverted spin (e.g., the Pitts Special can be pushed into an inverted spin with ease). To avoid the risk of this happenning it may be better to hold the elevator neutral.

Note 2: There is too the danger of entering a *Reversed Recovery* by leading with elevator, which then might blanket the rudder, as shown in Fig. 12.5b(2).

Note 3: Having said this, a conventionally sedate subsonic (e.g., light) aeroplane with heavy wing-mounted engines and A > B may stand a better chance of recovering from an inadvertent spin if elevator-DOWN is applied before rudder.

Rudder

Rudder is the primary source of out-spin yaw. The moment which is produced alters the wing-tilt, raising the outer wing in the spin above the horizon. The effect is anti-spin when B > A and pro-spin when A > B.

A number of fin and rudder combinations are sketched in Fig. 12.5, which also shows the way of constructing the way in a 45° nose-down spin, to check how much of the rudder is shielded and ineffective. The T-tail in Fig. 12.5a(4) is the best for recovery from an erect spin, but not from one which is inverted. There are tails around like those in Fig. 12.5a (5 and 6). Extreme caution is needed before attempting to clear tails of this kind for spinning. Fit an anti-spin parachute and wear a parachute.

Figure 12.5b(1) and (2) shows why the elevator is held up with full-back stick, while rudder is applied first in the Standard Spin Recovery. Reversed recovery, which is the situation in (2), occurs when the elevator leads, shielding the rudder and reducing its working area. Rudder becomes less effective with increasing angle of attack as a spin flattens and the rate of descent slows.

The effect of rudder upon anti-spin yawing moment, over a range of B/A values, is shown in Fig. 12.6c. The implication of this is that the longer the span of the wing, and the more powerful the inertia A is relative to B, the larger the area of unshielded rudder needed to recover from a spin.

Note 1: Careful attention must be paid to the arrangement of rudder surfaces to keep AT LEAST $\frac{1}{4}$ to $\frac{1}{3}$ of the rudder area working outside the wake of the tailplane and elevator. It is surprising how many designers still fail to do so, combining unfavourable arrangements (see Fig. 12.5a(2), (5) and (6)) with equally tricky (slippery and highly streamlined) rear fuselage sections.

Note 2: Look again at the Epsilon modifications in Fig. 11.14.

Note 3: Fin and rudder surfaces must never have too much taper, for the reasons given in Chapter 8 (see Figs 8.5d and 8.6). Narrow tip chord (low Reynolds number) and large angles of attack when spinning, precipitate fin and rudder tip-stalling.

Body damping contributions

☐ *Fuselage shape.* The principal aerodynamic yaw-damping moments come from the rear fuselage (shape of the sections and depth) and the fin and rudder surfaces. Curved, smooth and sweetly blended sections top and bottom encourage cross-flows to slip past unhindered with minimum resistance. There is no turbulent increase in drag to slow the spin rate and discourage flattening.

The most important part of the fuselage is the last third of the length between CG and rudder-hinge. This is the region on which to concentrate any modifications to the shape of the sections. Table 12–3 lists the effects of rear fuselage sections upon spin damping. The best (though least aesthetically pleasing) is the flat-topped section with strakes and a rounded bottom.

☐ *Strakes*. The fuselage strake is one of the simplest modifications to apply to a prototype. It belongs to the same family of flow-breaking and vortex-generating devices as leading-edge breaker strips, dorsal and ventral fins. When a strake is normal to a flow instead of tangential it trips random turbulence, encouraging turbulent reattachment of a separating boundary layer. A crossflow which is separating in an unsteady manner, as when Kármán vortices are being shed causes oscillatory spins.

When the flow meets a strake at an angle of attack a vortex is shed. This represents a form of controlled and directed turbulence. The fuselage strake then acts in a similar way to a dorsal fin, by inducing a favourable flow across otherwise critically ineffective areas of tail surface.

☐ *Forebody strakes* can be most useful on aeroplanes short in wingspan with heavy fuselages and relatively large values of B/A. By causing even reattachment of the flow in a spin the forebody wake is thinned, cross-flow drag and damping is reduced, the spin speeds up, becomes smoother and, as in Fig. 12.4a(2), inertia forces cause out-spin yaw.

Example 12–3: Anti-Spin Strakes – De Havilland 82A Tiger Moth

Following a number of spinning incidents and accidents between January and July 1941 several De Havilland Tiger Moths, the basic trainer of the RAF, were tested by the Aircraft & Armament Experimental Establishment (A&AEE), Boscombe Down, and RAE Farnborough. One took up to five turns to recover. RAE found that if ailerons were not centralized during the recovery, spins could be produced which took three to 13 turns to recover. Stick deflection of about 1 inch out of spin had an adverse effect upon recovery. The type was officially classed as borderline.

An A&AEE report in 1941 concluded that to reduce five-turn spins to not more than two-turn, using Standard Recovery Drill:

☐ *anti-spin strakes* should be fitted (to improve damping, slowing the spin rate);
☐ *aileron mass-balance weights* should be removed (to increase B/A).

These resulted in the manufacturer taking the following actions:

☐ *Modification 103* was introduced in late 1941 to delete aileron mass-balance weights, which necessitated a reduction in V_{NE} to avoid flutter.
☐ *Modification 112* introduced anti-spin strakes in 1942 to reduce the number of turns required to recover. This modification was classed as '*Desirable*'.

Nearly 50 years later a number of owners of Tiger Moths wanted to operate them and carry out aerobatics in air shows, without strakes fitted, as originally built. The dilemma was that the CAA quite rightly refused to issue or renew an aerobatic C of A for any aeroplane or type, without a spinning clearance. Further, it properly refused to challenge the manufacturer's recommendation that Mod 112 was '*desirable*'. Even so, every counter argument and piece of hearsay that could be scraped together by owners was deployed to secure

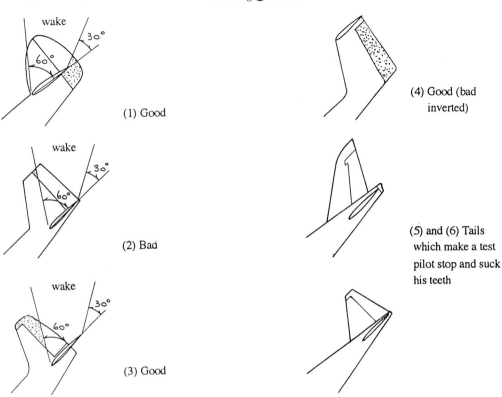

(1) Good

(2) Bad

(3) Good

(4) Good (bad inverted)

(5) and (6) Tails which make a test pilot stop and suck his teeth

As a rough rule have at least 1/3 of the rudder area
outside the wake of the stabiliser, approximated as shown

Fig. 12.5a   Tail configurations which are good, bad and likely to be dangerous ...

approval for spinning without strakes, while opposing the need for a CAA test flight to prove that an aeroplane, once classed as borderline, could now be relied upon to recover from any spin.

Eventually it was ruled that specific spin tests by a CAA certification test pilot must be carried out to clear an aeroplane, on an individual basis only, for operation in the Private Category. There would be no spin clearance without strakes for certification in the Transport Category, carrying a passenger for hire and reward.

Effects of Landing Gear, Flaps, Airbrakes and Stores

☐ *Landing gear* has no known adverse effect upon recovery, unless large spats are fitted to the wheels. But a story is told of the late (Sir) Geoffrey De Havilland, testing an aeroplane in his early career. He had flown it first with wire-spoked wheels, then later with them covered with fabric to improve their streamlining. This increased the side area forwards, causing the aeroplane to enter a spin, which fortunately he survived.

☐ *Flaps and slats* are neutral to adverse in their effects. Generally slats which open symmetrically cause no trouble. Asymmetric slats can have an adverse effect. If there is any doubt about flaps, retract them. If there are doubts about slats, lock them closed if possible.

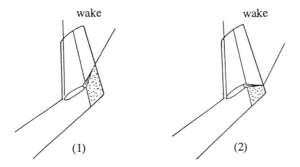

Effect of elevator deflection upon wake and, hence, shielding
of the rudder. This shows why it is wiser to attempt spin recovery
(particularly in a single-engined aeroplane) by applying corrective
rudder *before* attempting to unstall with elevator

Fig. 12.5b   ... all of which are made worse by smooth and slippery body cross-sections just
ahead of them.

☐ *Airbrakes* can be adverse, depending upon position, particularly if the thickened wakes
they leave degrade the authority of the fin and rudder surfaces.
☐ *Underwing and tip tanks and other stores* alter the B/A ratio in the direction of A. Their
aerodynamic effects M*ight* be small and indeterminate – but never assume this. They
must be cleared for spinning. For example, although a spin with wing tanks might be
satisfactory from the dynamic and aerodynamic point of view, the engine could suffer
fuel starvation. Further, the pilot must check the effects of inadvertent fuel asymmetry
upon spin entry and recovery of an aeroplane required for spinning.

Inverted Spinning

During tests a pilot must investigate the possibility of the aeroplane entering an inverted
spin accidentally. If it will, then he has the task of finding out what actions to take next. If
the aeroplane is for unlimited aerobatics, then inverted spins must be investigated as thoroughly as erect spins. If this is the case, make sure that the harness is adequate (five-point
attachment) and that all straps can be tightened individually.

In an inverted spin one must learn to think the other way around. The wing which drives
the spin is upside down. A spin to the pilot's left is really one to the right about the spin axis.
If the aeroplane has no inverted fuel system the engine will cut. Whether or not the propeller
will continue to windmill at low IAS under such conditions is another matter – probably
not.

If the spin flattens and the elevator trails with the wind the stick will move right forward.
If the harness is not tight it is easy to lose one's grip on the stick.

Recovery procedure is similar to that from an erect spin: throttle closed, ailerons neutral,
apply rudder opposite to the indicated direction of turn – PAUSE – and then this time pull
the stick back.

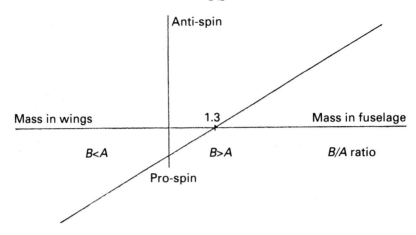

a. Yawing moment per degree of aileron into spin (in-spin stick)

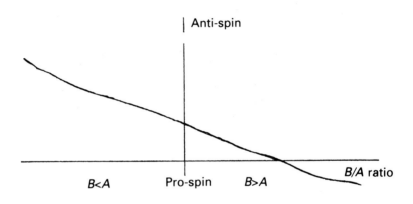

b. Yawing moment per degree of down elevator

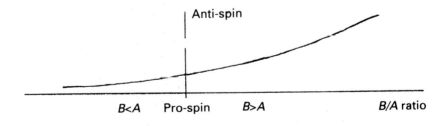

c. Yawing moment per degree of anti-spin rudder

Fig. 12.6   Sources of pro- and anti-spin yawing moment (Ref. 12.2).

Generally speaking recovery will be easier than when erect, because the fin and rudder are unshielded. However, it may be harder in a T-tailed aeroplane because, as we saw earlier, the fin and rudder could then be blanketed.

Inverted spinning should be approached with extra care and plenty of forethought.

TABLE 12–3

Body section contribution to yaw damping

Body cross-section	Damping effect*
Circular	1.0
Rectangular	2.5
Elliptical	3.5
Round top, flat bottom	1.8
Round bottom, flat top	4.2
Round bottom, flat top with strakes	5.8

* Damping of section/damping of circular section.

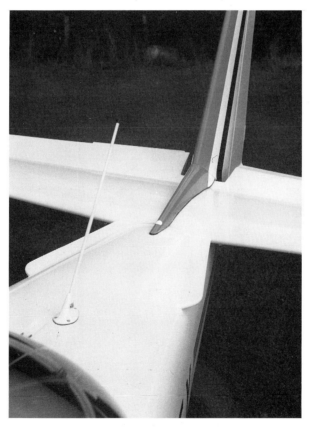

Plate 12–2   Final form of anti-spin strake on the Slingsby T-67.
(Author)

Recovery by releasing the controls

Sometimes it is said that all aeroplanes recover from spins on releasing the controls. I have found that it works for some and not for others. To recover from a spin usually requires positive actions and not a little determination. Also, one must give controls TIME TO BITE.

If the control column is left free to float about then the controls are not biting, they are trailing, being worked on by the air instead of doing work on the air. The De Havilland Chipmunk is one example.

Example 12–4: Stick-Force in the Recovery – De Havilland Chipmunk T 10

The De Havilland Chipmunk, which first flew in 1946, is shown in Fig. 12.7. A lively and delightful aeroplane, it needs positive handling in spite of having the sweetest flying controls found anywhere. Following a number of spinning accidents, particularly with

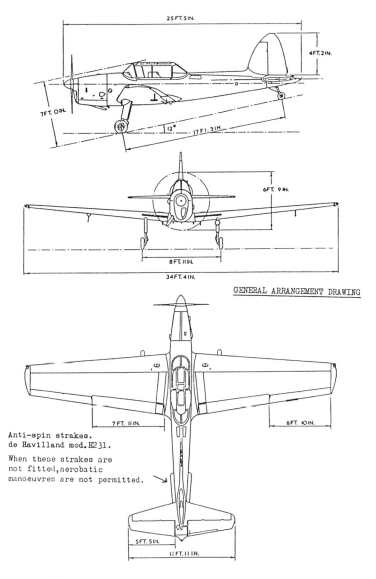

Fig. 12.7   De Havilland Chipmunk T 10 with broad-chord rudder (to keep the nose up in a slow roll) and anti-spin strakes (courtesy of former De Havilland Aircraft Company).

Chipmunks coming on to the Civil Register, it was said to be impossible to apply full rudder deflection against the spin if the handbrake had been left partially ON. This was investigated in detail. We measured rudder deflection obtainable between parking brake-OFF to full-ON and, apart from a modest increase in rudder force, found no restriction of rudder movement.

Look back to Standard Recovery Drill earlier in the chapter. Following a PAUSE, the stick is pushed progressively forward until the spin stops. If the stick of the Chipmunk was released from the fully back position (full nose-UP elevator), it would move forward only as far as neutral, or a little aft, with no sign of recovery from the spin. Flown dual, to push the stick forward from neutral needed about 28 lbf (125 N) to the front stop. Such a force, while not excessive, was close to the one-handed 30lbf of Table 9–1, which a military pilot of that period only cared to exert for a short while. Even so, it was three times higher than the 10 lbf (44.5 N) shown in Table 9–2, acceptable to pilots of civil light aircraft of the American school, designed to FAR 23.143(c).

We were convinced that the problem lay with the 28lbf push force. A light aeroplane pilot trained on Pipers and Cessnas, the commonest of club training aeroplanes, would possibly (not probably) think that he was pushing against the front stop, when the stick was only neutral. So Chipmunk cockpits and AFMs had warning placards incorporated for civil operations. To the best of my knowledge, from there on spinning problems ceased.

What I am certain of is that the Chipmunk cannot be relied upon to recover from a spin if the stick is released. I know that some other aeroplanes are also doubtful.

Therefore: BEWARE OF ADVICE TO RELEASE THE CONTROLS TO RECOVER FROM A SPIN.

Changes to the Basic Configuration

Often during prototype testing, or investigating role changes for customers, one needs to investigate the effects upon spin characteristics of significant changes in B/A, weight, CG position, control deflection and the installation of a larger engine with more power. Table 12–4 summarizes the effects of a number of such changes in a form used in Ref. 12.2.

Fitting a more powerful engine can be a mixed blessing. One benefits from livelier performance but:

☐ Pitching inertia B may grow much larger than rolling inertia A, changing a smooth spin into one which is oscillatory, or even erratic.
☐ Power is destabilizing and increased '*P*'-*effects* can make a departure into spin entry more likely.
☐ If the CG moves forward the spin may change from smooth to oscillatory.
☐ If a pilot accidentally takes reversed recovery action, having been caught by a sly, undiagnosed spin, a more powerful engine can make a recovery in the height available more of a gamble.

Example 12–5: Secondary Spin Mode – Piper Tomahawk

The Tomahawk, shown in Fig. 12.8, is an honest spinning aeroplane with a T-tail and effective rudder. The primary spin mode in a Tomahawk is so consistent that, unless one reads the registration placard on the instrument panel, it is almost impossible to tell one aeroplane

TABLE 12–4

Effects upon spin characteristics of changes to the basic configuration

Parameter	Spiral dive	Oscillatory spin	Normal smooth spin	Flat spin
Increasing B/A	*	←	←	←
Decreasing B/A	→	→	→	*
Increasing weight	* (slight)	←	←	←
Moving CG aft, B/A > 1	---------------------- (effects are small on the whole) --------------------------			
B/A < 1	→	→	→	*
Increasing wing span	→	→	→	*
Increasing fin size	*	←	←	←
Increased body damping (strakes, change of section)	→	→	→	*
Fitting a larger engine	*	←	←	←
Opening the throttle	→	→	→	*

Plate 12–3   Piper PA 38-112 Tomahawk, an honest spinning aero-plane with an interesting secondary spin mode. Ref.: Examples 12–5 and 12–6. (Courtesy of Piper Aircraft Corporation)

from another. Average recovery is in little more than $\frac{1}{4}$ turn, using either the manufacturer's Recommended Recovery Drill, or the Standard Recovery Drill.

When using the manufacturer's recommend spin recovery drill – which involves pushing the control column fully forward as the rudder pedal hits the stop – the spin stops and the aeroplane pitches smartly nose down. This, identified as 'the Tomahawk-bunt', is disliked.

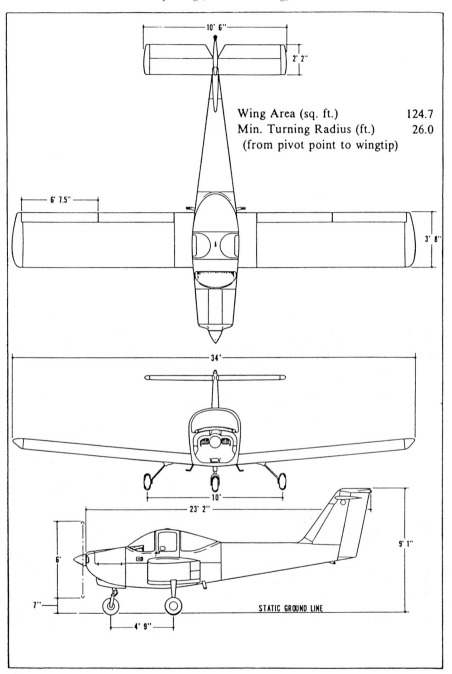

Wing Area (sq. ft.)   124.7
Min. Turning Radius (ft.)   26.0
(from pivot point to wingtip)

Fig. 12.8   Piper PA 38-112 Tomahawk (Ref.: 12.3, courtesy of Piper Aircraft Corporation).

An instructor demonstrating spin recovery technique during a formal investigation into the possible cause of a spinning accident, ran through the drill in the Pilot's Operating Handbook precisely, but on reaching the point in the Emergency Procedure where '(c) As the rudder hits the stop, push the control wheel fully forward. As the stall is broken, relax forward pressure to prevent an excessive airspeed build up.' (Ref. 12.3, POH), the rudder hit the stop, but the control wheel was hardly moved off its back stop (the same was noticed with a number of other instructors). The aeroplane recovered with ease in less than one turn, in spite of the action failing to correspond with the required procedure.

In the original accident there was forensic evidence of the control column having been held fully back. As both of us wore parachutes, an elevator-UP abused recovery was tried. After the spin stabilized, full rudder was applied for recovery with the control column held fully back. The Tomahawk continued to rotate in the direction of the spin, with throttle closed, ailerons neutral and elevator-UP. After about two noisy turns at 70 to 80 KIAS, with the skin popping and panting, the aeroplane pitched sharply nose down against elevator and doubled its rate of rotation to an estimated 1 turn/s *against* anti-spin rudder. Recovery was effected by simply moving the elevator fully nose-DOWN and waiting. After three to four further turns, during which one sensed a speeding up in rotation (a sign that it is on the way to recovering), the aeroplane recovered smartly with a strongly felt and seen yaw in the direction of the out-spin rudder.

The high rate of the secondary spin mode – which it is – was unexpected and disorientating. Far more interesting was the subsequent discovery that, although the primary spin mode was similar among several Tomahawks, the secondary was not. One would high-rotate only to the left, another only to the right. Yet another would do it both ways, and another would not do it at all.

Note: In my view the secondary spin mode is not a dangerous feature because one can recover. When clearing an aeroplane for spinning, if the manufacturer's Recommended Recovery Drill differs from the Standard Recovery Drill, then both are explored for the purpose of Type Certification. The reason is that pupils are taught the Standard Drill in training. Under stress pilots often revert to earlier learned patterns of behaviour (Chapter 4) and, if caught out with an unexpected spin, many automatically use the Standard Recovery Drill without further thought.

Example 12–6: Control column-free spin recovery – Piper Tomahawk

Further tests were carried out stalling and then spinning the Tomahawk with full nose-UP trim (FNUT), followed by releasing the control column for the recovery. It was found that with FNUT, throttle closed, the aeroplane settled into a gentle nodding descent on the point of the stall, elevator-free, at about 52 to 54 KIAS. Looking back through the rear of the canopy the elevator horns ahead of the hinge were fractionally lower than the leading edge of the tailplane, showing that the elevator was slightly UP from neutral.

On entering and stabilizing a spin for at least two to three turns, rudder for recovery was applied and, following the PAUSE in the Standard Recovery Drill (i.e. not the Emergency Drill in the POH), the control column was released from the fully back position. It moved forward to near NEUTRAL and the yoke rotated in the direction of the spin through half to full aileron. The elevator was then slightly UP, having returned to the FNUT position. Recovery occurred almost immediately in a rolling spiral.

This was found to apply to numerous Tomahawks in subsequent tests, all of which recovered from spins in rolling spirals when the control column was released – even with full nose-up trim set.

Care when Selecting an Historic Aeroplane for Reproduction

It follows that detailed research is needed to find out as much as possible about the spinning and other history of an aeroplane before deciding to build a replica. Because it has plenty of character in photographs and drawings, and an attractive shape which could fit an available engine, does not mean that if it looks pretty it will fly right. Look critically at the drawings and photographs – as many as you can find. Look for highlights which are a clue to shapes of sections. Look for low aspect ratio fins and rudders, with plenty of rudder area beneath the tailplane and elevator. Look too for long tail arms and deep rear fuselages with sharpish corners, which favour the generation of favourable vortices in a cross-flow.

Example 12–7: Spin Recovery – Fairey Flycatcher and Sopwith Camel Replicas

Two historic aeroplanes I found to recover fastest from developed spins – each within $\frac{1}{4}$ turn, to right and left – were John Fairey's replica of the biplane Fairey Flycatcher, designed to meet a 1922 specification for operation from Royal Navy aircraft carriers, and the Sopwith

Plate 12–4 Replica of Fairey Flycatcher. Note deep rear fuselage and ample rudder area below tailplane. (Courtesy of Richard Wilson via John Fairey)

Camel. Both had plenty of unshielded rudder area, long tail arms and favourable rear fuse-lage forms – the Flycatcher especially. So good was the Camel in its day, that a spin was advised to be the safest way of getting down through solid cloud to find out where you were. The Flycatcher reflected its original reputation for sweetness, flying like an extension of one's hands and feet.

The third aeroplane in the following example set in train a valuable body of British research into spinning and those physical characteristics which affect spin recovery.

Example 12–8: Warnings from the 1920s and 1930s (Ref. 12.4)

Caution is needed when considering replication of some of the most attractive historic ma-chines with character and performance. These are the European and American interceptor fighters of the 1920s and 1930s. Their technology was advancing fast. They span transition between the small metal and fabric-covered biplane and monoplane and include the period of winning the Schneider Trophy outright. Engines were powerful, increasingly so. Aero-dynamic, structural and propulsive research and development for what was to come finally in 1939 was being established. The highly aerobatic fighters were light aeroplanes by our standards, weighing less that 6000 lb (2730 kg).

One decade after World War I there were voices saying that '*The bomber will always get through*' and the military strategists foresaw the need for fast-climbing, agile, target-defence interceptors to meet the bomber threat. The policy which produced the interceptor fighter was vindicated in the Battle of Britain as far as Britain was concerned. From the point of view of the German Luftwaffe the shortcomings of such a policy were also clear. Essentially short-range Bf 109s had to escort bombers from French bases to London and other targets, leaving insufficient fuel to linger in protacted dogfights.

BLACKBURN LINCOCK III (1930)
To build a replica of an interceptor of the 1920s and early 1930s is to risk trouble when it comes to tail design. Figure 12.9a is a reproduction of an original general arrangement drawing of the Blackburn Lincock III of 1930 which, although it never saw service, is typical of a high performance fighter of the period. It was clean, had carefully balanced controls, and had the main loads – including the 35 gallon fuel tank – concentrated near the CG. It was powered by an air-cooled, radial, 270 hp Armstrong Siddeley Lynx Major, and weighed 2082 lb (946 kg). Although nothing is kown of Lincock flying qualities as this is written, all three marks were used for lively aerobatic displays by well-known pilots. But, in spite of compact and robust features which make it seem ideal for aerobatics, look at the tail configuration and compare it with tails shown in Fig. 12.5. What might a replica be like in a spin – assuming of course that one might find a comparable engine?

BOEING P-26A (1933) AND GRUMMAN F3F (1935–36)
On the other side of the Atlantic, tail design had similar troubling features. Two examples are shown in Fig. 12.9b(1) and (2). The first is the Boeing P-26A 'Peashooter' which en-tered service around 1933. The second is the Grumman F3F of 1935–36. While nothing is known as this is written of spinning problems with the P-26A, the early XF3F-1 had them, and the F3F was fitted with a larger rudder during development test flying.

Tail configurations of all three aeroplanes appear now to lack adequate rudder area out-side the wake of the tailplane and elevator, constructed for a 45° nose-down spin. Further-more, in the interests of streamlining for high speed, all had beautifully curved fuselage

Plate 12–5   The Blackburn Lincock of 1930 invites reproduction as an aerobatic aeroplane. However, careful research would be needed to ensure that the tail configuration would satisfy tests for a spinning clearance today. Ref.: Example 12–8. (Courtesy of former Blackburn Aircraft Ltd via the author)

cross-sections behind the wing, which would have provided little in the way of rotary tail-damping. But with protruding radial engines, wing–body junctions showing little skill in fairing, together with struts and bracing wires to leave dirty wakes behind them, hindsight would now cause us to expect trouble with any replica.

VICKERS JOCKEY (1934)
The Vickers Jockey was a small single-seat interceptor fighter being developed to meet a later Air Ministry Specification F5/34 (see Fig. 12.10). Like the three preceding examples it had a broadly similar tail configuration. It had a protruding Bristol Jupiter VIIF radial engine fitted with a Townend drag-reducing ring, driving a fixed pitch LH tractor propeller, an open cockpit, fixed undercarriage and spats fairing the wheels. During spinning tests the aeroplane, J 9122, was destroyed after failing to recover from a flat spin to the right, with the CG near the aft limit.

It is reported that after a fast three and a half turn spin to the left, which the aeroplane entered following a steep nose drop, it recovered in one and a half turns, losing a little under 2000 ft. A spin to the right was entered. The first turns were uneven, but it then steadied. The stick, when released, stayed back and to the right, in the direction of spin. Thereafter the spin began to flatten and seemed faster. The rudder was reversed and

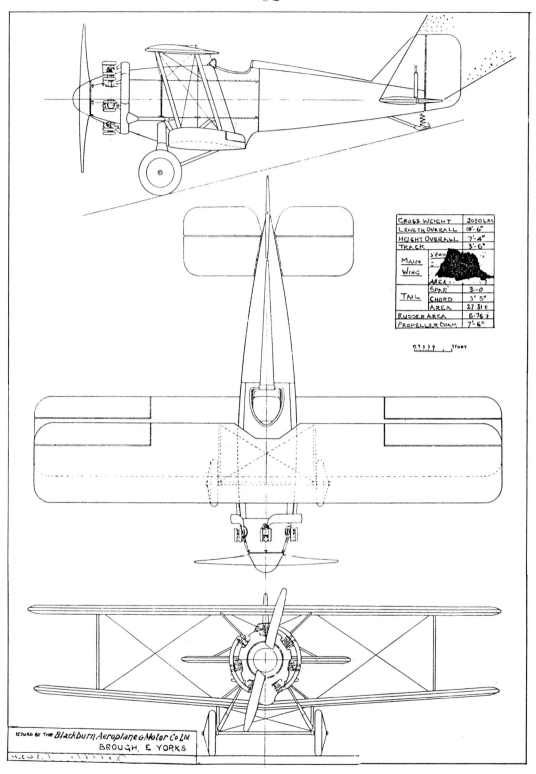

GROSS WEIGHT	2080 LBS
LENGTH OVERALL	19'-6"
HEIGHT OVERALL	7'-4"
TRACK	3'-6"
MAIN WING	SPAN	
	AREA	
TAIL	SPAR	3'-0
	CHORD	3' 5"
	AREA	27 31 E
RUDDER AREA	6·76 □
PROPELLER DIAM	7'-6"

ISSUED BY THE *Blackburn Aeroplane & Motor Co Ltd*
BROUGH, E YORKS.

Fig. 12.9a Blackburn Lincock III of 1930, a lively high performance fighter prototype, used by Sir Alan Cobham in 1933 for aerobatics in his National Aviation Day Displays.

Fig. 12.9a – Continued. Even so, a tail configuration of this kind on a replica should be treated with caution in the light of research, development and experience accumulated over the past 61 and more years. Note: The arrangement of the fin and rudder with the tailplane and elevator in the plan view does not correspond with the side view. Remember the warning in Chapter 1. Here, the side elevation is correct.
(Courtesy of the former Blackburn Aircraft Ltd)

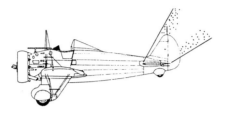

(1) *Boeing P-26A 'Peashooter'* (1933)

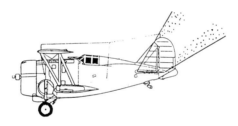

(2) *Grumman F3F* (1935-36)

Fig. 12.9b Two American aeroplanes of broadly the same period as in Fig. 12.9a, with tail configurations which would raise the same doubts about spinning in their replicas today.

the stick pushed forward. The aeroplane continued to spin flat, in spite of trying the effects of power and other control movements. At 5000 ft the pilot abandoned the Jockey successfully, and the aeroplane passed into history as having features which made it flat-spin prone.

Flat spinning research was subsequently carried out by the National Physical Laboratory NPL and the RAE, the latter establishment publishing two reports in 1933 of 1/22 model tests of the Jockey in the free-spinning tunnel at Farnborough.

The drawing shows features of the Vickers Jockey which would now alert a test pilot to the existence of likely unpleasant spinning qualities:

☐ A large, long and internally braced cantilever monoplane wing.
☐ A compact concentration of masses in the forward fuselage.

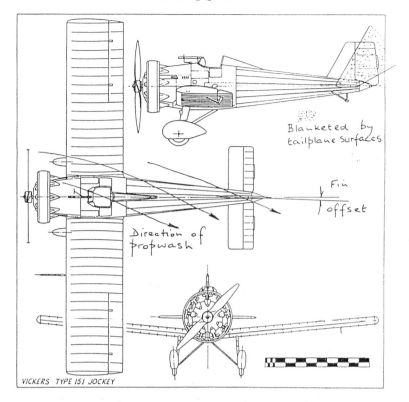

Annotations on figure: "Blanketed by tailplane surfaces", "Fin offset", "Direction of propwash", "VICKERS TYPE 151 JOCKEY"

Fig. 12.10   Vickers Type 151 Jockey (destroyed during full-scale spin tests in 1932) which became the subject of searching model spin tests by the National Physical Laboratory and the Royal Aircraft Establishment. Annotations are with hindsight from results and subsequent experience.

☐ A large radial engine which looks as if it would cause a broad turbulent wake, especially at low EAS and large angle of attack, degrading the authority of tail surfaces affected by it.

☐ Relatively small tailplane and elevator surfaces compared with the wings.

☐ An 'eyeballed' inertia combination of wing and fuselage mass distributions suggesting the probability that A > B (i.e. $(1 - B/A) > 1$).

☐ Propwash which would cause yaw to starboard (hence the fin offset, leading-edge to starboard to counteract this at the design speed).

☐ Rudder almost completely blanketed by wake from horizontal tail surfaces in a 45° nose-down spin.

A number of points arise from these last examples. The first is that enough has been learned about spinning since the 1930s to enable designers to steer clear of unfavourable tail arrangements. There is no excuse, more than one half-century later, for poor spin-prone tail design. It is for the test pilot to be aware of this, so that he can make warning noises in plenty of time.

Second, neither aeroplanes nor pilots are always what they might seem to be. An inexperienced or out-of-practice pilot may think one thing and do another without realizing it. One aeroplane may have a trick not shared with another of the same type, which remains hidden until triggered by a particular circumstance. Fortunately rogue aeroplanes are rare.

Finally, handling characteristics of different examples of one type of aeroplane, when the controls are used normally, may be so consistent that a pilot is lulled by docility of the type. But, when mishandled, an aeroplane which appears to be amiable and tolerant can become a tiger. For this reason aeroplanes belonging to the Tiger Club have their cockpits placarded:

'ALL AEROPLANES BITE FOOLS'

Example 12–9: Rudder Authority Reduced by Elevator on a T-tail

The point was made in Chapter 9 that a yawing component due to elevator, N_η, although rare is not unknown. During type certification of a T-tailed American light aeroplane we found that as the control yoke was pulled aft to rotate the aeroplane into the take-off attitude a point was reached at which full right rudder was applied before we were airborne. Later, in approaches to 75% power stalls, the same occurred again.

After lengthy discussions with the manufacturer it was concluded that the suction in-duced beneath the tail by the up-elevator diminished the favourable pressure distribution around the fin and rudder (the aerodynamicist would say that the elevator acted like a 'sink'). The aircraft had a RH tractor propeller which needed right rudder to maintain a straight heading, power-ON. At low speed and high angles of attack in that condition, full right rudder was needed before the stall was reached. This was a non-compliance with FAR 23.143(a) and (b), which required safe controllability and manoeuvrability throughout the flight envelope. Had the aircraft departed in yaw to port for any reason, the pilot would have had no right rudder available to correct matters. It would not be enough to argue: '*Well – he only has to close the throttle to have enough rudder again!*'. What if he or she is low, slow, and caught with a go-around in poor visibility? An unexpected departure into a spin would be possible.

Anti-Spin Parachutes

Aerobatic aeroplanes are usually light. One or two historic jet fighters weigh around 18 000 lb (say, 9000 kg). If a spinning clearance is sought fit an anti-spin parachute. The best attach-ment point is at the tail, where deployment causes the parachute pack to be ejected rear-wards, centrifugally.

When clear of the tail the parachute opens and produces powerful anti-spin yawing and nose-down pitching moments. It should incorporate a jettison device and a weak link, in case the pilot's jettison system fails.

The towing cable must not be too short. The minimum length should be no less than the length of the aeroplane, otherwise the parachute becomes trapped in the wake where it is useless or, exceptionally, may collapse back on to the aircraft.

As the cable is increased in length the effect of the parachute is to apply an increas-ingly nose-down pitching moment as against a yawing moment. Therefore, if there is a lurking danger of a deep-stall, use of a longer rather than a shorter cable should be considered.

Wing tip parachutes have been fitted to all-wing aeroplanes, but installations tend to be more complicated and less reliable than a parachute in the tail.

Anti-Spin Rockets

The anti-spin rocket, fitted to *both* wing tips and/or tail can be effective. It is less popular than the mechanically simpler anti-spin parachute, because it combines explosives with electrics, and still needs safety devices.

Reference 12.1 suggests that comparing rockets of equal thrust used in roll, pitch and yaw:

$$\text{Anti-spin effectiveness in (yaw/roll)} = \text{approximately 8} \qquad (12\text{--}5)$$
$$\text{in (yaw/pitch)} = \text{approximately 15} \qquad (12\text{--}6)$$

Abandoning an Aeroplane in a Spin

The point was made earlier that a test crew should be equipped with parachutes. Thin back-pack glider parachutes are available and these are often designed for high speeds. One should always check the certificated speed, stated in the Specification.

The primary danger when baling out of a light aeroplane is that of catching a propeller-bite. It used to be recommended to evacuate the cockpit on the inside of the spin – until tests in spinning tunnels showed that the pilot tended to follow a path close to that of an advancing propeller.

Therefore it became policy to recommend leaving the aircraft on the outside of the spin, as there was then less risk through hitting the tail than being bitten by the propeller.

When carrying out a spinning sortie, especially with another pilot or an observer:

- ☐ Pick airspace which is away from built-up areas or other hazards on the ground.
- ☐ Brief him/her thoroughly on your intentions and actions in the event of an emergency.
- ☐ Pick a decision height at which, without heroics, the aeroplane shall be abandoned if there is no sign of spin recovery. Two-thousand five-hundred feet (760 m) is a minimum, as it can take more than 1000 ft to get clear of a spinning light aeroplane – and
- ☐ Aim to get the other pilot/observer out of the aeroplane *first*.
- ☐ If the parachute is ready and waiting in the seat for the pilot or observer to sit on it, check that a parachute strap has not been wrongly routed. CHECK before flight, even though it will be awkward, by unfastening your *seat harness* to make sure that both of you can free yourselves from your seats without hindrance.
- ☐ As things can become hectic, may I remind you again of advice given much earlier: make sure that you carry a knife that is easily reached. Also that knife, knee-pad, test-cards, tape-recorder and pencil are tied on to you.

Note: Take heart: what you prepare for rarely happens.

References and Bibliography

12.1 Van Mansart, M., *Influence de la Geometrie des Avions Legers sur leur Vrille.* L'Association Aeronautique et Astronautique de France: Colloque, *Aerodynamique et Aviation Legere*, 1987
12.2 Empire Tests Pilots' School notes.
12.3 Report: 2126, *Pilot's Operating Handbook, PA 38-112, Tomahawk.* Revised edn, Piper Aircraft Corporation, 18 December 1978
12.4 Andrews, C. F., *Vickers Aircraft Since 1908*, London: Putnam, 1969
12.5 Mason, S., *Stalls, Spins and Safety,* New York: McGraw-Hill, 1982

GROUND AND WATER HANDLING
AND EARLY TESTS

CHAPTER 13

Ground and Water Handling and Early Test Flights

'Now lose height by a series of S turns. Never turn your back on the field. Don't do the turns too close to it or too far away. Finally, at about five hundred feet, do your last turn in and land well up the field in the usual way ...
... because it's far better to run a risk of trickling gently into the far hedge than it is to under-shoot and to stall into the near hedge through holding up the nose in an endeavour to get in.'
Pilot's Summer, an RAF Diary (c 1935?)   Frank D. Treadrey (1939)

It is said that if line and other pilots can be trained on flight simulators, then flying qualities of aircraft yet unflown can be dealt with in the same way, especially those with fly-by-wire and fly-by-light control systems. So, why waste time and money on flight test programmes? So runs the argument. The only response might not be positive but it is practical. Would you wish to buy a seat for yourself, or your wife and children, in an aeroplane that had never been flown by a flesh and blood pilot?

The test pilot goes out looking for trouble and expecting to find what is wrong far more often than what is right. For example, my first real flying instructor was Frank Bullen, a Blackburn Aircraft Company test pilot in the late 1940s, before joining Hawkers. His attitude was: '*Always keep your eye open for a field and on the wind direction, your engine could fail in the next minute*'. That saved my neck and other people's aircraft on some 13 occasions when testing civil aircraft one quarter of a century later, to the point of never carrying out stall tests beyond comfortable gliding range of the active runway threshold of an airfield.

Test flying has many facets, from the tentative first flight of an unknown prototype, to clearance of modifications, to airborne accident investigation into loss(es) of a tested and certified type with a history of previously substantial safe operation.

First flight

The attitude of the test pilot is defensive and doubting, regardless of the task in hand. This goes without saying when dealing with a prototype, whereas impartiality might not be quite so easy when the type is a known quantity. Even so, the test pilot distances himself and suspends partiality, from signing the pre-flight acceptance to completing the snag-sheet afterwards.

Before the first and subsequent test flights are carried out there are final checks of the cockpit, instrumentation and equipment, followed by taxying trials on land or on water. Their purpose is to clear ground or water handling, and they are more important than rushing into the air. They must not be skimped, because some unexpected or unpleasant characteristic may remain undiscovered which leads to damage on the first landing. More than once odd behaviour when taxying has revealed mis-loading of ballast.

Pre-flight – documents and actions

☐ *Weight and balance* (see Chapter 5). If the aeroplane is not new, or if the data is more than five years out of date, it may be necessary to reweigh the machine before flight. Only a careful check of weight and balance during pre-flight calculations will reveal this. Check too the ballast when loaded into the aeroplane. Before a test flight I once discovered the rear seat-buckets to be full of loose broken concrete, under an equally loose, squashed, cardboard carton for a household refrigerator.

Note 1: Carry no passenger on a test flight.

Note 2: A mid-range CG should be selected if possible. As no passengers should be carried expect an aeroplane not only to be lightly loaded but, if it is used or second hand, then the CG might be more forward than aft. If so then when the time comes you will find it easy to lift the tail. But, be ready for it being too easy. With a CG that is too far forward it may be hard to rotate into the take-off attitude, or to get the tail down on landing.

Note 3: A convenient form of ballast is to use plastic jerry-cans, containing different amounts of sand, or gravel, with the weight illegibly dye-lined on the outside. The handles are strong, making it easy to lash them down.

☐ *Fuel state*. Carry enough fuel to give yourself time to breathe and think in the event of an emergency. Many modern 'automobile-type' fuel gauges in light aeroplanes are inaccurate. Think of getting back on the ground regardless when $\frac{1}{3}$ fuel is indicated. So:
☐ *Do not aim to carry out too many tests* on one flight.
☐ *Pick airspace* where you have radio (and if possible radar) cover, plenty of altitude and good visibility.
☐ *Aim to stay close to the airfield*, within gliding range, just in case the engine stops.
☐ Where relevant, *historic or other background information* should be accessible, together with textbook and any other technical data.
☐ A current *Certificate of Fitness for Flight, C of A* or other equivalent will be needed to make the flight legal. If it is not, then:
☐ A current *Certificate of Insurance*, which is necessary, will be invalidated.

Ground Handling Tests, Including Pre-Night Flying

☐ *On walking out* to the aircraft look carefully at the way it sits: are the wing-tips equal heights above the ground – what about oleo compression, is it even – is the aircraft excessively nose or tail-down?

Note 1: If the aeroplane is well used and has stood around in the open, expect it to be tail-heavy. There is always the danger of water having seeped into the rear fuselage and tail-surfaces. Look for it. If you find water then DO NOT FLY before it has been properly dried out and inspected. There was the tragic case of a wood and glue replica of a Schneider Trophy seaplane which, sometime in its life had been immersed. Much later the tail came off in flight. What is not known is whether immersion contributed to the accident.

Note 2: If you have not checked the sit of the aeroplane on the ground before flight, you will not be able to refute any allegation that *you* have caused damage done later by someone else in flight or on the ground. There are those who will try this trick when money is an issue, putting a solo pilot without witnesses into an untenable position! This was one reason why I always invited the owner of an aeroplane to fly with me on an airworthiness test flight, and asked before finally shutting down if he was satisfied.

Note 3: *Briefing before a test flight.* If an owner, or his company pilot, accompanies you on a test flight as the P2 then make sure that he or she is thoroughly briefed on what you intend to do and why. Also point out that you, as the test pilot are the pilot-in-charge only for the purpose of the tests. Outside of the tests, as on normal take-off and landing, or in an emergency, the properly qualified owner or his pilot is the captain.

A briefing avoids the difficulty of having the owner, or his pilot, interfere at the wrong moment in a flight which is normally crammed with action to enable it to be as economically short as possible. Years ago, when checking the stick-force needed to lift-off during a deliberately mis-trimmed take-off in a light aeroplane, the hand of the company pilot appeared out of nowhere, just before rotation, and altered the elevator trimmer-setting to that given in the take-off checklist. It spoilt the test-point and investigation of an important feature. The fault was mine for not having told him beforehand that this minor item (as far as the main test was concerned) would be checked in passing, on a flight to look at something else.

☐ *Check entrance to and exit from the cockpit* for pilot and crew (and from the cabin, if the aeroplane is to carry passengers). How will this be accomplished in an emergency? There are more emergencies than just those in the air. In the event of a wheels-up landing, will it be possible to evacuate the aircraft safely, without obstruction?

☐ *Can the seats and rudder pedals be adjusted* adequately and easily, so that all system controls are reached by pilots in the normal (i.e. accepted) percentile range (see Fig. 4.3 *et seq.*, and Example 4–1) when strapped in tightly and wearing full flying clothing?

☐ *Is the pilot's view satisfactory?* One must consider flight in adverse conditions when the main windscreen is obscured by ice, oil, or a messy bird-strike. A direct vision panel which can be opened and looked through when landing, and through which a fully clothed arm can be thrust so as to wipe the windscreen, is essential. To look through a DV panel may well involve a curved approach, which is acceptable as long as airspeed can be maintained with ease.

☐ *If designed for night or instrument flying,* will internal reflections from internal instrument lights and faces prevent the pilot from seeing out through the windscreen and other cockpit glazing. A simple test is to cover the cockpit glazing with a thick sheet in the hangar (e.g., a rick-cover) and then to turn up the cockpit lights, normal and then emergency. More than once a white emergency flood light in the cockpit roof has caused such intense reflections that it was impossible to see out, and a dimmer control had to be fitted.

☐ *Is cockpit lighting* arranged to enable a pilot (left- or right-hand seat) to see and reach everything of significance in any corner of the instrument panel(s). It is of no use tucking circuit breakers away, unlit, at the RH-side of the panel, where the P1 cannot reach them. Yet, it happens frequently. Rheostats are needed to dim lamps selectively, either individually or more usually in banks or clusters.

☐ *Can all system controls and switches be reached and operated* when wearing gloves and other protective clothing, and can all flying controls be operated fully and freely, without fouling knees and thighs or other parts of the body when in flying kit?

☐ *Are the flying (and other) controls connected correctly?* This fault is unthinkable but it has happened. The Chief Test Pilot, father of one of my friends, together with the Chief Designer of the company were lost in the crash of a large aeroplane in the UK, many years ago, because the ailerons had been connected the wrong way round. Unfortunately the ailerons could not be seen and checked from the the cockpit by the pilot, even so, they had passed more than one separate ground inspection without the error being noticed.

Note: Arrange with the groundcrew, or the last member of the test crew due to climb on board, to check control movements with the pilot before start-up. Here a DV panel is useful.

Early engine running

Engine runs start long before the first flight of a prototype. But, in any case one should check that the engine, cowlings and airframe have no excessive or unusual vibration. Engine and propeller response to their controls should be noted. The 'gear ratio' between movement of a lever and response is important. Large response for small control movement results in overcontrol and lack of precision (this I discovered when first attempting to complete an intense package of tests during a necessarily short first flight in a McDonnell Douglas F-4 Phantom. Then I was bedevilled by what seemed to me large changes in thrust with small movement of throttle, compared with those of British aircraft then in service).

In the next chapter we shall have more to say about engine handling tests.

Taxying (testing the landing gear)

Ground handling tests explore three areas of concern. Unless there is cause for technical doubt, a used aeroplane is taxied with normal care and good airmanship, paying particular attention to the normal and parking-brake systems (wear which causes the parking brake to slip is one of the commonest snags with used aeroplanes). A prototype, on the other hand, warrants rougher handling to thoroughly test the landing gear units and shock absorption system. It is an airworthiness requirement that the shock-absorbing mechanism may not damage the aeroplane when taxied on the roughest ground that can reasonably be expected in normal operation. Requirements are defined in Refs 5.8 and 5.9.

☐ *Longitudinal stability and control,* depend upon wheelbase, brakes, shock absorber rates and the effects of brakes, none of which must make the aeroplane vulnerable to nosing over, or to excessive pitching.

(1)   With a tailwheel aeroplane nose-over is caused by the CG being too far forward, or the mainwheels too far back. The lines drawn from the point of contact of the mainwheels, leaning aft at 10° to 13° from the vertical in Fig. 10.21, were to check this point specifically. Figures 13.1 and 13.2 (see Ref. 1.5) show both types of landing-gear geometry. Rules are added to help reduce troublesome features. If the aeroplane feels sensitive longitudinally, with the tail lifting when taxying and braking, then it is nose-heavy and the overturning coefficient is too small, where:

Plate 13–1 McDonnell Douglas F-4 Phantom, a naval aeroplane basically with a remarkably long service life on land and at sea. (Courtesy of Ian Black via Julian Stinton)

$$\text{overturning coefficient, } K = Rh/Wa = R/W \tan \theta \qquad (13\text{–}1)$$

Fix: by moving the CG further aft with ballast (see *Postscript* to Example 2–1: Hawker Fury 1). Alternatively, and if the aerodynamics can be juggled safely, move the mainwheels forward. This was necessary when a replica of a De Havilland DH2 was changed into a Vickers Gunbus 'for a film, by lengthening the body nacelle to accommodate a front gunner (rough sketch in Fig. 13.1d). The mainwheels were moved forward 4 in (10 cm) following investigation of CG–CP relationships (see Figs 10.13 and 10.14). Additional fin surface ahead of the rudder was needed before a Permit to Fly was issued.

(2) If the wheelbase is too short, then with both tailwheel and nosewheel units an aircraft will pitch readily, with risk of grounding propeller tips, and the tail-bumper striking the ground when the static sit of the aeroplane is tail-down. If the CG is well forward there is also the risk of a roll-over when taxying fast and turning sharply. This will be appreciated from Fig. 13.2b.

(3) There must be no tendency to rebound into the air on take-off or landing. If this occurs it could be due to a short-coupled wheelbase and over-powerful shock absorbers.

(4) Wheelbrakes must operate smoothly. If they are harsh they may induce nose-over. If the wheelbase is short-coupled they may also cause uncomfortable nodding especially when jabbed.

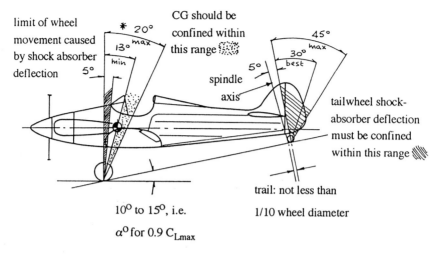

a. Summary of rules derived from various British and
American sources. The larger the angle marked ✳ the
easier it will be to groundloop. If the angle is too small
it will be too easy to nose over

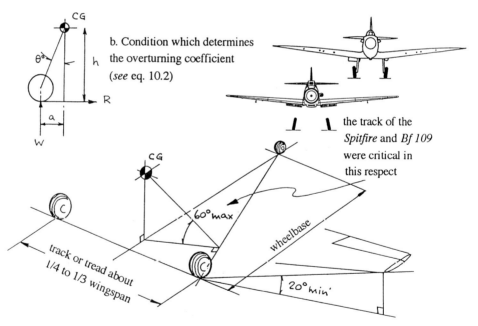

c. If the track is too narrow wing tips will be damaged by
tipping sideways when groundlooping or in a crosswind

Fig. 13.1a to c   The tailwheel undercarriage (Ref. 1.5).

Fixes for (2), (3) and (4) are along the lines of those suggested in (1), in that undercarriage
geometry is deficient. Some adjustment can be made by shifting the CG with ballast, but
one should check that there is no tendency to sit on the tail-bumper when loading. If there

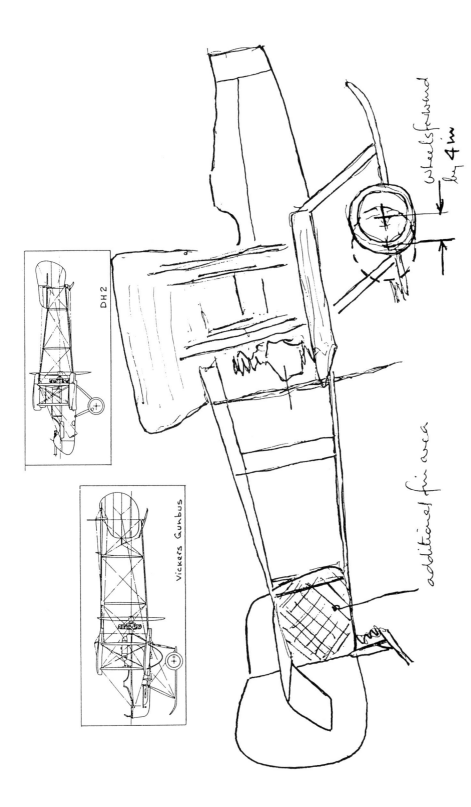

Fig. 13.1d DH 2 replica transformed into a 'Vickers Gunbus' by lengthening the nose of the nacelle. The mainwheels were moved forwards 4.0 in (about 10 cm), following investigation of the CG–CP relationship described in Chapter 10 (see Figs 10.13 and 10.14). Additional fin area was also needed. The scale insets are for comparison of the original aircraft.

Plate 13–2a   Replica of a De Havilland DH 2 which was then modified for film-work to resemble ...
(Courtesy of Ian Black via Julian Stinton)

Plate 13–2b   ... the Vickers Gunbus of 1914.
(Courtesy of Philip Jarrett Collection)

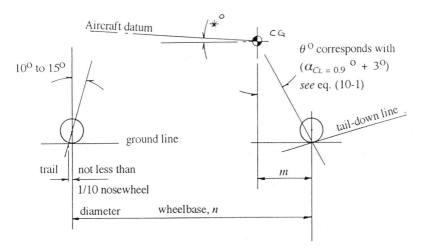

a. The angle ✱ should not be too large otherwise the nosewheel will touch first in a tail-high (too-fast) landing, causing
wheelbarrowing and loss of directional stability

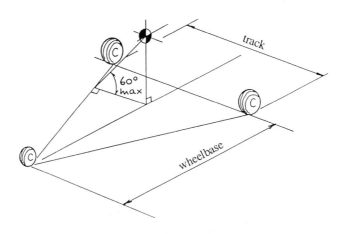

b. If the track is too narrow wing tips may be damaged

Fig. 13.2   The nosewheel undercarriage (Ref. 1.5).

is such a risk, then a tail-bumper strut is needed – clearly marked for easy removal before
start-up (taxying out with a strut still in place has happened)! Improvements might also
be achieved by alteration of oleo pressures and/or shock absorber rates.

☐ *Directional stability and control* requirements are concise. They do not distinguish between directional and lateral stability because they are interconnected.

(1)   *Directional handling* tests must demonstrate freedom from any tendency to
ground-loop:

$$\text{in a } 90° \text{ crosswind component} = 0.2 \, V_{S0} \tag{13-2}$$

(2)   It is out of the question to expect above-average piloting skill or alertness in power-
off landings at normal landing speed. It must be possible to keep straight without

use of brakes or engine power until the speed has decreased to less than 50% of the speed at touchdown. This implies the need for adequate authority of the aerodynamic controls when acting alone.

(3)   When taxying there must be adequate directional control. Interpretation of the requirements is that below one half of the touchdown speed, the aerodynamic controls acting alone will not suffice.

(4)   *Track of the undercarriage* (see Figs 13.1 and 13.2). The wider the track the shallower the overturning-angle, subtended between the CG and the line joining main and nose- or tailwheel, and the less likely is the aeroplane to suffer a ground-loop. In the case of historic tailwheel aeroplanes, those to be watched with particular care are all marks of Supermarine Spitfire and Messerschmitt Bf 109. Both had relatively thin wing sections and their main landing gear joints were located inboard, adjacent to the fuselage. Both had high landing accident rates, but:

> 'The principal deficiency of the Bf 109 was the narrow track undercarriage which was marginally adequate from the strength point of view. Throughout its career the aircraft suffered an unusual number of undercarriage failures, sometimes during test flights before delivery to the Luftwaffe. The problem was exacerbated by the aircraft's tendency to ground loop on landing ... this usually resulted in the aircraft coming to rest on one wing tip with a bent propeller.'   (Ref. 7.1)

(5)   *Toe-in* and *toe-out* are shown in Fig. 13.3. Their effects become apparent when caught by a side-gust, or touching down with drift in a crosswind, e.g., from starboard which, as drawn in Fig. 13.3c, causes the port wheel to be the more heavily loaded of the two. Toe-out helps to keep the loaded wheel running straight. Toe-in increases a tendency to ground-loop. Although toe-out has an advantage over toe-in, it should not be more than 1° or 2° at the most.

Both toe-in and toe-out cause excessive tyre-scrubbing.

Tractor and pusher propeller, ground and water handling

Where the tractor gains is in ground and water handling. Then propwash over the tail enables the pilot to use power with rudder to blow the tail round in the desired direction. In the past some single-engined seaplanes with twin fins and rudders have weathercocked into wind without difficulty, but needed an additional central rudder to enable the pilot to turn out of wind.

Example 13–1: HM 293 Flying Flea Derivative

The control configuration of the French tandem-winged, tail-dragging HM 293, is derived from the Flying Flea (Pou-du-Ciel) designed in the 1930s by Henri Mignet. Pitch control was by increasing the lift of the foreplane, linked by bell crank and push-rod to the stick. There were no ailerons and to turn one pushed the rudder and relied upon dihedral to produce skid-induced bank. Climb, glide and level flight depended upon throttle setting. The mainwheels had heel-brakes. The whole system was delightfully simple.

The test flight of the HM 293 was from an owner's short airstrip, cut through an apple orchard. It had to wait from morning to late evening for the wind to drop. The owner cut the grass while the aeroplane was airborne. On landing the aeroplane sat down easily and the tail-dragger pilot's instinct took over: chopping the throttle, pulling back on the stick, keeping straight with rudder and applying the brakes. But, pulling back on the stick lifted

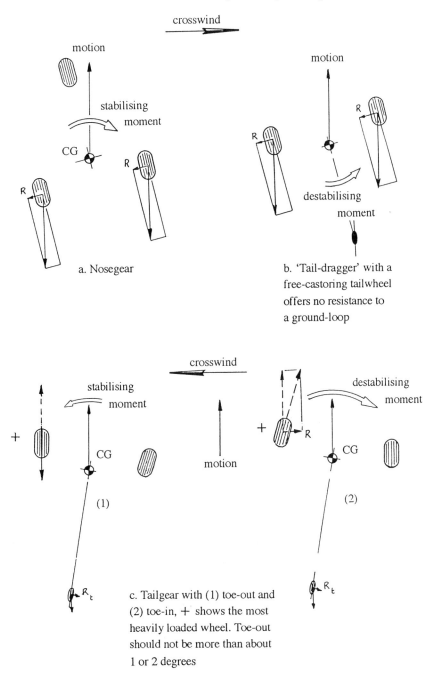

Fig. 13.3 The effect of wheel alignment on directional stability and instability (tendency to ground-loop). This is one of the first things to check if an aeroplane is hard to keep straight when taxying.

the mainwheels off the ground, making the brakes useless. The tailwheel, mounted at the bottom of the rudder, slavishly followed its every movement. The aeroplane left the strip backwards and came to rest, without damage, in the pile of cut grass.

Note: That event reflects on Fig. 1.7, in which the Système 'D' aeroplane was thought likely to land tailwheel first at times. A strong locking device would be needed, isolated from the rudder.

Example 13–2: Tailwheel Replacing Skid – De Havilland DH 88 Comet Replica

An aeroplane with a tailwheel that cannot be locked for take-off and landing is tricky. The original DH 88 Comet, built in 1934 for the Robertson race from England to Australia, had a gun-metal tipped tailskid, as shown in Fig. 13.4 (Ref. 13.1). The now restored original G-ACSS, owned by the Shuttleworth Trust at Old Warden, was given a castoring tailwheel (which could not be locked) for operation from modern paved surfaces. This led to the aeroplane suffering a ground-looping accident. The aeroplane now has a lockable tailwheel.

Shuttleworth pilot John Lewis after flying the restored Comet was quoted by 'Uncle Brian' in *Aerospace*, of the Royal Aeronautical Society, as saying:

'Very interesting, but I hope they don't build any more ....'

Water Handling

The flavour of water handling requirements for seaplanes is similar to that for ground handling, of landplanes and amphibians, with these additions:

☐ *Porpoising* tendencies must be neither dangerous nor uncontrollable at any normal operating speed on water.

☐ There must be *no uncontrollable tendency to waterloop* in 90° crosswinds up to $0.2\,V_{S0}$ (see Equation (13–2)), at any speed at which the aircraft is expected to operate on the water.

   The same applies to ground-looping in the case of an amphibian operating from an airfield.

Suffice it to say that on those occasions when the air, the attitude and the water are right, there is nothing in flying to equal the satisfaction felt on bringing a seaplane back out of the air to a landing on water.

Appendix A provides a background to seaplane handling. It is an extract from a paper to the Royal Aeronautical Society and carries its own captions and references.

Fast Hops and Lift-Off

The first test flight of an aeroplane must be planned with care, and the advice of Wilbur to Orville Wright at the head of Chapter 4 is apposite. For obvious reasons the test programme depends upon what is to be investigated. If the aeroplane is a new prototype with an untried engine and airframe combination, then there is both engine and airframe handling to be looked at. If, on the other hand, it is an untried engine in a testbed airframe, then attention will be focused primarily upon the engine. If the engine is more powerful than before then, of course, the test crew will be interested in the effects of power upon previous known handling qualities. Finally, the airframe may be new, or the subject of major modifications, while the powerplant is old and trusty. In this case primary interest will be in flying qualities of the airframe, and secondarily in the functioning of the powerplant.

In the next chapter we discuss the testing of powerplants and assume that the aeroplane is a new prototype, about which nothing is known.

Plate 13–3a   Rebuilt original De Havilland DH 88 Comet with the added tailwheel which caused the trouble. Ref.: Example 13–2. (Courtesy of British Aerospace via Shuttleworth Collection)

Plate 13–3b   The wooden, twin-engined record-breaking DH 88 led eventually to the fast wooden, twin-engined De Havilland Mosquito of World War II, the Mk II night-fighter version of which is shown here. (Courtesy of the former De Havilland Aircraft Company)

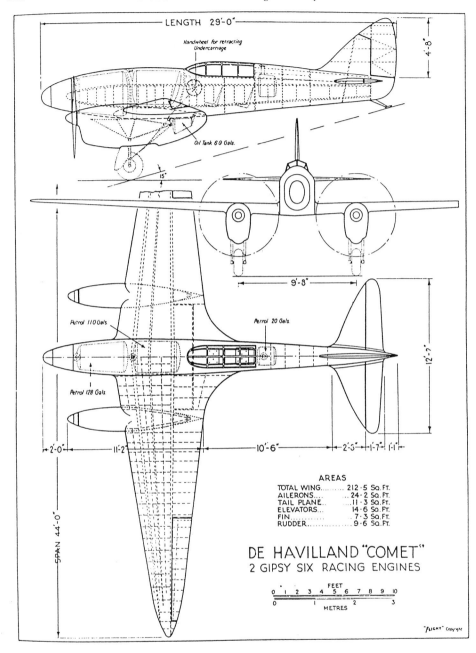

Fig. 13.4 A contemporary drawing of the De Havilland DH 88 Comet which won the MacRobertson race from Mildenhall in England to Melbourne, Australia, in 1934. Note the *tailskid* (a restored original Comet with *tailwheel* has flown, Ref.: Example 13–2, and there is a similar replica in the USA). Although fast and one of the most beautiful aeroplanes ever, the American entry, the functionally elegant Douglas DC 2, carried a load of passengers and reshaped the future (courtesy of *Flight* and IPC Ltd).

Plate 13–4 Amphibians, like the Russian Beriev Be-40 Albatros, shown here with high lift slats and flaps deployed and main gear travelling, require stringent tests for land and water operations, the latter in various sea-states, to determine water conditions. Note nose-up trim position of the tail to cope with increased nose-down pitching moment from high-lift devices and landing gear. (Courtesy of Simon Watson)

It is prudent to attempt one or two fast runs and hops before attempting to fly. These provide an opportunity to feel and assess response to each control in turn. They also help a pilot to gain confidence in himself and in the machinery strapped to him.

At the end of Chapter 6 a rule of thumb was given for an attempted first flight from a short field or strip. Apply it to the first flight of your prototype, especially if it is a home-built. Take into account the state of the surface and obstacles ahead of and within, say, a minimum of 20° on both sides of your selected path.

The initial ground runs and hops are invaluable for assessing whether or not the half-way mark (TORA/2) is reached with speed enough to spare.

Having decided to fly, check again the engine speed, pressures, temperatures and all flying controls:

☐ Set all trimming controls to neutral.
☐ Record the flap setting.

(a)

(b)

Plate 13–5   Testing the AMF Chevvron seaplane, with floats designed using the methods of Ref. 1.5:
(a) on the approach,
(b) on the step.
(Courtesy of Angus M. Fleming)

☐ Note engine conditions and that all flying controls are full and free.
☐ Note the time.

As the aeroplane accelerates:

☐ Note the speed at which the nose or tailwheel lift, and then V_{LOF}, the speed at which the main wheels leave the ground. With a nosewheel unit the speed at which the aeroplane rotates often coincides with V_{LOF}.
☐ Note trim changes while accelerating to the speed selected for the climb, say 1.3 V_{LOF}, which is around 1.4 V_S.
☐ Was there any airframe buffet on rotation, or hint of premature stall warning?

A Coarse Look at Flying Controls

If the aeroplane has rectractable gear it is often wise to leave it down on a first flight, until it has been proved that it can be lowered again after being first raised in flight. There are plenty of other things to be found out first in the take-off and landing configurations. Also there is no need to rush the raising of take-off-flap, but do remember to note the magnitude and direction of any trim changes when you do so.

It will be obvious during acceleration to the selected climb speed whether a push or a pull on the stick is needed to maintain the climb attitude. Note the changes of stick position and force needed to maintain it. If all is well the nose should rise with increasing airspeed and vice versa. Note too any changes in trim position to remove residual forces.

When settled in the climb try the controls to assess, crudely, the harmony of the aileron, elevator and rudder forces, A : E : R, for more or less equal rates of roll, pitch and yaw. Large control movements are unnecessary.

When settled in the climb (and assuming that the aeroplane can be trimmed), a convenient check of the longitudinal stability at constant power is to ease up the nose and reduce airspeed by 10%. Hold the new airspeed briefly and then release the stick (with your hand held close in front, just in case). The nose should fall and may tend to overshoot the trimmed position. Then push the stick forward to increase speed by 10%, hold briefly and release. If longitudinally stable the nose will rise.

The remaining time to a safe altitude – not less than about 3000 ft (1000 m) – should be spent noting altimeter readings, times, airspeed and engine conditions, and outside air temperature.

When level and certain of position (which, even with full radar control, should never be beyond gliding range from an airfield with engine idling) reduce power and trim at a gliding speed which feels comfortable (the speed will be not far from 1.3 V_S, or about V_S + 20 knots for a light aeroplane). This is done before attempting to approach the stall, just in case the engine stops and you have to make a forced landing. Time the steady height loss through about 500 ft. Use the short descent to look at the effect upon engine cooling – is it excessive?

Now try an *approach to the stall*, flap-UP at 1 knot/s speed reduction, BUT DO NOT STALL a new prototype. The reason for this is that the authority of the controls must be explored at low speed, just in case they are inadequate should the stall become a spin. Note the airspeed and:

☐ Any warning or hint of an impending stall.
☐ Any tendency to pitch nose-UP or nose-DOWN (any hint of self-stalling?).
☐ Deflect each flying control in turn, especially the rudder, through small angles. This is to ensure that a departure is unlikely to be provoked before the controls are tested more thoroughly later:

☐ Any loss of altitude.

Repeat with flap-DOWN, first in the take-off position, then fully for landing so that you know how the aeroplane is likely to behave in these configurations. As the flaps (and landing-gear) are lowered note the trim changes and stick forces involved.

It is useful to look for adverse effects of power, wrong tail-setting, and need for up- or down- or side-thrust in the recovery from the lowest speeds attained in the stall-approaches. Apply full power, or MCP, from throttle-CLOSED and maintain 1.4 V_{S1} (flap-UP, gear-DOWN) noting changes of control force. Repeat, maintaining 1.3 V_{S0} (flap-UP, gear-DOWN, then flap-DOWN and gear-DOWN as in the landing configuration). Some aeroplanes suffer from ridiculously large stick-forces when full power is applied and flaps are raised or lowered.

Following the approaches to the stall increase power and regain altitude and speed in level flight. The airspeed should not exceed the limitation with gear down and locked, V_{LE}, or V_A, the design manoeuvring speed. Try each control in turn at the higher speed. Has there been a marked alteration in feel or response to each?

Elevator and rudder control force ratios normally feel about n times heavier at V_A than at V_{S1}, where n is the manoeuvring load factor because if:

$$V_A = V_{S1} (n^{0.5}) \tag{13-3}$$

the forces vary as: $(V_A/V_{S1})^2 = n =$ say, 4 to 6, depending upon aircraft type while at V_D/M_D the control force ratios will be 2 to 4 times heavier than at V_A, depending upon the presence of aeroelasticity.

Controls should be well coordinated and sweetly harmonized, i.e. with elevator twice as heavy as aileron, and rudder twice as heavy as elevator, for broadly similar rates of response (see Equation (2–5)). Their existing harmony (or lack of it) and control forces multiplied by 4 to 6 at higher speed enable you to come to conclusions about the need for balancing: horns for lightening, cord or heavying strips at trailing edges for bias and increased hinge moments.

It is easiest to start with the ailerons, which must have authority at the stall and lightness at high speed. Providing them with enough area takes care of the former, and a Frise balance is a good start for the latter. Knowing that the rudder should be about four times heavier than the ailerons helps one to adjust what should be the heaviest of the controls. The elevator is then balanced so that the control forces lie mid-way between those of the ailerons and rudder.

Great care is needed to avoid overdoing balancing. Approach it a small step at a time. Better to be over-heavy than suddenly over-light and skittish, with risk of over-control and a PIO.

Fin area and dihedral

Although an aeroplane must be stable, the degree varies. Designers rarely have doubts about the effectiveness of the wings and engines. The portion of the airframe with the greatest number of imponderables is the combination of the fuselage surfaces aft of the CG – and the tail.

The fin and rudder unit is of paramount importance. Fin area must ensure at all times that the directional centre of pressure, CP, remains behind the CG when a side-gust strikes or, put another way, that weathercock stability remains positive anywhere in the flight envelope.

If after take-off you have encountered turbulence it will have given an indication of how pleasantly or otherwise the aeroplane responds both directionally and laterally. Straight steady sideslips, as described early in Chapter 11, with modest amounts of rudder are useful at this point to provide hard evidence. If top rudder is needed to keep the nose on or a little below the horizon to maintain speed in a straight sideslip, as shown in Fig. 13.5, then fin area is sufficient. No need of rudder means that fin is not quite big enough. But if bottom rudder is needed there is an excess of keel surface area ahead of the CG and fin area aft is deficient.

Next, trim the aeroplane in a glide, flaps-UP and stick-free, around an estimated 1.2 V_S, or 1.2 V_{S0}. Establish a slight bank with aileron and a touch of rudder to hold a straight path. First release the control column while holding the rudder fixed and see if the lower (up-wind) wing rises or falls. Then repeat in each direction, only this time keep hold of the control column and release the rudder.

If the fin and dihedral are in correct proportion, the lower wing rises when the control column is released, reducing the angle of bank. But if the lower wing remains down and the aeroplane turns towards it, entering a spiral dive, then the fin area is excessive for the dihedral, or the dihedral is insufficient.

Repeat the test with flaps-DOWN – and if the gear has been raised, then repeat the test with both gear-DOWN and flaps-DOWN.

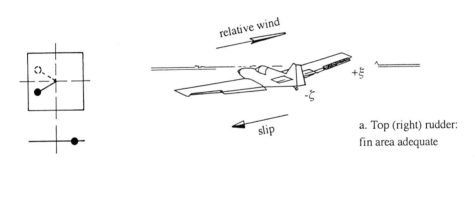

a. Top (right) rudder: fin area adequate

For code *see* table 4-5

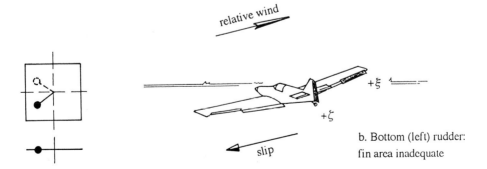

b. Bottom (left) rudder: fin area inadequate

Fig. 13.5 Assessment of fin authority from the amount of top or bottom rudder to maintain straight steady sideslips. Here there is enough dihedral because left aileron is needed to hold 10 degrees port bank – or the maximum angle obtainable with full rudder deflection or 150 lbf (34 N), according to type of aircraft – against the relative wind or airflow. Nose attitude is held relative to the horizon with elevator.

The amount of dihedral may be estimated from the lateral deflection of the control column to hold a steady sideslip. This is proportional to the aileron deflection needed to neutralize the rolling moment – the larger the deflection the more powerful the rolling moment. Of course, much depends upon the size and type of lateral control surfaces. The answer helps you to assess whether the dihedral is enough, or the fin and rudder are too large, or vice versa.

However, response of the aeroplane on releasing the control column is not always a clear measure of dihedral effect, especially when an aeroplane is very small because, as we saw much earlier, if the wing stays down in spite of the appearance of adequate dihedral, it could be due to friction in the aileron circuit. You will recall the requirement that dihedral effect – which inevitably involves some loss of efficiency due to friction – should not be worse than neutral. If friction is present and the angle of bank neither increases nor decreases, then dihedral could be enough. The *fix* is to do what one can to reduce control circuit friction.

Note: Expect dihedral effect to be reduced, more often than not, with flap-DOWN and gear-DOWN (see Fig. 11.2c(1)).

Effectiveness of the rudder(s) and other flying controls has already been examined at low speed in tentative approaches to the stall, in the clean and landing configurations. Having checked control forces, fin and rudder effectiveness at low speed, and dihedral one may draw tentative conclusions, and make recommendations about possible adjustments to improve fin and rudder authority, control-gearing, dihedral effect – and any need for trim or other tabs.

Longitudinal stability and tail-setting angle

An opportunity was taken to carry out a short check of longitudinal stability during the initial climb. This is the first longitudinal stability test in serial 10 of Table 4–6. Exactly the same is repeated in the approach configuration with gear and flap down and sufficient power for a standard 3° approach path (rate of descent about 300 ft/min), at an estimated 1.4 V_{S0}, and again between 1.3 V_{S1} and V_A, gear and flap-UP. For most piston-engined light aeroplanes the power for a standard approach is around 1700 to 1900 RPM. Note whether it is possible to trim the aircraft, movement of the control column and the stick-forces.

Tail-setting angle, the angle of incidence of the tailplane, is arranged to counter changes in pitch with downwash, and with power. These are greatest at low airspeeds when flap, landing-gear and power combinations can be extreme. If the tailplane is fixed, without scope for adjustment, the tests usefully measure the authority of the elevator trimmer. They are carried out as shown in Table 4–6, serial 6, 7, 8 and 9. The results help you to conclude whether or not side, up or down-thrust is needed to cope with changes in power.

If the thrust line is above the drag axis then applying power causes nose-down pitch, while throttling back causes the nose to rise. This is a problem with seaplanes especially, which have high-mounted engines. Throttles are often mounted on a roof-console and it is better to simply push them forward with the flat of the hand. If roof-mounted throttles are gripped with the fingers, pitching caused during a bumpy take-off, or change of power, accelerate and decelerate the hand, causing surges of thrust, a possible PIO and porpoising (see Appendix A).

It is better to omit checks of minimum trim speed from the first test flight of a prototype (see Table 4–6, serial 13), until behaviour at the stall has been explored. They may be included when checking a used aeroplane with published handling characteristics.

Full-rudder sideslips

Full-rudder sideslips are best left until you feel more confident about other aspects of handling, which is why they have been separated from other sideslip tests. They are carried out with the greatest of care, applying rudder slowly (feeling it and the aeroplane all the way), at estimated 1.2 V_{S1} and 1.2 V_{S0}, as shown in Table 4–6, serial 12. Their purpose is to look for rudder lightening or reversal. Results give clues to the need for a dorsal fin.

Functioning Checks

Functioning tests of the plethora of systems and equipment associated with the modern aeroplane have been omitted, to keep the book within limits. But, on a first flight you must note:

☐ Oddity of behaviour and any strange sounds and smells (see Example 13–3, below).
☐ Signs of control vibration or flutter.
☐ Satisfactory working of:

  (1) Electrics.
  (2) Suction and pressure systems.
  (3) Instruments.
  (4) Fuel supply.
  (5) Engine conditions: RPM, temperatures and pressures.
  (6) If an autopilot is fitted, check that it can be overpowered manually and 'killed' if necessary.

Example 13–3: Take-Off – Powerchute Raider

The parawing of the Raider is similar to that of a paraglider (French: parapente) and is a navigable high-aspect ratio parachute, comprised of 11 linked cells of zero-porosity ripstop nylon, coated with polyurethane. Each cell was open at the leading-edge, stitched along the trailing-edge and joined laterally inside by a cross-port to the adjacent cells. These ensured that all cells inflated with the engine running and propwash filling only one or two initially.

A powered tricycle unit hung beneath.Throttle – like the Flying Flea and HM 293 – was used to control climb, level flight and the glide. Flying controls were by two cords, one to the trailing-edge of each wing tip, which acted as airbrakes or 'flaperons'. Pull on one cord and the increased drag of the wing on that side caused the Raider to turn in the same direction. A pull on both enabled one to flare on landing.

The trike unit had a steerable nosewheel, but could not be taxied. It was pointed into wind manually, the wing deployed behind it and the engine started. Propwash inflated the cells one by one and the wing lifted in the wind. With equal amounts of wing in view above and behind on each side, the foot throttle was opened up for take-off, while keeping straight with the nosewheel.

On the test flight for a CAA Permit to Fly, just as the unguarded nosewheel reached the point of lift-off it ran through a patch of fresh cow-dung, filling my mouth and nose, covering my spectacles. For those who wish to know, in the interests of scientific enquiry, it tastes exactly the same as new-mown grass.

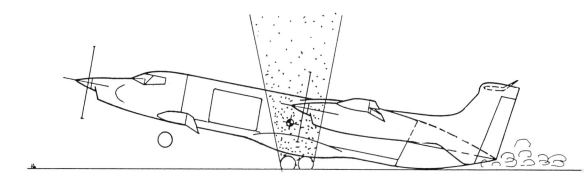

Fig. 13.6 Three-surface transport in Example 13–4 over-rotated on ground. Impact with rough and wet area showers debris and water into the wing-mounted propellers and engine intakes. With three surfaces more or less in line there is the danger of interference from the foreplane with the mainplane, and from both with the tailplane. Aircraft attitude might have to be restricted to avoid ground-stalling of any surface.

There is value in testing the simplest of flying machines. While I recommended the fitting of either a mudguard, or a dental-spittoon (one articulated for both left and right-handed pilots), the point was that a stone flung by the nosewheel could incapacitate a pilot.

On seeing the effects of the dung the applicant, a droll ex-paratrooper, wryly observed that: 'It couldn't have happened to a better organization'.

Example 13–4: Take-Off and Landing – Tri-Motor Transport (Example 6–8)

At an early stage the project test pilot and engineer voice reservations about the tri-motor in Fig. 6.16, and tests to be run. Appreciating that it is shaped for a wide CG range and flexibility in loading they anticipate difficulties during STOL tests to find out what happens when the aeroplane is over-rotated on take-off and landing, especially in a crosswind, see Fig. 13.6.

Their main concern is that the plane of the propellers and intakes of the wing-mounted engines appear to be too close to the point of contact of the mainwheels with the ground. Whether the propellers are in normal operation or in reverse there is a risk of picking up stones, sand, dirt, solid water, ice and slush at high power. Moreover propwash from the centre-engine could make matters much worse with solids (not only do stones and sand cause damage to the leading edges and tips of the propellers, I have had trouble with beads of ice lifted from a frozen surface). The gear is retractable and one possible solution is to design undercarriage doors which serve as stone or mudguards, without risk of them becoming clogged or distorted.

A further fix could involve lengthening the wing-nacelles, moving the engines further forward. But if that is done the port propeller comes much closer to the loading door on the port side of the fuselage; and also the CG is moved forward. Although the port engine would be shut-down for loading and unloading (leaving the other engines running), the propeller will need a brake to prevent it windmilling, so as to reduce loading and unloading time.

Moving the CG forward increases the distance between it and the point of contact of the mainwheels (which is, in effect, their centroid). What must then be watched is any reduction in lateral overturning coefficient as the CG shifts towards the apex of the triangle formed by the nosegear.

A CG further forward increases the pitching moment about the mainwheels, increasing the speed, V_R, at which the pilot is able to rotate on take-off; and also the decision speed, V_1. The latter shortens respectively the required emergency distance, take-off run and take-off distance (EDR, TORR and TODR).

There is concern about side area ahead of the mainwheels, tending to turn the aeroplane out of a crosswind, while the forward CG loads the nosegear heavily. Steering then causes severe tyre-scrubbing and rapid wear.

Note: One tandem-winged aeroplane I flew – a smaller scale version of the larger freighter – suffered this problem during tests intended to find the maximum distance that could be accepted between the CG and the maingear. Almost half-right rudder (and nosewheel steering) was needed to hold the runway centre-line in a moderate crosswind from starboard. A touch-and-go was impracticable, for on rotating the nose was blown downwind, with insufficient rudder authority left to keep straight.

After Landing

☐ Have the aeroplane checked for oil and other leaks, signs of rubbing, chafing, loosening, things flapping about and unusual noises when controls and cocks are moved.
☐ Write down your impressions while they are still vivid and reliable.

Subsequent Flights

Here one explores in greater detail features discovered or suspected earlier, with a view to eliminating or curing vices and developing good qualities.

☐ *Investigating high speed handling*: as in all things this must be approached with the greatest care, looking for flutter, buzz, adverse compressibility effects, control reversal (aeroelasticity) and other undesirable features.

(1) The aim is to achieve the design diving speed, V_D, but do not be surprised if you cannot achieve it, because of some feature that is off-putting. For example, the aeroplane may start snaking because the propeller is behaving like a windmill and beginning to dam the flow over the tail surfaces, disturbing the wake. Instead you will settle for a demonstrated V_{DF}.

(2) Once a V_{DF} is found, this is factored by 0.9 to produce V_{NE}, the never exceed speed (see Equation (3–68)).

Compressibility Effects

Although most of the aeroplanes in this book are subsonic, a number of jet-engined home-builts, like the Bede-10, will encounter compressibility effects. Pilots of other ex-military aeroplanes, operated privately with Permits to Fly, experience compressibility as a matter of course. During World War II Spitfires were regularly tested to M 0.86 and Martindale, an RAE test pilot, reached M 0.92 on one occasion (Ref. 13.4). Such effects often begin to appear around M 0.6 at which, with wear and tear on a conventional subsonic airframe, local airflow reaches the speed of sound somewhere on the skin.

High Mach numbers cause compressibility effects which can vary between the mildly irritating to serious. High airspeeds cause high air loads at low altitude when compressibility and high dynamic pressures combine. Therefore, when testing an aeroplane with high enough performance to encounter such effects one should arrange to investigate its qualities at high altitude, where EAS is reduced for a given TAS and Mach number. For example, using Figs 3.21 and 3.22, M 1.0 at 36 000 ft (11 000 m) is about 300 KEAS, whereas at sea level it is 660 KEAS. But the dynamic pressure and forces on the airframe at that altitude are only in the ratio of the square of (300/660), i.e. about 1/5 those at sea level, making overstressing less likely. The same is true for lower Mach numbers. At a constant 150 KEAS the dynamic pressure at 25 000 ft (7600 m) is around one half of its sea level value, but the true airspeed has risen to 225 KTAS, about M 0.4, somewhat less than M 0.6, where compressibility begins.

As we saw in Chapter 9, response to control at altitude is more lively, because of relative aircraft density, which also worsens the effects of mishandling and departures. Fortuitously the situation improves as the aircraft descends into denser air.

The commonest symptom of compressibility is buffet caused by the formation of shock-waves, wafer-thin pressure fronts at right angles to local airflows, through which a previously supersonic component decelerates back to subsonic speed. Shock-waves are accompanied by violent separations of the flow and adverse effects upon control surfaces and tabs downstream, which have caused control-reversal. Hinge-moments change sharply. Stability is degraded and there may be gross, unpredictable, changes of trim accompanied by loss of control. Better to encounter such phenomena at higher rather than lower altitude. As an aircraft descends the air becomes denser and warmer, the Mach number falls and compressibility effects are reduced or disappear altogether.

Compressibility effects occur earlier, at lower Mach numbers, when manoeuvring at high speed. The reason is that the angle of attack is increased and with it the velocity of airflows over humps and crests on the skin.

The airworthiness requirements with which we are concerned are published in FAR and JAR 23.253. The maximum operating limit speed or Mach number, V_{MO}/M_{MO}, whichever is critical at a particular altitude, must be established, such that it is not greater than the design cruising speed or Mach number, V_C/M_C. Further, it must be sufficiently below V_D/M_D to make it highly improbable that the latter speed will be exceeded in operations.

Once the maximum operating limit speed and Mach number have been established flight tests are needed to simulate upsets in pitch and roll, and inadvertent increases in airspeed. For the tests the aeroplane is trimmed at any likely speed up to V_{MO}/M_{MO}. The conditions to be explored include gust upsets, inadvertent control movements, low stick-force gradient in relation to control friction, passenger movement (when possible), and levelling off from climb and descent, from Mach to airspeed limit altitude.

Allowance must be made for pilot reaction time after the occurrence of an effective inherent or artificial speed warning. The aeroplane must be recoverable to a normal attitude without:

☐ Exceptional piloting strength or skill.
☐ Exceeding V_D/M_D, or the structural limitations.
☐ Buffeting that would impair the ability of a pilot to read the instruments, or to control the aeroplane for recovery.
☐ There may be no control reversal about any axis at any speed up to V_D/M_D.

☐ Any reversal of elevator control force or tendency of the aeroplane to pitch, roll or yaw must be mild and readily controllable using normal piloting techniques.

Example 13–5: Areas of Difficulty – Semi-Scale Su-35 (Example 1–1)

The semi-scale twin-jet aeroplane sketched in Example 1–1 took the idea of the jet-powered home-built a step further than those introduced in Ref. 1.5. Having a configuration inspired by the high performance Russian Sukhoi Su-35, the company test pilot is tasked with planning tests to establish its high-speed characteristics before it is launched as a kit.

Figure 13.7 shows that the thrust/weight ratio would lie somewhere near 0.35 to 0.40, with the weight between 3750 lb and 4285 lb, averaging say 4000 lb (1820 kg). As F/W_0 M for a trainer averages about 0.45, the achievable Mach number in level flight could be 0.83 – at lighter weights, M 0.9.

At the heart of the concept lies the question of reliability of twin (experimental) engines. Testing of the engines to establish their acceptability involves a comprehensive, expensive programme, quite apart from testing of the airframe and systems.

A weight of 4000 lb (1820 kg), a high approach speed with a wing-plus-body lifting area of 120 ft^2 (11 m^2), giving a touchdown speed in exceed of 100 KEAS, place the aircraft in a different class from other home-builts. Quality control of the kit, the materials it contains, and construction manuals must be to a high and detailed standard.

The test pilot will therefore be dealing with a fast and potentially tricky aeroplane, intended by the manufacturer to be constructed and flown by amateurs. Early in the project the pilot pointed out the need for oxygen, adequate instrumentation and avionics. Although not considered then, further study points to a requirement for power-operated flaps, airbrakes and landing-gear, and a healthy amount of fuel to cope with two engines while providing a reserve for emergencies. Because of the high take-off and landing speeds and small wheels, operations are limited to adequately paved airfields. This also means smooth paved surfaces, because small wheels are more sensitive and respond to small stones and other discontinuities, triggering overcontrol and PIOs.

In addition to subsonic handling, compressibility introduces complications. There is always the risk of degradation of control by compressibility effects. These are exacerbated by elasticity of the aerodynamically thin surfaces, which lack structural depth. Use of materials must be watched closely, and the machine could need considerably more metal than the carbon and glass composites already planned. The aeroplane is not large and small inertia radii of gyration, coupled with powerful controls at high speed, will make it lively in response at higher altitudes.

Tests must explore low-speed and stall handling; field performance; spinning if aerobatics are intended (and there is little point in contemplating such an aeroplane if they are not); asymmetry (one-engine-inoperative); weight, balance and ease of misloading (the second occupant is a long way ahead of the CG); adequacy of and response to manual control (PIOs and relative aircraft density effects); longitudinal stability, directional stability, roll with sideslip; buffet boundaries; range and endurance; and adequacy as an instrument platform – inevitably some pilots will be caught out by weather, with plenty of scope for mishandling and pilot error in operation; functioning of the systems – power, fuel and other – and the compatibility of the engines with the semi-scale airframe.

In other words, a jet home-built – especially a scale replica of a supersonic fighter – has a number of tricky, unique, corners into which a test pilot must poke and shed light while gathering numbers, impressions and searching for dangerous features. With a semi-scale

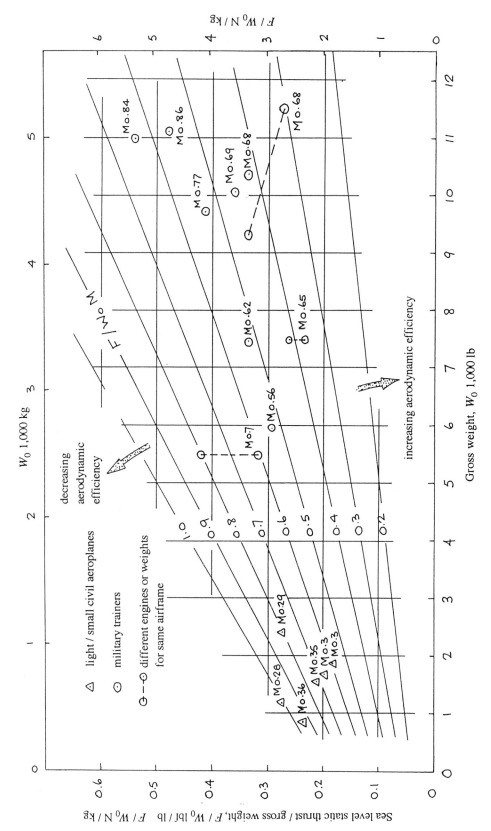

Fig. 13.7 The diagram is extracted from Ref. 1.5 to show thrust/weight ratios of a number of light and small civil jets and military trainers against constant 'aerodynamic efficiency', measurable in terms of $[1/(F/W_0M)] = W_0M/F$ (see Equation (2–11a)).

Su-35 especially there is nothing which might be taken for granted with any confidence. Every point will need proof that the level of safety is acceptable if it is to be marketed, either as a set of drawings alone, or as a kit.

Insurance

In any project requiring test flying a pilot must ALWAYS question what is being done and why – and then seek further advice if the answers do not satisfy him or her:

- □ especially where potential performance might appear to be be unusual or excessive;
- □ especially where spinning is concerned;
- □ especially before poking around in the low-and high-speed corners of the flight envelope where normally one does not go.
- □ So as to keep an eye on everything that the technical staff get up to.

If you are the test pilot or flight test engineer MAKE SURE that you have insurance cover. Pilots are the first and easiest people to blame when something goes wrong on a test flight – which is what a test pilot is conditioned to expect.

It is a tradition of the profession to bring an aeroplane back – if at all possible (but, of course, one must never push one's luck to stupid lengths). To quote again the ever-practical Frank Bullen and his repeated prodding from the front seat of the Tiger Moth:

'If it goes wrong in the next minute, where will you put it down?'

References and Bibliography

13.1 *De Havilland, The Golden Years 1919–1939, Flight International* Special, Sutton, Surrey: IPC Transport Press, 1981

13.2 Stinton, D., *The Approach to the First and Subsequent Test Flights*. Paper read before the Light Aeroplane Group, the Royal Aeronautical Society, 1987

13.3 Uwins, C. F., *Experimental Test Flying*. Paper read before the Bristol Branch of the Royal Aeronautical Society, 21 March 1929, *Aeronautical Journal*, May 1929

13.4 Derry, J. D., High-speed flying, *Journal of the Royal Aeronautical Society*, October 1951

13.5 Duke, N. F., *Test Pilot*. London: Alan Wingate (Publishers), 1953

CHAPTER 14

Dealing with Powerplants

'The job of the test pilot is to anticipate other people's mistakes and by reproducing them to find out what happens. Inevitably he makes mistakes of his own too, but this is also part of the learning process. Any mistake which can possibly be made will, unless designed out, sooner or later be made by someone.'
Captain Hadley G. Hazelden, DFC, formerly Chief Test Pilot of Handley Page Ltd

The machinery packed into the engine-room of an aeroplane has its own brand of humour. In this final chapter we tie together a number of near random and often troublesome loose ends, caused by engine, propeller, exhaust and other systems, with which the test pilot has to contend.

Powerplant Requirements

Stringent civil requirements for the safety of engines and their systems are published in JAR 23.901 and FAR 23.901 *et seq.* (Refs 5.8 and 5.9), while Ref. 10.3 lists extensive tests of powerplant systems needed to show compliance with FAR 23. The tests are substantially the same for the European requirements. Military requirements are in the appropriate Military Specifications.

There is little value in picking out more from the civil requirements than those parts which affect primarily the measurement of performance. Compliance with airworthiness requirements is checked by the test pilot, using his common sense. He should have knowledge of what is applicable to both piston and turbopropeller powerplants.

Only in the home-built world is one likely to encounter an engine which has not been subjected to extensive development testing. When a production engine is installed in a prototype or production aeroplane the handling characteristics, controls and instrument displays are investigated in detail. First, to confirm that the matching of engine and airframe is acceptable for all anticipated flight conditions in the design role. Second, to demonstrate compliance with airworthiness and any other requirements.

The Specification

Specifications for both engine and aeroplane set the standard of engine performance and handling required. Taken together they are the basis for engine testing. A specification:

☐ Forms a rational statement of what is wanted in an aeroplane or engine so that it may best satisfy the requirements in terms of what is possible.

☐ Must always take account of the current state-of-the-art in technology and good practices.

☐ Is the yardstick against which the resultant aircraft and/or engine is measured.

☐ In theory is drawn up by the customer. In practice it is written as the aircraft and/or engine proceeds through the different stages of design to production. This is inevitable as lead-times to produce satisfactory airframes and engines are long.

Note: What is said for engines and airframes applies equally to avionics and other specialized equipment.

There are penalties for failure to keep within these typical tolerances:

(1) Range within ±3% to 5% at optimum range speed.
(2) Maximum speed within ±3%.
(3) Equipped airframe and structure weights within ±2% to 3%.
(4) Take-off and landing distances within ±5% to 7%.
(5) Noise levels no more than 2 or 3 decibels above that specified.

Note: Tolerances usually have upper and lower limits, one of which presents a bonus for the operator. But if the bonus is too large it points to the possibility of the design having suffered some shortfall or a deficiency.

Tests carried out with a prototype and during development, must be repeated in part at least during production, to ensure that quality is maintained, and that modifications and production methods have not degraded engine performance and handling.

Modifications can have adverse cumulative effects upon flying qualities, even though each may appear innocuous and not worth the cost of testing. If there is reasonable cause for doubt then do not guess, carry out a proper check, no matter how brief your time airborne (test pilots have fewer flying hours in their logbooks than flying instructors and airline pilots, because test flights are expensive and only last long enough to gather essential information).

Engine tests can often be combined with other tests. For example, a timed climb produces important engine and airframe results; while a stall, or a run to V_D or V_{NE} may reveal every bit as much about powerplant behaviour as about the airframe.

Test Equipment

Basic cockpit equipment includes: ASI, altimeter, engine RPM, temperatures (JPT, TGT, EGT(s) in the case of a turbine installation), pressures, fuel pressure and flowmeter, stop watch and OAT. If testing is to be extensive and in depth, then additional test equipment would also include: a manual event marker with a transmission facility, normal accelerometer, mounted as close as possible to the CG. Instrument errors should be not more than ±1%.

Powerplant Tests in General

As a general rule, engine tests cover whatever is appropriate. Turbojet, turbofan and turboprop engines are simpler to operate and more reliable than older piston-propeller units. The aircraft in which they are used are (with certain exceptions, like those used in agricultural operations, or the Bede-10 kit-built jet) subject of extensive professional testing to a high standard. This chapter is a broad-brush sweep, mainly with simpler and older aircraft and engines in mind:

☐ *Starting and stopping the engine.* It must be possible to start and stop an engine under any conditions in which starting is permitted, both on the ground and in flight, with the minimum risk of fire or mechanical damage:

(1) Starting and stopping techniques and limitations must be established and included in the Aircraft Flight Manual, or other advice to pilots, including operating placards.

(2) An altitude and airspeed envelope must be established for restarting or relighting the engine in flight.

(3) If the minimum windmilling speed does not produce enough electrical power, then an independent ignition source must be provided for in-flight starting.

(4) Turboprop engines windmill freely. The risk of a prop-strike and the need for a propeller-brake must be considered for some operations.

☐ *Engine behaviour taxying and on take-off.* The test pilot should look at the effects of taxying into and out of wind, and in crosswinds. In the absence of a crosswind limitation stated in the specification, it is reasonable to work to the lesser of, say, $0.4\ V_{s1}$ or 25 KTAS, unless the aeroplane is so light that the crosswind test would be hazardous. The pilot should then recommend a crosswind speed which he thinks is safe.

☐ *Engine behaviour in the climb.* Cooling problems, RPM creep at fixed throttle settings.

☐ *Engine behaviour in general.* Operating characteristics must be explored to ensure that nothing giving rise to hazard (with a turbine, e.g., stall, surge or flame-out) occurs during normal and emergency engine handling, including:

(1) *Deceleration* to and at flight idle.

(2) *On opening up* again and accelerating to full power.

(3) *During a balk and go-around.* The tests should include the effects of fuel booster pumps, and first checks should be carried out at a safe altitude.

(4) There must be *no adverse excitement of the vibration characteristics* of components, the failure of which could be catastrophic.

(5) *Under negative g* there must be no hazardous malfunction of the engine or any component or system of the powerplant within the prescribed flight envelope.

(6) On landing, *when using power for retardation* (e.g., reverse thrust: see *propellers*).

☐ *Turbocharger and intercoolers.* (Sir) Stanley Hooker, responsible for the Rolls-Royce Merlin and most British engines since, is quoted as saying:

> 'If one thinks of the cylinders and pistons of an engine as the heart which converts the force of the burning air and petrol mixture into mechanical power by the downward motion of the piston, then the supercharger is the lungs of the engine, and by its efficiency controls the power output.'   (Ref. 14.1)

The efficiency of the supercharger depends upon the efficiency of its rotor and the diffuser, which converts the centrifugal energy imparted by the rotor into the high pressure feed of air to the combustion stage. The temperature of the compressed air is high and cooling is needed to make the explosive air–fuel mixture manageable.

The points which follow are obvious when a type certificated engine is fitted. But homebuilders are drawn to stalls in the various 'Flea-Markets' which spring up at airshows and conventions, and it is not unknown for used and repaired parts of dubious provenance to appear subsequently beneath cowlings, and to creep through the inspection nets of a few engineering organizations. If nothing is known about a part, assume it to be a source of potential danger.

(1) *Turbochargers* must be of approved types, with cases that are strong enough to contain a burst compressor or turbine at the highest speed obtainable with normal speed-control devices inoperable. Figure 14.1 shows a typical turbocharger (with intercooler omitted). Figure 14.2 lists a number of handling points (both, Ref. 1.5).

(2) *Intercoolers* must withstand in-flight loading and vibration. Should one fail it must be impossible for fragments to be ingested by the engine, or for the airflow to discharge harmfully on to any part of the aeroplane under all operating conditions.

(3) *Check that a turbocharger wastegate and control system works.* Simple tests reveal what is at fault: e.g., a sticking wastegate, or the wastegate actuator is leaking. On the other hand it could be the controller which is at fault.

There was a case which almost came to court as a result of a turbocharged twin failing to meet its scheduled climb performance. The wastegate on the starboard engine was found to be disconnected – two checks later!

(4) *Functioning at altitude.* Tests to a prescribed altitude might be required on the first test flight of an engine, and certainly during development testing of an aeroplane. On engines with automatic pressure controllers the pilot ensures that, with throttles fully forward the manifold pressure does not exceed the maximum permissible. It should be possible to maintain manifold pressure close to the maximum permissible (typically, within 1 in (2.5 cm) Hg). Stability of RPM, manifold or fuel flow pressure (booster-pump(s)-ON and -OFF) should be noted, together with engine temperatures and pressures.

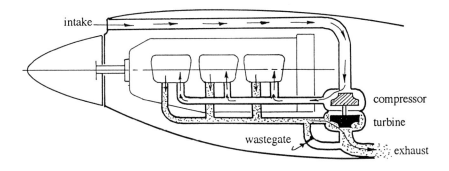

With the wastegate closed exhaust gases drive the turbine and the compressor - which turbo-charges the air from the intake to the induction system. When the wastegate is opened, exhaust gases by-pass the turbine and there is no supercharging.

Fig. 14.1 The basic turbo-supercharger.

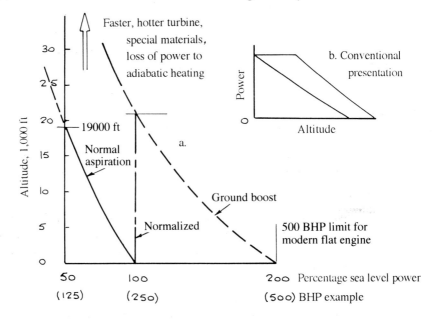

Some points on turbocharger handling:

1. High altitude = high turbocharger rpm and cylinder head temperature (CHT).
2. CHT is at least 30° F (20° C) higher at high altitude.
3. High CHT increases risk of detonation and damage to engine.
4. Turbocharged engines need at least 100 Octane fuel.
5. Open throttle slowly to avoid surge and over-boost.
6. Observe power-change sequence (rich mixture = cooler CHT)

   Increasing power: enrich -- increase rpm -- open throttle to increase boost pressure.

   Decreasing power: reduce throttle -- reduce rpm -- adjust mixture.
7. *Take-off*: consult Owner's Manual, Flight or other approved Manual for minimum oil

   temperature, to so as to avoid overboosting an engine that is too cool.
8. *Climb*: unless specified in the appropriate manual use full rich mixture to improve

   cooling and to avoid detonation.
9. *Cruising*: observe CHT, turbine inlet temperature (TIT) and exhaust gas

   temperature (EGT) limits.
10. AVOID THERMAL SHOCK.

Fig. 14.2 Turbocharger handling.

☐ *Fuel flow.*

(1) If the time taken to regain power after switching to a tank containing minimum
     fuel exceeds 10 s, then Ref. 10.3 states that suitable instructions and information

should be in the Aircraft Flight Manual (or other sources of information for pilots).

(2) Time taken to regain power when selecting another tank through a cross-feed system, or from a fuel-transfer system, should be investigated.

Propellers with pitch-changing devices raise a number of important points to be checked:

(1) Each featherable propeller must have a means to unfeather it in flight.

(2) Where it is possible to reverse the pitch of a propeller in flight then:

   (a) A commuter aeroplane must be able to continue to fly safely and land with the reverse-pitch control in any possible position.

      Note: Reverse-pitch *control*, not propeller!

   (b) A reverser system for ground operations must have been designed in such a way that no single failure, or reasonably likely combination of failures or malfunction of the system, will result in unwanted reversed thrust under any unexpected operating condition.

   (c) It must also be demonstrated that, in the event of a likely malfunction of the reversing system, it is possible to prevent divergence from the intended landing path by more than 30 ft (9 m) on both wet and dry runways in crosswinds up to 10 knots (Ref. 14.2).

   Note: In the UK it is unacceptable to move propeller pitch to less than the flight idle with one action alone of the pilot. *Two separate and distinct actions are needed to enter the reverse range.* While this is entirely reasonable for aeroplanes carrying innocent fare-paying, or other passengers, because inadvertent reversal of thrust in flight is usually catastrophic, it has caused difficulties (see Example 14–1).

(3) Safe vibration characteristics of any type of metal or composite propeller must be demonstrated throughout the flight envelope.

Tests of an aeroplane, like that in Example 1–1, fitted with twin turbojet engines designed originally for use in RPVs, would concentrate upon specifically:

☐ Engine relighting, hot and cold, to determine the envelope.
☐ Hot and cold slam accelerations.
☐ Engine accelerations at high *g* and angle of attack.
☐ Checks for engine surge.
☐ Operation of (variable geometry or secondary (upper surface)) air intakes and airflow controls, to ensure that no sympathetic interference between the engines is likely. For example, if one engine surges, then it should not cause the other engine to surge and flame-out.
☐ Checks of reheat lighting and burning, when appropriate.

Points to Watch

Turbine engines

Turbine engines include:

☐ *Turbojets:* which have relatively small mass flows of air accelerated by combustion to high velocities. They are thirsty and noisy, but useful at high altitude and airspeed. For the range of aircraft dealt with in this book they are too specialized and more novel than practical.

☐ *Turbofans:* have moderate mass flows and accelerations. They produce moderate thrust and are not unduly thirsty, at all airspeeds and altitudes. They are quiet.

☐ *Turboprops:* which have high mass flows accelerated to lower (somewhat quieter) velocities, except for those passing through the small gas-turbine core. They are good at low altitudes and low (subsonic) airspeeds, up to about 400 knots, and are the engines with which we are concerned here.

As a general rule problems with jet engines are largely caused by abuse, by not observing their limits, and by mishandling. Exhaust gas temperatures and engine speeds must be watched closely. An incorrectly designed intake, or attempting to increase thrust too swiftly, causes flow to the compressor to stall and the engine to surge and/or flame-out.

TURBOPROPS

The turbo-propeller engine combination combines the simplicity of the gas turbine with the power-conversion properties of the propeller. Most small turboprops consist of a hot gas-generator unit, which is a small turbojet, and a separate power unit, consisting of a (free) power turbine driving a shaft to the propeller, via reduction gears. Other 'fixed-shaft' units consist of a gas generator with an extra power turbine wheel, connected to the propeller by shaft and gearbox. Fixed shaft units tend to have lower specific fuel consumption than free.

Management of the power output of a turboprop involves both fuel and propeller control. Engine parameters are:

☐ *RPM*, the engine rotational speed needed to produce the power, indicated in percent.

☐ *Fuel flow* in pounds, or kilogrammes, per hour.

☐ *Torque*, the power output, indicated by torquemeter in horsepower, foot-pounds, or percent power, depending upon the aircraft.

☐ *Temperature*, in degrees Celsius.

Both torque and temperature have red-lined limits.

Once the required engine speed has been selected, operation of a *power lever* (as against a throttle of a piston engine) enables the pilot to control power between maximum and flight-idle in flight, and from ground-idle into the reverse range on the ground. The ability to direct thrust forwards or to reverse it with one hand on a single lever has the greatest attraction for the pilot (others are smoothness, quietness and reliability). Although there is a stop which forms a flight idle-gate, it is often possible to override this unintentionally in flight. For that reason much care and effort is spent in UK certification test programmes to ensure that *two separate and distinct actions* are needed by the pilot to enter the reverse range.

Turbines are hot and operate at high RPM. Because the entire rotating assembly is activated in the starting cycle, a battery which is low, or a weak electrical power source can cause thermal damage, or even lead to replacement of the hot section of the engine.

Typical of the lengths to which engine manufacturers go to protect their engines from abuse is the fixed-shaft Garrett TPE 331 turboprop. This is fitted with an Integrated Engine Computer (IEC) to provide auto-start, torque and temperature limiting, data logging, trend monitoring and fault detection/isolation. The IEC simplifies trouble-shooting, which is said to reduce time and cost of repair.

Example 14–1: Agricultural Aeroplanes

Agricultural aeroplanes, although relatively simple, are a special case. Airframe designs

Plate 14–1 Agricultural aeroplanes, especially those like the former
NAC 6 Fieldmaster modified for fire-fighting, need special tests.
(Courtesy of Desmond Norman)

tend to be old and latest variants of many types in use world-wide have been re-engined
from piston to turboprop, and fitted with dual controls for training. Turboprops have brought
a number of problems. The increase in power is often destabilizing. Yet the lighter and
smaller scantlings of a turboprop engine allows the hopper to be enlarged. With AG-engine
modifications they can run on either jet A1 or diesel fuel. They are capable of carrying loads
around half their gross weights. With fuel in the hopper a single-engined ag-aeroplane can
have a 2500 nm ferry range. Unladen rates of climb of 3000 ft/min reduce to around 1000 ft/
min with full load. For example, the Fletcher FU24 (Cresco 600, re-engined and built under
licence in New Zealand), has a crew of two in front of the hopper, which can be replaced
with a module for passenger/utility roles. It is then said to be capable of carrying up to ten
passengers, or acting as a stretcher-bearing ambulance. The hopper can dump up to 1800
litres of water or fire-retardant chemicals in 3 s.

Succcssful and safe operation over such a number of roles places heavy demands upon
the pilot and upon controllability of aircraft and engine. Pilot fatigue is a constant hazard.
The pilot must have everything to hand, because AG-operations involve similar integration
between man and machine as in a fighter. Split-second control of reverse pitch, either to
hold the now feather-light aircraft down and stop, with half of its take-off weight gone – or
to go around again – takes skill and highlights conflict between the opposing requirements
of both Authority and operator.

The requirement for two separate and distinct actions by an AG-pilot to bring the power-
lever back through the *beta-range* (propeller-pitch altered by movement of the power-
lever), and then clear the flight-idle and ground-idle stops so as to enter the reverse-pitch
range, might be too much for some at times. The Fletcher is said to have got around the

problem by the power-lever being rigged for a wide arc in the beta-range. This makes it less easy for the pilot to inadvertently get into reverse.

Trouble-Shooting with Piston-Propeller Engines

Two aspects of performance which give cause for concern as aeroplanes grow older are shortfall in rate of climb and in level speed – rate of climb most of all, because it is a sensitive indicator of malaise in numerous areas. In Table 7–2 we saw that the dominant source of poor rate of climb lies with the powerplant, which accounted for 11 out of the 30 causes listed.

Table 14–1 is useful for trouble-shooting with piston engines (see Ref. 14.3). The way to use it is to note the symptoms, which may be one or more, and find their combination in the right-hand columns. On the left-hand side is the probable cause of defective running.

It is the reduction in excess of power output available (translated into insufficient thrust, too much drag, or both) which causes losses of climb and level speed performance, range and endurance, and agility, as we saw in Chapter 7.

Two-stroke engines

Microlight aircraft rely upon two-stroke engines, which are light and easily obtainable power units, used for chain-saws, outboard motorboats, snowmobiles and much else. From the point of view of the test pilot they are contrary. They are not as flexible in operation as four-stokes, either liking to run flat-out and objecting to idling, or idling happily while refusing to run comfortably flat-out. Because a two-stroke fires every revolution and runs at high speed, between 5000 and 9500 RPM, they are noisy. Silencing, when adequate, is often elaborate and heavy. Many two-strokes are mounted as pushers, making them prone to CG problems which can be solved at the expense of some contortion of pipes and momentum losses. Being high-revving, noise is a particular problem to be addressed during flight testing.

Note: If fatigue failures are to be avoided, care is needed in the vibration mounting and matching of two-stroke engines and exhausts. An exhaust leak will not only make an engine run roughly and lose power, it can also stop it by disrupting the back-pressure. While no cause was discovered at the time, the forced landing which ended the flight in Example 9–13, began when the two-stroke engine stopped for no accountable reason at V_{NE}.

Modern rotary combustion (Wankel) engines

The Wankel engine is a modern development of an idea which goes back as far as a water pump in 1588. It is a compact unit, with a triangular rotor, fitted with seals at each corner, running inside a figure-of-eight trochoidal chamber. In spite of lightness (roughly two-thirds that of a four-stroke of equivalent power), the ability to run on different fuels, and potentially smooth and quiet running, development has been sluggish because of high temperatures, centrifugal forces, oscillation of the rotor and vibration, all of which cause excessive wear on the seals.

Engines run at speeds around 7000 RPM. Propeller gearing is needed to reduce tip-speed and noise. Flight tests must pay particular attention to noise requirements.

Exhausts and apertures

Exhaust gases are dangerous because they contain carbon monoxide, CO. While the gas by

TABLE 14–1

(Copy of original Ref. 14.3)

82. AERO ENGINE FAULT FINDING TABLE

1075

[TO FOLLOW END OF CHAPTER XVI.]

Probable Cause of Defective Running.		The engine will not start.	The engine stalls.	Oil Pressure Low	Oil Pressure High	Boost Pressure Low	Boost Pressure High	Rough running	Loss of power	Misfiring	Popping-back	Black exhaust smoke	Pre-ignition	Overheating	Abnormal noises
Defective starting apparatus	Starter defects	●	—	—	—	—	—	—	—	—	—	—	—	—	—
Distributor wrongly timed															
IGNITION DEFECTS															
Spark plugs oiled up	No spark at plugs	◐	●	—	—	—	—	●	●	●	—	—	—	—	—
Electrodes incorrectly set															
High tension wires broken or detached															
Defective insulation															
Magneto requires adjustment	Intermittent spark	◐	◐	—	—	—	—	●	●	●	—	—	●	—	●
To spark plugs	Wires wrongly connected.	◐	◐	—	—	—	—	●	—	●	●	●	—	—	—
On magneto															
Defective magneto	No current from magneto.	◐	—	—	—	—	—	—	—	—	—	—	—	—	—
Defective switch or in " off " position															
Faulty insulation on earth wire															
Too much advance	Magnetos incorrectly timed.	●	◐	—	—	—	—	●	●	●	●	●	●	—	—
Insufficient advance															
CARBURATION DEFECTS															
Tanks empty	No supply of fuel to carburettor.	◐	◐	—	—	—	—	—	—	—	—	—	—	—	—
Tank air vents stopped up															
Cocks in " off " position															
Leaks in circulation system															
Obstruction in circulation system															
Defective fuel pump															
Defective fuel pump drive															
Foreign matter in fuel	Intermittent fuel supply.	—	◐	—	—	—	—	—	—	●	—	—	—	●	—
Float needle valves sticking, due to foreign matter															
Defective pipe joints															
Piping system or filters dirty or obstructed															
Altitude control open															
Pressure insufficient	Insufficient fuel supply	—	◐	—	—	—	—	●	◐	●	—	—	●	●	●
Leaks in the circulation system															
Carburettor tuned too weak															
Carburettor tuned too rich															
Excessive pressure	Excessive fuel supply	—	◐	—	—	—	—	●	●	—	—	●	—	●	—
Air intake obstructed															
Excessive priming		◐	—	—	—	—	—	●	●	—	●	●	●	●	—
Unsuitable fuel		◐	◐	—	—	—	—	●	●	●	●	●	●	●	—
Water or dirt in fuel		◐	◐	—	—	—	—	●	●	●	●	—	●	●	—
Faulty or broken pressure pipes	Automatic	—	—	—	—	●	—	◐	●	●	—	—	—	●	—
Control incorrectly adjusted	Throttle control	—	◐	—	—	—	●	◐	●	◐	—	—	—	●	—
LUBRICATION SYSTEM DEFECTS															
Tanks empty	Insufficient oil	—	—	●	—	—	—	—	●	—	—	—	●	●	●
Cocks closed															
Oil leaks															
Abnormal clearances															
Incorrect setting of oil relief valve															
Dirty filters, or obstructed system															
Oil too viscous															
Badly fitting piston rings	Excessive oil	—	—	—	◐	—	—	—	●	—	—	—	—	—	—
Incorrect setting of oil relief valve															
Insufficient time to warm up	Oil too cold	—	—	—	●	—	—	—	—	—	—	—	—	—	—
MECHANICAL DEFECTS															
Spark plugs or non-return valves loose	Loss of Compression	●	●	—	—	—	—	●	●	●	—	—	—	◐	●
Piston rings stuck, broken or badly scored															
Bad scores in cylinder bore															
Valves not seating properly		●	◐	—	—	—	—	●	●	●	●	—	—	●	●
Valves distorted or stuck in guides															
Valve springs broken or too weak															
Excessive clearances		—	—	—	—	—	—	◐	—	—	—	—	—	●	◐
Airscrew hub loose															
Engine loose on mounting		—	—	—	—	—	—	◐	—	—	—	—	—	—	●
Airscrew out of track or loose on hub															
COOLING SYSTEM DEFECTS															
Insufficient coolant	Insufficient cooling	—	—	—	—	—	—	—	●	—	—	—	●	●	—
Air locks in system															
Pump not circulating coolant															
Radiator not wound out sufficiently or shutters not opened.															
Insufficient time to warm up	Engine too cold	—	—	—	—	—	—	●	—	—	●	—	—	—	—
Radiator out or shutters open when warming up															

itself is undetectable by the average nose and palate, its presence can be deduced from the smell of accompanying exhaust products. Quite small concentrations may cause headaches and can make one feel sick. Larger concentrations kill insidiously.

Exhaust fumes in the cockpit are dangerous. There is usually a lower pressure there than elsewhere so that gases are sucked in and are hard to blow out. The slower the build-up the harder contamination is to detect.

Example 14–2: Carbon Monoxide Contamination Through a Rivet Hole

There was one particular single-engined aeroplane which suffered from a slow build-up of CO in the cockpit. Eventually its path was traced from the exhaust stack beneath the nose cowling, down the outside of the fuselage, to entry through a hole left by a missing rivet in the tail-cone. Static pressure was higher at the tail-end of the fuselage than in the cabin, causing the gas to be sucked slowly forward inside the fuselage.

Many light and small aeroplanes are modified for aerial survey, photography, and despatch of anything from supplies to parachutists. This often involves: the cutting of ports and other apertures (in the belly), flying with a door removed, and the fitting of deflectors to the door-frame which make it easier to clear the aircraft in flight without snagging the parachute harness or damaging oneself.

Because modern single-engined light aircraft with flat four and six engines have short exhaust stubs, ejecting on the underside of the cowling in the vicinity of the firewall, inevitably their ends vent upstream of belly apertures and cabin doors. It is vitally important to check for the presence of CO contamination, no matter how well sealed a hole appears to be, or how far away from the exhaust plume a cabin door is located.

FAR and JAR 23.1121(a) require nothing more explicit than *safe disposal of exhaust gases*, without CO contamination or fire hazard, by each separate exhaust system. BCAR Section K-3, 5.2.1, on the other hand, defines a safe level by stating that concentration *shall not exceed one part in 20 000 parts of air in any flight or ground condition or aeroplane configuration which is likely to be maintained for more than 5 minutes.*

A deflector plate or pad fitted to the frame of an open door may cause no handling difficulties but, being aerodynamic in shape, it may well generate an upright vortex, with suction enough in its core to draw exhaust products from beneath the belly of the aircraft, upwards and into the cabin.

Camera ports are usually open, with no transparent cover and with the camera mounted inside the aircraft, a few inches or centimetres clear of instead of level with the skin, to afford protection from oil, insects and grass. The camera mounting is often made locally from bent plate. Inevitably there are small gaps at the joints and where it fits the skin. These need careful sealing. What tends to be forgotten is that a camera is removable and an equally well sealed blanking-plate must be fitted in its place because, by 'Sod's Law', the aeroplane will next fly without it.

Checks for contamination

Checks are carried out using a standard gas test equipment. Samples are taken when taxying, up, down and crosswind, before and after flight; during the climb and level cruise; in a clean configuration in high speed flight; at low speed with gear and flaps down for landing. It is not an exercise to be rushed. Take thirty minutes at least, because CO can be very slow to build up to an unacceptable level.

Old and Historic Engines

Ideally, replicas and rebuilds of historic aeroplanes should be fitted with their proper engines, where these still exist. Early engine types were air-cooled rotary, and water-cooled in-line. Later, with improving technology, came the classic middle range of air-cooled fixed engines, both radial and in-line, with liquid cooled for higher power, until the advent of the big air-cooled radial engines of World War II. Like eggs, all need great care in handling.

Never assume that because a particular type of engine had a history of sound and solid service that it will continue where it left off. There were different limits to the materials then in use, which lacked the benefit of modern methods and technology. There is always the possibility of fatigue failure of parts which cannot be replaced. This may mean cannibalizing other engines or adapting parts from elsewhere. If a part has to made it can be difficult to find the equivalent material, with the result that there is risk of incompatibility when ancient and modern parts are set to work together in the same assembly. This may result in unexpectedly different kinds of wear, and patterns of failure not previously encountered.

Plate 14–2 Albatros DVa replica of Example 14–4. (Origin unknown, author's collection)

Fatigue and lack of adequate reserve strength is a continuing problem. A modern replica of an SE5A, constructed to Royal Aircraft Factory drawings and fitted with an original, old, in-line engine, is banned from spinning because of risk to the engine structure from stresses imposed by autorotation of the aeroplane. It is also limited to low positive *g* manoeuvres only, to preserve the engine in working order into the forseeable future.

World War I rotaries

The Sopwith Camel in Example 10–4 had a Clerget rotary engine. Rotaries were unique, and no other engine then equalled their weight/power ratio, which was around 3 to 1 (Ref. 14.4). Figure 14.3 shows where they fit now within the spectrum of piston engine technology. Being heavy by modern standards, here is the source of CG problems with full-scale replica aeroplanes fitted with lighter, fixed, air-cooled radial engines (see Example 10–5).

Rotary engines were limited to an output around 230 BHP (172 kW), because of high rotational stresses, they had high fuel and (castor) oil consumption, were costly to maintain (15 hours between overhauls), and had low cylinder mean effective pressure (MEP). They had no exhaust pipes and discharged to atmosphere through the valve port, past the valve stem, spring and rocker arm which, running in hot gases, contributed to frequent failures.

The engine cylinders rotated on a fixed shaft to which the pistons were fixed. The shaft was fastened into the airframe. The propeller was bolted on to the crankcase, to which the cylinders were locked by various methods.

The crankcase was the chamber in which fuel and air were mixed, before being fed through the rear end of the hollow crankshaft and into the cylinders by induction pipe or port, depending upon engine type. Lubrication was by means of castor oil because this did not mix with petrol.

Note: A modern thought. Flying behind a tractor rotary leaves the pilot smothered in a mist of castor oil. The illness, consumption, was regarded in World War I as Flying Sickness D (from which Victor Yeates died after writing *Winged Victory*). In the light of what we now know, might it have been instead a then undiagnosed form of lung cancer, caused by breathing a burnt mixture of carcinogens in a cocktail of exhaust products and castor oil?

Power of a rotary is controlled by means of a separate throttle lever, governing the air supply, and a smaller, fine adjustment fuel lever working on a concentric shaft in the same quadrant, corrected the fuel-air mixture. Ignition timing is fixed. One constantly fine-tunes the fuel lever with changing altitude, to avoid sluggishness in operation, reduction in power and sooting of plugs.

When flying the Camel my briefing was to correct the fine-tuning of the Clerget after climbing through about 200 ft (some 60 m) after take-off, to avoid the risk of stalling the engine. To cut the engine the aeroplane had a 'blip-switch' on the stick, operated by the right hand, because the air and fuel levers, which were adjusted almost continuously by the fingers of the left hand, left none free for much else. If the levers were not adjusted, then fuel and oil consumption became prohibitively high, reducing endurance.

RPM were low. Maximum for the Camel were in the region of 1100 to 1200. But the Gnome-powered Pup cruised sweetly on only about 600, with correspondingly low noise.

Advice on handling a rotary engine, which is every bit as valuable for modern pilots flying such aircraft as it was then, is copied in Table 14–2 from data believed to be from Ref.14.5. It is thought to refer to a BR1-engined Camel, of the Royal Naval Air Service (RNAS).

The rotary is best remembered for its inertia, causing gyroscopic precession. The RH tractor rotary causes a Camel to keep its nose up in a steep turn to port, and needed some left

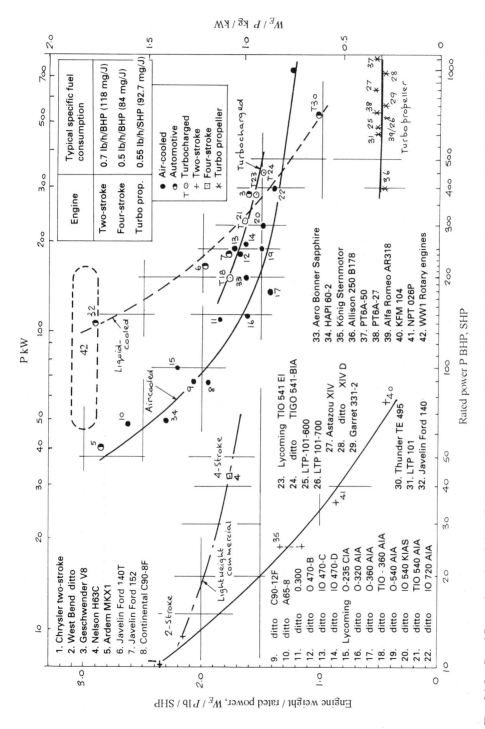

Fig. 14.3 Specific equipped weights of piston and turbopropeller engines with World War I rotary engine data added (taken from various published sources and author's estimates).

TABLE 14–2

Advice on flying a rotary
Naval Squadron No. 8
13th November 1917

DON'T open up STRAIGHT AWAY, it does not give the oil a chance to circulate, and ruins the
      obdurators.
DON'T run your engine TOO LONG on the ground. It is only necessary to open fully for a FEW
      SECONDS.
DON'T forget to test BOTH magnetos when running on the ground, and occasionally in the air.
DON'T exceed 1250 revolutions at any time. It causes the ballraces to "creep", and other unpleas-
      ant things.
DON'T allow your engine to 'pop' and 'bang'. It is caused by TOO MUCH petrol and damages
      the valves, in addition to overheating.
DON'T "blip" except when throttled right down. It is extremely bad flying and puts unnecessary
      strain on the whole machine.
DON'T switch off at any time in the air or the plugs will oil up.
DON'T close the throttle when the petrol is turned off. Allow a cool draft to blow right through the
      engine by keeping it wide open.
DON'T miss the chance to let your engine cool down, by a short glide after a long stiff climb.
DON'T always set the control levers by the figures on the bracket, but by the SOUND of the
      engine. That ever-changing density of the air requires an ever-changing mixture.
DON'T forget there will be no pressure in the tank after a long glide with the engine off.
DON'T condemn an engine immediately you are "let down".
DON'T be too ready to blame your mechanics.
DON'T forget that SYMPATHY and thorough knowledge of all "work", especially "carburation"
      is very important.
DON'T forget the oil pulsator.

---

rudder to hold it down. A steep turn to starboard needed full left (top-)rudder, because this time gyroscopic precession caused the nose to fall.

When in service stunting in Camels was banned during flights over England, after questions were asked in the House of Commons about the number of accidents during what we now call aerobatics. For example, when performing a simple loop, pulling back on the stick and applying power precesses the nose to starboard and, when erect, the right way up, this is seen from above as the aeroplane turning to the right. But as the aeroplane passes through the vertical, still rotating to starboard, an eye-witness above would begin to see the underside and the nose now rotating to the left – which is still to starboard from the Camel's point of view.

To the pilot of the Camel, head craned back, looking for the horizon coming into view, there is the surprise of failing to see any expected landmark. The aeroplane, which is now inverted and probably slow with insufficient rudder for the manoeuvre, is by then wandering away in another direction. Kick in more rudder too late in an effort to correct matters, and the observation of Hill's, quoted from Ref. 10.12 in Example 10–3, ceases to be just words.

Water-cooled engines

Water-cooled engines are always heavier than air-cooled, because of the water jacket, radiator pipes and pump – and water. Beyond the demands of good airmanship and engine handling, the main point to be watched is control of temperature.

Example 14–3: Replica – SE5A

In-house notes on flying the replica of an SE5A in the Shuttleworth Collection contain the following advice:

'COOLANT
60° for opening up (full open shutters when reach). Aim for 80°–90°. Must keep above 60° in flight. Normally runs 85° in flight shutters full open, but NEED watch shutter position at all times, especially on landing or glides.
    Coolant overflow/boil off is starboard pipe on upper plane, fuel is port – They are two good reasons (!) for wearing goggles even when taxying, let alone if you are brave enough to apply neg "*g*" ....

'CLIMB
60 mph ideally, but up IAS if coolant (*temperature*) starts to creep up ....

'AEROBATICS
Avoid –ve *g* to preserve engine and to avoid mush (*mouth, face*) full of fuel and gunk from cockpit fuel pipes and from upper plane overflows ....'

Difficulties arise with rebuilds and replicas of old aircraft when there are no original engines available. In the period between the wars the upright, liquid-cooled, in-line engine was replaced by the lighter, air-cooled, inverted in-line engine. When building with the intention of flying a full-scale replica of, say, an Albatros DVA, or Fokker D VII, both of which had upright water-cooled engines, one is forced to look for an in-line engine of a later date, which not only looks 'right' when mounted the other way up, but runs properly when mounted upside-down.

Example 14–4: Replica – Albatros DVa

Various problems were encountered with a full-scale flying replica which formed part of a museum collection. Unlike the original Mercedes DIIIa engine, the replica was fitted with a Ranger. The normally inverted Ranger engine, mounted cylinders upwards as a 'look-alike', was much lighter than the original, putting the CG further aft. But the Ranger did not like running the wrong way up. It was designed with the ignition system out of the way, above the engine, and now it was at the bottom. Oil, running downwards under gravity, contaminated the ignition, causing rough and erratic running.

Also there were doubts about comparable and total accuracy of drawings from German, Australian and UK sources. Most showed the wheel axle lying in the same vertical plane as the leading-edge of the much smaller bottom wing. Yet, when the replica was measured to find the actual CG it was discovered that the axle was 3.25 in (8.26 cm) ahead of the lower plane leading-edge. This finding cast doubts upon the drawings, the build-standard, or both.

There was uncertainty about the wing section, which resembled the thin and highly undercambered USA 5 more than, say, RAF 15. If correct, then the Albatros would, like the USA 5, have a CP-coefficient around 50% of the chord – and also the calculated equivalent plane chord. But was such a guess reliable enough to approve test flying, in view of the unsorted other problems? It was decided not to start until better answers, or firmer confirmation of those to hand already, had been obtained.

With the axle 3.25 in forward of where the drawings showed it to be. Ballasting to put the CG near the guessed 50% equivalent plane chord risked ground-looping, as reported with the Sopwith Triplane replica. Weight and balance calculations pointed to a need to move the CG forward, to nearer 35% of the equivalent plane chord, and this involved the addition of

Plate 14–3   'Big Iron' the Chance-Vought F4U-4 of The Old Flying Machine Company, with a 'civilianized' version of a military Pratt and Whitney R-2800 two-row radial engine. To all intents and purposes an unadulterated aeroplane – except in the eyes of those tasked with clearing it for operation on the Civil Register, around the air-show circuits. (Courtesy of John Rigby via The Old Flying Machine Company)

at least 178 lb (81 kg) ballast fastened to the engine bearer/cowl former, 41 in (104 cm) ahead of the lower-plane leading-edge. A recommendation was made to re-stress the structure, because the ballast was equivalent to hanging the weight of a man beneath the engine of an aeroplane weighing slightly less than 2000 lb (901 kg).

Following test flights longitudinal stability was said to be improved, with no reports of ground-looping.

Example 14–5: Fuel Starvation – Westland Lysander Rebuild/Replica

The Lysander in Example 10–3 suffered around nine forced and emergency landings because of fuel starvation. The Bristol Mercury III engine always had a reputation for being tricky, care being needed to re-open the throttle slowly after closing it.

Starting was not a problem. But, even though the throttle was set for the cruise and then left alone, the engine popped and banged and occasionally ran down – after about 45 minutes on two, quite unrelated, occasions. On four other flights the engine ran down after shorter, equally unrelated, intervals. On the last 45 minute occasion (in spite of the plugs, carburetter and fuel lines having been so thoroughly serviced and cleaned that the engineers swore the problem could not recur) the engine misfired and ran down, leaving nothing for it but to land up-hill, with a tailwind. The flight ended with the Lysander on its back in a bone-dry newly harrowed field – the only one without burning stubble – its spats filled with earth.

The fuel tank was behind the pilot's shoulders. The original fuel filler and overflow system could not be reproduced. Instead, to save money the tank was fitted with a fuel filler-cap cannibalized from a light aeroplane. If I remember correctly this had three vent holes which were meant to align with a kidney-shaped hole in the gasket fitted inside. What we did not know was that the filler-cap and gasket turned separately each time the cap was replaced. This meant that after several refuellings the vent holes became misaligned to almost any degree until one, two, or three holes were covered by the gasket. Without adequate venting suction inside the fuel tank starved the engine at different rates and intervals.

Postscript: The Lysander was repaired and returned to flying condition. Overturning during the forced landing, because the spats turned into dredger-buckets, revealed why they – or their side-panels at least – were removed during the war when operating from sand, snow and soft ground.

This is an appropriate point to leave the subject of flying qualities and flight testing, having questioned drawings, examined consequences, and heard something of what aeroplanes tell us about themselves. The main thing is to have the wit to listen, record it and bring back an accurate account – together with conclusions and recommendations.

Just because something appears in print after being sifted, checked and recounted with care, does not mean that it is right in all places and at all times for everyone. Quality of information, science, technologies, philosophies, policies and attitudes based upon events and experience move onwards and change. Flight testing draws from many disciplines and the experience and advice of many different people, some of whom are no longer around. Keep your critical faculties sharp. Never fly until you are satisfied with the answers you have been given. There are many so-called experts around, willing to baffle brains with the intestinal products of bulls – as long as *they* take no responsibility for the consequences.

What is here is written in good faith from the most reliable sources, past and present, that I found at the time. Those are the ones which, to my ear and eye and gut, contain the soundest ring of truth. The way in which I saw things and have reported them may not suit every reader, but that is how they were and I can only tell it as it was. I have not said anywhere that this or that is the best way of doing something. Whether or not what is done works is a matter of technique, experience, judgement – and above all good airmanship. A technique which works for one person might not work for another.

Remember what was said in the Preface, and:

WATCH IT, LIKE THE REST OF US YOU ARE MORTAL, AND JUST AS PRONE TO MAKING MISTAKES AS I AM.

While I can accept no responsibility for changes in sources, techniques and policies, mistakes and misquotations in the text are mine. Please tell me, through my publishers, if you discover anything needing correction.

References and Bibliography

14.1 Hooker, Sir S., an autobiography, *Not Much of an Engineer*. Shrewsbury: Airlife Publishing, 1984

14.2 CAP 531, *BCAR 23 Light Aeroplanes, ACB 23.933(c)*. London: Civil Aviation Authority, December 1987

14.3 Air Publication 1081, *Royal Air Force Pocket Book*. London: Air Ministry, 1937

14.4 Macmillan, N., Rotaries remembered, *Shell Aviation News*, No. 312, 1964

14.5 Sykes, R., Flying the Camel in 1918, *Flight International*, 2 May 1968

APPENDIX

Aero-Marine Design and Flying Qualities of Floatplanes and Flying-Boats

This paper was originally given by **Dr** *Darrol Stinton as a lecture on 8 October 1986 at The Royal Institution of Naval Architects. Subsequently it was published in* The Aeronautical Journal of the Royal Aeronautical Society, *March 1987.*

Prologue

Quite by chance after this lecture had been written I saw at first hand part of the fire-fighting operation in the Tanneron in the South of France using flying-boats. This provided an excellent point from which to begin.

On 23 August 1986 a vast fire was started deliberately, it was alleged, in the valley of the River Siagne, between Grasse and Mandelieu. There was a strong mistral blowing at the time. It was brought under control two days later with the aid of a small force of around nine 'Pelicans': Canadair CL-215 flying-boats (Plate A–1). Each aeroplane can scoop up over 5000 litres (more than five tonnes) of water, in 10 seconds while planing at 70 knots (Plate A–2). At times three aircraft would land in loose formation to do so. In the end the fire, said to be the worst since 1706, was reported to have destroyed 10 000 hectares (say 25 000 acres) of farm, forest and homes. Four people died, 700 suffered from burns and asphyxia, 8000 were left homeless.

Such fires seriously disturb the eco-system for years. Topsoil washes away. Regeneration of a forest is delayed for decades. Large areas become semi-desert, covered with scrub.

The operation was fast. In my opinion no other class of aeroplane could have done the job so effectively in the *block* time available. Landplanes take longer to turn around. But airframes and engines of the flying-boats are wearing out and the French Government is searching for replacements.

Introduction

This lecture on aspects of seaplane design, operation and flying expresses my own views as an aero-marine consultant and as a qualified test pilot, and not those of the Civil Aviation Authority, or of Her Majesty's Government. It falls naturally into two parts. The first is concerned with hydrodynamic and aerodynamic features needed to operate in a marine environment. The second concentrates upon seaplane flying qualities when operating from water.

Plate A–1   Canadair CL-215 firebombing. Note pitch-up as load goes.
One of the aircraft is said to have made 225 drops in one day, totalling
1350 tonnes water. (Canadair)

Seaplane is a generic term for both a floatplane or a flying-boat, with or without wheels.
It is an aeroplane which can operate from water, and is not simply a displacement craft that
can fly.

Up to the Second World War the seaplane, in the form of the long range flying-boat, was
in a more advanced state of development and opened up more air routes for passengers
and cargo than the land plane. The Schneider Trophy was won outright by Great Britain
with the Supermarine S6B floatplane. This gave us the speed record in 1931; the Merlin
engine, the Spitfire, the Hurricane, high speed aerodynamics and control surface design,
advanced engine gearbox and cooling design, and useful steps forward in structures and
aviation medicine.

It was the need for round-the-clock long range artillery over hundreds, not just tens of
miles which later brought in the heavy land-based bomber to displace the big seaplane. The
attendant substantial runways and pans, maintenance facilities and factories led to a plethora

Plate A–2 Canadair CL-215 in planing attitude for picking up water.
(Canadair)

of main and dispersal airfields; to the big landplane manufacturers, like Boeing; and to the many, extended, ready-made airports post war.

The result is that someone talking seriously about this neglected species must tread somewhat warily, so as not to arouse out-of-date notions, predilections and prejudices in an audience.

The hardest question to answer is 'well, it *looks* alright – but what can you *do* with it?' Here, the Prologue has given one example. But, to provide a more extensive answer the point must be made that one cannot compare seaplanes with 'equivalent' landplanes. There are none. It is like comparing a Range Rover with, say, a saloon car of similar capacity. One has to have a reason for buying the Range Rover: because it has features and qualities that the roadable saloon has not.

Operational Qualities

The most substantial argument advanced in favour of seaplanes is that the surface of the globe is $\frac{3}{4}$ covered with water (Fig. A1) which is virtually limitless for take-off and landing and therefore, size. The biggest argument against them is that this vast amount of water is hardly useable. Even so, useable water is more extensive than all of the aerodromes put together. While a water base favours the construction of very big and heavy aeroplanes – far bigger and heavier than can be catered for on land – sea-states are an important factor (Table A1, Fig. A2 and Plate A–3) (Refs 1, 2, 21).

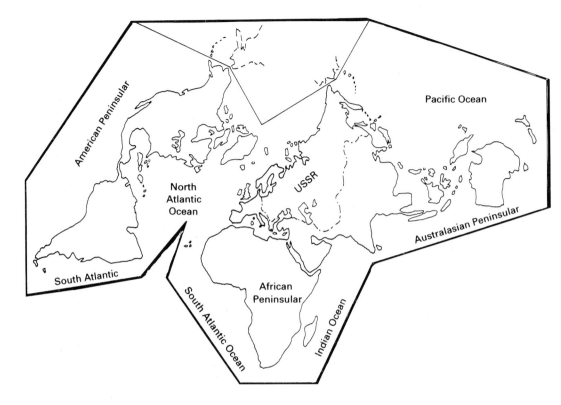

Fig. A1   Distribution of land masses showing Eurasian 'heartland' with radial peninsulars. Less than one third is land.

The following tentative specification from a French source for a water-bomber lends point:

Specification for water-bomber

- To carry 6 to 8 tonnes water
- Two turbopropeller engines
- Two man crew
- Rate of climb 1000 to 1500 ft/min
- To pull 3 g applied
- Endurance $3\frac{1}{2}$ hours with 45 min reserves
- Range speed 180 knots
- Full IFR nav/comm equipment
- To be amphibious
- Light (power assisted) ailerons
- Retractable tip floats
- The ability to stop quickly (free turbine engines)
- Airbrakes
- Materials resistant to salt water corrosion
- The ability to operate in waves of 2 metres from trough to crest

TABLE A1

Beaufort wind scale and wave height relevant to seaplane operations (from Ref. 1)

Beaufort Number	Descriptive term	Mean velocity in knots	Specifications		Probable wave height* in metres	
			Sea	Coast		
0	Calm	Less than 1	Sea like a mirror	Calm	—	(—)
1	Light air	1–3	Ripples with the appearance of scales are formed, but without foam crests	Fishing smack just has steerage way	0.1	(0.1)
2	Light breeze	4–6	Small wavelets, still short but more pronounced; crests have a glassy appearance and do not break	Wind fills the sails of smacks which then travel at about 1–2 knots	0.2	(0.3)
3	Gentle breeze	7–10	Large wavelets. Crests begin to break. Foam of glassy appearance. Perhaps scattered white horses.	Smacks begin to careen and travel about 3–4 knots	0.6	(1)
4	Moderate breeze	11–16	Small waves, becoming longer; fairly frequent white horses	Good working breeze, smacks carry all canvas with good list	1	(1.5)
5	Fresh breeze	17–21	Moderate waves, taking a more pronounced long form; many white horses are formed. (Chance of some spray.)	Smacks shorten sail	2	(2.5)
			(limit of Fig. A2b)			
6	Strong breeze	22–27	Large waves begin to form; the white foam crests are more extensive everywhere. (Probably some spray.)	Smacks have double reef in mailsail; care required when fishing	3	(4)
7	Near gale	28–33	Sea heaps up, and white foam from breaking waves begins to be blown in streaks along the direction of the wind	Smacks remain in harbour and those at sea lie-to	4	(5.5)

* Significant wave height is defined as the average value of the height of the largest one-third of the waves present.

Note: Use the lower code figure if the observed wave height is shown on two lines of the table, e.g. a height of 4 metres is coded as 5.

Table A1(a)

Sea state code (Ref. 1)

Sea state code		
Code figure	Description of sea	Significant wave height* (metres)
0	Calm (glassy)	0
1	Calm (rippled)	0–0.1
2	Smooth (wavelets)	0.1–0.5
3	Slight	0.5–1.25
4	Moderate	1.25–2.5
5	Rough	2.5–4
6	Very rough	4–6
7	High	6–9
8	Very high	9–14
9	Phenomenal	Over 14

* Significant wave height is defined as the average value of the height of the largest one-third of the waves present. *Note:* Use the lower code figure if the observed wave height is shown on two lines of the table, e.g. a height of 4 metres is coded as 5.

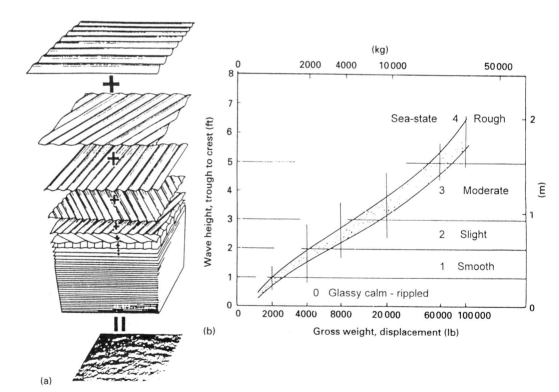

Fig. A2 (a) The sum of simple sine waves makes a sea (from Ref. 2); (b) Limiting wave heights for hulls of different displacements (after Ref. 3).

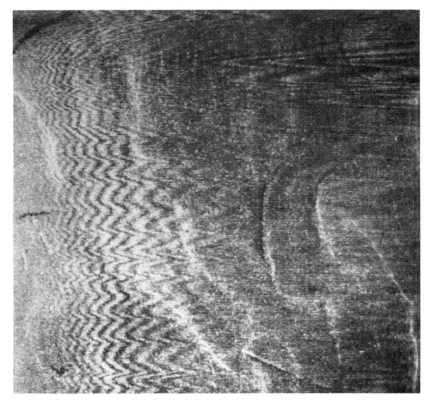

Plate A–3   Wave spectra seen on radar. (Intradan)

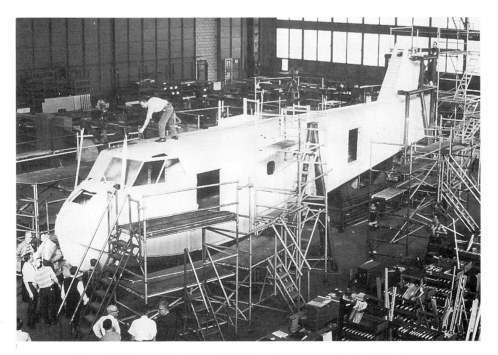

Plate A–4 Conventional metal hull under construction. (*Flight International*)

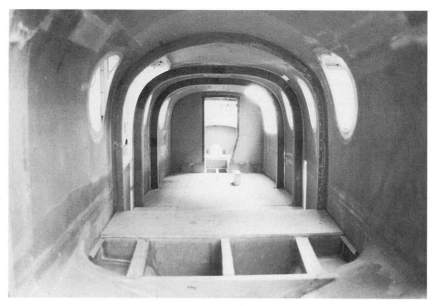

Plate A–5   GRP construction of hull (top) and wing (bottom) of
Dornier Seamaster. (Dornier)

The last named is critical. Figure A2b shows the aircraft would have to weigh around
100 000 lb (45 000 kg), which means it would need more than two turbopropeller engines
for a start – and it could lift many more than 6 to 8 tonnes of water. Quite huge dock and
harbour facilities would be needed for seaplanes, in the form of floating platforms, with
their infrastructure of supply and communications. Fortunately, the logistics are eased
by modern technology; by the positive public demand for airports to be out of sight
and sound of people; and by the heavy engineering now available from the oil-related
industries.

There are a number of strategic factors which make water-based aircraft a reasonable subject for military interest.

☐ Much heavier loads can be moved about the world off water than land. Thus, seaplanes can be more productive than landplanes where productivity is measured in terms of weight carried and block speed, i.e.

$$\text{Productivity} = \text{weight carried} \times \text{block speed}$$
$$= \text{ton nautical miles per hour} \tag{A1}$$

☐ The geo-strategic problem for the West is that it consists of an alliance of nations which are peninsular-hoppers. This is in the sense that the central Euro-Asian landmass (which includes Mackinder's Heartland) is defined by the watersheds of the main rivers and is occupied mainly by their opponent, the USSR. The Heartland has radiating from it the Americas, the African and South-East Asian/Australasian peninsulars of continents and islands, separated, as we have seen, by wide oceans.

☐ The fact of such peninsulars and long established ports – which grew into major capital cities – assisted the development of airlinks world-wide before World War II, using water-based aircraft.

In recent history it takes little wit to realise that military operations in the Gulf, or in the Falklands, might have been materially assisted by the availability of one or two big flying-boats with bow loading.

Big flying-boats existed and were operationally successful as long ago as the latter half of World War I (Fig. A3). Their technology, which took the direction of metal construction between the wars (because a big wood and fabric structure could soak up 600 lb water while afloat), was immediately applicable to landplanes (Fig. A4) which will be recognisable by any naval architect or aircraft designer. The high speed Schneider Trophy floatplanes were metal (Fig. A5, Plates A–6, A–19).

Today the seaplane is attractive in a number of areas: as a flying-ship, for global air transport and logistic supply. Recent papers by Dr Claudius Dornier, Professors Dennis Howe and John E. Allen (Refs 4, 5, 6) discuss this and point a way forward (Fig. A6). A fair comparison with the Dornier design is that of Japanese Shinmeiwa project for about 1200 passengers, around 1977 (Fig. A7). The aircraft would have had a supercritical wing spanning some 256 ft (78 m) with a hull of 300 ft (91.4 m), six turbofans and lift-augmentation by means of upper surface blowing.

A different approach is shown in Fig. A8 (Plate A–7) employing the principles of Dr-Ing A. M. Lippisch in the United States. Such machines or aerofoil boats – 'ekranoplan' in the Soviet Union – would have the potential to carry freight, 1000 passengers or more, cruising in surface (air cushion) effect at 200/300 knots at altitudes equal to one half wing span. At such low altitudes there is a dramatic drop in lift-induced drag, which increases range by half as much again. Projected figures for such machines suggest 50 ton-miles per gallon on 50% power at cruising speeds around 200 knots, depending upon the propulsion system. Very heavy and awkward loads, like generating gear, or a super-tanker propeller shaft, might be carried externally beneath (or within) the centre section, by straddling a pontoon or jetty for loading and unloading.

On a smaller scale seaplanes (especially amphibians) are useful for:

☐ *Surveillance:* large sea areas around our coasts contain Warsaw Pact shipping. In Indonesia and other countries there is the need to find and stop fast gun- and drug-runners. There is also the need to watch for tankers flushing out offshore – this is a common problem off Orkney and Shetland.

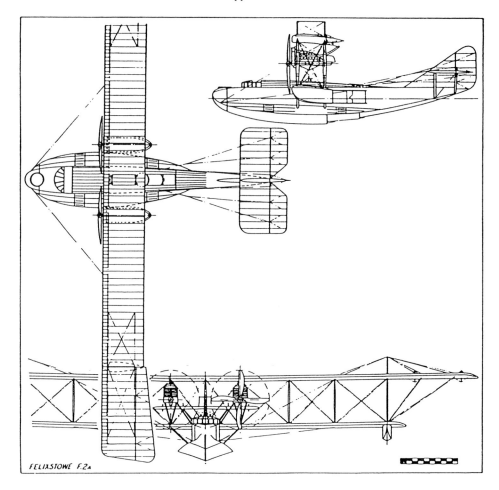

FELIXSTOWE F.2ₐ

Fig. A3 Felixstowe 'Large America' (1917) (Duval, G. R., *British Flying-Boats and Amphibians 1909–1952*, Putnam).

- □ *Air Sea Rescue:* On many occasions seaplanes are cheaper than helicopters – which windsurfers refuse because, unlike lifeboats, helicopter crews will not pick up the surfboard too.
- □ *Environmental Control:* Flight over large areas of coastline and river estuaries reveals contamination by effluents, which can only be seen from the air. Seaplanes can often land for sampling and further investigation.
- □ *Earth and Ocean Research.*
- □ *Passenger and Freighting* (Fig. A9, Plates A–8, A–9): getting to outback settlements in remote areas, like N Canada; as well as for tourists to Pacific and Caribbean Islands.
- □ *Water-bombing forest fires* (Fig. A10, Plates A–1, A–2); in this latter case quite large aeroplanes are needed, with three engines for safety and the ability to taxy after engine failure in narrow rivers, like the Rhone. Pilots are specialized and often ex-Navy (France).
- □ *Cleaning up of oil spills* (foaming and other chemicals being added to water uplifted while on board).

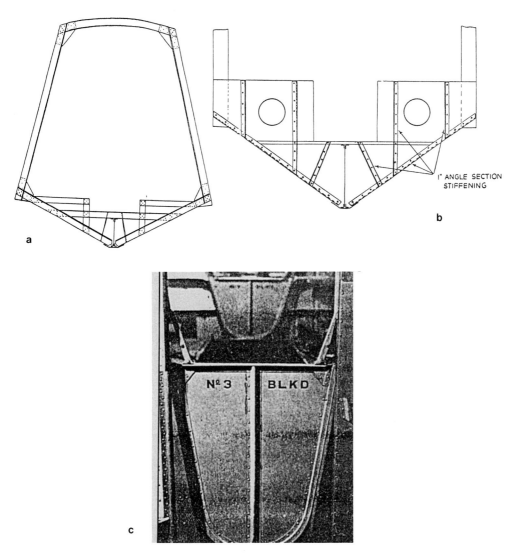

Fig. A4 Examples of metal seaplane construction, early 1930s: (a) How main frames are often made up. 'Zed' section was generally employed; (b) Vertical stiffening at main frames; (c) The swashplates in position, forming a watertight compartment.

For fire-fighting one must think in terms of carrying loads of water of 50% to 55% take-off weight, and scooping it up while planing. Vertical hoppers are needed to avoid big longitudinal trim changes when manoeuvring in flight.

To carry, say, 5000 lb water in a flying boat weighing 9500 lb, a hopper of about 80 cubic feet capacity would be required. A base cross section of 3 ft × 3 ft (say 1 metre square) needs a hopper height of 9 ft (about 3 metres). This would occupy the full depth of a hull, and more. Therefore, we are looking at quite a deep hull structure, which means weight and drag.

Size is an important point. Anything weighing less than about three tonnes (6000 lb, 3000 kg) tends to be a rich man's toy, in spite of being a five to six seater. The limiting sea

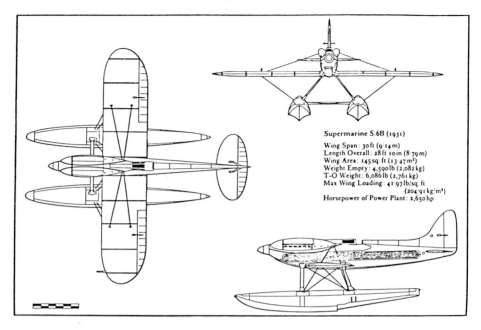

Fig. A5a   The Supermarine S6B was the outright winner of the Schneider Trophy and was hard to handle through propeller torque (Ref. 7).

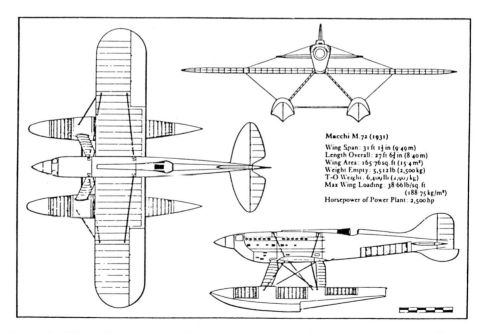

Fig. A5b   The Italian Macchi M72 had twin engines and counter-rotating propellers to eliminate torque effects (Ref. 7).

state for this weight (Fig. A2b and Table A1) is a wave crest to height around only 2 ft (0.6 m) – which corresponds with sea-state 3. Size costs money, so one has to have a good reason for building such a machine. Within this brief argument lies the reason why

Plate A–6 Replica Supermarine S5 the first finale British winner. (*Flight International*)

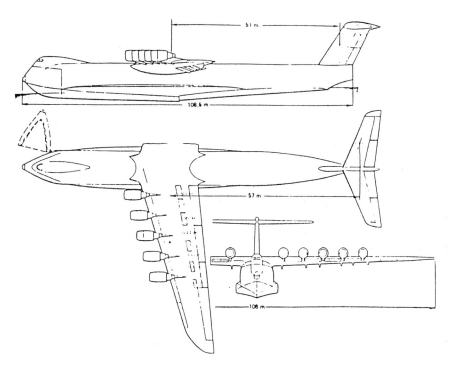

Fig. A6 Dornier 1000 ton Flying Ship (*c*1985) (Dr Claudius Dornier, Ref. 4).

small seaplanes are rare in Britain today: there is too little sheltered water for them to be of use.

This is not the same elsewhere. Large parts of Canada and North America, the East and West Indies, and Mediterranean coast of Europe, the Arctic, Antarctic, Soviet Russia,

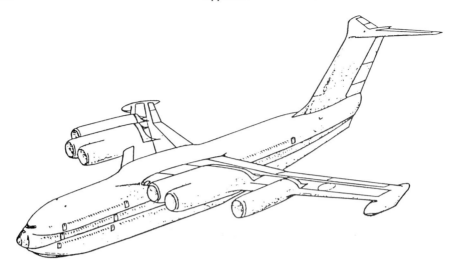

Fig. A7   Shinmeiwa Model GS (c1977) 500 ton 1200 passenger flying ship.

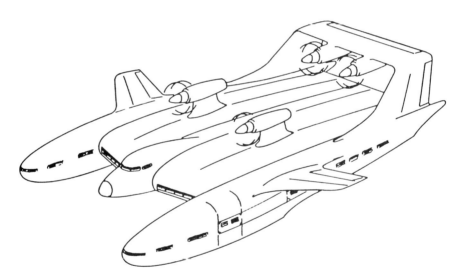

Fig. A8   500 ton ram-wing aerofoil boat to cruise at an altitude equal to about half-beam (after Dr-Ing A. M. Lippisch).

Pacific Islands and parts of Australasia are well suited to seaplane operations. Furthermore seaplanes, flying-boats in particular, are regularly operated from ice.

Figure A11 summarizes simply the steps from operational needs (which one can relate to Fig. A1, and which are framed in the form of operational regiments) to the final shape of an aircraft.

Aero-Hydrodynamics

Hulls and floats

Here is a conflict between the optimum shape of a seaplane for aerodynamic efficiency and

Plate A–7   Rhein Flugzeugbau X-114 aerofoil-boat which applies the principles of Dr Lippisch. (Rhein Flugzeugbau)

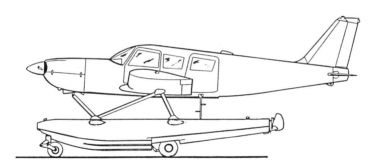

Fig. A9   Piper Cherokee 6 amphibian on Pee Kay Model 3500A floats.

that needed for good hydrodynamics. Figure A12 shows the generalized hydrodynamic form, which has the object of:

- Buoyancy and Static Stability.
- The ability to generate hydrodynamic-lift at low speeds, while being strong enough to cope with forces increasing as $(speed)^2$ from a medium 800 times denser than air.
- Reduction of the tendency of any convex-curved and streamlined body to stick to the water (*Coanda effect*).
- Reduced area of wetted surface, which causes friction drag.
- Suppression of spray reaching propellers, intakes and other working parts.
- Dynamic stability on water.
- Manoeuvrability and control on water.
- Adequate performance and versatility over all – which includes loading, unloading, replenishment and maintainability. Maintainability: accessibility and ease, needs special consideration, not only because tools refuse to float when dropped.

Plate A–8   Cessna Caravan 1 floatplane. (Cessna)

Plate A–9   Forebody warp of Grumman Widgeon (Author)

Coanda effect, named after Henri Coanda (1886–1972), a Rumanian engineer who discovered its importance, is the primary reason for special shaping of hulls and floats.

Viscosity of water makes it stick to a surface so that relative flows adhere to and follow curves (Fig. A13b). When water is caused to flow past a curved surface a pressure gradient is established by centrifugal reaction, lowest pressure on the inside of the curve and highest on the outside. The phenomenon can be demonstrated by dangling a spoon by the handle and bringing its convex back into contact with a jet flowing from a tap. The spoon is drawn vigorously into the jet in direct proportion to the rate of mass flow. The same thing happens when immersed in air, which is why aerofoil sections are shaped as they are, enabling aircraft to fly. But water, being 800 times denser than air, and with a free surface of discontinuity along which a body is trying to skim, produces more violent suction at much lower speeds.

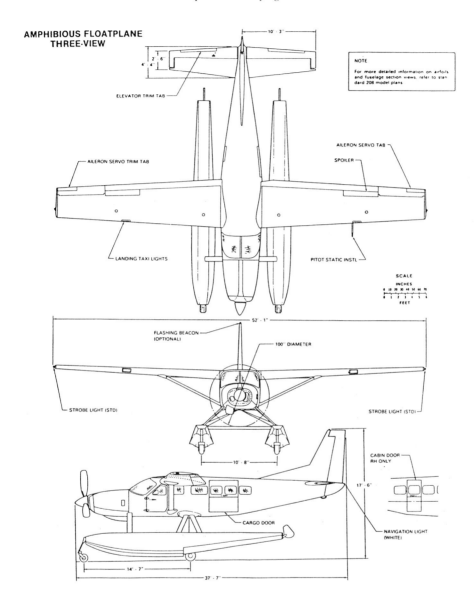

AMPHIBIOUS FLOATPLANE THREE-VIEW

NOTE

For more detailed information on airfoils and fuselage section views, refer to standard 208 model plans

Fig. A10   Cessna Caravan 1 amphibian (Cessna Aircraft Company).

The trick is to introduce a layer of air between the body and the water by means of sharp corners (which water cannot negotiate) formed by the main step of a hull or float, and by chines more or less at right angles to spray displaced on either side. The step is positioned approximately where the normal pressure from the water is at its peak, before decreasing to a suction further aft. Suction aft is adverse and causes porpoising during the take-off run, and increases the load to be lifted. The layer of air introduced by the step acts as a lubricant because air is over 60 times more slippery than water. But in flight, where kinematic viscosity counts (i.e. the ratio of dynamic viscosity to density), air is 13 times more able to turn

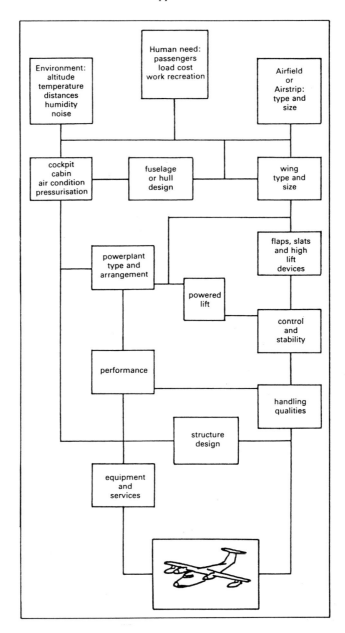

Fig. A11a   Steps from initial operational needs to final aircraft.

sharp corners than water. So the same steps and chines, which caused water flows to separate, make the relative airflow slow down and stagnate, to be carried along in 'draggy' regions of intense energy-absorbing vorticity. Lumps of such vorticity then become detached and shed into the wake to cause airframe buffeting, which can be damaging.

   The dilemma for the designer is how to get rid of aerodynamic steps and chines in flight. The problem has never been adequately solved. Solutions have been attempted along the lines of:

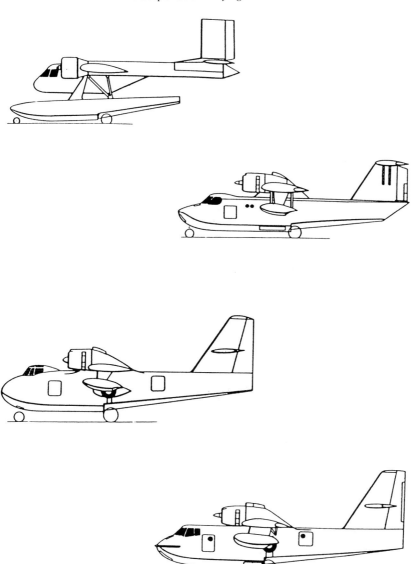

Fig. A11b   Steps taken by Canadair to increase adaptability of the CL-215.

☐ Hydroskis
☐ Hydrofoils
☐ Integrated aero-hydrodynamics with blended hulls ('Skate' project in the USA c1950).

and these will be dealt with shortly.

The shape of planning surfaces is critical and Fig. A14 shows ten sections with notes of their basic properties. Figure A14i is the worst, being most vulnerable to Coanda, closely followed by d.

Figure A14g, the Porte-type, appeared in Fig. A3 of the Felixstowe F2A. Plate A–24 shows a hybrid form of rearstep: part Porte, part Linton-Hope(?) of a Blackburn Iris c1924, *see* too Fig. A24a: Supermarine NIB Baby (1918).

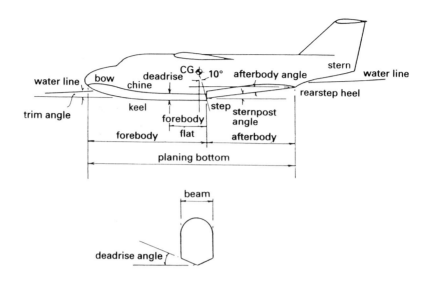

Fig. A12   Parts of a seaplane hull (and floats).

The softest riding sections are vee-d. Figure A15 (Plate A–9) shows the effect of deadrise angle on impact loads (Ref. 8) and also illustrates forebody warp. Such warp reduces the tendency of a hull to porpoise. But the steeper the deadrise angle towards the bow the deeper runs the forefoot. This is de-stabilizing directionally and encourages waterlooping: the nautical, and very wet equivalent of a ground loop.

Figure A16 (Plate A–10) shows the Japanese PS-1 maritime patrol flying-boat. The aeroplane has a yaw-vane mounted on a mast ahead of the windscreen, and a large dorsal addition to the fin and a large side area ahead of the wing (due in part to the deep forefoot and high-set cockpit). It is fair to surmise that the designer has had directional problems both in the air and on water, because of deficient aerodynamic weathercock stability with flaps down. There is no sign of a water-rudder.

In the displacement regime, before the aircraft begins to plane, energy is wasted in the form of wave making. Figure A17 (after Ref. 9) is typical. (See also Plates A–11, A–13). As the machine accelerates it must be encouraged to lift by up-elevator which increases the trim-angle, so enabling the aircraft to climb its own bow wave. In the case of an ordinary boat a sawn-off transom terminating a straight run aft avoids the Coanda effect of Fig. A13b and results in a shallower trim-angle than would be the case in Fig. A13c. Seaplanes must accelerate quickly to reach the planing regime beyond the hump. Figure A18 illustrates the general form of the combined resistance of a seaplane up to the unstick speed, V_{US}.

Spray

Spray is caused by the peak pressure developed along the stagnation streamline in the area where the planing-bottom enters the water, and occurs in two forms (Fig. A19, Plates A–14, A–16, A–20). The first, *ribbon* or *velocity spray*, is flung sideways in a flat trajectory from the line of forward contact of the planing-bottom with the surface of the water. Being light

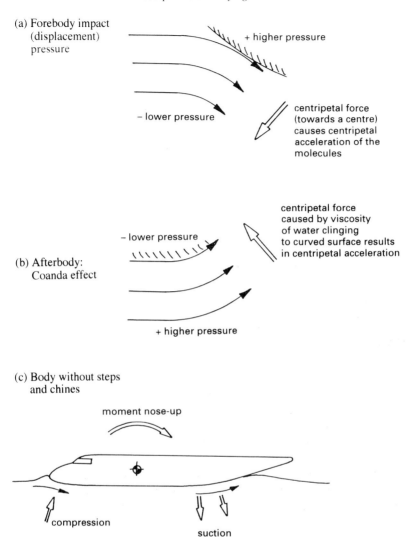

(a) Forebody impact
(displacement)
pressure

+ higher pressure

– lower pressure

centripetal force
(towards a centre)
causes centripetal
acceleration of the
molecules

centripetal force
caused by viscosity
of water clinging
to curved surface results
in centripetal acceleration

– lower pressure

(b) Afterbody:
Coanda effect

+ higher pressure

(c) Body without steps
and chines

moment nose-up

compression

suction

Fig. A13   Displacement and Coanda effects.

it causes few problems, apart from misting of windscreens. The second kind, *blister spray*, is heavy and far more damaging because it tends to be thrown upwards and rearwards by the chine in a heavy cone. The height to which blister spray rises determines the heights of wings, engines and tail-surfaces (Plates A–13 and A–14).

Spray is suppressed by deflection (i.e. reflection by the planing surface) and its damaging effects are ameliorated by aeration, which reduces the solidity of its mass. Spray control is most necessary with vulnerable, hot and precisely made, expensive jewels, like turbopropeller engines. These need spray separators, in the form of plenum chambers, between air intake and engine. Piston engines do not seem to suffer quite as badly and have gulped, but continued to run, after being submerged momentarily. Hollow-grinding the forebody helps as shown in Fig. A19b, but the concavity of the curve is critical.

(a) Rectangular section. Basically a buoyant water-ski. Planes well but wet at low speeds below 'hump'.

(b) Straight vee. Clean running. Planes well, and reduces alighting loads. Simple to make.

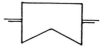

(c) Inverted vee. Works as well as b, except for tendency to bounce.

(d) Rounded bottom. Will plane reasonably, but water flows up around curves, causing excessive spray.

(e) Hollow concave. Said to plane well but sensitive and harsh, said to generate more drag in flight than c.

(f) Concave vee. Planes well, but more drag on take-off than b, and generates much spray.

(g) First World War Porte-type, similar hydrodynamics to b.

(h) Dornier step and hull section. Clean running, but harder to make than b.

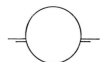

(i) Circular section is poor, sucking itself deeper into water.

(j) First World War Linton-Hope false planing bottom: heavy, but used by Supermarine flying-boats.

Fig. A14 Hull and float sections and properties.

Too little affects only the boundary layer and not the main mass. Too much causes the boundary layer to rebound off the main mass and so rise higher. Figures A19c and d show fixed spray dams (Plate A–9). That in d is also illustrated in Fig. A16c and d, which takes the form of an inverted gutter around the chine, from which spray is ejected aft.

For a spray dam to work successfully it must make an included angle with the spray direction of not more than 90°. Spray is also suppressed by increased fineness of the planing bottom, which also decreases aerodynamic drag. Knowler in his Fifth Louis Bleriot

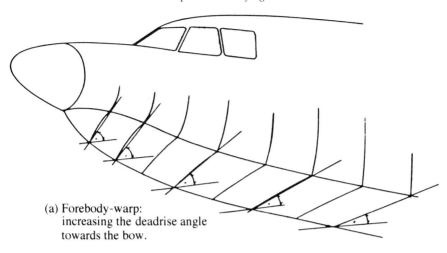

(a) Forebody-warp:
increasing the deadrise angle
towards the bow.

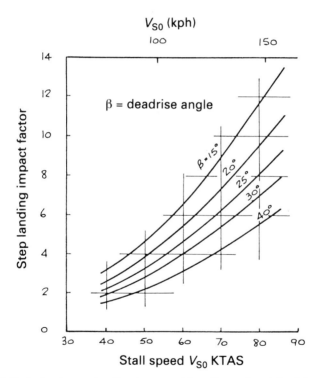

(b) Variation in step landing impact with stall speed and deadrise
angle for a 4000 lb (1818 kg) seaplane (after Ref 8)

Fig. A15   Deadrise and forebody-warp.

lecture (Ref. 10) discussed the effects of fineness and recorded points made by observers
about the following terms in a useful equation for load coefficient, which could be used to
improve hydrodynamics without spoiling the aerodynamics:

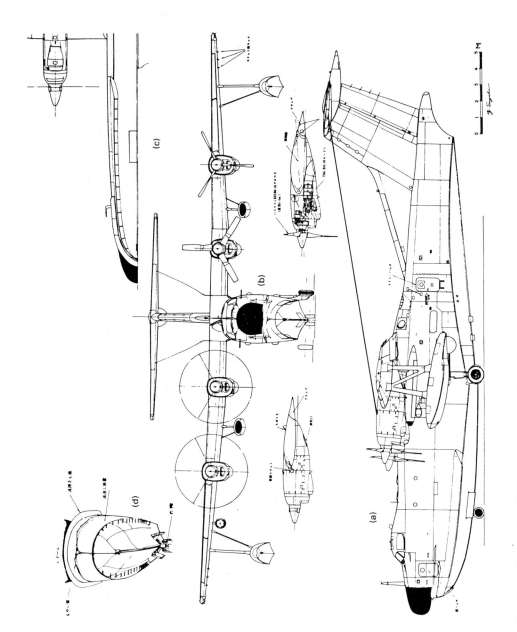

Fig. A16   Shinmeiwa PS-1 anti-submarine flying-boat.

Plate A–10 Shinmeiwa PS-1 short take-off using powerful flaps and plenty of propwash. (Shinmeiwa)

$$K = \frac{\text{displacement of aircraft}}{\text{unit weight of water} \times (\text{length})^2 \times \text{beam}}$$

$$= \frac{\Delta}{(wl^2b)} \tag{A2}$$

It had been found that if K as defined was kept constant, then hulls with varying length/beam ratios had equivalent resistance and spray characteristics. For example, hull fineness (l/b) could be improved (and aerodynamic drag reduced) without affecting the spray height as long as Δ/l^2b was kept constant. But if the product (lb) was fixed while (l/b) only was altered, this caused the spray height to decrease, because:

$$l^2b = \text{constant} \, x \, (l/b)$$

$$= f \, (\text{fineness ratio}) \tag{A3}$$

The practice with planing bottom design has been to make the forebody ahead of the step 3 to 3.5 times the maximum beam. The overall length/beam is roughly double, around 6 or 7 making:

$$\text{overall } l^2/b = 36 \text{ to } 49 \text{ in general}$$
$$\text{and forebody } l_f^2/b = 9 \text{ to } 12.25$$

Dornier's solution to spray suppression is a combination of the laterally stepped, almost flat, planing bottom shown in Fig. A14h and sponsons (Stumeln), which deflect spray while providing buoyancy and static stability. Certainly, recent tests with the Dornier ATT (Fig. A28b, Plate A–17) have shown very flat spray profiles. A flat bottom planes earlier than a vee.

Aerodynamic drag

But much finer bodies can be designed than this with lower aerodynamic drag (Ref. 11). Knowler also records, for example, a forebody/beam ratio of 6.5 ($lf^2/b = 42.25$) which had 5% less hull drag than when $lf/b = 3.5$.

Table A2 gives a broad idea of drag increments introduced by hull geometry.

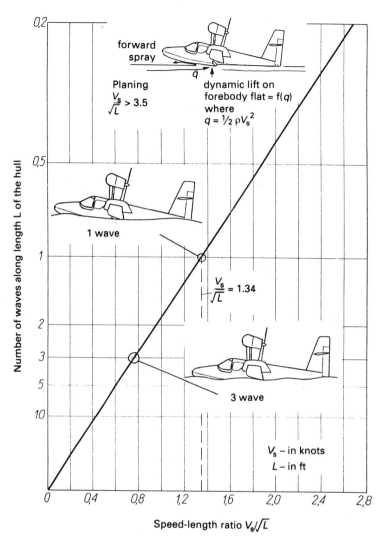

Fig. A17   Wave development by a hull (after Ref. 9).

Skin friction drag (proportional to surface area/volume and, hence, to the adverse effect of the square-cube law upon craft of smaller sizes), and parasite drag caused by increasing acreage of junctions, militates against multi-hull flying-boat configurations. The *ekranoplan* is an exception when flying in surface effect.

A badly shaped forebody chine causes considerable aerodynamic hull drag. The earliest, simplest, steps increased the drag of the basic streamlined body upon which the hull was based by about 48%. An elliptical step (Fig. A20) increases drag by around 15%, but the latest seaplane hulls can be built with a total drag increment around 12%, compared with a value of 4% to 5% for a similar landplane. Ideally, complete ventilation of the hull on the hovercraft principle, by using a cushion of air, would provide the greatest reduction in drag, but the weight penalty of such a mechanical system would be very high. None of these

Plate A–11   Two-wave sequence showing aircraft to be taxying at $V_s/\sqrt{L} \sim 0.9$. (Canadair)

Plate A–12   Martin Seamaster experimental jet flying-boat needing water-flaps for manoeuvring and braking. (Professor J. E. Allen)

problems has been fully and satisfactorily solved because of a lack of advanced seaplane research in recent years.

Hydroskis

Conventional hulls and floats are bulky, but not as much as the original buoyant box incorporating a flat planing bottom which was, essentially, a hydroski that would float. With the

Plate A–13   Short Sunderland making blister spray. (Professor J. E. Allen)

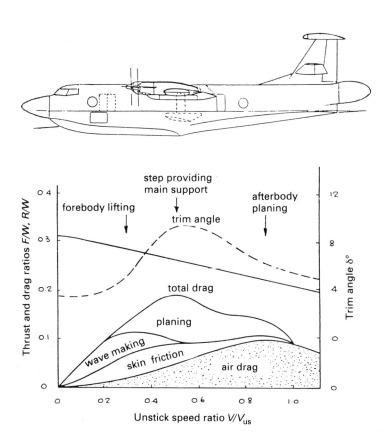

Fig. A18   Combined resistance with thrust and trim angle during take-off.

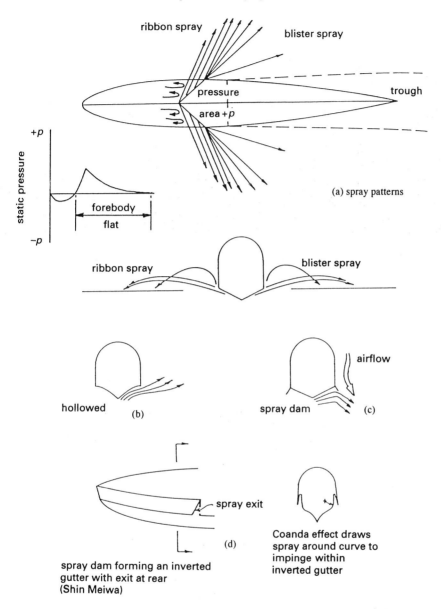

Fig. A19 Spray formation and devices for its suppression.

coming of the jet engine research turned back to the hydroski, fitted to a clean slender hull, which lifts the aeroplane bodily out of the water and can be used for landing on rough water at sea. However, hydroskis, although tried on a number of aeroplanes, including the Convair Sea-Dart (1953) (Plate A–15) have not been developed further, simply because seaplane development has been overtaken by other events – except in Japan and maybe in Soviet Russia.

The ski is good at absorbing landing shocks, as shown in Fig. A21 which is derived from data published years ago by Saunders-Roe. The curves apply to a conventional rough-water hull, with and without a retractable ski.

Plate A–14   Dornier Do24 ATT making blister spray. (Dornier)

TABLE A2

Aerodynamic drag increments
(after Ref. 10)

Geometry	Description	Drag increase (%)
	body of revolution	0
	cambering	3 to 5
	addition of fin and cabin	2 to 3
	addition of planing bottom	8 to 9
	squaring mid-body chine	1
	turning downwards bow chine	1 to 8
	addition of main step	20 to 38
	Total	35 to 64
	Average	50

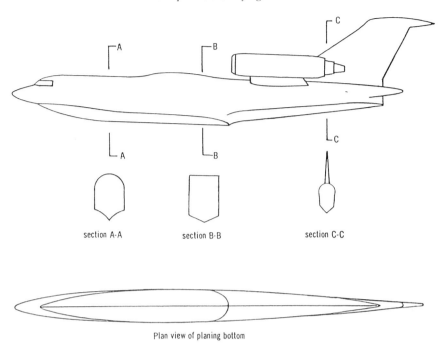

section A-A   section B-B   section C-C

Plan view of planing bottom

Fig. A20   Elliptically faired step and planing tail.

Plate A–15   Convair Sea Dart with hydroskis. (Professor J. E. Allen)

The disadvantage of the ski is that it generates much hydrodynamic drag before it planes, more than a conventional hull, as may be seen by comparing Figs A18 and A22. This demands more installed power. This need not be too detrimental with a jet-engined aeroplane (which usually has power in hand) because the lower accelerations for which the hull is designed result in lighter structure and gross weights, in spite of the additional weight of the skis and their mechanisms.

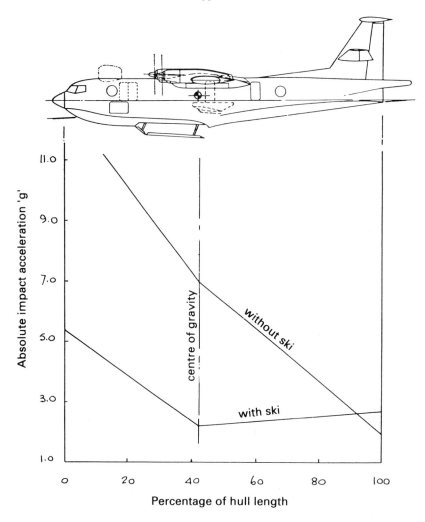

Fig. A21   Hydroski shock absorbing properties.

However, hydroskis do not fit well with propeller driven aeroplanes which, being designed for lower cruising speeds cannot afford surplus installed power to cope with the additional drag on take-off.

Hydrofoils

The hydrofoil is, in effect, a small water-wing that remains completely immersed until lift-off, and which is capable of generating lift/drag ratios around 30/1. While the attraction of such an arrangement is that relatively small, retractable, surfaces can be used to lift the hull in the displacement regime, like hydroskis, power demands are high. Some attempts have been made to support aeroplanes completely on hydrofoils, but the operating speeds are so fast that the suction over the upper surfaces is too intense and the water 'boils', a phenomenon known as cavitation. Cavitation, which may occur unpredictably, causes an immediate loss of lift/drag and longitudinal instability. Further, hydrofoils are structurally vulnerable and also vulnerable to damaging effects of debris, which, by spoiling the hydrodynamic cleanliness, precipitate cavitation.

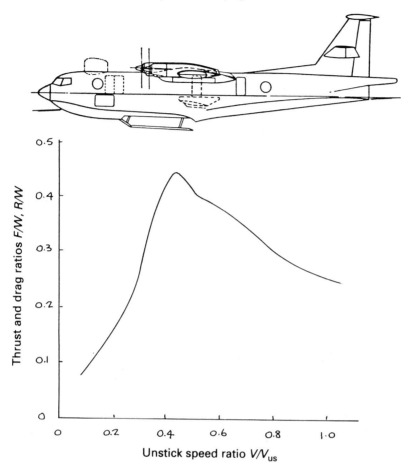

Fig. A22 Typical combined resistance curve for a slender hull with hydroski. Compare with Fig. A18.

If surplus power can be afforded, then experiments indicate that the best arrangement is a main lifting foil slightly aft of the CG, stabilized by a forward canard foil. Such an arrangement reduces the hump resistance/weight by about one third, from 0.18 to 0.12. The main purpose of hydrofoils would be to lift an aircraft beyond the hump then, after cavitation, they would be retracted to leave it planing on the hull surfaces. In this way, foils allied with spray dams might allow shallower hulls to be designed.

Even so, those who have experienced the jet foil to France in a rough sea will appreciate just how vulnerable and limited such devices can be in reality.

Summarized Design Penalties

Increased drag and increased structure weight reduce the fuel and payload carried by any aeroplane. Table A3 shows a typical breakdown of weight for a light twin engined landplane (Ref. 12).

Table A4 shows, conservatively, calculated penalties of adaptations of the basic aeroplane in Table A3 to a flying-boat, a floatplane, and to a float-amphibian. In these calculations the accent is upon increased wetted area and the interference drag of bracing and junctions.

Table A3

Breakdown of all-up weight

Item	Approximate % weight
Powerplant	18
(Fuel)	(22–31)
(Payload) } Disposable load	40
	(9–18)
Structure	28
Equipment and services	14
Total	100

Table A4

Penalties compared between landplane, flying-boat, floatplane, float-amphibian
(5000 lb (2273 kg) design weight)

Aircraft	Increase in structure weight	Increase in parasitic drag	Reduction in cruise speed	Reduction in payload × (block/speed) for given range	Overload to achieve payload and range
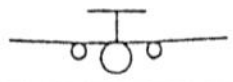	0	0	0	0	0
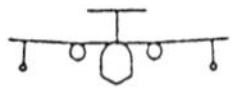	+14%	+15%	−7%	−34%	+12%
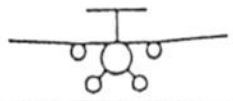	+25%	+22%	−10%	−55%	+19%
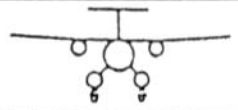	+43%	+28%	−12%	−87%	+27%

Numerous light landplanes are fitted with floats, especially in countries like the USA and Canada in spite of penalties like those shown in Table A4. Simply, one can generally accept overload and longer take-off and landing runs on water than on land. Even so Fig. A23 shows graphically the effect of drag upon airspeed and rate of climb of a small light seaplane – and the way in which added keel surface is needed in the form of a ventral fin, to provide weathercock stability in flight.

Effects of Engine Location

One of the most significant aspects of seaplane configuration, which affects both handling and performance, is the necessarily high mounting of engines, putting thrust-lines high and centres of gravity and centres of drag low. Thus, seaplanes tend to suffer nose-down pitching when power is applied. This is the reverse of what is desirable for a pilot. It is wise not to grip the throttles on take-off, because thumping and pitching on water jolts the hand causing fluctuations in power, which can introduce pilot-induced pitching oscillations. Figure A24a and Plate A–16 illustrated the point about high-set engines.

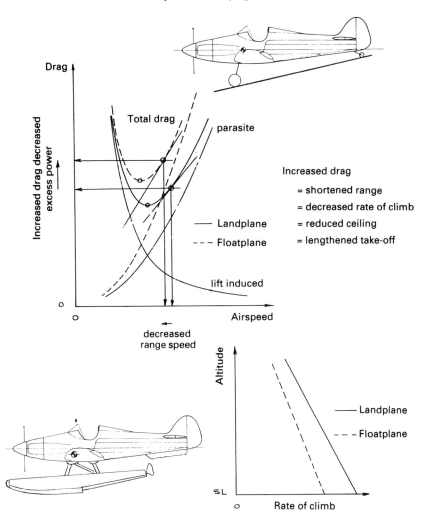

Fig. A23 Small landplane showing additional keel surface in form of ventral fin; and performance penalties when fitted with floats.

Figure A24b shows an additional factor, namely that of the effect of drag, caused by interference and other parasite sources ahead of the tail. The dynamic pressure: which affects all aerodynamic forces and, hence, the authority of the flying control surfaces:

$$q = \tfrac{1}{2}\, \rho V^2 \qquad\qquad \text{((A4) (and Equation 3–26))}$$

(where ρ is air density and V is true airspeed) is often badly reduced. This means that seaplanes often need bigger tail surfaces than landplanes. The effect is especially marked when the engine is throttled back and the windmilling propeller dams the airflow to an extent. The aeroplane pitches nose up and one feels the tail becoming less effective as larger rudder movements are needed to achieve the desired responses.

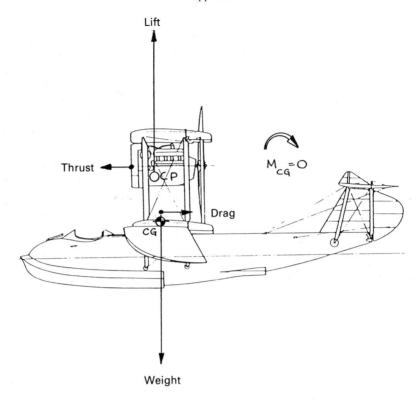

Fig. A24a   Typical arrangement of lift, drag, thrust and weight to produce zero moment about the centre of gravity when cruising. Note too the under-cambered tailplane to generate additional nose-up pitch with increased power (i.e. working in propwash); and the Linton-Hope type bottom Supermarine N1B Baby (1917) (Fig. A14j).

Flying-boats have relatively long forebodies. A single engine mounted high and well aft (Figs A17 and A24, Plate A–16) interferes badly with the tail when the propeller is windmilling. The long nose generates aerodynamic lift at large angles of attack, like a javelin. Such lift can induce a secondary stage in the stall, and loss of directional stability.

One light flying-boat tested by the author pitched up in a secondary stall and this was accompanied by lateral 'slicing' of the nose as the tail became less effective. Admittedly, this was a result of an attempt to investigate a tendency to deep-stall in an aeroplane with a high-mounted tailplane and elevator. Stall warning would normally have prevented this happening.

Finally, and as mentioned earlier in passing, the installation of turbopropeller engines involves special design to avoid spray ingestion.

Flying Qualities

This brings us to the matter of safe flying qualities. These embrace:

☐ Performance
☐ Handling
☐ Functioning

Plate A–16 Lake LA-4 planing – note large up-elevator tabs to augment authority of longitudinal trimmer. (Alan Deacon)

both in the air *and on water*. Table A5 shows the various factors which affect seaplane flying qualities. Here, we are not particularly concerned with functioning. Effect upon performance has been hinted at in Fig. A23, and in discussion of drag.

Water Handling

Handling breaks into two parts

- Control
- Stability

each of which tends to oppose the other in flight to a much greater extent than on water. Too much stability, i.e. the tendency to return to the undisturbed state, and control authority is reduced. Overpowerful controls and an aeroplane becomes too lively and tiring because the benefit of stability is reduced.

Static stability on water

Buoyance is proportional to the volume of water displaced: the weight of the displaced volume being equal to the weight of the aircraft. Buoyancy acts upwards through the centre of buoyancy, CB, while weight acts downwards through the CG. Static stability when heeled is measured in terms of the distance between the *metacentre* M and the CG. The metacentre is the point of intersection of the line of action of the buoyant reaction in the plane of symmetry of the aircraft. The distance between the metacentre and the CG is the *metacentric height*. If the CG lies below the metacentre when heeled the aircraft is statically stable; and vice versa (Fig. A26).

In a similar way static stability in pitch is measureable in terms of the metacentric height of the intersection of the buoyant reaction with the lateral plane through the CG.

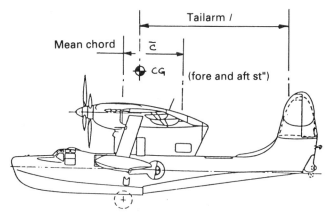

The Supermarine Seagull (1948) had to 'grow'
a third fin and rudder after the first flights

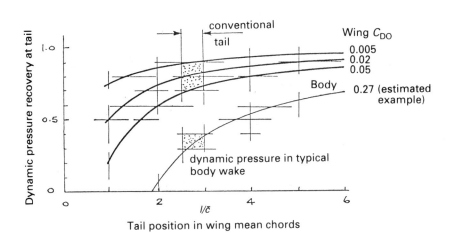

Tail position in wing mean chords

Fig. A24b Another factor: drag slows the flow and reduces dynamic pressure recovery at
the tail, making one tail less effective than another.

Blended hulls

A promisingly neat way of providing static stability, which satisfies both aero and hydrodynamic criteria, is by blending wings and bodies. This is illustrated in Fig. A27: a conversion
on paper of a small landplane into a flying-boat. The fairings are extensions of the planing
bottom. Even so, spray-dams are needed. These prevent as much drag reduction as one
might hope for initially. Because this aeroplane is small, with smooth laminar surfaces, a
tailplane has been added on top of the fin to augment control in pitch. This is because the
foreplane is badly placed for being covered with spray and salt deposits which would spoil
lift by breaking down laminar flow (for example, drops of rain on the smooth wings of
motor-gliders have delayed take-off through lost lift and increased drag by destroying laminar
flow). This is one of the hardest problems to get around with high performance water-based
aircraft.

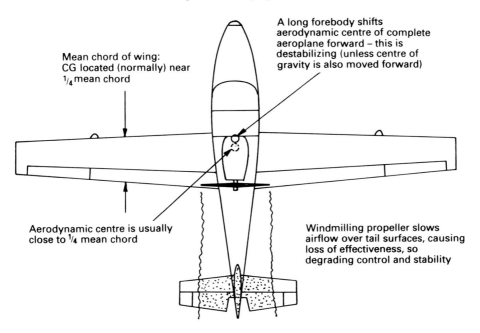

Mean chord of wing:
CG located (normally) near
$\frac{1}{4}$ mean chord

A long forebody shifts
aerodynamic centre of complete
aeroplane forward – this is
destabilizing (unless centre of
gravity is also moved forward)

Aerodynamic centre is usually
close to $\frac{1}{4}$ mean chord

Windmilling propeller slows
airflow over tail surfaces, causing
loss of effectiveness, so
degrading control and stability

Note: the <u>aerodynamic centre</u> is that point on the aeroplane about which the resultant
pitching moment of all aerodynamic forces remains more or less constant,
within the working range of angles of attack, at a given speed – it is a <u>fulcrum</u> of sorts

Fig. A25 Effect of long forebody and windmilling propeller set well aft.

While attractive, the blended hull, an American 'skate' concept from 1947 (Ref. 13) –
has not been developed further, in spite of the apparent advantages of large payload vol-
ume/surface area, and better lift/drag ratio than any other forms of seaplane.

Floatplanes

Floats are large and relatively heavy items. The American FAR 23.751 requires floats to
have buoyancy of 80% in excess of the total required displacement to support the aircraft in
fresh water (i.e. 1.8Δ). Floats have powerful aerodynamic effects that are invariably
destabilizing.

By comparison with a hull, floats suspended beneath a fuselage act like a pendulum,
resisting roll initially, while building up inertia.

Controllability and performance are never as good as with a landplane, because of the
additional weight of floats and the weight of a stronger supporting structure. Larger control
movements are needed to achieve the same response rates with floats as when fitted with
wheels. Motions in pitch roll and yaw take longer to start: to get the aeroplane 'wound up',
and longer to stop for the same reason, because of their 'flywheel' effects. The centre of

TABLE A5

Factors affecting seaplane flying qualities

Quality	Embracing	Factors affecting
Performance	take-off and landing distance	see state, tide flow, wind
	climb	excess weight
	manoeuvre	stall speeds too fast
	ceiling	insufficient excess power
	range	excessive drag (step, chine, etc.)
	endurance	insufficient lift to spare
	speed	high water drag
Handling	In air:	sea state, tide flow, wind
	control	centre of gravity wrong
	stability (stick and rudder fixed and free)	flying surfaces wrongly rigged
	trimability	control surfaces lacking authority
	On water:	bad fairing and rigging
	Control, stability, trim} porpoising,	control moments wrong
	pattering, skipping	flying and engine controls:
	manoeuvrability – plough and step turns	backlash, friction and ease of operation
	aileron, sailing	thrust line and drag alignment
	On land:	
	ground handling (amphibian)	
Functioning	Cockpit comfort and safety	seats, harness and their adjustment
	instrument presentation	hydraulic, pneumatic, electrical, fuel, oil
	flying controls	pressurization and de-icer systems
	engine controls (location)	cockpit and external lighting
	flap, slat and gear travel	Nav/Comm system and equipment
	doors and canopies	serviceability: corrosion, lubrication,
	communication and navigation	protection
	marine equipment	adequacy of marine equipment

gravity is altered, often adversely. Of all of the controls the rudder is most important, for spin prevention and recovery. Floatplanes generally need more rudder than landplanes to provide the considerable outspin yaw forces needed to effect spin recovery. As noted earlier, landplanes need additional fin and rudder area when fitted with floats.

Single engined seaplanes generally need water rudders. These are relatively fragile and arc used only at low speed. They must be retracted on take-off and landing. Water rudders are most useful when used in conjunction with the normal rudder, and are thus smallest and most cost effective when the normal rudder operates in a propeller slipstream.

Multi-engined machines can often manage without water rudders.

On water all seaplanes are affected by:

☐ Wind
☐ Tidal stream
☐ Waves
☐ Propulsive thrust (under the control of the pilot)

Floatplanes, which are in general more lightly loaded than flying-boats with more surfaces sticking up high into the air, are more susceptible to the effects of wind than tide. Some

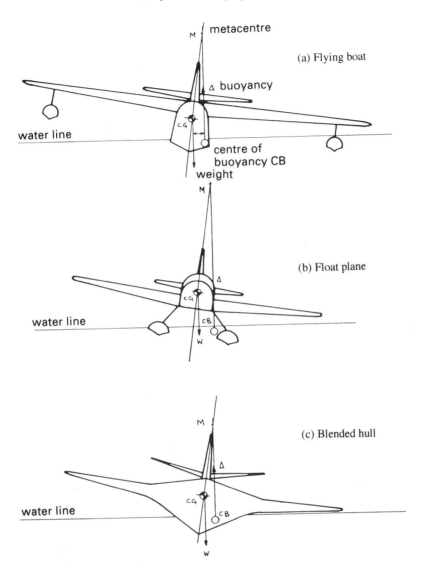

Fig. A26 For static stability on water the metacentre of a seaplane is above the centre of gravity. The greater the metacentric height the more powerful the righting moment.

additional control over the often adverse effect of wind can be gained by lowering the landing gear of an amphibian when taxying. This can be especially useful when testing the slope and condition of the bottom before beaching.

Flying-boats

The body or hull of a flying-boat supports the whole of the machine, and also provides accommodation. Lateral stability on the water is provided either by stub planes, e.g. sponsons, or by wing-mounted floats (usually at the tip). The attachment to tip floats must not be so

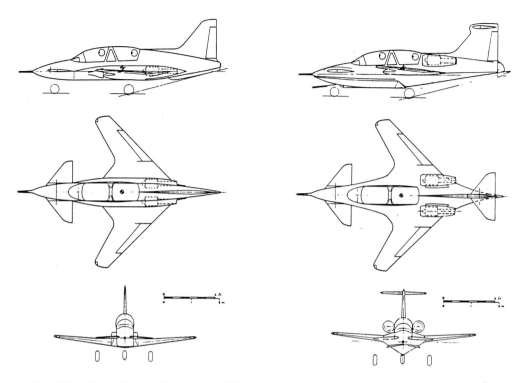

Fig. A27   Canard lightweight jet modified to a small flying-boat by means of a blended-hull.

strong as to cause a wing to break off when striking an obstacle, nor should they be so weak as to come off too easily. Sponsons are shown in Fig. A28a and b (Plate A–17). Wing-mounted floats cause much drag and attempts have been made in the past to retract and fair them (Fig. A29, Plate A–18).

Flying-boats, in their larger sizes, are generally more seaworthy than floatplanes. While wing-tip floats are economically small and light, they must not become detached too easily – nor must they be attached so firmly as to cause the wing to break off before a float.

Sponsons are seaworthy devices and have the added advantages of spray suppression, stowage of amphibious landing gear, and provision of easy access between the water and the hull (useful for rescue missions). But, sponsons must be rigged correctly, otherwise they cause much more drag and in extreme conditions can be swamped more dramatically than tip floats.

On the point of landing gear stowage: doors in a hull area are a problem, needing special care and attention to structural design, functioning and sealing. A simple failure of a door when planing can lose the aircraft.

Forces and Effects of Motion

Because hydrodynamic forces are high, advantages can accrue from short take-off and landing (STOL) features. High lift flaps are a good example, coupled with propeller slipstream and leading edge devices, to help the aircraft to take off and touchdown at slower speeds in more or less level attitudes (small trim-angles). All these devices tend to cause drag and

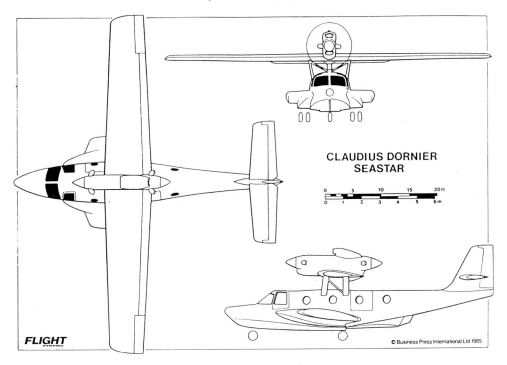

Fig. A28a   Claudius Dornier Seastar.

certainly add weight. Furthermore, they inevitably reduce cruise performance. Thus, they are best used on machines designed for slower cruising speed, rather than long range at high speed. This has an inhibiting effect upon the designer who hopes to produce a fast world-beating seaplane today. If you want a STOL seaplane that is cost effective it will probably be a dray-horse.

Seaplanes can never be stopped on water whenever engines are running. It is always prudent to start the engine(s) as soon as possible after launching unless the aircraft is to be moored immediately. With engines stopped and adrift, if there is any wind or current, a seaplane will always move in relation to the land, the bottom, and to buoys or shipping at anchor.

A seaplane tends to weathercock into wind. When drifting with the current it will point into the relative wind. When moored it will take up a resultant heading between wind (acting on surfaces above water) and current, acting upon submerged keel surfaces. A lightly loaded biplane, with much surface area and aerodynamic drag will be more *wind-rode* when moored than a clean monoplane, which will be more *tide-rode* (i.e. swung to the tidal stream).

It follows that, in addition to basic airmanship, a seaplane pilot needs a sound grasp of the principles of seamanship. Further a seaplane must also carry a load of marine equip-ment, like line and anchor at least, for when it is a boat (Table A6).

When moving under the influence of wind and current the pilot can change the heading of his aeroplane by means of aileron deflection, in effect aileron-sailing (see later), increas-ing the drag more on one side of the CG than the other. Fin bias and rudder deflection are useful in much the same way. The stronger the wind the more useful are the flying control surfaces – and deflected flaps.

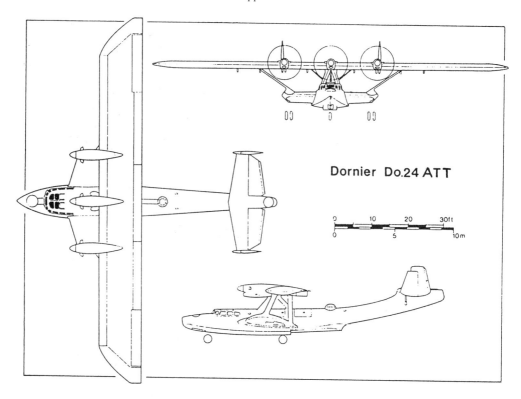

Fig. A28b   Dornier Do24 ATT.

Drogues (in effect water-brakes) are conical canvas tubes like windsocks and are useful for checking drift or speed. The pilot can run his engine at a higher RPM, increasing aero-dynamic control and manoeuvrability, without increasing forward speed. But, as with all such devices which involve the use of lines at sea, a drogue cannot simply be heaved over-board. Not only must the free end be tied in-board, but the heaving line must be free to run with the drogue attached to the line by a swivel and a spring hook.

The Hump and Planing

On opening the throttle to accelerate on water there is a change of longitudinal trim which is the resultant of the hull, or floats, tending to pitch nose up with the bow wave; and the thrust line, which is usually above the CG, causing the aeroplane to pitch nose down. Figure A17 (Plate A–11) shows what takes place with a boat hull. The water thrust aside by dis-placement sets up a wave system along the length of the hull, running away diagonally from bow to stern. This generates a system of transverse waves which travel at the same speed as, and at right angles to, the hull. The transverse waves absorb energy which is the biggest factor in wave-making resistance at speed. The length between the crests of a standard sinusoidal wave is:

$$L = (V/1.34)^2$$

i.e.

$$V/\sqrt{L} = 1.34 \tag{A5}$$

Plate A–17 Dornier Do24 ATT showing spray-suppression by sponsons. (Dornier)

and the normal maximum speed of a boat fits the same formula.

When V is measured in knots and L in feet, then at

$$V/\sqrt{L} = 1 \tag{A6}$$

the wave is shorter than the waterline length, and the hull is supported at bow and stern. This represents the speed for best economy in the displacement regime (Refs 18, 20).

As $V/\sqrt{L}$ increases beyond 1 the wave length grows longer than the hull and the craft begins to climb its own bow wave. The stern squats and additional power is needed to climb

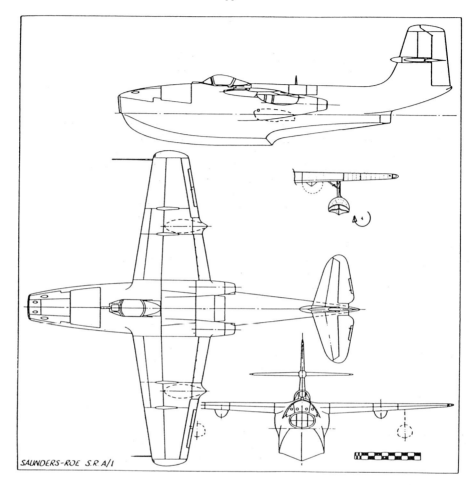

Fig. A29   Saunders-Roe SR/A1 (c1947) with retractable wing floats.

Plate A–18   Consolidated Catalina (Canso) with retractable tip-floats.
(Author)

TABLE A6

Equipment needed for operation from water

Item	Note	Approx weight (lb)
1. Ground anchor and chain with line plus float	Folding non-rigid for aircraft < 12 500 lb, plus 15 to 20 fathoms of chain for large aircraft. Small: 3 times depth in fresh and 5 times in salt, with length of chain between line and anchor to prevent chafing.	30 up to 200
2. Boat hook	Folding pole long enough to reach water, with detachable spring-loaded hook, tied to adequate line, incorporating spliced loop to pass over bollards. Stow within reach of bow.	3.5
3. Drogue/sea anchor	Two or three	2–3
4. Bilge pump	With hose to reach from port to lowest drain point (usually at main step)	4–5
5. Towing pennant	Fixed item: 3 steel cables spliced to a ring, all able to take load 1.5 × weight of aircraft. Stowage in flight within reach of pilot and best carried externally. Shock absorber cord to tension and served to prevent chafing hull/float.	3–4
6. Swashplates	Normally kept in position. Divide hull into watertight compartments. Tops above load waterline. Check before flight	5–8
7. Warpline	Carried in bow on revolving drum	3–5
8. Dinghy & Paddles	Inflatable, CO_2 bottle, with manual valves for topping up by mouth	8–10
9. Leak repair outfit	To include tapered rubber plugs	2–3
10. Mooring and out of control lights	Only if over-night. Mooring white light, visible all directions over radius of 1 nml. Out of control: two red lights, one 6ft above the other, visible radius 2 nml.	3 to 6
11. Signal lamp/torch		2
12. Distress pyrotechnics	Must include hand guard. Be kept watertight at highest point possible.	2
13. Signal pistol	Desirable, with selection of cartridges: red, white, green. Pistol & cartridges stowed separately, pistol 'broken'.	4–5
14. Life jackets	Enough for each person. CO_2. Plus manual inflation.	say 5 each
15. Sharp knife	With slip-stone. Keep only for emergency.	1
Warning: All items of marine equipment must be checked regularly, washed off in fresh water after contact with salt water and lightly greased as required.		90–265

uphill and surmount the 'hump' into the planing regime. Eventually a favourably shaped hull will plane completely when $V/\sqrt{L}$ reaches at least 3.5, i.e. it will be supported by impact lift on the specially shaped planing bottom. In the planing regime total resistance is reduced (Fig. A18). The total resistance R is shown as a ratio of drag/weight, R/W, against unstick-speed ratio, V/V_{us}. Two other curves have been added: the thrust/weight, T/W, and trim-angle. All seaplanes have reduced acceleration in the vicinity of the 'hump' where $(T/W-R/W)$ is least. However, note that aerodynamic resistance continues to increase with speed. Skin friction would also increase, were it not for aerodynamic lift raising the hull and so reducing the wetted surface area. Improved supercritical hulls (sea knives) (Ref. 22) and swept back steps of the Dynaplane-type (Refs 23, 24) cannot be adapted easily from surface craft to aircraft because of the poor resulting lift/drag ratio in flight. Sea-knife, for example, is a straight-edged, broad transomed-wedge.

Dynamic stability on water

There are three kinds of dynamic longitudinal instability: *porpoising*, *skipping* and *pattering*. Porpoising is the most dangerous and can occur at both small and large angles of trim, mainly small.

At small trim-angles porpoising is reduced by the forebody-flat. The forebody-flat extends 1.5 beam-widths forward of the step and, being flat, sustains more or less constant pressure over the whole surface. Curvature in this area would cause a variation in longitudinal pressure distribution with trim and alter the longitudinal metacentric height with any disturbance so that pitching motion would be aggravated. Later hulls with refined slender lines do not have marked forebody-flats, instead they employ forebody warp (Fig. A15a).

Porpoising at large trim-angles is caused by the afterbody dipping into the water. This is prevented by maintaining large afterbody keel and sternpost angles; and by decreasing the length of the afterbody by introducing a rear step. Increasing both angles increases aerodynamic drag. Porpoising at high speeds results in skipping, the aeroplane being thrown clear of the water before stalling back again. Porpoising is also caused by the step centroid being too far in front of or behind the CG. Skipping is caused by the step being too shallow and, therefore, insufficient ventilation of the planing-bottom. Tests indicate that the depth of the step should be 6% to 10% of the beam.

Manoeuvrability and control

Fine hulls have long forebodies and deep-running keels that move the centre-of-lateral-area forward relative to the CG, like forebody warp. This decreases directional stability making such hulls more prone to ground (water) looping. Careful judgement is needed to balance adequate control for the pilot, when taking off and landing, against hull shapes which, while providing directional stability, do not suffer loss of sea-kindliness when moving fast on water.

Long fine hulls had their origins with W. Soltorf at the Hamburg 27 tank in the late 1920s. The Blohm and Voss Company used this design of hull for the BV222 and BV238 flying-boats. After World War II NACA and the US Navy Bureau of Aeronautics are alleged to have handed the German research to the Japanese, where the technology has been applied by Shinmeiwa.

The fine hull (with length/beam around 10/1) cannot be used effectively with a small aircraft, because there is not enough beam for stowage of disposable load and equipment. Hull sections are forced to bulge outboard beyond the chines, and such curvature of the hull sides can cause yaw if spray strikes one side before the other.

Directional instability may be cured by a skeg, a small fin, protruding into the water from the afterbody keel, but its effectiveness is limited by the range of trim-angles at which it runs in solid water.

Directional control is by water-rudder (as we have already mentioned) or by water-flaps. Water-flaps (which can be used in the air as air-brakes), open differentially under water for turning, or together for braking. They are fitted either side of the afterbody keel and are most necessary for jet aircraft that do not have the beneficial effects of propeller slipstream to help with manoeuvre and control (Plate A–12).

Flying from Water

Flying a seaplane is little different from flying a landplane – except that there is more to handling on water than on land. All aeroplanes have minds of their own and airworthiness requirements are intended to make their minds predictable.

Perhaps the most striking differences between seaplanes and landplanes are:

☐ On water, when no longer planing, the pilot is aware that he is in charge of quite another sort of animal. He feels newly vulnerable and sometimes at a considerable disadvantage when his aircraft is converted into an unwieldy boat, with bits that stick out more on either side than fore and aft. Control can feel inadequate when juggling with aerodynamic and hydrodynamic forces which are often working against one another.

☐ In the air seaplanes are often ungainly in response to control. Pendulous effects are introduced because centres of mass (CG) and centres of aerodynamic effort of lifting, stabilizing (and destabilizing) and control surfaces may be far apart, and in less than optimum locations.

Here are some random observations to give a flavour to what follows:

☐ A landplane loses about 10% cruising speed when fitted with floats. The loss is less than might be expected because floats fly, being designed to lift their own weight aerodynamically.

☐ A floatplane may be lifted off water by rocking from side to side, lifting one float out before the other.

☐ Suction can be broken by rocking a seaplane fore and aft, to shorten the take-off.

☐ No seaplane should be landed into the face of a swell.

Seaplane testing

When testing such aircraft much depends upon whether a seaplane is a floatplane or a flying-boat, and whether or not it is amphibious. One must expect longitudinal, lateral and directional stability to be degraded. There might be marked changes in handling qualities at the stall. Response to control will be less lively. If the machine is to be cleared for aerobatics (which must include spinning in the UK) then spin and spin recovery characteristics will almost certainly be affected adversely. Take-off distance, rate of climb and ceiling suffer.

Apart from the normal testing of stall characteristics, the ability to overshoot from a balked landing, V_{DF} (the demonstrated maximum diving speed in flight), control, stability, and rate of climb as an indication of performance, will also be investigated. The tests also involve take-off from and landing on water – and in the safe transition from one to the other.

Assessment of wind and water conditions

As in all forms of flying, meteorological conditions must be right. Seaplanes have the advantage of being able to operate on water under a cloud ceiling that would be prohibitive for a landplane. Navigation by shore line is generally much easier. There is also the relative peace of mind in knowing that, in an emergency, the aeroplane can be put down on land without much difficulty, as well as on water.

If there is no windsock available – and this is usually the case where there is no seaplane base (as in the UK), one has to find the wind and use the wind-line for take-off direction. As we have seen, a seaplane weathercocks naturally into wind, and there is no point in making things harder than necessary by attempting a take-off in some other direction. Wind always blows from a region of flat calm water. A narrow band of flat water along a shoreline is an indication of a strong wind. A wide band shows that the wind is slack. Wind direction is also marked by white streaks of foam, or wind-lanes, if the wind is strong enough; but these must not be confused with similar lines made by a current. When picking take-off direction the pilot must also make a critical assessment of tidal flow, seastate and obstacles, leaving plenty of sea-room.

Special checks before flight

Of all the things checked pre-flight seven are outstanding with a seaplane:

- *Watertightness:* Drain plugs, and see that there is no more than, say, a cupful of water in the bilges already. Quite apart from the presence of water indicating a leak, water is heavy (10 lb weight for 1 imperial gallon, 1 kg per litre) and it can make a large difference to rate of climb, as well as to stability.
- *Water-Rudder Controls:* Check that water rudder(s) can be retracted. If this cannot be done before take-off and after landing, there is the danger that the water rudder(s) will be damaged.
- *Elevator trim setting:* This should be set for neutral stick pressure during the planing phase of take-off, otherwise porpoising may result.
- *Ease of starting the engine – and idling that is not too fast.*
- *Rope, anchor, float and baler on board* (as a minimum): Make sure that the rope is tied securely inboard (allow enough rope to pay out: at least three times the depth in fresh, and five times in salt water, to allow the anchor flukes to bite).
- *Carry a paddle.*
- *Wear a lifejacket and carry a knife:* The only time a pilot will need a knife is the moment that he *really* needs it. This is even more likely to occur in an emergency on, or in, water than on land.

Taxying

In the displacement phase a seaplane is a boat and the rules and principles of seamanship must be observed in addition to good airmanship. Taxying must be carried out slowly, with plenty of sea room. At low speed and without a strong tailwind, the stick must be held back to lift the nose, which reduces spray and improves manoeuvrability. Momentum built up in a turn can carry the seaplane through the desired heading, so plenty of anticipation is needed with rudder, and asymmetric throttling with multi-engines. Further, to maintain direction when taxying the pilot should get into the habit of picking a fixed reference point on land. Short bursts of throttle, which help to keep speed low, give better control when turning tightly at low speed. Depending upon the direction of rotation of the propeller a seaplane turns better one way than another. A right hand tractor propeller causes propwash to spiral clockwise towards the tail, making the machine more prone to turn left than right, and vice versa.

- *Ploughing*
  In a strong wind a technique called a '*plough turn*' is often needed to turn downwind, Fig. A30 (Plate A–20). Such a turn is started into wind with plenty of power, nose high and tail low. This is shown in Fig. A31. Note the use of aileron into wind. When ploughing (Plate A–19) the centre of buoyancy moves aft, increasing the lateral area forward, reducing weathercock stability and assisting the turn out of wind.

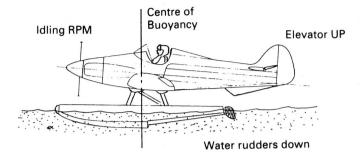

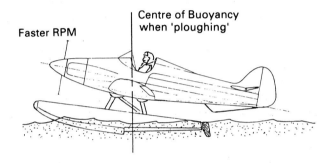

Fig. A30 Attitude when idling and when 'ploughing'.

□ *Step Taxying and Turning*

When there is a long clear run to be covered on water it is convenient to step taxy, placing the seaplane 'on the step'. To do this the stick is pulled fully back and power applied until the floats or hull rise out of the water, climbing the bow wave as we saw at the top of Fig. A17. At this point the back pressure on the stick is eased and the stick moved forward to around neutral (Plate A–21). The seaplane will be felt to accelerate and thump along on the surface. If the stick is pushed too far forward the forebody keel digs in and the consequent loss of directional control and possible porpoising can be embarrassing. When on the step and planing the correct attitude can be felt by small stick movements fore and aft. If the stick is moved back too far the afterbody touches and a deceleration is felt. When planing comfortably the throttle setting can be reduced so that a constant speed is maintained. Step taxying is fast and the water rudder(s) must have been raised beforehand. Directional control is now by means of the main air-rudder. When turning at high speed the two primary forces which affect the resulting path are centrifugal force and the wind. Fast skidding turns can be made on the step using rudder alone, with aileron into the turn. When turning crosswind, from upwind to downwind, centrifugal and wind force tend to oppose one another, as shown in Fig. A32. But, when step turning into wind from downwind centrifugal and wind forces combine to make the aeroplane unstable, Fig. A33. So, when turning from downwind to upwind in windy conditions, use minimum speed in a displacement condition, water rudder(s) down if

Appendix

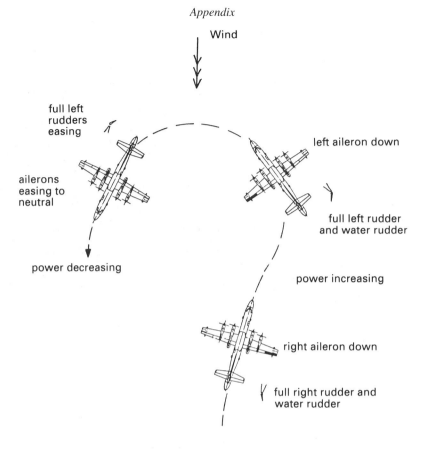

Wind

full left
rudders
easing

left aileron down

ailerons
easing to
neutral

full left rudder
and water rudder

power decreasing

power increasing

right aileron down

full right rudder and
water rudder

Fig. A31 Plough turning to downwind.

Plate A–19 Supermarine S5 ploughing (c1931). (Royal Aircraft
Establishment)

Plate A–20   Planing with tail controls suggesting the start of a plough turn. (*Flight International*)

Plate A–21   De Havilland Sea Tiger. Elevator position shows it is in transition over the 'hump', from ploughing to planing. (*Flight International*)

necessary or outboard engines opened up more than inboard, and aileron as required into the turn. If the pilot does not take these actions a floatplane may bury the downwind float, and a flying-boat submerge the downwind tip-float. In either case the downwind wing may then dig into the water, leading to a capsize.

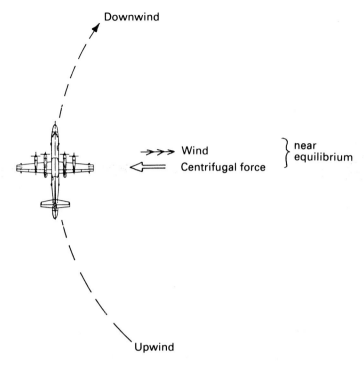

Fig. A32   Step turning out of wind.

Take-off

Taking off for the first time in a seaplane that he does not know the pilot must be prepared for marked differences between the longitudinal stability, and 'feel', between water and air. Big changes may occur instantaneously on lift-off. Therefore, one must concentrate upon nose attitude. After lift-off the take-off attitude should be maintained for the initial climb.

   With the water-rudder raised for take-off and pointing into wind, the throttles are opened wide and the stick pulled fully back, while correcting yaw with rudder. The object is to get through the spray and on to the step as quickly as possible. As soon as the nose attitude has stabilized the pilot lets the stick move forward until, if trimmed correctly, the elevators are about neutral and planing begins, as shown in Fig. A34 (Plate A–22). Sometimes a slight forward pressure must be applied to plane. The elevator is shown neutral in the figure. Sometimes a slight backward pressure is needed on the stick to plane correctly. If the nose is pushed down too much porpoising might occur. When planing, turns, using the air-rudder normally, can be made to adjust the take-off direction. The best planing angle is that where the afterbody keel *almost* touches the water (Plate A–23). If the seaplane is run too tail up, then too much of the forebody is wet and take-off is delayed. Also the craft needs more rudder to correct the line of take-off. The pilot becomes aware of a tendency to nose over. But if the pilot then attempts to pull the seaplane off early, the afterbody digs in, speed is lost and the aeroplane may begin to proceed in a series of jarring hops, becoming airborne, and then dropping back again.

   Glassy water take-offs may involve ruffling the surface first, by making a wide sweeping turn. The object is to get air under the planing surfaces. The beauty of seaplane flying is

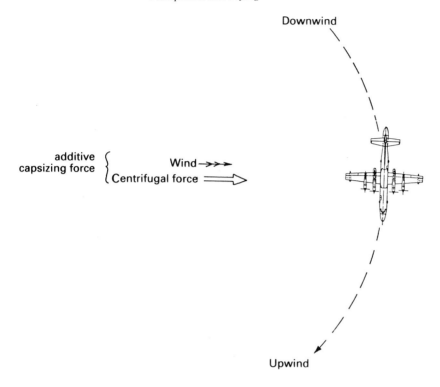

Fig. A33   Step turning into wind.

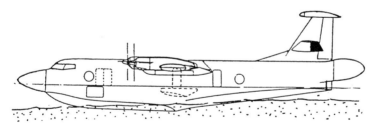

Fig. A34   Flying boat planing, note position of elevator: stick almost neutral.

that the pilot can take off in a short distance by accelerating around a stepped turn first, then straightening for the final leap into the air. Another useful technique with a floatplane is to use aileron on the take-off run to lift one float out of the water before the other. Dropping a small amount of flap on lift-off is effective if used with care. But, flaps can cause quite strong nosedown pitching moments and the seaplane to adopt a nosedown attitude. If used too soon flaps *can* delay the seaplane from getting on to the step.

☐ *Roughwater take-offs* need more nose-up trim than when the water is smooth. This helps to reduce spray and any tendency to bury the bow. However, there is a tendency to bounce and stall, and a concentrated effort must be made to flatten the attitude as soon as possible to avoid being flung into the air prematurely.

☐ *Taking off out of wind* can be tricky. There is the risk of the downwind float being buried as the nose rises. Then the pilot must cut the throttle and turn downwind, which allows

Plate A–22 De Havilland Sea Tiger planing. Note water-rudders are up and that there is less spray than in Plate A–21, and elevator is neutral. (*Flight International*)

Plate A–23 Cessna floatplane on lift-off, afterbodies of floats paralleling the water. (J. M. Ramsden)

the downwind float to surface. If the pilot is losing control and has not submerged a float, he must chop the throttle and allow the machine to weathercock into wind. The safest technique for a crosswind take-off is to start into wind and then turn crosswind when planing. The upwind wing is held down with aileron and only just enough rudder applied to prevent turning.

☐ *Taking off in a swell* can involve out of wind take-off, so as to run for as far as possible along the crest of the swell. By this means it may be possible to become airborne between the crests of two or three swells, so minimizing the chance of being bounced into the air prematurely. It is wise not to push for such a take-off regardless if it is hard to get airborne. Stop and try again from a slightly different direction.

☐ *Racing (Schneider) Seaplanes (Ref. 14) Figs A5a and A35, Plate A–19* Although what follows is historic, it makes such fascinating points that it is worth recording.

The Supermarine S6A and S6B were observed to be easy to fly, but were longitudinally unstable on take-off. They had no water rudders and, when taxying, yaw was slow and hard to stop without a burst of throttle. If the engine was opened up the aeroplane swung left and the pilot was blinded by spray.

To take off the aeroplane was:

> 'Put well to the right of the wind and held there without yaw while the pilot ensures he has a clear run into wind. He then applies full right rudder and full elevator control, bends his head well forward and down under the windscreen to shield his goggles, and then opens the throttle full out fairly rapidly. In this way he can go through the first second or two while the aircraft is swinging left and the cockpit is covered with spray, and he can get a clear view with clean goggles afterwards to apply rudder control as soon as it becomes available.'

It was for this reason that the Macchi M72 (Fig. A5b) was fitted with counter-rotating propellers on its pair of twin-tandem engines.

☐ *Porpoising*

As we have noted earlier, porpoising is most common on take-off, but it can be encountered landing, and is caused by the planing bottom running at the wrong angle. It usually occurs when the nose, or bow is too low. Quick reactions are needed to oppose pitching motions. But, if in doubt, chop the throttle and start the take-off again (Refs 15, 17).

Landing (Fig. A36) (Plate A–24)

There is always a trap in landing. Water looks calmer from the air than it really is – and a seaplane can land safely in conditions in which it would be dangerous to attempt to take off again.

Much of what has been said of handling on take-off applies equally to landing. Attitude of the nose is all important. Seaplanes glide more steeply than landplanes (more drag). So it is easier to land power on than off, because there is less chance of error. Seaplanes have no shock absorbers. Before alighting the pilot must check (Ref. 15).

☐ Conditions of wind and sea.
☐ Presence of floating obstructions, moving ships and their wash.
☐ Depth of water and presence of rocks and shoals (these can often be seen quite clearly through the water when overflying the landing area).
☐ The path along which it is desired to taxy after alighting.

In a calm or choppy sea it is wise to touch down in a normal flying position to avoid putting excessive strain on hull or floats through landing to tail down (Plate A–24).

☐ *Glassy surfaces* are deceptive. The pilot should aim to land near some object. On a calm

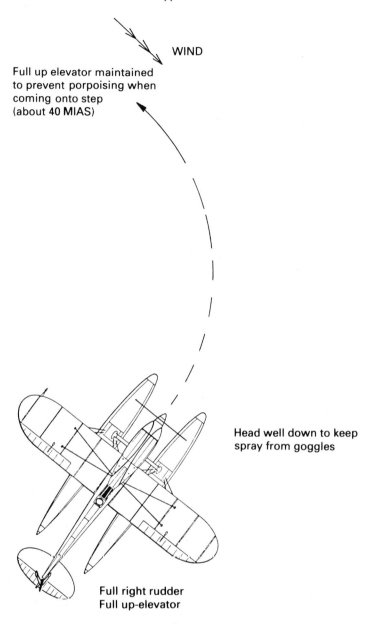

WIND

Full up elevator maintained
to prevent porpoising when
coming onto step
(about 40 MIAS)

Head well down to keep
spray from goggles

Full right rudder
Full up-elevator

Fig. A35   Start of take-off path, Schneider Trophy Seaplane (Supermarine S6A and S6B).

lake in haze it has been advised to throw overboard a couple of cushions during a low
pass, so as to provide a line and a way of estimating height above the surface. The
seaplane must be flown on in a level attitude, closing the throttle and easing back slowly
on the stick at touchdown, to avoid being thrown back into the air. It is also advisable not
to change one's mind on nearing the water, so as to make a normal touchdown, because
judgement of height is deceptive.

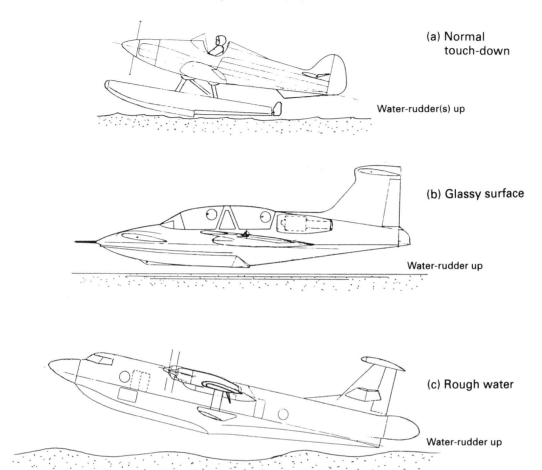

(a) Normal
touch-down

Water-rudder(s) up

(b) Glassy surface

Water-rudder up

(c) Rough water

Water-rudder up

Slow and stalled

Fig. A36   Touchdown attitudes: normal, glassy surface, and rough water.

☐ *Rough sea landings* must be made at low speed rounding out higher than usual to touch tail-down in a stalled attitude. This is to keep the hydrodynamic drag forces behind the CG. On no account must the forebody be allowed to touch first – or one float touch first. Quite apart from there being a danger of losing a float, a water-loop under such conditions can be disastrous.

☐ *Landing crosswind* is tricky and often necessary. A crab technique is best, straightening out nose-up, with wings level at the beginning of touchdown; and without drift.

Aileron sailing (Fig. A37)

Sailing is a way of moving sideways across water while keeping a seaplane substantially head into a strong wind. To do this the pilot adjusts power and uses asymmetric aileron, together with rudder – and wing flaps if necessary – to change the drag axis of the machine.

When drifting backwards the seaplane is little affected by the wind – except in producing drift – and it travels backwards more or less in line with the keel. When propulsive thrust is

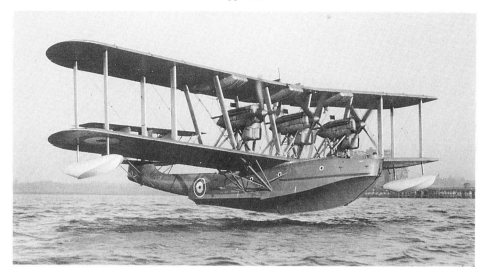

Plate A–24 'The moment of truth'. Blackburn Iris c1930, on touch-down, elevator up. The rear step appears to be related to the Linton-Hope form in Fig. A14j. (J. M. Ramsden)

introduced and drift is checked, the wind acts on the airframe surfaces to move the seaplane sideways in the direction in which the nose is pointed.

When sailing backwards water rudders oppose the air rudder and must be retracted. Also, the stick must be held forwards to keep the elevator down.

Amphibians

An undercarriage and retraction mechanism is heavy, adding around 5% to the basic sea-plane weight. However, the fact of such gear can be most useful, not only for feeling the bottom as when approaching the shore, but also as a source of water-drag; allowing the engine to be run at higher RPM, making tail controls more effective, without gaining speed.

Flying from Ice

The strength of seaplane hulls and floats enable them to be flown from ice. In Canada, Major F. J. Steven (Ref. 16) has given much useful information on the quality of fresh and salt water ice and the conditions which determine what a frozen surface will be like for operations using Grumman Albatross flying-boats.

Required minimum ice thickness is given as:

$$T = \frac{27}{8} W \tag{A7}$$

where T is thickness in inches and W the weight of the aircraft. Steven recommends that no aeroplane should land on lake ice less than six inches (16 cm) thick.

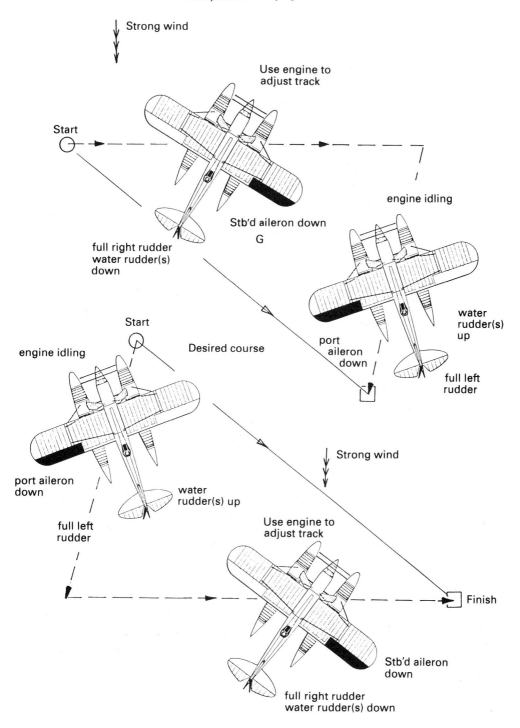

Strong wind

Use engine to
adjust track

Start

engine idling

Stb'd aileron down
G

full right rudder
water rudder(s)
down

water
rudder(s)
up

port
aileron
down

full left
rudder

Start

engine idling

Desired course

port aileron
down

water
rudder(s) up

full left
rudder

Strong wind

Use engine to
adjust track

Finish

Stb'd aileron
down

full right rudder
water rudder(s) down

Fig. A37 Aileron sailing.

Even though ice may be thick enough in theory, *resonance waves* associated with movement of the aircraft over the surface, and dependent upon water depth and radius of influence (proportional to ice thickness), have been known to cause it to break prematurely. Points made in Ref. 16 are that lakes used for such operations should not be too long and narrow, unless the ice is very thick, or an emergency exists.

When operating from ice covered with snow care must be taken when using reverse thrust. Not only because snow will obscure visibility but because any imbalance in RPM can aggravate a swing, and lumps of ice can damage airframe surfaces.

Conclusions

Seaplane design, operation and flying qualities require unusual expertise beyond that needed for conventional landplanes of similar size, for similar tasks. Marine aircraft are both advantaged and disadvantaged by water and its particular properties, which conflict with air, while obeying the same fundamental physical and fluid laws. Nevertheless, water bestows the operational advantages of cheap room needing little preparation (provided one can accept the design penalties of increased structure and equipment weights, and increased drag as well as operational problems of loading, unloading, maintaining and servicing on water). Seaplanes are not off-shore machines – except in very large sizes. They are best suited to operations to and from lakes and relatively sheltered waters. The size of aircraft, as indicated by weight (hence height of flying surfaces, propellers and hot parts of engines above water) is limited by wave height. The greater the wave height, the bigger the seaplane that is needed to cope. Such size is useful in that it provides volume for lifting heavy loads that might not be moved in other ways except, maybe, by airship.

Because of the unique features needed to float and to deflect spray, seaplanes are stronger and heavier and tend to be clumsier and more ungainly than comparable land planes. It is hard to achieve lift/drag ratios that are equal to those of well-designed landplanes. Even so, they can fly into and out of remote areas without costly pre-prepared surfaces, or expensive STOL equipment, often in shorter block times – as in the case of fire bombing.

Seaplanes, while limited in many ways, are the only aircraft that are able to operate in many remote areas, doing work that land planes cannot do. They should be seen as necessary tools for specialized work, and as complementary in every way to the land-based aeroplane. The seaplane may have limitations in performance and sweetness of handling compared with more elegant landplane designs; but the landplane cannot compare with the 'go-almost-anywhere-at-any-time' features the seaplane can offer.

A good case can be made for continuing research and development, albeit on a small scale in university and college, so as to improve their usefulness. I also believe that learned Societies, like the RAeS and RINA, should help to keep a balanced overview and provide moral encouragement for the continuing study of such aircraft.

Acknowledgments

I am indebted to Dornier GmbH for its time, photographs, films and opportunity to examine its flying-boats (and to Herr von Meier and Herr Lucas especially); Professor John E. Allen for material and slides; Professors John Stollery and Dennis Howe of the College of Aeronautics for the copy of the discussion papers 'Is there a future for the Flying Boat' (22 February 1984); and Dr Paul Franceschi for technical information about French

'Pelican' fire-bombers. Also to Canadair and Rhine Flugzeugbau for the provision of data and photographs; Bill Barnhouse of H. W. Barnhouse Incorporated, a consultant and test pilot, slides; to Alan Deacon, Short's Chief Test Pilot in Belfast, for the LA-4 photograph; to Captain Keith Sissons and Tom Freer of the Tiger Club for Sea Tiger and other floatplane photographs; and to J. M. Ramsden, Editor in Chief of *Flight International* for the use of his rare photographic archive.

References

1. BR67(2) Admiralty Manual of Seamanship 2 HMSO, 1981.
2. Gerritsma, Professor Ir. J. and Kenning, Ir. J. A. Winglet Keels in waves. Delft University of Technology (Ocean Racing: Seahorse No. 93, April 1986).
3. BCAR G4-10. Appendix. Civil Aviation Authority.
4. Dornier, Dr Claudius *The 1000 ton Flying Ship – The Potential*. Aero Marine Committee RAeS, and RINA discussion, 22 February 1984.
5. Howe, Professor D. The Long Range Transport Flying Boat. Some Design and Operational Problems (see Ref. 4).
6. Allen, Professor John E. Is There a Future for the Flying Boat? Past Innovations and Future Possibilities (see Ref. 4).
7. Stinton, D. *The Design of the Aeroplane*, Collins, 1983, 85.
8. Thurston, D. B. *Design for Flying*. New York, London: McGraw Hill Book Company, 1978.
9. Marchaj, C. A. *Sailing Theory and Practice*. Adlard Coles/Granada, 1982.
10. Knowler, H. The Fifth Louis Bleriot Lecture, The Future of the Flying Boat, *Journal RAeS*, May 1952.
11. McLarren, R. New Hull Fineness Slashes Drag, Research Review, *Engineering Aviation Week*, 14th March 1949.
12. Stinton, D. *The Anatomy of the Aeroplane*, Collins, 1966, 80, 85.
13. Stout, Ernest G. A Review of High-Speed Hydrodynamic Development, Royal Aeronautical Society, 1951.
14. Orlebar, A. H. Wing Commander, Flying and Water Characteristics of S6A and S6B Seaplanes. R&M 1575, 1934.
15. Air Publication 1098 – Royal Air Force Flying Training Manual, Part II – Seaplanes. Air Ministry, 1938.
16. Steven, Major F. J. Flying-boat Pilotage. *Shell Aviation News* Nos 382 to 385 incl. c1972.
17. Hoffsommer, A. *Flying with Floats*. Pan American Navigation Service, 1969.
18. Mudie, R. and C. *Power Yachts*. Granada Publishing in Adlard Coles Limited, 1977.
19. Collected in one binding:
    Macmillan, Captain N. Flying Floatplanes, c1939.
    Sutherland, Flight Lieutenant J. Handling of all Types of Seaplanes Afloat, c1939.
    Wilcockson, Captain A. S. Practical Notes on Flying the Mayo Composite Aircraft, c1939.
    Russell, Flight Lieutenant H. M. Piloting a Flying Boat, c1939.
    Macmillan, Captain N. Flying amphibians, c1939.
    Downer, Captain F. W. M. The Development of the Flying Boat, Aeronautics. George Newnes Limited, c1939.

20. Du Cane, P. *High Speed Small Craft.* Temple Press. 1951.

21. Bagg, M. and Thomas, J. O. The Detection of Internal Waves in the North Atlantic using Real Aperture Airborne Radar. *International Journal Remote Sensing*, 1984, **5**, 6, 969–974.

22. Payne, Peter R. Supercritical Planing Hulls, *Ocean Engineering*, 1984, **11**, 2, 129–184.

23. Clement, Eugene P. The Dynaplane-type Planing Boat. *The National Architect*, January 1980, 15–16.

24. Clement, E. P. Designing for Optimum Planing Performance. *International Shipbuilding Progress*, March 1979, **26**, 295, 61–64.

INDEX

Index